AF718093

GENERALIZED LINEAR MODELS

Biostatistics: A Series of References and Textbooks

Series Editor

Shein-Chung Chow

President, U.S. Operations
StatPlus, Inc.
Yardley, Pennsylvania
Adjunct Professor
Temple University
Philadelphia, Pennsylvania

1. *Design and Analysis of Animal Studies in Pharmaceutical Development*, edited by Shein-Chung Chow and Jen-pei Liu
2. *Basic Statistics and Pharmaceutical Statistical Applications*, James E. De Muth
3. *Design and Analysis of Bioavailability and Bioequivalence Studies, Second Edition, Revised and Expanded*, Shein-Chung Chow and Jen-pei Liu
4. *Meta-Analysis in Medicine and Health Policy*, edited by Dalene K. Stangl and Donald A. Berry
5. *Generalized Linear Models: A Bayesian Perspective*, edited by Dipak K. Dey, Sujit K. Ghosh, and Bani K. Mallick

ADDITIONAL VOLUMES IN PREPARATION

Medical Biostatistics, Abhaya Indrayan and S.B. Sarmukaddam

Generalized Linear Models

A Bayesian Perspective

edited by

Dipak K. Dey
The University of Connecticut
Storrs, Connecticut

Sujit K. Ghosh
North Carolina State University
Raleigh, North Carolina

Bani K. Mallick
Texas A&M University
College Station, Texas

CRC Taylor & Francis
Taylor & Francis Group
Boca Raton London New York

CRC is an imprint of the Taylor & Francis Group,
an informa business

Library of Congress Cataloging-in-Publication Data

Dey, Dipak.
Generalized linear models : a Bayesian perspective / Dipak K. Dey, Sujit K. Ghosh, Bani K. Mallick.
p. cm. — (Biostatistics ; 5)
Includes index.
ISBN 0-8247-9034-0 (alk. paper)
1.Linear models (Statistics) 2. Bayesian statistical decision theory. I. Ghosh, Sujit K., 1970- II. Mallick, Bani K., 1965- III. Title. IV. Biostatistics (New York, N.Y.) ; 5.

QA276 .D485 2000
519.5'35—dc21 00-024055

This book is printed on acid-free paper.

Headquarters
Marcel Dekker, Inc.
270 Madison Avenue, New York, NY 10016
tel: 212-696-9000; fax: 212-685-4540

Eastern Hemisphere Distribution
Marcel Dekker AG
Hutgasse 4, Postfach 812, CH-4001 Basel, Switzerland
tel: 41-61-261-8482; fax: 41-61-261-8896

World Wide Web
http://www.dekker.com

The publisher offers discounts on this book when ordered in bulk quantities. For more information, write to Special Sales/Professional Marketing at the headquarters address above.

Copyright © 2000 by Marcel Dekker, Inc. All Rights Reserved.

Neither this book nor any part may be reproduced or transmitted in any form or by any means, electronic or mechanical, including photocopying, microfilming, and recording, or by any information storage and retrieval system, without permission in writing from the publisher.

Current printing (last digit)
10 9 8 7 6

Series Introduction

The primary objectives of the Biostatistics series are to provide useful reference books for researchers and scientists in academia, industry, and government, and also to offer textbooks for undergraduate and/or graduate courses in the area of biostatistics. This series will provide comprehensive and unified presentations of statistical designs, analyses, and interpretations of important applications in biostatistics, such as those in biopharmaceuticals. A well-balanced summary will be given of current and recently developed statistical methods and interpretations for both biostatisticians and researchers/scientists with minimal statistical knowledge who are engaged in applied biostatistics. The series is committed to providing easy-to-understand state-of-the-art references and textbooks. In each volume, statistical concepts and methodologies will be illustrated through real examples.

Generalized linear models (GLMs) have been frequently used in pharmaceutical research and development, especially in clinical research and development for demonstration of the safety and efficacy of a pharmaceutical compound under investigation. However, the concept for the analysis of GLMs with mixed effects for categorical and/or longitudinal data is often misused or misinterpreted due to its complexity. This volume provides a comprehensive overview of key statistical concepts and methodologies including the Bayesian approach for analysis of GLMs including logistic regression and log-linear models from both a theoretical and a practical point of view. In addition, it includes important issues related to model diagnostics and variable selection in GLMs.

This volume serves as an intersection for biostatisticians, practitioners, and researchers/scientists by providing a good understanding of key statistical concepts and methodologies for analysis and interpretation of GLMs and GLMs with mixed effects as well. This volume is in compliance with good statistics practice (GSP) standards for good clinical practice (GCP) as required by most regulatory agencies for pharmaceutical research and development.

Shein-Chung Chow

Preface

Generalized Linear Models (GLMs) are widely used as flexible models in which a function of the mean response is "linked" to covariates through a linear predictor and in which variability is described by a distribution in an exponential dispersion family. These models include logistic regression and log-linear models for binomial and Poisson counts as well as normal, gamma and inverse-Gaussian models for continuous responses. Standard techniques for analyzing censored survival data such as Cox regression can also be handled within the GLM framework. Other topics closely related to GLMs include conditionally independent hierarchical models, graphical models, generalized linear mixed models (GLMMs) for estimating subject-specific effects, semi-parametric smoothing methods, pharmokinetic models and spatio-temporal models.

GLMs thus provide a versatile statistical modeling framework for medical and industrial applications, but questions remain about how the power of these models can be safely exploited when training data are limited. This volume demonstrates how Bayesian methodology allows complex models (ranging from simple logistic regression models to semi-parametric survival models for censored data) to be used without fear of the "over-fitting" that can occur with traditional GLM methods which are usually based on normal approximation theory. Insight into the nature of these complex Bayesian models is provided by theoretical investigations and practical implementations. Presupposing only basic knowledge of probability and statistics, this volume should be of interest to researchers in statistics, engineering and medicine.

This volume will serve as a comprehensive reference book for practitioners and researchers. Each part in the volume has chapters written by an expert in that particular topic, and the chapters are carefully edited to ensure that a uniform style of notation and presentation is used throughout. As a result, all researchers whose work uses GLM theory will find this an indispensable companion to their work and it will be the reference volume for this subject for many years to come. In particular, each chapter describes how to conceptualize, perform and criticize traditional GLMs from a **Bayesian perspective**. In addition, how to use modern computational methods to summarize inferences using simulation is elucidated.

The primary users of this volume include professionals in statistics and other related disciplines who work in the pharmaceutical industry, medical centers (including public health and epidemiology) and public and private research and academic institutions.

Our hope is that this volume will also help researchers identify areas of important future research and open new applications of generalized linear models using Bayesian approaches.

The papers in this volume are divided into six parts: General overview, extension of the GLMs, categorical and longitudinal data, semiparametric and nonparametric approaches, model diagnostics and variable selection and challenging problems.

In part I, Gelfand and Ghosh introduce Bayesian analysis of generalized linear models from its developments. Sun, Speckman and Tsutakawa describe random effects in generalized linear mixed model with fully explained examples. Ibrahim and Chen develop methods of prior elicitation and variable selection for generalized linear mixed models with an example of pediatric pain data.

Chapters in Part II of the volume describe several extensions of GLMs. Ferreira and Gamerman introduce dynamic modeling approach for GLMs. They also lay out computational steps with two applications. Dey and Ravishanker extend GLMs in the presence of overdispersion. Both parametric and nonparametric approaches to overdispersed GLMs are considered. Nandram proposes Bayesian GLMs for inference about small areas and describes an application with mortality data of U.S.A.

Part III concerns modeling categorical and longitudinal data. Modeling dichotomous, polychotomous and count data are quite useful and challenging in the presence of correlation. In this part, first Chib describes methods for the analysis of correlated binary data using latent variables. He also describes three algorithms for implementation. Chen and Dey extend this to correlated ordinal data and propose algorithms for analysis of such data. Bayesian methods for time series count data are described by Ibrahim and Chen with an application to the analysis of pollen count. Albert and Ghosh propose and analyze item response modeling for categorical data. This part concludes with a case study using Bayesian probit and logit models by Landrum and Normand.

Part IV describes GLMs using rich classes of nonparametric and semiparametric approaches. Semiparametric GLMs are considered by Mallick, Denison and Smith using Bayesian approaches. The chapter by Basu and Mukhopadhyay presents a semiparametric method to model link functions for the binary response data. Next, Haro-López, Mallick and Smith develop a data adaptive robust link function. In the last chapter of Part IV, Kuo and Peng present a mixture-model approach to the analysis of survival data.

Part V deals with important issues relating to model diagnostics and variable selection in GLMs. In this part, the chapter by Dellaportas, Forster and Ntzoufras presents Bayesian variable selection in using Gibbs sampler. Next, Ibrahim and Chen describe variable selection methods for Cox models. This part is concluded by Dey and Chen on Bayesian model diagnostics for correlated binary data.

Part VI concludes the volume with challenging problems. Wakefield and Stephens develop a case study by incorporating errors-in-variable modeling. Iyengar and Dey review parametric and semiparametric approaches for the analysis of compositional data. Denison and Mallick describe classification trees from a Bayesian perspective and apply the algorithm on a case study problem. In the next chapter, Gelfand, Ravishanker and Ecker develop a new modeling and inference method for point-referenced binary spatial data. The part closes with the chapter by Best and Thomas on graphical models and software for GLMs.

The cooperation of all contributors in the timely preparation of their manuscripts is greatly appreciated. We decided early on that it was important to referee and critically evaluate the papers which were submitted for inclusion in this volume. For this substantial task, we relied on the service of numerous referees to whom we are most indebted. Among those whom we wish to acknowledge are Sudipto Banerjee, Pabak Mukerjee and Kaushik Patra.

Finally we thank the editors at Marcel Dekker, Inc. for considering our proposal. Our special thanks go to Debosri, Swagata and Mou for their encouragements in this project .

Dipak K. Dey,	Sujit K. Ghosh	and	Bani K. Mallick
Storrs, CT, USA	Raleigh, NC, USA		College Station, TX, USA

Contents

Contributors

James Albert
Department of Mathematics and Statistics,
Bowling Green State University, U.S.A
albert@bgnet.bgsu.edu

Sanjib Basu
Division of Statistics,
Northern Illinois University, U.S.A
basu@math.niu.edu

Nicky Best
School of Medicine at St. Mary's,
Imperial College, U.K.
n.best@ic.ac.uk

Ming-Hui Chen
Department of Mathematical Sciences,
Worchester Polytechnic Institute, U.S.A
mhchen@wpi.edu

Siddhartha Chib
John M. Olin School of Business,
Washington University, U.S.A
chib@simon.wustl.edu

Petros Dellaportas
Department of Statistics,
Athens University of Economics and Business, Greece
petros@aueb.gr

David D.T. Dension
Department of Mathematics,
Imperial College, U.K
d.denison@ma.ic.ac.uk

Dipak K. Dey
Department of Statistics,
University of Connecticut, U.S.A
dey@stat.uconn.edu

Mark D. Ecker
Department of Mathematics,
University of North Iowa, U.S.A
ecker@math.uni.edu

Marco A. R. Ferreira
Instituto de Matematica,
Federal University of Rio de Janeiro, Brazil
marco@dme.ufrj.br

Jonathan J. Forster
Department of Statistics,
Athens University of Economics and Business, Greece
forster@aueb.gr

Dani Gamerman — Instituto de Matematica,
Federal University of Rio de Janeiro, Brazil
dani@dme.ufrj.br

Alan E. Gelfand — Department of Statistics,
University of Connecticut, U.S.A
gelfand@uconnvm.uconn.edu

Malay Ghosh — Department of Statistics,
University of Florida, U.S.A
ghoshm@stat.ufl.edu

Rubén A. Haro-López — Department of Mathematics,
Imperial College, U.K
haro@gauss.rhon.itam.mx

Joseph G. Ibrahim — Department of Biostatistics,
Harvard University, U.S.A
d.denison@ma.ic.ac.uk

Malini Iyengar — Clinical Biostatistics,
Smith Kline Beecham Pharmaceuticals, U.S.A
malini@stat.uconn.edu

Lynn Kuo — Department of Statistics,
University of Connecticut, U.S.A
lynn@stat.uconn.edu

Mary Beth Landrum — Department of Health Care Policy,
Harvard University, U.S.A
landrum@hcp.med.harvard.edu

Bani K. Mallick — Department of Statistics,
Texas A & M University, U.S.A
bmallick@stat.tamu.edu

Saurabh Mukhopadhyay — Clinical Biostatistics,
Merck & Co., Inc., U.S.A
saurabh_mukhopadhyay@merck.com

Balgobin Nandram — Department of Mathematical Sciences,
Worchester Polytechnic Institute, U.S.A
balnan@wpi.edu

Sharon Lise-Normand — Department of Health care policy,
Harvard University, U.S.A
sharon@hcp.med.harvard.edu

Ioannis Ntzoufras — Department of Statistics,
Athens University of Economics and Business, Greece
ioannis@aueb.gr

Fengchun Peng	Credit Marketing, Sears, Roebuck and Co., U.S.A `peng@stat.uconn.edu`
Nalini Ravishanker	Department of Statistics, University of Connecticut, U.S.A `nalini@stat.uconn.edu`
Adrian F.M. Smith	Queen Mary and Westfield College, University of London, U.K `AFMS.BSU_DOM`
David Stephens	Department of Mathematics, Imperial College, U.K `d.stephens@ma.ic.ac.uk`
Dong Chu Sun	Department of Statistics, University of Missouri Columbia, U.S.A `sun@stat.missouri.edu`
Paul L. Speckman	Department of Statistics, University of Missouri Columbia, U.S.A `speckman@stat.missouri.edu`
Andrew Thomas	School of Medicine at St. Mary's, Imperial College, U.K. `a.thomas@ic.ac.uk`
Robert K. Tsutakawa	Department of Statistics, University of Missouri Columbia, U.S.A. `tsutakawa@stat.missouri.edu`
Jon Wakefield	School of Medicine at St. Mary's Imperial College, U.K. `j.wakefield@ma.ic.ac.uk`

Part I

General Overview

1

Generalized Linear Models: A Bayesian View

Alan E. Gelfand
Malay Ghosh

ABSTRACT Generalized linear models (GLMs) offer a unifying class of models which are widely used in regression analysis. Though initially introduced into the community from a classical viewpoint, in the past decade the Bayesian literature employing these models has witnessed rapid growth. This is due in part to their attractiveness within familiar hierarchical modeling but as well to the wide availability of high speed computing to implement simulation based fitting of these models. The objective of this chapter is to provide a brief, somewhat selective, summary and overview of this recent literature. In particular, we focus upon the range of proposed GLMs, prior specification, propriety of the resultant posterior, semiparametric approaches and model determination, i.e., model adequacy and model choice.

1. Introduction

Generalized linear models (GLMs), originally introduced by Nelder and Wedderburn (1972), provide a unifying family of models that is widely used for regression analysis. These models are intended to describe non-normal responses. In particular, they avoid having to select a single transformation of the data to achieve the possibly conflicting objectives of normality, linearity and homogeneity of variance. Important examples include the binary and the count data. Over the years, GLMs have expanded much in scope and usage, and are currently applied to a very broad range of problems which include analysis of multicategory data, dynamic or state-space extensions of non-normal time series and longitudinal data, discrete time survival data, and non-Gaussian spatial processes.

By now, there are several excellent textbooks discussing inference for GLMs from a classical point of view (McCullagh and Nelder, 1989; Fahrmeir and Tutz, 1991, Lindsey, 1995). These books provide a rich collection of estimation and hypothesis testing procedures for various parameters of interest primarily from a frequentist point of view. Breslow and Clayton (1993) have extended these models further by introducing random effects in addition to the fixed effects. The resulting models, usually referred to as generalized linear mixed models (GLMM's) have further widened the scope of application of GLMs for data analysis. Software routines such as PROC MIXED in SAS have facilitated the computations involved in the classical implementation of GLMs.

Bayesian methods for analyzing GLMs are of more recent origin. A general hierarchical Bayesian (HB) approach for such analysis began with West (1985) and Albert (1988), although various special cases were considered earlier. Leonard and Novick (1986) considered Bayesian analysis of two-way contingency tables, while Albert (1985) considered simultaneous estimation of several Poisson means via a hierarchical log-linear model.

The present article aims to review the Bayesian perspective with regard to GLMs. This includes a range of hierarchical model GLM specifications, approaches for Bayesian model fitting and techniques for model checking and model choice. In Section 2, after a brief introduction to the model and the classical inference procedure, we proceed to the discussion of some of the Bayesian models that have appeared in the literature. We compare the various priors that have been proposed, and discuss also the methods required for their implementation. This includes the work of Albert (1988), Ibrahim and Laud (1991), Dellaportas and Smith (1993), Zeger and Karim (1991), Ghosh, Natarajan, Stroud and Carlin (1998) among others. Section 3 discusses the propriety of posteriors under the different priors. In particular, we discuss the results of Ibrahim and Laud (1991), Natarajan and McCulloch (1995), Hobert and Casella (1996), Gelfand and Sahu (1999), Ghosh, Natarajan, Stroud and Carlin (1998). Section 4 discusses semiparametric and nonparametric Bayesian procedures for GLMs. Section 5 briefly looks at overdispersed GLMs. Finally, Section 6 addresses the issues of model diagnostics and model selection.

2. GLMs and Bayesian Models

2.1 GLMs

Consider measurements (discrete or continuous) for n individuals. For the ith individual, the response variable is denoted by y_i, and the corresponding vector of covariates is denoted by $\boldsymbol{x}_i$. Responses may be continuous real variables, or counts or binary. Fahrmeir and Tutz (1991) contains many interesting examples of binary and count data. As an example of binary data, they consider infection from births by caesarean section. The response variable is the occurrence or non-occurrence of infection. They consider three dichotomous covariates: (a) planned or unplanned caesarean, (b) presence or absence of risk factors such as having diabetes or being overweight, and (c) use or nonuse of antibiotics as prophylaxis. An example of count data involves the effect of two agents of immuno-activating ability that may induce cell differentiation (Piegorsch, Weinberg and Margolin, 1988). As response variable, one considers the number of cells that exhibited markers after exposure. The covariates are the agents TNF (tumor necrosis factor) and IFN (interferon). It is of interest to know whether these agents stimulate cell differentiation independently or whether there is an interaction effect.

There are certain distributional and structural assumptions associated with GLMs. The key distributional assumption is that conditional on the θ_i, the y_i are independent with pdf's belonging to the one-parameter exponential family, that is,

$$f(y_i|\theta_i) = \exp[a^{-1}(\phi_i)\{y_i\theta_i - \psi(\theta_i)\} + c(y_i;\phi_i)], \tag{1}$$

where the θ_i are unknown, but the $a(\phi_i)(> 0)$ are known. The usual structural assumption is that $\theta_i = h(\boldsymbol{x}_i^T\boldsymbol{b})$, where h is a strictly increasing sufficiently smooth function, $\boldsymbol{b}(p\times 1)$ is the vector of unknown regression coefficients, and the $\boldsymbol{x}_i(p\times 1)$

are known design vectors of dimension p. The parameters θ_i are usually referred to as the canonical parameters. Important special cases include the binomial distributions with success parameters $p_i = \exp(\theta_i)/[1+ \exp(\theta_i)]$, $a(\phi_i) = 1$, and the Poisson distributions with means $\lambda_i = \exp(\theta_i)$, $a(\phi_i) = 1$. The $N(\mu_i, \sigma_i^2)$ distributions are also covered by (1) with $\theta_i = \mu_i, a(\phi_i) = \sigma_i^2$. The gamma and the inverse Gaussian distributions are other important special cases of (1).

The classical estimation procedure for GLMs is maximum likelihood. For simplicity, we assume that the ϕ_i are known and that $\boldsymbol{X}^T = (\boldsymbol{x}_1, \cdots, \boldsymbol{x}_n)$ has rank p. The likelihood function is given by

$$L(\boldsymbol{b}) \propto \exp\left[\sum_{i=1}^{n} a^{-1}(\phi_i)\{y_i h(\boldsymbol{x}_i^T \boldsymbol{b}) - \psi(h(\boldsymbol{x}_i^T \boldsymbol{b}))\}\right]. \tag{2}$$

The corresponding score vector is

$$\frac{d \log L(\boldsymbol{b})}{d\boldsymbol{b}} = \sum_{i=1}^{n} a_i^{-1}(\phi_i)\{y_i - \psi'(h(\boldsymbol{x}_i^T \boldsymbol{b}))\}h'(\boldsymbol{x}_i^T \boldsymbol{b})\boldsymbol{x}_i, \tag{3}$$

and the Fisher information matrix is

$$\boldsymbol{I}(\boldsymbol{b}) = E\left[-\frac{d^2 \log L}{d\boldsymbol{b}d\boldsymbol{b}^T}\right] = \boldsymbol{X}^T \boldsymbol{D}\boldsymbol{V}(\boldsymbol{b})\boldsymbol{\Delta}^2(\boldsymbol{b})\boldsymbol{X}, \tag{4}$$

where $\boldsymbol{D} = \text{Diag}\,(a^{-1}(\phi_i), \cdots, a^{-1}(\phi_n))$, $\boldsymbol{V}(\boldsymbol{b}) = \text{Diag}\,(\psi''(h(\boldsymbol{x}_1^T \boldsymbol{b})), \cdots, \psi''(h(\boldsymbol{x}_n^T \boldsymbol{b}))$, and $\boldsymbol{\Delta}(\boldsymbol{b}) = \text{Diag}\,(h'(\boldsymbol{x}_1^T \boldsymbol{b}), \cdots, h'(\boldsymbol{x}_n^T \boldsymbol{b}))$.

The maximum likelihood estimators are obtained as iterative solutions of the likelihood equations $\frac{d \log L(\mathbf{b})}{d\mathbf{b}} = \mathbf{0}$. If the log-likelihood $\ell(\boldsymbol{b}) = \log L(\boldsymbol{b})$ is concave, then the MLE is unique when there exists at least one $\hat{\boldsymbol{b}}$ within the admissible parameter set where $\ell(\boldsymbol{b})$ attains the local or global maximum.

The asymptotic theory of the MLE works in this situation as well. Under mild regularity conditions, the MLE $\hat{\boldsymbol{b}}$ of $\boldsymbol{b}$ is asymptotically $N(\boldsymbol{b}, n^{-1}\boldsymbol{I}^{-1}(\boldsymbol{b}))$.

2.2 Bayesian Models

For a Bayesian model associated with the likelihood (2), we require a prior for $\mathbf{b}$. A commonly used choice is $N(\boldsymbol{b}_0, \boldsymbol{\Sigma})$, where $\boldsymbol{b}_0$ and $\boldsymbol{\Sigma}$ are known. This prior appears for example in Dellaportas and Smith (1993). Then, writing $\boldsymbol{y} = (y_1, \cdots, y_n)^T$, the posterior of $\boldsymbol{b}$ is given by

$$\pi(\boldsymbol{b}|\boldsymbol{y}) \propto \exp\left[\sum_{1}^{n} a^{-1}(\phi_i)\{y_i h(\boldsymbol{x}_i^T \boldsymbol{b}) - \psi(h(\mathbf{x}_i^T b))\} - \frac{1}{2}(\boldsymbol{b} - \mathbf{b}_0)^T \boldsymbol{\Sigma}^{-1}(\boldsymbol{b} - \mathbf{b}_0)\right]. \tag{5}$$

The above posterior is not analytically tractable. In fact, there does not exist any closed form expression for the norming constant. Also, finding posterior means, variances etc. by numerical integration is not easy even for moderate p. The most convenient approach seems to be the Markov Chain Monte Carlo (MCMC) numerical integration techniques which require generating samples from the posterior. They can be implemented in general using the Metropolis-Hastings algorithm, but if the posterior is log-concave, then one can also use the adaptive rejection sampling approach of Gilks and Wild (1992).

With little or no prior information, an alternative is to use noninformative priors. This implies that the posterior distribution is essentially the likelihood that the Bayesian analysis will be close to a likelihood analysis, possibly attractive to frequentists. One of the commonly used noninformative priors due to Laplace (1812) is $\pi_L(\boldsymbol{b}) \propto 1$. However, the following example due to Laud and Ibrahim (1991), shows that such a prior can sometimes lead to an improper posterior.

Example 1. Suppose $n = 1$, $p = 1$, and

$$f(y_1|x_1) = (bx_1)^{-1} \exp[-y_1/(bx_1)]; \;\; y > 0, x_1 > 0, \;\; b > 0.$$

Then if $\pi_L(b) \propto 1$

$$\pi_L(b|y_1) \propto (bx_1)^{-1} \exp[-y_1/(bx_1)]. \tag{6}$$

Integration with respect to b over $(0, \infty)$ gives $\int_0^\infty \pi_L(b|y_1)db = \int_0^\infty z^{-1} \exp\left(-\frac{y_1}{z}\right) dz = +\infty$, so that the posterior is improper.

Laud and Ibrahim (1991) proposed Jeffreys' prior for this problem given by $\pi_J(\boldsymbol{b}) \propto |\boldsymbol{I}(\boldsymbol{b})|^{1/2}$ with $\boldsymbol{I}(\boldsymbol{b})$ given in (4). They provided sufficient conditions under which the resulting posterior $\pi_J(\boldsymbol{b}|\boldsymbol{y})$ is proper.

The likelihood function given in (2) generalizes the normal fixed effects model to the one-parameter exponential family. Breslow and Clayton (1993) have extended this further to include mixed effects models. For a fixed effects model, under the link function $g \equiv h^{-1}$, we have $g(\theta_i) = \boldsymbol{x}_i^T \boldsymbol{b}$. In contrast, for the random effects model, one incorporates the random effects as well, and writes $g(\theta_i) = \boldsymbol{x}_i^T \boldsymbol{b} + \boldsymbol{z}_i^T \boldsymbol{u}_i$, where the $\boldsymbol{z}_i$ are also known and the $\boldsymbol{u}_i$ are i.i.d. $N(\boldsymbol{0}, \boldsymbol{\Sigma}_u)$. Breslow and Clayton (1993) advocated penalized quasi-likelihood estimates (PQL) for estimating $\boldsymbol{b}$.

We note an example of generalized linear mixed effects model (GLMM), considered in Crowder (1978), and also in Breslow and Clayton (1993). This concerns data on the proportion of seeds that germinated on each of 21 plates arranged according to a 2×2 factorial layout by seed variety and type of root extract. It turns out in this example that the within-group variation exceeds that predicted by the binomial sampling theory. The heterogeneity due to plate-to-plate variability is accounted for by Crowder (1978) and Breslow and Clayton (1993) by means of a GLMM that employs the canonical link setting

$$\theta_i = \boldsymbol{x}_i^T \boldsymbol{b} + u_i,$$

$i = 1, \cdots, 21$, where $\boldsymbol{b}$ represents the fixed effects associated with seed and the extract, and the u_i, assumed to be iid $N(0, \sigma_u^2)$, represent random effects associated the plates.

The Bayesian procedure, as before assigns a $N(\boldsymbol{b}_0, \boldsymbol{\Sigma}_b)$ distribution to $\boldsymbol{b}$. More generally, a hierarchical Bayesian model is considered whereby one assigns distributions to $\boldsymbol{\Sigma}_b$ and σ_u. One option is to use an inverse Wishart distribution for $\boldsymbol{\Sigma}_b$ and inverse gamma distribution for σ_u^2. Such distributions can possibly be improper, but care must be exercised in order that the resulting posterior is proper.

Specifically, let $\boldsymbol{\Sigma}_b$ and σ_u^2 have independent priors

$$\pi(\boldsymbol{\Sigma}_b) \propto \exp[-\frac{1}{2} tr(\boldsymbol{\Psi}\boldsymbol{\Sigma}_b^{-1})]|\boldsymbol{\Sigma}_b|^{-\frac{1}{2}\nu}, \tag{7}$$

and

$$\pi(\sigma_u^2) \propto \exp(-\frac{a}{2\sigma_u^2})(\sigma_u^2)^{-\frac{g}{2}-1}. \tag{8}$$

We shall write symbolically $\boldsymbol{\Sigma}_b \sim IW(\boldsymbol{\Psi}, \nu)$ and $\sigma_u^2 \sim IG(\frac{a}{2}, \frac{g}{2})$.

The joint posterior of $\boldsymbol{\theta} = (\theta_1, \cdots, \theta_n)^T, \boldsymbol{b}, \boldsymbol{\Sigma}_b$ and σ_u^2 is now given by

$$\begin{aligned}\pi(\boldsymbol{\theta}, \boldsymbol{b}, \boldsymbol{\Sigma}_b, \sigma_u^2|\boldsymbol{y}) &\propto \exp[\sum_{i=1}^{n}\{a^{-1}(\phi)(y_i\theta_i - \psi(\theta_i))\}] \\ &\times (\sigma_u^2)^{-\frac{n+g}{2}-1}\exp[-\frac{\sum_{i=1}^{n}(\theta_i - \boldsymbol{x}_i^T\boldsymbol{b})^2 + a}{2\sigma_u^2}] \\ &\times |\boldsymbol{\Sigma}_b|^{-\frac{p+\nu}{2}}\exp[-\frac{1}{2}tr\{(\boldsymbol{b}-\boldsymbol{b}_0)(\boldsymbol{b}-\boldsymbol{b}_0)^T + \boldsymbol{\Psi}\}\boldsymbol{\Sigma}_b^{-1}]. \quad (9)\end{aligned}$$

This posterior is analytically intractable, and one needs numerical integration for posterior analysis. Gibbs sampling (Gelfand and Smith, 1990; Gelfand, Hill, Racine-Poon and Smith, 1990) has proved to be very useful for implementation of the Bayesian model fitting. This requires sampling from the full conditionals

$$[\boldsymbol{b}|\boldsymbol{\theta}, \boldsymbol{\Sigma}_b, \sigma_u^2, \boldsymbol{y}] \sim N((\sigma_u^{-2}\boldsymbol{X}^T\boldsymbol{X}+\boldsymbol{\Sigma}_b^{-1})^{-1}(\sigma_u^{-2}\boldsymbol{X}^T\boldsymbol{\theta}+\boldsymbol{\Sigma}_b^{-1}\boldsymbol{b}_0), (\sigma_u^{-2}\boldsymbol{X}^T\boldsymbol{X}+\boldsymbol{\Sigma}_b^{-1})^{-1}]; \quad (10)$$

$$[\boldsymbol{\Sigma}_b|\boldsymbol{\theta}, \boldsymbol{b}, \sigma_u^2, \boldsymbol{y}] \sim IW(\boldsymbol{\Psi} + (\boldsymbol{b}-\boldsymbol{b}_0)(\boldsymbol{b}-\boldsymbol{b}_0)^T, \nu + p); \quad (11)$$

$$[\sigma_u^2|\boldsymbol{\theta}, \boldsymbol{b}, \boldsymbol{\Sigma}_b, \boldsymbol{y}] \sim IG(\frac{\sum_{i=1}^{n}(\theta_i - \boldsymbol{x}_i^T\boldsymbol{b})^2 + a}{2}, \frac{n+g}{2}); \quad (12)$$

$$\pi(\theta_i|\theta_j(j \neq i, \boldsymbol{b}, \boldsymbol{\Sigma}_b, \sigma_u^2, \boldsymbol{y}) \propto \exp[a^{-1}(\phi)(y_i\theta_i - \psi(\theta_i)) - \frac{(\theta_i - \boldsymbol{x}_i^T\boldsymbol{b})^2}{2\sigma_u^2}], \quad (13)$$

where $\boldsymbol{X} = (\boldsymbol{x}_1, \cdots, \boldsymbol{x}_n)^T$. It is easy to generate samples from the normal, inverse Wishart and the inverse gamma distributions. The only nonstandard conditionals are the $\pi(\theta_i|.)$, $i = 1, \cdots, n$, which are known only up to multiplicative constants. One can use the Metropolis-Hastings algorithm to generate samples. Alternately, one can use the adaptive rejection sampling of Gilks and Wild (1992) since these posteriors are log-concave.

The above model should be contrasted to that of Albert (1988). Albert (1988) begins with the likelihood given in (1), but does not model the θ_i as $\theta_i = h(\boldsymbol{x}_i^T\boldsymbol{b})$. Instead, he considers independent conjugate priors for the θ_i at the first stage, namely

$$\Pi(\theta_i|m_i, \lambda) = \exp[\lambda(m_i\theta_i - \psi(\theta_i)) + k(m_i; \lambda)] \quad (14)$$

An easy calculation shows that $E[\psi'(\theta_i)] = m_i$. He models the prior means m_i as $m_i = h(\boldsymbol{x}_i^T\boldsymbol{b})$. Thus Albert moves the GLM to the second stage specification, that is, for the θ_i's rather than the customary first stage specification for the y_i's. The remaining prior parameter λ is a precision parameter that reflects the strength of one's prior beliefs about the means m_i. As λ approaches infinity, the prior distribution of the $\psi'(\theta_i)$ becomes increasingly concentrated about the mean m_i, and the Bayesian model approaches the first stage specification in (2).

To complete the prior specification at the second stage, a distribution needs to be assigned to $\boldsymbol{b}$ and λ. West (1985) assigns a normal distribution to $\boldsymbol{b}$, and a chi-squared distribution to λ. Albert (1988) assigns instead the prior $\Pi(\boldsymbol{b}, \lambda) \propto (1+\lambda)^{-2}$, that is a priori $\boldsymbol{b}$ and λ are independent with $\boldsymbol{b} \sim$ uniform (R^p), and $\Pi(\lambda) \propto (1+\lambda)^{-2}$, a heavy tailed prior with infinite first moment.

With Albert's model, it is possible to calculate

$$E[\psi'(\theta_i)|\boldsymbol{b}, \lambda, \boldsymbol{y}] = [a^{-1}(\phi_i)y_i + \lambda m_i]/[a^{-1}(\phi_i) + \lambda].$$

This implies

$$E[\psi'(\theta_i)|\boldsymbol{y}_i] = y_i - E\left[\frac{\lambda}{a^{-1}(\phi_i) + \lambda}(y_i - m_i)|\boldsymbol{y}\right]. \quad (15)$$

Again, explicit evaluation of the conditional expectation given in the right hand side of (16) is difficult. Albert uses three approximations: (a) Laplace's method, (b) Quasi-likelihood approach, and (c) Brooks's method. A detailed application is given for the binomial-logit hierarchical model.

Albert's model differs from the one given in (5), (7) and (8) in that he connects the regression variables not directly with the parameters of interest, but indirectly with the prior means. The approximations that he considers are attractive alternatives to direct numerical integration, especially when p is large. On the other hand, these approximations do not appear to be simpler than using MCMC integration methods.

A slight variant of the model given in (1), (5), (7) and (8) is due to Zeger and Karim (1991). They consider a stratified sampling situation where y_{ij} is the response of the jth unit in the ith stratum. (In a longitudinal data or repeated measurements situation, j denotes the jth measurement on the ith subject). The corresponding vector of auxiliary characteristics or covariates is denoted by $\boldsymbol{x}_{ij}(p \times 1)$. The one-parameter exponential family model is now given by

$$f(y_{ij}|\theta_{ij}) = \exp[a^{-1}(\phi_{ij})\{y_{ij}\theta_{ij} - \psi(\theta_{ij})\} + c(y_{ij};\phi_{ij})], \tag{16}$$

where the link function $h(.)$ yields $h(\theta_{ij}) = \boldsymbol{x}_{ij}^T\boldsymbol{b} + \boldsymbol{z}_{ij}^T\mathbf{u}_i, j = 1, \cdots, n_i; i = 1, \cdots, m$. The $\boldsymbol{u}_i$ are iid $N(\mathbf{0}, \boldsymbol{\Sigma}_u)$.

At the final stage of the hierarchical model, one assigns mutually independent priors to $\boldsymbol{b}$ and $\boldsymbol{\Sigma}_u$ with $\boldsymbol{b} \sim$ uniform (R^p) and $\boldsymbol{\Sigma}_u \sim IW(1\boldsymbol{\Psi}, \nu)$. This posterior is also analytically intractable, and Gibbs sampling can be used to generate samples from the necessary conditionals.

The Zeger-Karim formulation does not include possible error in misspecifying the model. Ghosh, Natarajan, Stroud and Carlin (1998) consider slightly more general modeling of $g(\theta_{ij})$, namely $h(\theta_{ij}) = \boldsymbol{x}_{ij}^T\boldsymbol{b} + \boldsymbol{z}_{ij}^T\boldsymbol{u}_i + e_{ij}$. The errors e_{ij} account for model misspecification. A special case of the latter model will be discussed in the next section for studying the propriety of posteriors.

3. Propriety of Posteriors

If only proper priors are used, then one necessarily gets proper posteriors. However, as mentioned earlier, Bayesian analysis often relies on diffuse and flat priors which are mostly improper. In such instances, it is imperative to verify the propriety of posteriors. Otherwise, descriptive measures such as moments, quantiles etc. of the posteriors do not carry any meaning. Checking the propriety of posteriors is all the more important when the Bayesian procedure is implemented via MCMC technique as it may so happen that all the full conditionals are proper distributions, and yet the posterior is improper. (See Casella and George (1992) for an elementary example).

As shown in the previous section, Laplace's prior does not necessarily lead to a proper posterior. The same comment applies to Jeffreys' prior. Ibrahim and Laud (1991) have investigated conditions under which Jeffreys' prior leads to proper posteriors for GLMs. We present their main result below.

THEOREM 1. Consider the likelihood function given in (2), and Jeffreys' prior $\Pi_J(\boldsymbol{b}) \propto |\boldsymbol{I}(\boldsymbol{b})|^{1/2}$, where $\boldsymbol{I}(\boldsymbol{b})$ is given in (4). Assume that $r(\boldsymbol{X}) = p$ and the likelihood function is bounded. Then, a sufficient condition that the posterior distribution $\Pi(\boldsymbol{b}|\boldsymbol{y})$ is proper is that the integrals

$$\int \exp[(y_i\theta - \psi(\theta))/a(\phi_i)][\psi''(\theta)]^{1/2}d\theta \tag{17}$$

are finite for all $i = 1, \cdots, n$.

Ibrahim and Laud (1988) also show that a necessary and sufficient condition for Π_J to be proper is that $\int_\Theta [\psi''(\theta)]^{1/2}d\theta$ is finite where Θ denotes the parameter space. Dey, Gelfand and Peng (1997) extend these results to the case of overdispersed GLMs (see Section 5). Hobert and Casella (1996) have provided conditions ensuring propriety of posteriors for normal models. Natarajan and McCulloch (1995) have considered a version of a hierarchical model for binary data which ensures the propriety of posteriors.

Ghosh et al. (1998) have considered the model given in (14), but $h(\theta_{ij}) = \boldsymbol{x}_{ij}^T\boldsymbol{b} + u_i + e_{ij} (j = 1, \cdots, n_i; i = 1, \cdots, m)$. The u_i and the e_{ij} are mutually independent with the u_i iid $N(0, \sigma_u^2)$ and the e_{ij} are iid $N(0, \sigma^2)$. Also, $\boldsymbol{b}, \sigma_u^2$ and σ^2 are mutually independent with $\boldsymbol{b} \sim$ uniform $(R^p), \sigma_u^2 \sim IG\left(\frac{1}{2}a, \frac{1}{2}g\right)$ and $\sigma^2 \sim IG\left(\frac{1}{2}c, \frac{1}{2}d\right)$.

Ghosh et al. (1998) provide sufficient conditions for the propriety of posteriors under this model. A slightly more general version of their theorem is proved in Ghosh and Natarajan (1998) which we present below.

THEOREM 2: Assume that $f(y_{ij}|\theta_{ij})$ is bounded for all $j = 1, \cdots, n_i; i = 1, \cdots, m$. Let $S = \{(i,j) : \int f(y_{ij}|\theta_{ij})d\theta_{ij} < \infty\}$, and $s =$ cardinality of S. Assume that $s \geq 1$. Then the posterior $\pi(\boldsymbol{\theta}, \boldsymbol{b}, \sigma_u^2, \sigma^2|\boldsymbol{y})$ is proper if $a > 0, c > 0, m + g > 0$ and $s + d > p$.

In a more recent article, Gelfand and Sahu (1999) have linked the issue of propriety of posteriors with Bayesian identifiability. Suppose the Bayesian model is denoted by the likelihood $L(\boldsymbol{\theta}; y)$ and the prior $\pi(\boldsymbol{\theta})$. Suppose $\boldsymbol{\theta} = (\boldsymbol{\theta}_1, \boldsymbol{\theta}_2)$. Following Dawid (1979), if $\pi(\boldsymbol{\theta}_2|\boldsymbol{\theta}_1, \boldsymbol{y}) = \pi(\boldsymbol{\theta}_2|\boldsymbol{\theta}_1)$, we say that $\boldsymbol{\theta}_2$ is not identifiable. This means that if observing data $\boldsymbol{y}$ does not increase our prior knowledge about $\boldsymbol{\theta}_2$ given $\boldsymbol{\theta}_1$, then $\boldsymbol{\theta}_2$ is not identified by the data. Noting that

$$\pi(\boldsymbol{\theta}_2|\boldsymbol{\theta}_1, \boldsymbol{y}) \propto L(\boldsymbol{\theta}_1, \boldsymbol{\theta}_2; \boldsymbol{y})\pi(\boldsymbol{\theta}_2|\boldsymbol{\theta}_1)\pi(\boldsymbol{\theta}_1), \tag{18}$$

$\boldsymbol{\theta}_2$ is nonidentifiable if and only if $L(\boldsymbol{\theta}_1, \boldsymbol{\theta}_2; \boldsymbol{y})$ is free of $\boldsymbol{\theta}_2$, that is $L(\boldsymbol{\theta}; \boldsymbol{\theta}_2; \boldsymbol{y}) = L(\boldsymbol{\theta}_1, \boldsymbol{y})$. Hence, Dawid's formal definition of non-identifiability is equivalent to lack of identifiability of the likelihood. We may also observe that if $L(\boldsymbol{\theta}_1, \boldsymbol{\theta}_2; \boldsymbol{y})$ is free of $\boldsymbol{\theta}_2$, then the posterior

$$\pi(\boldsymbol{\theta}_1, \boldsymbol{\theta}_2|\boldsymbol{y}) \propto L(\boldsymbol{\theta}_1; \boldsymbol{y})\pi(\boldsymbol{\theta}_2|\boldsymbol{\theta}_2)\pi(\boldsymbol{\theta}_1) \propto \pi(\boldsymbol{\theta}_1|\boldsymbol{y})\pi(\boldsymbol{\theta}_2|\boldsymbol{\theta}_1) \tag{19}$$

is proper if and only if both $\pi(\boldsymbol{\theta}_1|\boldsymbol{y})$ and $\pi(\boldsymbol{\theta}_2|\boldsymbol{\theta}_1)$ are proper.

To see how (19) works for GLMs suppose that rank $(\boldsymbol{X}) = r < p$. Then it is possible to make a one-to-one transformation from $\boldsymbol{b}$ to $(\boldsymbol{\delta}, \boldsymbol{\rho})$ such that (2) has the alternate representation

$$L(\boldsymbol{\delta}, \boldsymbol{\rho}) \propto \exp\left[\sum_{i=1}^n a^{-1}(\phi_i)\{\boldsymbol{u}_i h(\boldsymbol{x}_{0i}^T\boldsymbol{\delta}) - \psi(h(\boldsymbol{x}_{0i}^T\boldsymbol{\delta}))\}\right], \tag{20}$$

where $\boldsymbol{X}_0^T = (\boldsymbol{x}_{01}, \cdots, \boldsymbol{x}_{on})$ is a $r \times n$ matrix of rank r. Thus the likelihood does not depend on $\boldsymbol{\rho}$. Since propriety of $\pi(\boldsymbol{b}|\boldsymbol{y})$ is equivalent to the propriety of $\pi(\boldsymbol{\delta}, \boldsymbol{\rho}|\boldsymbol{y})$, it follows from (19) and (20) that $\pi(\boldsymbol{b}|\boldsymbol{y})$ is proper if and only both $\pi(\boldsymbol{\delta}|\boldsymbol{y})$ and $\pi(\boldsymbol{\rho}|\boldsymbol{\delta})$ are proper. If the latter holds, all we need to verify is that $\pi(\boldsymbol{\delta}|\boldsymbol{y})$ is proper.

Gelfand and Sahu (1999) have shown that with the canonical link $\theta_i = \boldsymbol{x}_i^T \boldsymbol{b}$, if $\text{rank}(\boldsymbol{X}) = r < p$, a sufficient condition for a proper $\pi(\boldsymbol{b}|\boldsymbol{y})$ is that the likelihood is bounded, and that at least r of the y_i's belong to the interiors of their respective domains.

4. Semiparametric GLMs

Sections 2 and 3 contain discussion of fully parametric Bayesian GLMs. In this section, we enrich this class of models by wandering nonparametrically near (in a suitable sense) this class. As a result, parts of the modeling are captured parametrically, in particular, the linear regression structure on some monotonically increasing transformed scale of the canonical parameters. Other aspects such as the link function or the distribution of the random effects are specified nonparametrically. This is an example of what has now become known as semiparametric regression modeling.

One of the difficulties with Bayesian modeling in the nonparametric case is that, unlike its parametric counterpart where the dimension of the parameter space is finite, nonparametric modeling requires an "infinite dimensional" parameter. Thus, the Bayesian approach, in assuming all unknowns are random, requires an infinite dimensional stochastic specification. However, significant advancement of research in this area during the past twenty five years or so has provided tractable ways to make such specifications. Further, more recent advances in Bayesian computation enable the fitting of models incorporating these specifications and even extensions of these specifications.

The various probabilistic specifications which yield Bayesian nonparametric modeling include discrete mixtures, Dirichlet processes, mixtures of Dirichlet processes, Polya tree distributions, Gamma processes, extended Gamma processes and Beta processes. Gelfand (1998) contains a review of all these. For brevity, we will discuss only the discrete mixtures and mixtures of Dirichlet processes in the context of conditionally independent hierarchical GLMs.

We may note that the basic object which we are attempting to model is an unknown function, say $g(\cdot)$. The parametric approach writes g as $g(\cdot, \boldsymbol{\theta})$, $\boldsymbol{\theta} \epsilon \Theta$ and then places a prior distribution over $\boldsymbol{\theta} \epsilon \Theta$. The nonparametric approach assumes only that $g \in \mathcal{G}$, where $\mathcal{G}$ is some class of functions. A Bayesian approach requires assigning a prior over the elements of $\mathcal{G}$. In what follows, we illustrate this with examples where the elements of $\mathcal{G}$ are monotone functions.

We begin our discussion with mixture models. First we notice that modeling a strictly monotone function g is equivalent to modeling a distribution function. For instance, if the range of g is R^1, then $T(g(\cdot))$ with $T(z) = k_1 \exp(k_2 z)/[1 + k_1 \exp(k_2 z)]$ with $k_1 > 0, k_2 > 0$ is a df. Similarly, if the range of g is R^+, then $T(g(\cdot))$ with $T(z) = k_1 z^{k_2}(1 + k_1 z^{k_2})^{-1}$ with $k_1 > 0, k_2 > 0$ is a df.

The mixture model approach models an unknown df using a dense class of mixtures of standard distributions. For instance, Diaconis and Ylvisaker (1985) observe that discrete mixture of Beta densities provide a dense class of models for densities on $[0, 1]$.

As a special case, consider modeling the link function h in a generalized linear model. Mallick and Gelfand (1994) suggested modeling $h(\theta)$ by

$$T(h(\theta)) = \sum_{\ell=1}^{r} w_\ell IB(T(h_0(\theta); \;\; c_\ell, d_\ell), \tag{21}$$

where r is the number of mixands, $w_\ell \geq 0, \sum_{\ell=1}^{r} w_\ell = 1$ are the mixing weights, $IB(u,c,d) = \int_0^u \{x^{c-1}(1-x)^{d-1}/B(c,d)\}dx$, and h_0 is a centering function for h. Then (21) provides a generic member of the dense class. Inversion of (21), provides a generic h. Since h is determined by specification of $r, \boldsymbol{w}^{(r)} = (w_1, \cdots, w_r)$, $\boldsymbol{c}^{(r)} = (c_1, \cdots, c_r)$ and $\boldsymbol{d}^{(r)} = (d_1, \cdots, d_r)$, introducing a distribution on g requires specification of a distribution of the form $f(r)f(\boldsymbol{w}^{(r)}, \boldsymbol{c}^{(r)}\boldsymbol{d}^{(r)}|r)$. Mallick and Gelfand (1994) suggest-fixing $c^{(r)}$ and $d^{(r)}$ to provide a set of Beta densities which "blanket" $[0,1]$, for example $c_\ell = \ell, d_\ell = r+1-\ell, \ell = 0,1,\cdots,r$, but treat $\boldsymbol{w}^{(r)}$ as random, given r. As shown by these authors, if $\boldsymbol{w}^{(r)} \sim$ Dirichlet $(\alpha \mathbf{1}_r)$, where $\mathbf{1}_r = (1, \cdots 1)^T$, then $E[T(h(\theta))] \approx T(h_0(\theta))$, that is h is roughly centered about h_0. This class of models is easy for inferential purposes. Since finding the posterior of h is equivalent to finding the posterior of $\boldsymbol{w}^{(r)}$, all one needs is to generate samples from the latter.

Next we turn to Dirichlet processes. Since the appearance of the classic paper of Ferguson (1973), such processes have been used quite extensively for Bayesian nonparametric inference. A probability measure G on $\mathcal{G}$ is said to follow a Dirichlet process with parameter αG_0, symbolically written as $G \sim DP(\alpha G_0)$, if for any measurable partition $B_1, \cdots, B_m$ of $\mathcal{G}$, $(G(B_1), \cdots, G(B_m)) \sim$ Dirichlet $(\alpha G_0(B_1), \cdots, \alpha G_0(B_m))$. Here, G_0 is a specified probability measures, and α is the "precision" parameter. This name for α is justified since $V[G_0(B_\ell)] = G_0(B_\ell)(1-G_0(B_\ell))/(\alpha+1)$ which decreases in α for every ℓ.

Computationally, it is most convenient to work with a family of Dirichlet mixture distributions. Let $\{f(\cdot|\boldsymbol{\theta}), \boldsymbol{\theta}\epsilon\Theta \subset R^P\}$ be a parametric family of densities with respect to some dominating measure μ. Consider the family of probability distributions $\mathcal{F} = \{F_G : G\epsilon\mathcal{G}\}$ with densities

$$f(y|G) = \int f(y|\boldsymbol{\theta})dG(\boldsymbol{\theta}). \tag{22}$$

Here $G(\boldsymbol{\theta})$ is viewed as the conditional distribution of $\boldsymbol{\theta}$ given G. It is assumed that $G \sim DP(\alpha G_0)$, whence $f(y|G)$ arises by mixing with respect to a distribution having a Dirichlet process.

Mukhopadhyay and Gelfand (1997) provide a general discussion of Bayesian inference based on Dirichlet process mixed models. Mixtures of Dirichlet processes were first introduced in Antoniak (1974), and have been considered subsequently in Lo (1984), Brunner and Lo (1989, 1994), (1991), Escobar (1994), Escobar and West (1995), MacEachern and Müller (1994), Gelfand and Mukhopadhyay (1995), and Newton, Czado and Chappell (1996) among others.

To illustrate the implementation of the semiparametric Bayesian procedure in the context of GLMs, we begin with conditionally independent observations y_i with pdf given in (1). In the next stage, we model $h(\theta_i) = \boldsymbol{x}_i^T\boldsymbol{b}+u_i$, where the u_i are iid from G with $G \sim DP(\alpha G_0)$. Integrating over G, the u_i have joint pdf $f(u_1, \cdots, u_n|G_0, \alpha)$. They are no longer independent, but this joint distribution can be written explicitly.

To complete the prior specification, one needs to specify G_0, and also specify a distribution for $\boldsymbol{b}$ and α. Typically, one assigns a $N(\boldsymbol{b}_0, \boldsymbol{\Sigma})$ distribution for $\boldsymbol{b}$ and, following Escobar and West (1995), a gamma distribution for α. (Empirical experience suggests setting $\alpha = 1$ may be preferable to adding a hyperprior.) A simple choice for G_0 is a normal distribution. However, the Bayesian model fitting can be implemented, at least in principle, for an arbitrary G_0 and α.

In order to implement the model fitting, one begins with the joint posterior of $\boldsymbol{b}$

and $\boldsymbol{u} = (u_1, \cdots, u_n)^T$ as

$$\begin{aligned} &\pi(\boldsymbol{b}, \boldsymbol{u}, \alpha | G_0, \boldsymbol{y}) \\ &\propto \exp\left[\sum_{i=1}^{n} \left\{y_i h^{-1}\left(\boldsymbol{x}_i^T \boldsymbol{b} + u_i\right) - \psi\left(h^{-1}\left(\boldsymbol{x}_i^T \boldsymbol{b} + u_i\right)\right)\right\}\right] \\ &\times \exp\left[-\tfrac{1}{2}\left(\boldsymbol{b} - \boldsymbol{b}_0\right)^T \boldsymbol{\Sigma}^{-1}\left(\boldsymbol{b} - \boldsymbol{b}_0\right)\right] f\left(u_1, \cdots, u_n | G_0, \alpha\right) \pi\left(\alpha\right). \end{aligned} \tag{23}$$

The MCMC implementation of the Bayesian procedure requires generating samples from the full conditionals

(a)
$$\begin{aligned} \pi\left(\boldsymbol{b} | \boldsymbol{u}, \alpha, G_0, \boldsymbol{y}\right) &\propto \exp\left[\sum_{i=1}^{n} \left\{y_i h^{-1}\left(\boldsymbol{x}_i^T \boldsymbol{b} + u_i\right) - \psi\left(h^{-1}\left(\boldsymbol{x}_i^T \boldsymbol{b} + u_i\right)\right)\right\}\right] \\ &\times \exp\left[-\tfrac{1}{2}\left(\boldsymbol{b} - \boldsymbol{b}_0\right)^T \boldsymbol{\Sigma}^{-1}\left(\boldsymbol{b} - \boldsymbol{b}_0\right)\right]; \end{aligned} \tag{24}$$

(b) $\pi\left(u_i | \boldsymbol{b}, u_j \left(j \neq i\right), \alpha, G_0, \boldsymbol{y}\right)$, a mixed density placing point masses proportional to $f(y_i | \boldsymbol{b}, u_j)$ at each $u_j (j \neq i)$ and continuous mass proportional to $\alpha f(y_i | \boldsymbol{b}, u_i)$ $f(u_i | G_0)$;

(c) $\pi(\alpha | \boldsymbol{b}, \boldsymbol{u}) \propto f(u_1, \cdots, u_n | G_0, \alpha) \pi(\alpha)$. If $\pi(\alpha)$ is gamma, Escobar and West (1995) have discussed how to generate samples from $\Pi(\alpha | \boldsymbol{b}, \boldsymbol{u})$ by introducing an additional parameter, say, γ. The reader is referred to their paper for details.

5. Overdispersed Generalized Linear Models

Being based on the one-parameter exponential family of distributions, GLMs assume a known functional relationship between the mean and the variance. This makes these models unsuitable for certain applications, especially those where the samples are too heterogeneous to be explained by such a simple functional relationship. In such instances, one is naturally led to a wider class of models.

A popular approach for creating a larger class has been through mixture models. For instance, the one parameter exponential family defining the GLM is mixed with a two parameter exponential family for the canonical parameter θ (or equivalently the mean parameter μ) resulting in a two parameter marginal mixture family for the data. The resulting overdispersed family of mixture models no longer belongs to the exponential family (e.g. beta-binomial, gamma-Poisson). More importantly, since the likelihood depends on the sample size, while the mixture distribution does not, the relative overdispersion of the resulting mixture family to the original exponential family tends to infinity as the sample size increases. An implication of such models is that taking additional observations within a population does not increase knowledge regarding heterogeneity across populations.

A second class of models, usually referred to as exponential dispersion models (EDM), arises when $a(\phi) = \phi$ so that ϕ behaves like a scale parameter. Jorgensen (1987) provides an extensive treatment of such models. The resulting two parameter family of distributions no longer belongs to the exponential family.

An alternative approach due to Efron (1986) models overdispersion through so-called "double-exponential" families. Such families are derived as a saddle point approximation to the density of an average of n^* random variables from a one parameter exponential family with large n^*. The parameter n^* written as $n\rho$ for actual sample size n introduces ρ as a second parameter in the model along with canonical parameter θ. Ganio and Schafer (1992) have shown that EDM's can be embedded within Efron's double exponential family, and the associated asymptotic inference applies. These asymptotics result in overdispersion relative to the original exponential family which tends to a constant as $n \to \infty$, unlike the mixture case.

A more general class of models was introduced by Gelfand and Dalal (1990). For a given one-parameter exponential family, they introduced a two-parameter exponential family where one parameter is the overdispersion parameter in addition to the canonical parameter. This model includes Efron's model as a special case, and also includes a family discussed in Lindsay (1986).

Dey, Gelfand and Peng (1997) adopted a Bayesian approach for fitting these models using Jeffreys' prior. Following the technique of Laud and Ibrahim (1991), they also proved the propriety of posteriors under such priors under certain conditions. Some of their results are presented below.

We begin with conditionally independent random variables y_i $(i = 1, \cdots, n)$ such that

$$f(y_i|\theta_i, \tau_i) = b(y_i) \exp[\theta_i y_i + \tau_i T(y_i) - \rho(\theta_i, \tau_i)]. \tag{25}$$

In the above f is a density with respect to some σ-finite measure μ. Assuming that (25) is integrable with respect to y_i, if $T(y_i)$ is convex, then for distributions with common means, $V(y_i)$ increases in τ_i. Let $\boldsymbol{y} = (y_1, \cdots, y_n)^T$, and define $\theta_i = h_1(\boldsymbol{x}_i^T \boldsymbol{b})$ and $\tau_i = h_2(\boldsymbol{z}_i^T \boldsymbol{\alpha})$, where h_1 and h_2 are strictly increasing. The resulting likelihood is

$$L(\boldsymbol{b}, \boldsymbol{\alpha}; \boldsymbol{y}) = \exp[\sum_{i=1}^n \{\theta_i y_i + \tau_i T(y_i) - \rho(\theta_i, \tau_i)\}]. \tag{26}$$

We shall use the notation $\rho^{(r,s)} = \partial\rho^{r+s}/(\partial\theta^r \partial\tau^s)$. Then straightforward calculations yield

$$E\left(-\frac{\partial^2 \log L}{\partial b_j \partial b_k}\right) = \sum_{i=1}^n \rho^{(2,0)}(\theta_i, \tau_i) x_{ij} x_{ik} [g'(\boldsymbol{x}_i^T \boldsymbol{b})]^2; \tag{27}$$

$$E\left(-\frac{\partial^2 \log L}{\partial \alpha_j \partial \alpha_k}\right) = \sum_{i=1}^n \rho^{(0,2)}(\theta_i, \tau_i) z_{ij} z_{ik} [h'(\boldsymbol{z}_i^T \boldsymbol{\alpha})]^2; \tag{28}$$

$$E\left(-\frac{\partial^2 \log L}{\partial b_j \partial \alpha_k}\right) = \sum_{i=1}^n \rho^{(1,1)}(\theta_i, \tau_i) x_{ij} z_{ik} g'(\boldsymbol{x}_i^T \boldsymbol{b}) h'(\boldsymbol{z}_i^T \boldsymbol{\alpha}). \tag{29}$$

Writing $\boldsymbol{X}^T = (\boldsymbol{x}_1, \cdots, \boldsymbol{x}_n)$, $\boldsymbol{Z}^T = (\boldsymbol{z}_1, \cdots, \boldsymbol{z}_n)$, $\boldsymbol{M}_\theta$ as the $n \times n$ diagonal matrix with $(\boldsymbol{M}_\theta)_{ii} = \rho^{(2,0)}(\theta_i, \tau_i)(g'(\boldsymbol{x}_i^T \boldsymbol{b}))^2$, $\boldsymbol{M}_\tau$ a diagonal matrix with $(\boldsymbol{M}_\tau)_{ii} = \rho^{(0,2)}(\theta_i, \tau_i)(h'(\boldsymbol{z}_i^T \boldsymbol{\alpha}))^2$, and $\boldsymbol{M}_{\theta,\tau}$ a diagonal matrix with $(\boldsymbol{M}_{\theta,\tau})_{ii} = \rho^{(1,1)}(\theta_i, \tau_i) g'(\boldsymbol{x}_i^T \boldsymbol{b}) h'(\boldsymbol{z}_i^T \boldsymbol{\alpha})$, one gets the Fisher information matrix

$$\boldsymbol{I}(\boldsymbol{b}, \boldsymbol{\alpha}) = \begin{pmatrix} \boldsymbol{X}^T \boldsymbol{M}_\theta \boldsymbol{X} & \boldsymbol{X}^T \boldsymbol{M}_{\theta,\tau} \boldsymbol{Z} \\ \boldsymbol{Z}^T \boldsymbol{M}_{\theta,\tau} \mathbf{X} & \boldsymbol{Z}^T \boldsymbol{M}_\tau \boldsymbol{Z} \end{pmatrix}. \tag{30}$$

Jeffreys' prior is then given by $|\boldsymbol{I}(\boldsymbol{b}, \boldsymbol{\alpha})|^{1/2}$.

Suppose we assume that $\boldsymbol{X}$ and $\boldsymbol{Z}$ are of full column rank, and that $L(\boldsymbol{b}, \boldsymbol{\alpha}; \boldsymbol{y})$ is bounded above. Then, the posterior of $(\boldsymbol{b}, \boldsymbol{\alpha})$ is proper if, for each $y_i (i = 1, \cdots, n)$,

$$\int_T \int_\Theta \exp[\theta y_i + \tau T(y_i) - \rho(\theta, \tau)] (\rho^{(2,0)}(\theta, \tau) \rho^{(0,2)}(\theta, \tau))^{1/2} d\theta d\tau < \infty. \tag{31}$$

This result is proved in Dey et al. (1997).

Recognizing the limitations of the one parameter exponential family, for example, an implicit mean-variance relationship and unimodality, Mukhopadhyay and

Gelfand (1997) introduced a Dirichlet process mixed GLM (DPMGLM). These models provide a more flexible first stage specification, while retaining linear structure on a transformed scale.

6. Model Determination Approaches

With the availability of a wide range of GLMs to consider in analyzing a dataset, the problem of model determination becomes critical. Model determination comprises model checking - is the model adequate? - and model selection - among a set of adequate models, which one is best?

First we consider model adequacy which has received much less attention in the literature than model choice. In providing the probabilistic components of a hierarchical model, we rarely believe that any of the distributions is correct. Those specifications further removed from the data are often intentionally made less precise, not because we *believe* them to be correct but in order to permit the data to drive the inference. However, what is *true* is apart from model checking. If we undertake model criticism we must examine the adequacy of what is specified and we must assume proper priors (or else the observed data could not have arisen under the model). High dimensional models, e.g., those having more parameters than data points, as well as very vaguely specified hierarchical models will be difficult to criticize.

A formal Bayesian model adequacy criterion (as in Box, 1980) proposes that the marginal density of the data be evaluated at the observations. Large values support the model, small values do not. Assessment of the magnitude of this value could be facilitated by standardizing, using the maximum value or an average value of this density (Berger, 1985). However, a high dimensional density ordinate will be difficult to estimate well and hopeless to calibrate. In addition, with hierarchical models, failures, such as outliers, mean structure errors, dispersion misspecifications and inappropriate exchangeabilities, can occur at each hierarchical stage. The formal procedure does not provide feedback regarding the adequacy of the stagewise specifications.

Chaloner and Brant (1988), Chaloner (1994) and Weiss (1995), focusing on outlier detection suggest posterior-prior comparison. Their strategy is to identify random variables whose distribution, a priori, is a standard one. In particular, they choose functions of so-called realized residuals. Given the data, the posterior distribution of each such function is obtained. If it differs considerably from its associated prior, using tail area comparison, a lack of model fit is claimed. For a realized residual itself, an outlying observation is asserted. If the entire model specification is correct, such comparisons will be successful on average but will fail to recognize the *variability* in the posterior.

A second approach, referred to as model expansion or elaboration, captures model failures by specifying a more complex model using mixtures. Though most often used to detect outliers, recently, Albert and Chib (1997) use this approach for other model failures, in particular, exchangeability in the direction of partial exchangeability. Regardless, the model of interest becomes nested within the expanded or full model, so model choice procedures replace model checking to criticize the adequacy of the reduced one. Recent work of Müller and Parmigiani (1995) and Carota, Parmigiani and Polson (1993) combines elaboration with posterior-prior comparison using the Kullback-Leibler distance between these two distributions for the elaboration parameter. This approach requires, for each sort of failure, a non-unique specification of an expanded model.

A third approach is taken up in Gelman, Meng and Stern (1995) who propose a posterior predictive strategy. These authors define a discrepancy measure as a function of data and parameters, treating both as unknown in one case, inserting the observed data in the other. They then compare the resulting posterior distributions given the observed data. Gelman, et al. dismiss prior predictive checking arguing that the prior predictive distribution treats the prior as a true "population distribution" whereas the posterior predictive distribution treats the prior as an outmoded first guess. However, model checking must examine the acceptability of the model fitted to the data. Model parameters must be generated from the prior prescribed under the model. Gelman, et al. can be criticized for using the data twice. The observed data, through the posterior, suggests values of the parameter which are likely under the model. Then, to assess adequacy, the observed data is checked against data generated using such parameter values, apparently making it difficult to criticize the model.

A fourth approach is developed in recent work of Hodges (1998). Limited to the case where all levels are Gaussian, he reexpresses linear hierarchical models as standard linear models with simple covariance structure. He then suggests the use of familiar linear models diagnostic tools, e.g., residual plots, added variable plots, transformations, collinearity checks, case influence, etc. Ad-hoc method is needed in tailoring some of these tools to the hierarchical structure.

Finally, Dey, Gelfand, Swartz and Vlachos (1998) suggest an approach which is entirely simulation based, requiring only the model specification and that, for a given data set, one be able to simulate draws from the posterior under the model. By replicating a posterior of interest using data replicates obtained under the model, the extent of variability in such a posterior can be seen. Then, the posterior obtained under the observed data can be compared with this medley of posterior replicates to ascertain whether the former is in agreement with them and accordingly, whether it is plausible that the observed data come from the proposed model. Such comparison can be implemented using a Monte Carlo test. Many such tests can be run, each focusing on a potential model failure.

Turning to model choice, for a collection of models $m = 1, 2, \cdots, M$, the formal Bayesian approach assumes that one is "true" but which is the true one is unknown. Assigning prior probabilities p_m that model m is true, the posterior probability of model m, is $Pr(m|\boldsymbol{y}) \propto f(\boldsymbol{y}|m)p_m$ where $f(\boldsymbol{y} \mid m)$ is the marginal or prior predictive density of $\boldsymbol{y}$ under model m. Hence, if $\boldsymbol{y}_{obs}$ denotes the realized data, the model which maximizes $f(\boldsymbol{y}_{obs}|m)p_m$ is selected. If $p_m = M^{-1}$ for all m, we choose the model with the largest $f(\boldsymbol{y}_{obs}|m)$, suggesting the use of this quantity as a general screening criterion. When models are compared in pairs, the Bayes factor emerges, $B = f(\boldsymbol{y}_{obs}|m_1)/f(\boldsymbol{y}_{obs}|m_2)$ for say models m_1 and m_2. B is viewed as a weight of evidence; $B > 1$ supports model $m_1, B < 1$ supports model m_2.

Bayes factors have a wide advocacy in the Bayesian community; see Kass and Raftery (1995) for a review. However, they lack interpretation in the case of improper priors which are frequently used in complex hierarchical specifications and they are difficult to compute for such models with large datasets (though there is much recent discussion, see, e.g., Raftery, 1995). The use of Schwarz's (1978) Bayesian information criterion (BIC) as an approximation to the Bayes factor requires the specification of model dimension. Unfortunately in the context of GLMs involving mixed effects, the dimension of the model is unclear. Moreover, the asymptotics associated with such approximation are invalid when the number of model parameters grows with sample size, as in random effects settings where the number of individuals grows large.

Recently, attractive alternatives have appeared. Gelfand and Ghosh (1998), noting that posterior prediction is often a primary use for a model, suggest a formal utility maximization approach for model selection. In particular, their approach amounts to obtaining a minimized expected posterior predictive loss for a given model and then selecting the model which provides the overall minimum.

For a version of log scoring (or deviance) loss, the minimization for a given model can be done explicitly yielding an expression which can be interpreted as a penalized deviance criterion. The criterion is comprised of a piece which is a Bayesian deviance measure and a piece which is interpreted as a penalty for model complexity. The penalty function arises without specifying model dimension or asymptotic justification.

Under the model in (1) with $a(\phi_i) = \phi/w_i, w_i$ known, the criterion becomes, for model m,

$$\begin{aligned} D_k(m) &= 2\textstyle\sum_{i=1}^n w_i(t_i^{(m)} - t(\mu_i^{(m)}))+ \\ &+2(k+1)\textstyle\sum_{i=1}^n w_i \left\{ \frac{t(\mu_i^{(m)})+kt(y_{i,obs})}{k+1} - t\left(\frac{\mu_i^{(m)}+ky_{i,obs}}{k+1}\right)\right\} \end{aligned} \tag{32}$$

In (32), $t(y) = y\theta(y) - \psi(\theta(y))$ where $\theta(\mu) = \psi'^{-1}(\mu), \mu_i^{(m)} = E(y_i|\boldsymbol{y}_{obs}, m)$ and $t_i^{(m)} =$
$E(t(y_i)|\boldsymbol{y}_{obs}, m)$, and k is a weight which typically does not affect the ordering of the models and so may be set to 1 for convenience. Since $t(y)$ is convex, Jensen's inequality ensures that each term in the right side of (32) is nonnegative. Gelfand and Ghosh clarify that the first term can be interpreted as a penalty function and the second as a goodness-of-fit term. The choice (1) determines $t(y)$, hence (32). For instance, in the Possion case $t(y) = y\log y - y$, in the binomal case, $t(y) = \log\left(\frac{y}{n}\right) + \frac{n-y}{n}\log\left(\frac{n-y}{n}\right)$. Usual continuity corrections are imposed to ensure that $t(y)$ can be calculated for any $y_{i,obs}$ and that $t_i^{(m)}$ exists.

Another, somewhat similar criterion has been discussed in Spiegelhalter, Best and Carlin (1998). Motivated by the work of Dempster (1974), their suggestion is to obtain the posterior distribution of the log likelihood at the observed data for each model and then compare these across models. In particular, for (1) they define the "Bayesian deviance" $D(\boldsymbol{\theta})$ to be $\sum_{i=1}^n D(\theta_i)$ where

$$D(\theta_i) = -2\log f(y_i|\theta_i) + 2\log f(y_i|\theta(y_i)). \tag{33}$$

Defining $\overline{D} = E(D(\boldsymbol{\theta})|y)$ and $p_D = \overline{D} - D(E(\boldsymbol{\theta}|\boldsymbol{y}_{obs}))$, Spiegelhalter, et al. propose the criterion

$$DIC = \overline{D} + p_D \tag{34}$$

where DIC denotes Deviance Information Criteria. They argue that $\overline{D}$, the posterior expected deviance, summarizes model fit while p_D, interpreted as the effective number of parameters, measures the complexity of the model. They show that DIC generalizes the familiar Akaike Information Criteria (AIC) (Akaike, 1973).

Both the Gelfand and Ghosh criteria and the DIC are readily computed from posterior samples. The BUGS software (Spiegelhalter, et al. 1996) provides a convenient package for fitting most Bayesian GLMs, and thus for providing posterior samples.

Finally, informal Bayesian model selection in the case of nested GLMs can be effected by obtaining the posterior distribution of the discrepancy parameter between the full and reduced models as in Albert and Chib (1997). Exploratory approaches using cross validation ideas, applicable to small or even moderate sized datasets are discussed in Gelfand, Dey and Chang (1992) and Gelfand (1995).

Acknowledgement

The work of the first author was supported in part by NSF grant DMS 96-25383. The work of the second author was supported in part by NSF grant SBR 98-10968.

References

Akaike, H. (1973). Information theory and an extension of the maximum likelihood principle. In *Proceedings of International Symposium on Information Theory*, ed. B.N.Petrov and F. Czaki, Budapest, Academia Kiado, 267-281.

Albert, J.H. (1985). Simultaneous estimation of Poisson means under exchangeable and independence models. *Journal of Statistical Computation and Simulation*, **23**, 1-14.

Albert, J.H. (1988). Computational methods using a Bayesian hierarchical generalized linear model. *Journal of the American Statistical Association*, **83**, 1037-1044.

Albert, J.H. and S. Chib (1997). Bayesian tests and model diagnostics in conditionally independent hierarchical models. *Journal of the American Statistical Association*, **92**, 916-925.

Antoniak, C.E. (1974). Mixtures of Dirichlet processes with applications to nonparametric problems. *The Annals of Statistics*, **2**, 1152-1174.

Berger, J.O. (1985). *Statistical Decision Theory and Bayesian Analysis*. Springer-Verlag, New York.

Box, G.E.P. (1980). Sampling and Bayes's inference in scientific modeling (with discussion). *Journal of the Royal Statistical Society, Series A*, **143**, 383-430.

Breslow, N.E. and D.G. Clayton (1993). Approximate inference in generalized linear mixed models. *Journal of the American Statistical Association*, **88**, 9-25.

Brunner, L.J. and Lo, A.Y. (1989). Bayes methods for a symmetric unimodal density and its mode. *The Annals of Statistics*, **17**, 1550-1566.

Brunner, L.J. and Lo, A.Y. (1994). Nonparametric Bayes methods for directional data. *The Canadian Journal of Statistics*, **22**, 401-412.

Carota, C., G. Parmigiani and N.G. Polson (1997). Diagnostic measures for model criticism. *Journal of the American Statistical Association*, **92**, 753-762.

Casella, G and George, E.I. (1994). Explaining the Gibbs sampler. *The American Statistician*, **46**, 167-174.

Chaloner, K. (1994). Residual analysis and outliers in Bayesian hierarchical models. In: *Aspects of Uncertainty*, eds. A.F.M. Smith and P.R. Freeman. Chichester, U.K., John Wiley, 153-161.

Chaloner, K. and R. Brant (1988). A Bayesian approach to outlier detection and residual analysis. *Biometrika*, **75**, 651-659.

Crowder, M. (1978). Beta-binomial ANOVA for proportions. *Applied Statistics*, **27**, 34-37.

Dawid, A.P. (1979). Conditional independence in statistical theory. *Journal of the Royal Statistical Society, series B*, **41**, 1-31.

Dellaportas, P. and Smith, A.F.M. (1993). Bayesian inference for generalized linear and proportional hazards models via Gibbs sampling. *Applied Statistics*, **42**, 443-459.

Dempster, A. (1974). The direct use of likelihood for significance testing. In: *Proceedings of Conference on Foundational Questions In Statistical Inference.* Eds: O. Barndorff-Nielsen, P. Blaesild and G. Schou, p. 335-352, Department of Theoretical Statistics: University of Aarhus.

Dey, D.K., Gelfand, A.E. and Peng, F. (1997). Overdispersed generalized linear models. *Journal of Statistical Planning and Inference,* **64**, 93-107.

Dey, D.K., Gelfand, A.E., Swartz, T. and Vlachos, P.K. (1998). Simulation based model checking for hierarchical models. *Test*, **7**, 325-346.

Efron, B. (1986). Double exponential families and their use in generalized linear regression. *Journal of the American Statistical Association*, **81**, 709-721.

Escobar, M.D. (1994). Estimating normal means with Dirichlet process priors. *Journal of the Statistical Planning and Inference*, **43**, 97-106.

Escobar, M.D. and West, M. (1995). Bayesian density estimation and inference using mixtures. *Journal of the American Statistical Association*, **90**, 577-588.

Fahrmeir, L and Tutz, G. (1994). *Multivariate Statistical Modelling based on Generalized Linear Models.* Springer-Verlag, New York.

Ferguson, T.S. (1973). A Bayesian analysis of some nonparametric problems. *The Annals of Statistics*, 1, 209-230.

Ganio, L.M. and Schafer, D.W. (1992). Diagnostics of overdispersion. *Journal of the American Statistical Association*, **87**, 795-804.

Gelfand, A.E. (1995). Model determination using sampling-based methods. In *Markov Chain Monte Carlo in Practice*, eds. W. Gilks et al., London, Chapman and Hall, 145-161.

Gelfand, A.E. (1998). Approaches for semiparametric Bayesian regression. In: *Asymptotics, Nonparametrics and Time Series.* Ed.: S. Ghosh, Marcel Dekker, Inc, New York. (to appear).

Gelfand, A.E. and Dalal S. (1990). A note on overdispersed exponential families. *Biometrika*, **77**, 55-64.

Gelfand, A.E., D.K. Dey and H. Chang (1992). Model determination using predictive distributions with implementations via sampling-based methods. In: *Bayesian Statistics 4*, eds. J.M. Bernardo et al., Oxford, U.K., Oxford University Press, 147-167.

Gelfand, A.E. and Ghosh, S.K. (1998). Model choice: a minimum posterior predictive loss approach. *Biometrika*, **85**, 1-11.

Gelfand, A.E., Hills, S.E., Racine-Poon, A. and Smith, A.F.M. (1990). Illustration of Bayesian inference in normal data models using Gibbs sampling. *Journal of the American Statistical Association*, **85**, 972-985.

Gelfand, A.E. and Sahu, S.K. (1999). On the propriety of posteriors and Bayesian identifiability in generalized linear models. *Journal of the American Statistical Association* (to appear).

Gelfand, A.E. and Smith, A.F.M. (1990). Sampling-based approaches to calculating marginal densities. *Journal of the American Statistical Association*, 85, 398-409.

Gelman, A., X-L. Meng and H.S. Stern (1995). Posterior predictive assessment of model fitness via realized discrepancies (with discussion). *Statistica Sinica*, **6**, 733-807.

Ghosh, M., and Natarajan, K. (1998). Small area estimation : a Bayesian perspective. *Multivariate, Design and Sampling*, Ed., S. Ghosh, Marcel Dekker, New York (to appear).

Ghosh, M., Natarajan, K., Stroud, T.W.F. and Carlin, B.P. (1998). Generalized linear models for small-area estimation. *Journal of the American Statistical Association*, **93**, 273-282.

Gilks, W.R. and Wild, P. (1992). Adaptive rejection sampling for Gibbs sampling. *Applied Statistics*, **41**, 337-348.

Hobert, J.P. and Casella, G. (1996). The effect of improper priors in Gibbs sampling in hierarchical linear mixed models. *Journal of the American Statistical Association*, **91**, 1461-1473.

Hodges, J. (1998). Some algebra and geometry for hierarchical models, applied to diagnostics. *Journal of the Royal Statistical Society, Series B*, **60** (to appear),

Ibrahim, J.G. and Laud, P.W. (1991). On Bayesian analysis of general linear models using Jeffreys's prior. *Journal of the American Statistical Association*, **86**, 981-986.

Jorgensen, B. (1987). Exponential dispersion models. (with discussion). *Journal of the Royal Statistical Society, Series B*,**50**, 150-?.

Kass, R.E. and Raftery, A.E. (1995). Bayes factors. *Journal of the American Statistical Association*, **90**, 773-795.

Leonard, T. and Novick, M.R. (1986). Bayesian full rank marginalization for two-way contingency tables. *Journal of Educational Statistics*, **11**, 33-56.

Lindsay, B. (1986). Exponential family mixture models. *The Annals of Statistics*, 14, 124-137.

Lindsey, J.K. (1995). *Modelling Frequency and Count Data.* London, Clarendon Press.

Lo, A.Y. (1984). On a class of Bayesian nonparametric density estimation:I, density estimates. *The Annals of Statistics*, 12, 351-357.

MacEachern, S.N. and Müller, P. (1994). Estimating mixture of Dirichlet process models. Technical Report, Institute of Statistics and Decision Sciences, Duke University.

Mallick, B.K. and Gelfand, A.E. (1994). Generalized linear models with unknown link functions. *Biometrika*, 81, 237-245.

McCullagh, P. and J.A. Nelder (1989). *Generalized Linear Models.* Chapman and Hall, London.

Meng, X-L. (1994). Posterior predictive p-values. *Annals of Statistics*, **22**, 1142-1160.

Mukhopadhyay, S. and Gelfand, A.E. (1997). Dirichlet process mixed generalized linear models. *Journal of the American Statistical Association*, **92**, 633-639.

Müller, P. and G. Parmigiani (1995). Numerical evaluation of information theoretic measures. Inc: *Bayesian Statistics and Econometrics: Essays in Honor of A. Zellner.* Eds: Berry, D.A., Chaloner, K.M., Geweke, J.F., John Wiley, New York, 397-406.

Natarajan, R. and McCulloch, C.E. (1995). A note on the existence of the posterior distribution for a class of mixed models for binomial responses. *Biometrika*, 82, 639-643.

Nelder, J.A. and Wedderburn, R.W.M. (1972). Generalized linear models. *Journal of the Royal Statistical Society, Series A*, 135, 370-384.

Newton, M.A., Czado, C. and Chappell, R. (1996). Semiparametric Bayesian inference for binary regression. *Journal of the American Statistical Association*, **91**, 142-153.

Piergorsch, W.W., Weinberg, C.R. and Margolin, B.H. (1988). Exploring simple independent action in multifactor table of proportions. *Biometrics*, **44**, 595-603.

Raftery, A.E. (1995). Hypothesis testing and model selection. In: *Markov Chain Monte Carlo in Practice.* Eds. W. Gilks, et al. London, Chapman and Hall, 163-187.

Schwarz, G. (1978). Estimating the dimension of a model. *The Annals of Statistics*, 6, 461-464.

Spiegelhalter, D.J., Best, N.G. and Carlin, B.P. (1998). Bayesian deviance, the effective number of parameters and the comparison of arbitrarily complex models. Preprint.

Weiss, R.E. (1995). Residuals and outliers in repeated measures random effects models. Tech. Rpt., Dept. of Biostatistics, UCLA.

West, M. (1985). Generalized linear models: scale parameters, outlier accomodation and prior distributions. In *Bayesian Statistics*,**2**, Oxford, Oxford University Press, 531-557.

Zeger, S.L. and Karim, M.R. (1991). Generalized linear models with random effects: a Gibbs sampling approach. *Journal of the American Statistical Association*, **86**, 79-86.

2

Random Effects in Generalized Linear Mixed Models (GLMMs)

Dongchu Sun
Paul L. Speckman
Robert K. Tsutakawa

ABSTRACT In this chapter, we examine the use of special forms of correlated random effects in the generalized linear mixed model (GLMM) setting. A special feature of our GLMM is the inclusion of random residual effects to account for lack of fit due to extra variation, outliers and other unexplained sources of variation. For random effects, we consider, in particular, the correlation structure and improper priors associated with the autoregressive (AR) model of Ord (1975) and the conditional autoregressive (CAR) model of Besag (1974). We give conditions for the propriety of the posterior distribution of the GLMM when the fixed effects have a constant improper prior and the random effects have a possibly improper conditional autoregressive prior. Several examples of exponential families as well as computational details for Markov chain Monte Carlo simulation are also presented.

1. Introduction

Traditional treatment of random effects in mixed linear and nonlinear models generally assumes that these effects are independent following some standard distributions such as normal or gamma. However, with the advent of Markov chain Monte Carlo (MCMC) methods and, in particular, the Gibbs sampler (cf. Gelfand and Smith, 1990), such restrictions are no longer necessary, and a much broader class of models, including those with correlated random effects, can be used in practice. (See Clayton (1996) for a general review of this recent development.)

In this chapter we consider generalized mixed linear models with random effects having the autoregressive and conditionally autoregressive properties commonly encountered in temporal and spatial covariates where one expects similarities among closely situated observations. Examples from disease mapping will be used to motivate these models.

The computational simplicity of MCMC methods enables one to extend the commonly used generalized linear mixed model (GLMM) to one that appends random residual effects to the linear term to account for lack of fit. These extra terms allow for the minor perturbations and occasional outliers commonly encountered in practice. However, the remarkable ease of application of the Gibbs sampler does not come without a price. There is potential nonconvergence and other annoying problems when using the algorithm, especially in situations where noninformative prior distributions are employed.

In Section 2 we formally define the GLMM with residual effects. Two examples are given. One has the normal distribution and the other the gamma distribution, with the choice depending on the nature of the observed data. For example when the data are Poisson, it is more natural to use the conjugate gamma distribution, although the normal may be just as appropriate and simple to use.

In Section 3 we discuss several forms of correlated random effects including the AR process of Ord (1974) and the CAR process of Besag (1974), which are useful in describing spatial correlations. We examine the joint distributions associated with these processes to get a better understanding of the underlying association implied by these models. Of particular interest are distributions that are improper and could create problems when used in the GLMMs.

In Section 4, we consider the incorporation of these spatial random variables into the GLMM setting and emphasize the special role of the link function in a Bayesian hierarchical framework. In the case where the residual effects are normally distributed, the fixed effects have a constant prior and random effects may have an improper prior, we give sufficient conditions for the existence of a proper posterior distribution of all parameters including the fixed and random effects and variance components.

In Section 5, we summarize the computational details including the full conditional distributions required for the implementation of the Gibbs Sampler.

2. The Model

Let $Y_1, \ldots, Y_N$ be the independent random observations, where Y_i has the probability density

$$f_i(y_i|\eta_i, \phi) = \exp[A_i(\phi)^{-1}\{y_i\eta_i - B_i(\eta_i)\} + C_i(y_i; \phi)]. \tag{1}$$

The function $A_i(\phi)$ is commonly of the form $A_i(\phi) = \phi w_i^{-1}$, where the w_i are prespecified weights. It is often assumed that the scale parameter ϕ is known. Consider, for example, the case when the population size in area i is m_i with unknown mortality rate p_i, and Y_i is Poisson distributed with mean $m_i p_i$. This is a special case of (1) with $\phi = 1, A_i(\phi) = 1, \eta_i = \log(m_i p_i), B_i(\eta_i) = \exp(\eta_i)$, and $C_i(y_i; \phi) = -\log(y_i!)$. When Y_i has a binomial distribution with parameters m_i and p_i, $\phi = 1, A_i(\phi) = 1, \eta_i = \log\{p_i/(1-p_i)\}, B_i(\eta_i) = m_i \log\{1 + \exp(\eta_i)\}$, and $C_i(y_i; \phi) = \log[m_i!/\{y_i!(m_i - y_i)!\}]$.

Generalized Linear Models. We wish to model the variability in η_i to account for various fixed covariates. The natural parameters η_i are modeled as

$$h_i(\eta_i) = \boldsymbol{x}_{1i}^t \boldsymbol{\theta}, \tag{2}$$

where the h_i are known monotone functions, $\boldsymbol{X}_1 = (\boldsymbol{x}_{11}, \ldots, \boldsymbol{x}_{1n})^t$ is an $N \times p$ design matrix and $\boldsymbol{\theta}$ is the vector of fixed effects. Such a model is commonly referred to as a generalized linear model (GLM) with canonical parameter η_i, scale parameter ϕ, and link function h_i (cf. McCullagh and Nelder, 1989). (Note that usually there is a single link function $h_i \equiv h$.)

Generalized Linear Mixed Models. We now extend the model to include random effects as follows. Let

$$h_i(\eta_i) = \boldsymbol{x}_{1i}^t \boldsymbol{\theta} + \boldsymbol{x}_{2i}^t \boldsymbol{Z}, \tag{3}$$

where h_i is a known monotone function, $\boldsymbol{X}_1 = (\boldsymbol{x}_{11}, \ldots, \boldsymbol{x}_{1n})^t$ and $\boldsymbol{X}_2 = (\boldsymbol{x}_{21}, \ldots, \boldsymbol{x}_{2n})^t$ are $N \times p$ and $N \times k$ design matrices, the $p\times$ 1 vector $\boldsymbol{\theta}$ represents fixed effects, and $\boldsymbol{Z}$ is a $k \times 1$ vector of random effects. Models given by (1) and (3) are often called generalized linear mixed models (GLMMs) and have been widely used in many problems such as disease mapping e.g., Breslow and Clayton (1993).

We can further extend the model to add additional residual effects by taking

$$h_i(\eta_i) = \boldsymbol{x}_{1i}^t \boldsymbol{\theta} + \boldsymbol{x}_{2i}^t \boldsymbol{Z} + e_i. \tag{4}$$

Here $\boldsymbol{e} = (e_1, \ldots, e_N)^t$ are residual effects satisfying some restriction such as $\mathbb{E}(e_i) = 0$ or $\mathbb{E}\exp(e_i) = 1$. In addition, $\boldsymbol{Z}$ and $\boldsymbol{e}$ are assumed mutually independent. We include random residual effects e_i to account for the lack of fit of (3) due to extra variation, outliers, and other unexplained sources of variation. Note that the random effect e_i is quite different from $\boldsymbol{Z}$ in the sense that $\boldsymbol{Z}$ often accounts for some special pattern such as random geographical effects and spatial correlation. In addition, the number of components of $\boldsymbol{Z}$ is often much smaller than N, the number of residual effects e_i. By a suitable choice of the design matrix, (4) may be encompassed under (3), but we do not do this in order to emphasize the separate roles of $\boldsymbol{Z}$ and $\boldsymbol{e}$. We will call the model given by (1) and (4) a GLMM as well.

There are many possible choices for the link functions h_i in models (2)–(4). For example, in the mortality setting cited earlier, Y_i has the Poisson distribution with mean $m_i p_i$ and $\eta_i = \log(m_i p_i)$. One possibility is to take $h_i(\eta_i) = \eta_i - \log(m_i) = \log(p_i)$, and a loglinear regression model may be applied. Alternatively, Y_i can be modeled with a binomial distribution. Then the logit link is canonical, and $\text{logit}(p_i) = \log\{p_i/(1-p_i)\} = \eta_i - \log(m_i - e^{\eta_i}) = h_i(\eta_i)$, resulting in logistic regression.

The random effects term $\boldsymbol{Z}$ in (3)–(4) is typically assumed to have a multivariate normal distribution. We will discuss in detail the choice of the distribution of $\boldsymbol{Z}$ in the next section.

Distribution of Residual Effects. We will assume that the residual effects e_i or some monotone functions of e_i have distributions belonging to an exponential family (1), with known common canonical parameter η but unknown scale parameter ϕ. For illustration, we will consider the following two classes of distributions for residual effects.

- *Normal Residual Effects.* Residual effects e_i are independent and identically normal with mean 0 and variance δ_0.
- *Gamma Residual Effects.* The $\exp(e_i)$ are iid gamma(R, R). Here a random variable W has the gamma(α, β) distribution if W has p.d.f.

$$f(w) = \alpha^{\beta}\{\Gamma(\alpha)\}^{-1} w^{\alpha-1} \exp(-\beta w).$$

Special cases of these models have appeared previously. Clayton and Kaldor (1987) and Waller *et al.* (1997) use a Poisson-normal model (Poisson for Y_i and normal for $\boldsymbol{Z}$) but without the residual term $\boldsymbol{e}$. This a special case of (3). Ghosh *et al.* (1998) use $\boldsymbol{e}$ in the binomial-normal model and treat spatial effects by taking $\boldsymbol{X}_2\boldsymbol{Z} = \boldsymbol{U}$, with $\boldsymbol{U}$ having a distribution defined by the conditional auto regressive CAR(1) model of Besag (1974). This is a special case of (4). In Sun, Tsutakawa, Kim and He (1997) and Sun, Tsutakawa and He (1998), $\boldsymbol{Z}$ consists of block-wise independent random effects, where each block contains random effects and the e_i are independent random variables with mean 0 and a common variance. West and

Aguilar (1997) give another interesting example analysing hospital quality monitors with an extra residual term in (1.4).

Special cases of Poisson-gamma models are found in Clayton and Kaldor (1987) and Tsutakawa (1988). Specifically, in Tsutakawa (1988), $\boldsymbol{Z}$ contains independent random effects, and the $\exp(e_i)$ are independent gamma variables with mean 1 and a common variance.

The general model (1) and (2) can be used for both continuous and discrete data. A discrete example of (4), which motivated much of this work, is studied in Sun, Tsutakawa, Kim and He (1998), where a spatio-temporal model for cancer mortality data is proposed. For a given gender, let Y_{ijk} denote the frequency of deaths from some specific cause in the ith region and jth age group during the kth time period, $i = 1, \ldots, I$; $j = 1, \ldots, J$; $k = 1, \ldots, K$. Conditionally on the fixed and random parameters, assume the Y_{ijk} are independent and Poisson with means $m_{ijk}p_{ijk}$, where m_{ijk} is the size of the ijkth target population. The model of Sun *et al.* takes the form

$$\log(p_{ijk}) = \theta_j + Z_i + (\mu_j + W_{ij})(t_k - \bar{t}) + e_{ijk},$$

where θ_j is the effect of the jth age group, Z_i is the effect of the ith region, t_k is the midpoint of the kth time period, and $\bar{t} = \sum_{k=1}^{K} t_k/K$. The rate of change over time is represented by $(\mu_j + W_{ij})$ for the jth age group in the ith region. Both θ_j and μ_j are treated as fixed effects, while Z_i and W_{ij} are random. The residual effects e_{ijk} are also random. A detailed description of the distributions of the random effects and prior distributions are given in Sun *et al.* (1998), where disease mapping and interpretation of numerical results for male lung cancer in the state of Missouri can be found.

3. Random Effects

3.1 Independent Random Effects

Historically, it was common to assume independent random effects for linear mixed models, i.e., $Z_1, \ldots, Z_N$ are independently and identically $\mathrm{N}(0, \delta_1)$ distributed. (See Harville (1977).) Typical examples include one-way ANOVA and two-way ANOVA models with random effects. Hobert and Casella (1996) gave necessary and sufficient conditions for the propriety of the posterior distribution for a class of noninformative priors for variances components assuming independence of random effects.

3.2 Correlated Random Effects

There are many important situations where the random effects should be modeled as correlated. Correlated models are especially appropriate for spatial effects. A number of related methods are commonly used.

Direct specification of correlation matrix. If the random effects are linearly ordered, as for example with longitudinal data, it may be convenient to specify a correlation structure directly. For example, to model correlation decreasing with distance, $\boldsymbol{Z} = (Z_1, \ldots, Z_k)^t$ can be taken to have the $\mathrm{MVN}(\mathbf{0}, \boldsymbol{\Sigma})$ distribution, where $\boldsymbol{\Sigma} = (\sigma_{ij})$ is the $k \times k$ matrix with elements

$$\sigma_{ij} = \tau\rho^{|i-j|}, \tag{5}$$

and $\tau > 0$ and $\rho \in (-1, 1)$ are constants. For MCMC methods with modest size k (say, $k < 100$ or so), it is sometimes feasible to generate $\boldsymbol{Z}$ from the joint conditional distribution directly. A number of authors including Cressie and Chan (1989) have used the distance between area i and area j to introduce spatial correlation.

AR models. Again assuming a linear ordering for the components of $\boldsymbol{Z}$, a commonly used structure is the AR(1) model with

$$Z_i = \rho Z_{i-1} + \epsilon_i, \; i = 2, \ldots, k, \tag{6}$$

where ρ is a constant in $(-1, 1)$, and the ϵ_i are independent and identically $N(0, \delta_1)$ distributed. If $Z_1 \sim N(0, \delta_1/(1 - \rho^2))$, the distribution of $\boldsymbol{Z}$ is given by (5) with $\tau = \delta_1/(1 - \rho^2)$.

Ord (1975) proposed a generalized AR(1) model by defining

$$Z_i = \rho \sum_{j=1}^{k} C_{ij} Z_j + \epsilon_i, \tag{7}$$

where the C_{ij} are fixed constants satisfying $C_{ii} = 0$, and $\epsilon_1, \ldots, \epsilon_k$ are iid $\mathrm{N}(0, \delta_1)$. Here ρ is a "correlation coefficient," measuring the correlation among Z_i in the sense that the larger $|\rho|$ is, the stronger the correlation among the components of $\boldsymbol{Z}$. For example, if the Z_i are linearly ordered, one can define their joint distribution by assuming

$$\begin{aligned} Z_1 &= \rho Z_2 + \epsilon_1, \\ Z_i &= \rho(Z_{i-1} + Z_{i+1}) + \epsilon_i, \; i = 2, \ldots, k-1, \\ Z_k &= \rho Z_{k-1} + \epsilon_k. \end{aligned} \tag{8}$$

One advantage of (7) is that the formulation generalizes easily to two or more dimensions. Taking $\boldsymbol{C} = (C_{il})$ to be the $k \times k$ matrix of coefficients, $\boldsymbol{I}$ the $k \times k$ identity matrix, and

$$\boldsymbol{W}_\rho = \boldsymbol{I} - \rho \boldsymbol{C}, \tag{9}$$

model (7) is equivalent to $\boldsymbol{W}_\rho \boldsymbol{Z} = (\epsilon_1, \ldots, \epsilon_k)^t$. If $\boldsymbol{W}_\rho$ is nonsingular, $\boldsymbol{Z}$ has a multivariate normal distribution with mean zero and covariance matrix $\boldsymbol{\Sigma} = \delta_1(\boldsymbol{W}_\rho^t \boldsymbol{W}_\rho)^{-1}$. A common choice of $\boldsymbol{C}$ is the adjacency matrix $\boldsymbol{A} = (a_{ij})_{k \times k}$, defined by

$$a_{ij} = \begin{cases} 1, & \text{if } j \text{ is adjacent to } i, \\ 0, & \text{otherwise.} \end{cases} \tag{10}$$

The class of distributions for $\boldsymbol{Z}$ when $\boldsymbol{W}_\rho = \boldsymbol{I} - \rho \boldsymbol{A}$ has been used in modeling random regional effects in disease mapping by Sun, Tsutakawa, Kim and He (1997) and random county effects in hunting success rates from a turkey hunting survey in the State of Missouri by He and Sun (1998).

One appealing way to view the prior for $\boldsymbol{Z}$ is through the conditional distributions of Z_i given $\boldsymbol{Z}_{-i} = (Z_j, j \neq i)$. For the simple AR(1) prior (6), it can be shown that $\boldsymbol{B} = \delta_1 \boldsymbol{\Sigma}^{-1}$ is a tridiagonal matrix with diagonal elements $(1, 1 + \rho^2, \ldots, 1 + \rho^2, 1)$ and off-diagonal elements $-\rho$. It follows easily that $\boldsymbol{Z}$ has the Markov property

$$Z_i | \boldsymbol{Z}_{-i} \sim N\Big(\frac{\rho}{1 + \rho^2}(Z_{i-1} + Z_{i+1}), \frac{\delta_1}{1 + \rho^2}\Big), \text{ for } i = 2, \ldots, k-1,$$

that is, the conditional distribution of Z_i given the rest depends only on adjacent variables. Curiously, the generalized AR prior specified through the adjacency matrix in (8) does not have a similar Markov property. This follows from the fact that $\boldsymbol{\Sigma}^{-1} = \delta_1^{-1}\boldsymbol{W}_\rho^t\boldsymbol{W}_\rho$ is a banded matrix but is not tridiagonal. Instead, $Z_i|\boldsymbol{Z}_{-i}$ depends on $(Z_{i-2}, Z_{i-1}, Z_{i+1}, Z_{i+2})$ for $3 \le i \le k-2$.

CAR(1) model. In an effort to use priors with the appealing first-order Markov property in spatial modeling, many authors have adopted conditional autoregressive or CAR models, which are developed by specifying the conditional distributions directly in a (presumably) consistent manner. One popular model takes

$$Z_i|\boldsymbol{Z}_{-i} \sim N\Big(\frac{\rho}{d_i}\sum_{j\neq i}^{k} C_{ij}Z_j,\ \frac{\delta_1}{d_i}\Big), \tag{11}$$

where C_{ij} and $d_i > 0$ are constants satisfying $C_{ii} = 0$. This is a special case of Besag's (1974) model with

$$f(Z_i|\boldsymbol{Z}_{-i}) = \Big(\frac{\alpha_i}{2\pi\delta_1}\Big)^{\frac{1}{2}}\exp\Big\{-\frac{\alpha_i}{2\delta_1}\Big(Z_i - \sum_{j\neq i}^{k}\beta_{ij}Z_j\Big)^2\Big\}, \tag{12}$$

$i = 1, \ldots, k$. Suppose $\boldsymbol{B}$ is the $k \times k$ matrix with diagonal elements α_i and ijth off-diagonal elements $-\alpha_i\beta_{ij}$. Besag proved that if $\boldsymbol{B}$ is symmetric and positive definite, these conditional distributions lead to the joint probability density of $\boldsymbol{Z}$,

$$f(\boldsymbol{Z}) = (2\pi\delta_1)^{-k/2}|\boldsymbol{B}|^{1/2}\exp\Big(-\frac{1}{2\delta_1}\boldsymbol{Z}^t\boldsymbol{B}\boldsymbol{Z}\Big), \tag{13}$$

i.e. $\boldsymbol{Z} \sim \text{MVN}(\boldsymbol{0}, \delta_1\boldsymbol{B}^{-1})$. In the context considered here, suppose

$$\boldsymbol{B} = \boldsymbol{B}_\rho = \boldsymbol{D} - \rho\boldsymbol{C}, \tag{14}$$

where $\boldsymbol{D}$ is a $k \times k$ diagonal matrix with positive elements $(d_1, \ldots, d_k)$, and $\boldsymbol{C}$ is a symmetric matrix with $C_{ii} = 0$. If $\boldsymbol{B}_\rho$ is positive definite, then the joint distribution of $\boldsymbol{Z}$ is (13), and the conditional distributions of Z_i given $\boldsymbol{Z}_{-i}$ are (11).

In practice, these models are important because the simple conditional distributions depending only on neighboring values for the Z_i are desirable for Bayesian analysis using Markov chain Monte Carlo methods. Here are two important cases.

Case 1. Assume that $\boldsymbol{C} = \boldsymbol{A}$, the adjacency matrix, and $d_i = \sum_j C_{ij}$. If $\rho \in (-1, 1)$, then $\boldsymbol{B}$ is positive definite and the conditional distribution of Z_i given $\boldsymbol{Z}_{-i}$ is $\text{N}(\rho\bar{Z}_i, \delta_1/n_i)$, where n_i is the number of neighbors of location i, and $\bar{Z}_i$ is the mean of the n_i neighboring $Z_j s$. (This corresponds to $\alpha_i = n_i$ and $\beta_{ij} = \rho/n_i$ if j is adjacent to i and zero otherwise.) This model was studied in Besag (1975) and Ripley (1981).

Case 2. Assume that $\boldsymbol{C} = \boldsymbol{A}$, the adjacency matrix, and $\boldsymbol{D} = \boldsymbol{I}$. Let λ_1 and λ_k be the smallest and largest eigenvalues of $\boldsymbol{C}$. If $\lambda_1^{-1} < \rho < \lambda_k^{-1}$, then $\boldsymbol{B}$ is positive definite and the conditional distribution of Z_i given $\boldsymbol{Z}_{-i}$ is $\text{N}(\rho\sum_{j\neq i} a_{ij}Z_j, \delta_1)$. This model was used in Ripley (1988).

However, there are potential problems in modeling the dependence among the Z_i through the choice of (α_i, β_{ij}) in (12). One problem, singularity of $\boldsymbol{B}$, is addressed further in the next section. Another possible problem is with specifying a symmetric matrix $\boldsymbol{B}$. The specification in Case 1 with conditional variance δ_1/n_i seems unrealistic when ρ is small, since in the limiting case $\rho = 0$, the Z_i are independent but the variances still depend on the number of neighbors. This may not make sense near boundaries or in nonregular cases. On the other hand, in Case 2 the conditional variance of Z_i does not depend at all on the number of neighbors. As an alternative, suppose we let the conditional distribution of Z_i given $\boldsymbol{Z}_{-i}$ be

$\mathrm{N}(\rho\bar{Z}_i, \delta_1(1+\rho/n_i))$ and assume $0 \le \rho \le 1$. Formally, this is equivalent to a CAR model with $\alpha_i = (1+\rho/n_i)^{-1}$ and

$$\beta_{ij} = \begin{cases} \rho/n_i, & \text{if } j \text{ is adjacent to } i, \\ 0, & \text{otherwise.} \end{cases}$$

Unfortunately, the ijth off-diagonal element of $\boldsymbol{B}$, $\alpha_i\beta_{ij} = \rho/(\rho+n_i)$, is not equal to $\alpha_j\beta_{ji}$ unless $n_i = n_j$. Even in the linearly ordered case, this fails at the boundary where $n_1 = 1$ and $n_j = 2$ for $1 < j < k$. Thus care must be taken in specifying a CAR model.

3.3 Strongly Correlated Random Effects

If the determinant $|\boldsymbol{B}|$ is zero, the set of full conditional distributions given by (12) is not "compatible," a definition used by Arnold and Press (1989), in the sense that there is no joint density of $\boldsymbol{Z}$ consistent with the corresponding conditional densities. However, there are situations in practice where a nonpositive definite $\boldsymbol{B}$ is desirable. For example, if $\rho \to 1$ in (14) when $\boldsymbol{C} = \boldsymbol{A}$, the adjacency matrix, and $\boldsymbol{D}$ is the diagonal matrix of row sums of $\boldsymbol{C}$, the model is a Markov random field. Clearly $\boldsymbol{B}$ is singular.

When the matrix $\boldsymbol{B}$ is nonpositive definite, there are two possible interpretations. One way is to consider a lower dimensional distribution, in the sense that it is proper in certain directions but degenerate in some other directions. For example, let r be the rank of $\boldsymbol{B}$, and let $\lambda_1, \ldots, \lambda_r$ be the positive eigenvalues of $\boldsymbol{B}$. Write $\boldsymbol{B} = \boldsymbol{\Gamma\Lambda\Gamma}^t$, where $\boldsymbol{\Gamma} = (\boldsymbol{\gamma}_1, \ldots, \boldsymbol{\gamma}_k)$ is an orthogonal matrix, and $\boldsymbol{\Lambda} = diag(\lambda_1, \ldots, \lambda_r, 0, \ldots, 0)$. Let $\boldsymbol{\Gamma}_1 = (\boldsymbol{\gamma}_1, \ldots, \boldsymbol{\gamma}_r)$ and $\boldsymbol{\Lambda}_1 = diag(\lambda_1, \ldots, \lambda_r)$. Then $\boldsymbol{B} = \boldsymbol{\Gamma}_1\boldsymbol{\Lambda}_1\boldsymbol{\Gamma}_1^t$. Now let $\boldsymbol{U}_1 = (U_1, \ldots, U_r)^t$ be a vector of independent random variables where $U_i \sim N(0, \delta_1\lambda_i^{-1})$. Then $\boldsymbol{Z} = \boldsymbol{\Gamma}_1\boldsymbol{U}_1$ has a singular normal distribution with mean $\boldsymbol{0}$ and covariance matrix $\delta_1\boldsymbol{B}^-$, where $\boldsymbol{B}^-$ is a pseudo-inverse of $\boldsymbol{B}$. We often write this distribution as $\mathrm{MVN}(\boldsymbol{0}, \delta_1\boldsymbol{B}^-)$. The joint distribution has the form

$$f(\boldsymbol{Z}) \propto (2\pi\delta_1)^{\frac{r}{2}}|\boldsymbol{B}|_+^{\frac{1}{2}} \exp\Big(-\frac{1}{2\delta_1}\boldsymbol{Z}^t\boldsymbol{B}\boldsymbol{Z}\Big), \tag{15}$$

where $|\boldsymbol{B}|_+$ is defined to be $\prod_{i=1}^r \lambda_i$, the product of all positive eigenvalues of $\boldsymbol{B}$. Note that based on such a singular normal distribution, the full conditional distribution of Z_i given $\boldsymbol{Z}_{-i}$ is degenerate instead of a normal distribution. The distribution of $\boldsymbol{Z}$ is essentially proper on a lower dimensional space, so $\boldsymbol{Z}$ is a vector of strongly correlated random effects.

Alternatively, we can sample an additional random sample $\boldsymbol{U}_2 = (U_{r+1}, \ldots, U_k)^t$ from a flat constant density over a $k-r$ dimensional Euclidian space. Now define $\boldsymbol{Z} = \boldsymbol{\Gamma}(\boldsymbol{U}_1^t, \boldsymbol{U}_2^t)^t$. We can see that the joint density of $\boldsymbol{Z}$ has the form (15), which is improper because $\boldsymbol{B}$ is singular. However, we can formally relate (15) to (12) by noting that

$$f(Z_i|\boldsymbol{Z}_{-i}) = f(Z_1, \ldots, Z_k) \Big/ \int_{-\infty}^{\infty} f(Z_1, \ldots, Z_k) dZ_i.$$

Hobert and Casella (1998) have called this type of relationship "functionally compatible," in contrast to one being "compatible."

Markov random field models. In (14), if $\boldsymbol{C} = \boldsymbol{A} = (a_{ij}), d_i = \sum_{j \ne i} a_{ij}$, and $\rho = 1$, then the distribution of $\boldsymbol{Z}$ is often called a Markov random field model (cf.

Kindermann and Snell (1980)). Such models have been used for modeling spatial correlations in disease mapping and other contexts by Besag, York, and Mollié (1991) and used by Bernardinelli and Montomoli (1992), Bernardinelli *et al.* (1995), Carlin and Louis (1996), Waller *et al.* (1997) and Ghosh *et al.* (1998) among others.

Autocorrelated random effects. We next give a class of strongly correlated distributions. Define the backwards difference operator $\boldsymbol{H}_k$ to be the $k \times k$ matrix

$$\boldsymbol{H}_k = \begin{pmatrix} 1 & 0 & 0 & \cdots & 0 & 0 \\ -1 & 1 & 0 & \cdots & 0 & 0 \\ \cdot & \cdot & \cdot & \cdots & \cdot & \cdot \\ 0 & 0 & \cdot & \cdots & -1 & 1 \end{pmatrix},$$

and let

$$\boldsymbol{G}_{kd} = [\boldsymbol{0}_{(k-d)\times d} \,|\, \boldsymbol{I}_{k-d}]$$

be the lower $(k-d)$ rows of the k-dimensional identity matrix I_k. Now let the structural matrix $\boldsymbol{B}$ in (15) be

$$\boldsymbol{B}_{kd} = (\boldsymbol{G}_{kd}\boldsymbol{H}_k^d)^t \boldsymbol{G}_{kd}\boldsymbol{H}_k^d. \tag{16}$$

Because $\boldsymbol{B}_{kd}$ has rank $k-d$, $\boldsymbol{Z}$ has a singular distribution, which we write as $\text{MVN}(\boldsymbol{0}, \delta_1 \boldsymbol{B}_{kd}^-)$, where $\boldsymbol{B}_{kd}^-$ is a pseudo-inverse of $\boldsymbol{B}_{kd}$.

AR(1) (The first order difference model). When $d = 1$, the prior on $\boldsymbol{Z}$ is called the first order difference or random walk prior. See Clayton (1996). In this case the structural matrix has the form

$$\boldsymbol{B}_{k1} = \begin{pmatrix} 1 & -1 & 0 & 0 & \cdots & 0 & 0 & 0 \\ -1 & 2 & -1 & 0 & \cdots & 0 & 0 & 0 \\ \cdot & \cdot & \cdot & \cdot & \cdots & \cdot & \cdot & \cdot \\ 0 & 0 & 0 & 0 & \cdots & -1 & 2 & -1 \\ 0 & 0 & 0 & 0 & \cdots & 0 & -1 & 1 \end{pmatrix}_{k\times k}.$$

AR(2) (The second order difference) Model. When $d = 2$, the prior on $\boldsymbol{Z}$ is called the stochastic-trend or second order difference prior. See Clayton (1996). In this case the structural matrix has the form

$$\boldsymbol{B}_{k2} = \begin{pmatrix} 1 & -2 & 1 & 0 & 0 & \cdots & 0 & 0 & 0 & 0 & 0 \\ -2 & 5 & -4 & 1 & 0 & \cdots & 0 & 0 & 0 & 0 & 0 \\ 1 & -4 & 6 & -4 & 1 & \cdots & 0 & 0 & 0 & 0 & 0 \\ \cdot & \cdot & \cdot & \cdot & \cdot & \cdots & \cdot & \cdot & \cdot & \cdot & \cdot \\ 0 & 0 & 0 & 0 & 0 & \cdots & 1 & -4 & 6 & -4 & 1 \\ 0 & 0 & 0 & 0 & 0 & \cdots & 0 & 1 & -4 & 5 & -2 \\ 0 & 0 & 0 & 0 & 0 & \cdots & 0 & 0 & 1 & -2 & 1 \end{pmatrix}_{k\times k}.$$

Note that this model can be obtained by the iterated formula

$$Z_i = 2Z_{i-1} - Z_{i-2} + \epsilon_i, \quad i = 3, \ldots, k.$$

Such a second order random effects prior has been used for patient monitoring by Berzuini (1996).

3.4 Some Examples of the AR(d) Model

To see the differences among the AR(d) models when we change d, in Figure 2.1 we graph three sample paths of the AR(d) process prior for $\boldsymbol{Z}$ using $k = 100$, $\delta_1 = 1$ and $d = 1, 2, 3$. Note that the rank of $\boldsymbol{B}_{kd}$ is $k - d$. When $d = 1$, the sample paths are simple random walks and locally rough, but sample paths are smoother when $d \geq 2$.

4. Hierarchical GLMMs

Bayesian analysis for the GLMs (1)–(2) and the GLMMs given by (1) and (3) is studied in Clayton (1996). We will discuss the GLMMs given by (1) and (4). For illustration, we will consider normal residual effects e_i. A full hierarchical Bayesian approach requires the specification of prior distributions for $\boldsymbol{\theta}$, the variance δ_0 of the distribution of e_i, the variance δ_1 of the distribution of random effects $\boldsymbol{Z}$, and the scale parameter ϕ.

Although the commonly used prior of the fixed effects $\boldsymbol{\theta}$ is normal, we will assume a noninformative prior for $\boldsymbol{\theta}$, in particular, one having a constant density. We will not give a specific form for the priors of $(\delta_0, \delta_1, \phi)$. Since the prior for $\boldsymbol{\theta}$ is improper, and the prior for $\boldsymbol{Z}$ is also improper for a singular $\boldsymbol{B}$, the joint posterior distribution may still be improper. As noted by Hobert and Casella (1996) and Sun, Tsutakawa and Speckman (1997), the propriety of the posterior is very important in Bayesian computation, especially when Markov chain Monte Carlo methods are used. Sun, Tsutakawa and Speckman (1997) considered a one-parameter distribution family where the prior for the parameters follows a linear mixed model and found conditions for a proper posterior distribution. Here we extend the results to the GLMM model, where the observations follow the densities (1) with canonical parameters η_i and a common scale parameter ϕ. We will only consider the case where there is a common variance component δ_1 for the whole vector of $\boldsymbol{Z}$. Some generalizations to block random effects can be found in Sun *et al.* (1998). We use the following notation. Note that for $B_i(\cdot)$ defined in (1), the first derivative B_i' is a strictly increasing function. Let H_i be the inverse function of B_i'. Note that for any fixed ϕ, the likelihood function $f_i(y_i|\eta_i, \phi)$ is bounded by

$$\begin{aligned} M_i(\phi) &\equiv \sup_{\eta_i} f_i(y_i|\eta_i, \phi) \\ &= \exp[A_i(\phi)^{-1}\{y_i H_i(y_i) - B_i(H_i(y_i))\} + C_i(y_i; \phi)]. \end{aligned} \tag{17}$$

Theorem 4..1 *Consider the GLMMs (1) and (4) with normal residual effects* e_i *iid* $\sim N(0, \delta_0)$. *Assume that*

(a) there exists a subset of $\{1, \ldots, N\}$, *say* $\mathcal{J}_n = (i_1, \ldots, i_n)$, *such that*

$$\int \prod_{j \notin \mathcal{J}_n} M_j(\phi) \left\{ \prod_{j \in \mathcal{J}_n} \int f_j(y_j|\eta_j, \phi) h_j'(\eta_j) d\eta_j \right\} F(d\phi) < \infty, \tag{18}$$

where $F(\cdot)$ *is the prior distribution for* ϕ;

(b) the design matrix $\boldsymbol{X}_1^* = (\boldsymbol{x}_{1,i_1}, \ldots, \boldsymbol{x}_{1,i_n})^t$ *has full rank* p, *and* $\boldsymbol{X}_2^* = (\boldsymbol{x}_{2,i_1}, \ldots, \boldsymbol{x}_{2,i_n})^t$ *has the same rank as the matrix* $\boldsymbol{X}_2 = (\boldsymbol{x}_{2,1}, \ldots, \boldsymbol{x}_{2,N})^t$;

(c) the prior for $\boldsymbol{\theta}$ *is a constant and* $\boldsymbol{Z}$ *follows the density (15);*

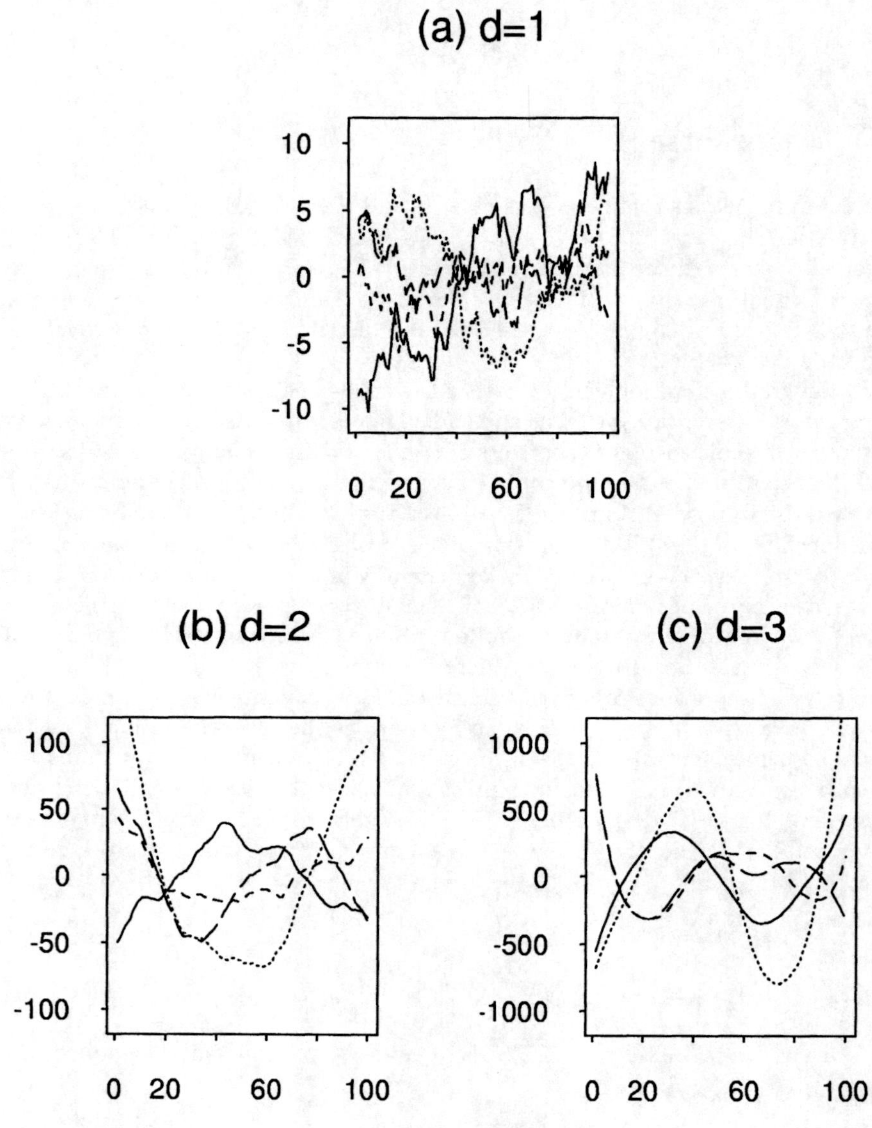

FIGURE 2.1. Sample paths of CAR (d) models for $\boldsymbol{Z}$ when $n = 100$.

(d) the rank of $(\boldsymbol{X}_2^{*t}\boldsymbol{R}_1\boldsymbol{X}_2^* + \boldsymbol{B})$ *is* k*, where* $\boldsymbol{R}_1 = \boldsymbol{I}_n - \boldsymbol{X}_1^*(\boldsymbol{X}_1^{*t}\boldsymbol{X}_1^*)^{-1}\boldsymbol{X}_1^{*t}$*;*

(e) the prior for (δ_0, δ_1) *satisfies the moment condition,*

$$\mathbb{E}\{\delta_0^{-\frac{1}{2}(n-p-k)}\delta_1^{-\frac{1}{2}k} + \delta_0^{-\frac{1}{2}(n-p)}\} < \infty. \tag{19}$$

Then the posterior distribution of $(\boldsymbol{\eta}, \phi, \boldsymbol{\theta}, \boldsymbol{Z}, \delta_0, \delta_1)$ *given* $\boldsymbol{Y} = (y_1, \ldots, y_N)$ *is proper.*

Proof. Without loss of generality, assume that $\mathcal{J}_n = \{1, \ldots, n\}$ in assumption (a) and that δ_i has a prior density g_i. The posterior density of $(\boldsymbol{\eta}, \boldsymbol{\theta}, \boldsymbol{Z}, \delta_0, \delta_1)$ given $(\boldsymbol{Y}, \phi)$ is

$$p(\boldsymbol{\eta}, \boldsymbol{\theta}, \boldsymbol{Z}, \delta_0, \delta_1 | \boldsymbol{Y}, \phi) \propto \prod_{i=1}^{N} f_i(Y_i|\eta_i, \phi)h_i'(\eta_i)\delta_0^{-\frac{1}{2}(N-n)} \times$$
$$\prod_{i=n+1}^{N} \exp\Big[-\frac{1}{2\delta_0}\{h_i(\eta_i) - \boldsymbol{x}_{1i}^t\boldsymbol{\theta} - \boldsymbol{x}_{2i}^t\boldsymbol{Z}\}^2\Big] \prod_{j=0}^{1} g_j(\delta_j)G.$$

where

$$G = \frac{1}{\delta_0^{\frac{n}{2}}\delta_1^{\frac{k}{2}}} \exp\Big\{-\frac{(\boldsymbol{V}^* - \boldsymbol{X}_1^*\boldsymbol{\theta} - \boldsymbol{X}_2^*\boldsymbol{Z})^t(\boldsymbol{V}^* - \boldsymbol{X}_1^*\boldsymbol{\theta} - \boldsymbol{X}_2^*\boldsymbol{Z})}{2\delta_0} - \frac{\boldsymbol{Z}^t\boldsymbol{B}\boldsymbol{Z}}{2\delta_1}\Big\}.$$

Here $\boldsymbol{V}^* = (h_1(\eta_1), \ldots, h_n(\eta_n))^t$. Let $\boldsymbol{\eta}^* = (\eta_1, \ldots, \eta_n)^t$. Using inequality (17) and integrating with respect to $(\eta_{n+1}, \ldots, \eta_N)^t$,

$$p(\boldsymbol{\eta}^*, \boldsymbol{\theta}, \boldsymbol{Z}, \delta_0, \delta_1 | \boldsymbol{Y}, \phi) \propto \prod_{i=n+1}^{N} M_i(\phi) \prod_{j=1}^{n} f_j(y_j|\eta_j, \phi)h_j'(\eta_j)\, G.$$

Using arguments similar to those in Sun, Tsutakawa and Speckman (1997), we get

$$\int_{\mathbb{R}^p}\int_{\mathbb{R}^k} G d\boldsymbol{\theta} d\boldsymbol{Z} \le \{\delta_0^{-\frac{1}{2}(n-p-k)}\delta_1^{-\frac{1}{2}k} + \delta_0^{-\frac{1}{2}(n-p)}\}.$$

Therefore, from assumption (e),

$$\int\int p(\boldsymbol{\eta}^*, \phi | \boldsymbol{Y}) d\boldsymbol{\eta}^* F(d\phi)$$
$$\propto \int \prod_{i=n+1}^{N} M_i(\phi) \Big\{\prod_{j=1}^{n} \int f_j(y_j|\eta_j, \phi)h_j'(\eta_j)d\eta_j\Big\} F(d\phi),$$

which is finite by (18). □

Remark 4..1 *A common prior for the variance components* δ_i *is inverse gamma*(a_i, b_i)*, whose density is*

$$g_i(\delta_i) \propto \frac{1}{\delta_i^{a_i+1}} \exp(-b_i/\delta_i). \tag{20}$$

Clearly, when $b_i > 0$, $n - p - k + 2a_0 > 0$ *and* $k > 2a_1$*, condition (19) holds.*

Remark 4..2 *When the prior of ϕ is degenerate, i.e., a known constant as in the Poisson or binomial cases, condition (18) becomes*

$$\int f_j(y_j|\eta_j,\phi)h_j'(\eta_j)d\eta_j < \infty, \textit{ for } j \in \mathcal{J}_n,$$

which is equivalent to the condition,

$$\int \exp[A_j(\phi)^{-1}\{y_j\eta_j - B_j(\eta_j)\}]h_j'(\eta_j)d\eta_j < \infty, \textit{ for } j \in \mathcal{J}_n. \tag{21}$$

A condition similar to (21) was required for all j in Ghosh et al. (1997) for propriety of the posterior distribution.

Example 4..1 Suppose $f_i(y_i|\eta_i,\phi)$ is Poisson with mean $\mu_i = m_i p_i$. This is a special case of (1) with $\phi = 1$ and $\eta_i = \log(m_i p_i)$. Let $h_i(\eta_i) = \eta_i - \log(m_i) = \log(p_i)$. Here $\phi = 1$ and $h_i(\eta_i)$ follows the linear structure (4). Then $f_i(y_i|\eta_i,\phi)$ is bounded for any $y_i \geq 0$, and

$$\int f_i(y_i|\eta_i,\phi=1)h_i'(\eta_i)d\eta_i = \int_0^\infty \frac{e^{-\mu_i}\mu_i^{y_i-1}}{y_i!}d\mu_i,$$

which is finite for $y_i > 0$. Under assumptions (b)-(e) of Theorem 4..1 , the joint posterior distribution of $(p_1,\ldots,p_N,\boldsymbol{\theta},\boldsymbol{Z},\delta_0,\delta_1)$ is proper.

Example 4..2 Suppose $f_i(y_i|\eta_i,\phi)$ is binomial with parameters m_i and p_i. This is a special case of (1) with $\phi = 1$ and $\eta_i = \log\{p_i/(1-p_i)\}$. Assume $h_i(\eta_i) = \eta_i$ has structure (4). Then $f_i(y_i|\eta_i,\phi=1)$ is bounded in η_i for any $0 \leq y_i \leq m_i$, and

$$\begin{aligned}\int_{-\infty}^\infty f_i(y_i|\eta_i,\phi=1)h_i'(\eta_i)d\eta_i &= \int_{-\infty}^\infty \frac{e^{y_i\eta_i}}{(e^{\eta_i}+1)^{m_i}}d\eta_i \\ &= \int_0^1 p_i^{y_i-1}(1-p_i)^{m_i-y_i-1}dp_i,\end{aligned}$$

which is finite if and only if $0 < y_i < m_i$. Under assumptions (b)–(e) of Theorem 4..1, the joint posterior distribution of $(p_1,\ldots,p_N,\boldsymbol{\theta},\boldsymbol{Z},\delta_0,\delta_1)$ is proper.

Example 4..3 When $Y_i|(\mu_i,\sigma^2) \sim \mathrm{N}(\mu_i,\sigma^2)$, we have $\eta_i = \mu_i$, $\phi = \sigma^2$, $A_i(\phi) = \phi$, $B_i(\eta_i) = \eta_i$ and $C_i(y_i,\phi) = -0.5\log(\phi) - y_i^2/(2\phi)$. If $h_i(\eta_i) = \eta_i$, this is a typical example of a normal hierarchical model. It is easy to see that $M_i(\phi) = 1/\sqrt{2\pi\phi}$ and $\int f_i(y_i|\eta_i,\phi)d\eta_i = 1$. Condition (18) becomes

$$\int_0^\infty \phi^{-\frac{1}{2}(N-n)}F(d\phi) < \infty,$$

which always holds when $N = n$ and F is a proper prior for ϕ. In addition, assumptions (b)–(e) of Theorem 4..1 hold. Then the joint posterior distribution of $(\mu_1,\ldots,\mu_N,\sigma^2,\boldsymbol{\theta},Z,\delta_0,\delta_1)$ is proper.

Example 4..4 Suppose $Y_i|(\mu_i, \alpha) \sim \text{gamma}(\alpha, \alpha/\mu_i)$, with density

$$f_i(y_i|\mu_i, \alpha) = \frac{\alpha^\alpha y_i^{\alpha-1}}{\Gamma(\alpha)\mu_i^\alpha} \exp\{-\alpha y_i/\mu_i\}.$$

Here α is the common shape parameter and μ_i is the mean of Y_i for given (μ_i, α). This is a special case of (1) with $\phi = \alpha$, $\eta_i = 1/\mu_i$, $A_i(\phi) = -1/\phi$, $B_i(\eta_i) = \log(\eta_i)$ and $C_i(y_i, \phi) = \alpha\log(\alpha) + (\alpha - 1)\log(y_i) - \log\{\Gamma(\alpha)\}$. Choose $h_i(\eta_i) = \log(\eta_i) = -\log(\mu_i)$. Then

$$M_i(\phi) = y_i^{-1}\alpha^\alpha e^{-\alpha}/\Gamma(\alpha) \text{ and } \int_0^\infty f(y_i|\eta_i, \phi)\frac{1}{\eta_i}d\eta_i = y_i^{-1}.$$

If $N = n$ and ϕ has a proper prior, condition (18) holds. If assumptions (b)–(e) in Theorem 4..1 hold, the joint posterior distribution of $(\mu_1, \ldots, \mu_N, \alpha, \boldsymbol{\theta}, \boldsymbol{Z}, \delta_0, \delta_1)$ is proper.

When ϕ is unknown but Y_i has a continuous distribution, as in the normal and gamma examples, we often choose $n = N$ and $\mathcal{J}_n = \{1, \ldots, N\}$, so that condition (18) becomes

$$\int \left\{\prod_{j=1}^N \int f_j(y_j|\eta_j, \phi)h_j'(\eta_j)d\eta_j\right\} F(d\phi) < \infty.$$

Remark 4..3 *Assumption (d) in Theorem 4..1 is crucial. Otherwise, the results may not hold. On the other hand, it is easy to see that the rank of the matrix* $(\boldsymbol{X}_2^t\boldsymbol{R}_1\boldsymbol{X}_2 + \boldsymbol{B})$ *equals* k *if either the rank of* $(\boldsymbol{X}_1, \boldsymbol{X}_2)$ *is* $p + k$ *or the rank of* $(\boldsymbol{B}) = k$. *The following results can be proved similarly.*

Theorem 4..2 *Assume that the rank of* $(\boldsymbol{X}_2^t\boldsymbol{R}_1\boldsymbol{X}_2 + \boldsymbol{B}) < k$, *where* $\boldsymbol{R}_1 = \boldsymbol{I}_N - \boldsymbol{X}_1(\boldsymbol{X}_1^t\boldsymbol{X}_1)^{-1}\boldsymbol{X}_1^t$, *and* $\boldsymbol{X}_1$ *and* $\boldsymbol{X}_2$ *are design matrices based on the full data. Under assumption (d), for any proper prior of* $(\delta_0, \delta_1, \phi)$, *the posterior is improper.*

Proof. Let G be defined as in the proof of Theorem 4..1, and replace n by N and $(\boldsymbol{X}_1^*, \boldsymbol{X}_2^*)$ by $(\boldsymbol{X}_1, \boldsymbol{X}_2)$. ¿From Sun, Tsutakawa and Speckman (1997), we know that for any given (δ_0, δ_1),

$$\int_{\mathbb{R}^p}\int_{\mathbb{R}^k} G d\boldsymbol{\theta} d\boldsymbol{Z} = \infty.$$

The results follows. □

The following result can be proved using the same argument as that of Theorem 4..1.

Theorem 4..3 *Given assumptions (a), (b) and (c) of Theorem 4..1, suppose that either condition (d1) or (d2) below holds:*

(d1) the rank of $(\boldsymbol{X}_2^{*t}\boldsymbol{R}_1\boldsymbol{X}_2^*)$ *is* k *and (19) is replaced by*

$$\mathbb{E}\{\delta_0^{-\frac{1}{2}(n-p-k)}\delta_1^{-\frac{1}{2}k}\} < \infty. \tag{22}$$

(d2) the rank of $\boldsymbol{B}$ is k and (19) is replaced by

$$\mathbb{E}\{\delta_0^{-\frac{1}{2}(n-p)}\} < \infty. \tag{23}$$

Then the result of Theorem 4..1 still holds.

5. Bayesian Computation

Bayesian inference for hierarchical GLMMs can be implemented via Markov chain Monte Carlo methods such as Gibbs sampling and/or the Metropolis algorithm. We assume that the prior for the variance components δ_i follows an inverse gamma(a_i, b_i) distribution with density (20). The proof of the following fact is omitted.

Fact 5..1 The full conditional distributions are as follows.

1. $\boldsymbol{\theta}|(\boldsymbol{\eta}, \phi, Z, \delta_0, \delta_1) \sim \text{MVN}_p((\boldsymbol{X}_1^t\boldsymbol{X}_1)^{-1}\boldsymbol{X}_1^t(\boldsymbol{V} - \boldsymbol{X}_2\boldsymbol{Z}), \delta_0(\boldsymbol{X}_1^t\boldsymbol{X}_1)^{-1})$.

2. $\boldsymbol{Z}|(\boldsymbol{\eta}, \phi, \boldsymbol{\theta}, \delta_0, \delta_1) \sim \text{MVN}_k(\boldsymbol{M}_1\boldsymbol{X}_2^t(\boldsymbol{V} - \boldsymbol{X}_1\boldsymbol{\theta}), \delta_0\boldsymbol{M}_1)$, where $\boldsymbol{M}_1 = (\boldsymbol{X}_2^t\boldsymbol{X}_2 + \delta_0\delta_1^{-1}\boldsymbol{B})^{-1}$.

3. $\delta_0|(\boldsymbol{\eta}, \phi, \boldsymbol{\theta}, \boldsymbol{Z}, \delta_1) \sim$ inverse gamma$(a_0 + \frac{n}{2},\ b_0 + \frac{1}{2}(\boldsymbol{V} - \boldsymbol{X}_1\boldsymbol{\theta} - \boldsymbol{X}_2\boldsymbol{Z})^t\ (\boldsymbol{V} - \boldsymbol{X}_1\boldsymbol{\theta} - \boldsymbol{X}_2\boldsymbol{Z}))$.

4. $(\delta_1|\boldsymbol{\eta}, \phi, \boldsymbol{\theta}, \boldsymbol{Z}, \delta_0) \sim$ inverse gamma$(a_1 + \frac{k}{2},\ b_1 + \frac{1}{2}\boldsymbol{Z}^t\boldsymbol{B}\boldsymbol{Z})$.

5. Given $(\phi, \boldsymbol{Z}, \delta_0, \delta_1)$, the η_j (or $v_j = h_j(\eta_j)$) are independent. In fact, since η_j and v_j are related by a one-to-one transformation, we can simulate from either η_j or v_j, depending on simplicity. The density of η_j given $(\phi, \boldsymbol{\theta}, \boldsymbol{Z}, \delta_0, \delta_1)$ is

$$s_j(\eta_j) \quad \propto \quad \exp\Big[\frac{y_j\eta_j - B_j(\eta_j)}{A_j(\phi)} - \frac{\{h_j(\eta_j) - \boldsymbol{x}_{1j}^t\boldsymbol{\theta} - \boldsymbol{x}_{2j}^t\boldsymbol{Z}\}^2}{2\delta_0}\Big]h_j'(\eta_j),$$

and the density of v_j given $(\phi, \boldsymbol{Z}, \delta_0, \delta_1)$ is

$$\tilde{s}_j(v_j) \quad \propto \quad \exp\Big[\frac{y_jh_j^{-1}(v_j) - B_j\{h_j^{-1}(v_j)\}}{A_j(\phi)} - \frac{\{v_j - \boldsymbol{x}_{1j}^t\boldsymbol{\theta} - \boldsymbol{x}_{2j}^t\boldsymbol{Z}\}^2}{2\delta_0}\Big],$$

where h_j^{-1} is the inverse function of h_j.

6. If the prior for ϕ is degenerate, so is its posterior. If ϕ has the prior density $g(\phi)$, then its posterior density given $(\boldsymbol{\eta}, \boldsymbol{\theta}, \boldsymbol{Z}, \delta_0, \delta_1)$ is

$$g^*(\phi) \propto g(\phi)\prod_{i=1}^{N}\exp[A_i(\phi)^{-1}\{y_i\eta_i - B_i(\eta_i)\} + C_i(y_i;\phi)].$$

Sampling from a normal or inverse gamma distribution is very simple. In Part 5 of Fact 5..1, the conditional density of η_i or v_i is often log-concave. For sampling from a log-concave density, Gilks and Wild's (1992) adaptive method or Berger and Sun's (1993) direct method can be used. Here are Poisson and binomial examples.

Example 4..1 (continued). When $h_i(\eta_i) = \eta_i - \log(m_i) = \log(p_i)$,

$$s_i(\eta_i) \propto \exp\Big[y_i\eta_i - e^{\eta_i} - \frac{1}{2\delta_0}\{\eta_i - \log(m_i) - (\boldsymbol{x}_{1i}^t\boldsymbol{\theta} + \boldsymbol{x}_{2i}^t\boldsymbol{Z})\}^2\Big].$$

Therefore

$$\frac{\partial^2}{\partial\eta_i^2}\log\{s_i(\eta_i)\} \quad = \quad -e^{\eta_i} - \delta_0^{-1} < 0.$$

Consequently, the conditional density of η_i given $(\phi, \boldsymbol{Z}, \delta_0, \delta_1)$ is log-concave. Since v_j is a linear transformation of η_i, the conditional density of v_j is also log-concave.

Example 4..2 (continued). When $h_i(\eta_i) = \eta_i = \log\{p_i/(1-p_i)\}$, we have

$$s_i(\eta_i) \propto \exp\Big[y_i\eta_i - m_i\log(1+e^{\eta_i}) - \frac{\{\eta_i - \log(m_i) - (\boldsymbol{x}_{1i}^t\boldsymbol{\theta} + \boldsymbol{x}_{2i}^t\boldsymbol{Z})\}^2}{2\delta_0}\Big].$$

We can show that

$$\frac{\partial^2}{\partial\eta_i^2}\log\{s_i(\eta_i)\} \quad = \quad -m_ie^{\eta_i}(1+e^{\eta_i})^{-2} - \delta_0^{-1} < 0.$$

So the conditional density of $\eta_i = v_i$ given $(\phi, \boldsymbol{Z}, \delta_0, \delta_1)$ is log-concave.

Example 4..3 (continued). When $h_i(\eta_i) = \eta_i$, we have

$$s_i(\eta_i) \propto \exp\Big[-\frac{(y_i-\eta)^2}{2\sigma^2} - \frac{\{\eta_i - (\boldsymbol{x}_{1i}^t\boldsymbol{\theta} + \boldsymbol{x}_{2i}^t\boldsymbol{Z})\}^2}{2\delta_0}\Big].$$

Clearly, the conditional distribution of η_i given others is normal with mean $\delta_i(\sigma^2 + \delta_i)^{-1}y_i + \sigma^2(\sigma^2+\delta_i)^{-1}(\boldsymbol{x}_{1i}^t\boldsymbol{\theta} + \boldsymbol{x}_{2i}^t\boldsymbol{Z})$ and variance $\sigma^2\delta_i(\sigma^2+\delta_i)^{-1}$.

Example 4..4 (continued). When $h_i(\eta_i) = log(\eta_i)$, we have

$$s_i(\eta_i) \propto \exp\Big[-\alpha y_i\eta_i + \alpha\log(\eta_i) - \frac{\{\log(\eta_i) - (\boldsymbol{x}_{1i}^t\boldsymbol{\theta} + \boldsymbol{x}_{2i}^t\boldsymbol{Z})\}^2}{2\delta_0}\Big].$$

This conditional density is not necessary logconcave. However, its transformation $\xi_i = \log(\eta_i)$ has the conditional density

$$\tilde{s}_i(\xi_i) \propto \exp\Big[-\alpha y_ie^{\xi_i} + (\alpha+1)\xi_i - \frac{\{\xi_i - (\boldsymbol{x}_{1i}^t\boldsymbol{\theta} + \boldsymbol{x}_{2i}^t\boldsymbol{Z})\}^2}{2\delta_0}\Big].$$

It is easy to verify that $\frac{\partial^2}{\partial\xi_i^2}\log\{\tilde{s}_i(\xi_i)\} = -\alpha y_ie^{\xi_i} - \delta_0^{-1}$, which is negative. Consequently, we can simply sample from the logconcave density of ξ_i, then make the transformation $\eta_i = e^{\xi_i}$.

For numerical illustrations of the Gibbs sampler discussed here, see the binomial application used in He and Sun (1998) and the Poisson example given in Sun *et al.* (1998).

References

Arnold, B.C. and Press, S.J. (1989). Compatible conditional distributions, *Journal of the American Statistical Association*, **84**, 152–156.

Berger, J.O. and Sun, D. (1993). Bayesian Analysis for the Poly-Weibull Distribution. *Journal of the American Statistical Association*, **88**, 1412–1418.

Bernardinelli, L., Clayton, D. and Montomoli, C. (1995). Bayesian estimates of disease maps: how important are priors? *Statistics in Medicine*, **14**, 2411–2431.

Bernardinelli, L. and Montomoli, C. (1992). Empirical Bayes versus fully Bayesian analysis of geographical variation in disease risk. *Statistics in Medicine*, **11**, 983–1007.

Berzuini, C. (1996). Medical Monitoring. In *Markov Chain Monte Carlo in Practice*, ed. by W.R. Gilks, S. Richardson, and D.J. Spiegelhalter. Chapman and Hall, 321–337.

Besag, J. (1974). Spatial interaction and the statistical analysis of lattice systems (with discussion). *J. Roy. Statist. Soc. Ser. B*, **36**, 192–236.

Besag, J. (1975). Statistical analysis of non-lattice data. *The Statistician*, **24**, 179–195.

Besag, J., York, J. & Mollié, A. (1991). Bayesian image restoration, with two applications in spatial statistics (with discussion). *Ann. Inst. Statist. Math.*, **43**, 1–59.

Breslow, N.E. and Clayton, D.G. (1993). Approximate inference in generalized linear mixed models. *Journal of the American Statistical Association*, **88**, 9–25.

Carlin, B.P. & Louis, T.A. (1996). *Bayes and Empirical Bayes Methods for Data Analysis,* London: Chapman and Hall.

Clayton, D. (1996). Generalized linear mixed models. In *Markov Chain Monte Carlo in Practice*, ed. by W.R. Gilks, S. Richardson, and D.J. Spiegelhalter. Chapman and Hall, 275–301.

Clayton, D. & Kaldor, J. (1987). Empirical Bayes estimates of age-standardized relative risks for use in disease mapping. *Biometrics*, **43**, 671–681.

Cressie, N. & Chan, N.H. (1989). Spatial modeling of regional variables. *Journal of the American Statistical Association*, **84**, 393–401.

Gelfand, A.E. & Smith, A.F.M. (1990). Sampling based approaches to calculating marginal densities. *Journal of the American Statistical Association*, **85**, 398–409.

Ghosh, M., Natarajan, K., Stroud, T.W.F., & Carlin, B.P. (1998). Generalized linear models for small area estimation. *Journal of the American Statistical Association*, **93**, 273–282.

Gilks, W.R. & Wild, P. (1992). Adaptive rejection sampling for Gibbs sampling. *Applied Statistics*, 41, 337–348.

Harville, D.A. (1977). Maximum likelihood approaches to variance component estimation and to related problems. *Journal of the American Statistical Association*, **72**, 320–338.

He, Z., and Sun, D. (1998). Hierarchical Bayes estimation of hunting success rates with spatial correlations. Revised for *Biometrics.*

Hobert, J.P. & Casella, G. (1996). The effect of improper priors on Gibbs sampling in hierarchical linear mixed models. *Journal of the American Statistical Association*, **91**, 1461–1473.

Hobert, J.P. & Casella, G. (1998). Functional compatibility, Markov chains and Gibbs sampling with improper posteriors. *J. of Computational and Graphical Statistics*, **7**, 42–66.

Kindermann, R. and Snell, J.L. (1980). *Markov Random Fields and Their Applications.* Amer. Math. Soc., Providence. RI.

McCullagh, P. and Nelder, J.A. (1989). *Generalized Linear Models*, Chapman & Hall, London.

Ord, K. (1975). Estimation methods for models with spatial interaction. *Journal of the American Statistical Association*, **70**, 120–126.

Ripley, B.D. (1981). *Spatial Statistics.* Wiley, New York.

Ripley, B.D. (1988). *Statistical Inference for Spatial Processes.* Cambridge University Press, Cambridge.

Sun, D., Tsutakawa, R.K., and He, Z. (1998). Propriety of posteriors with improper priors in hierarchical linear mixed models. Submitted

Sun, D., Tsutakawa, R.K., Kim, H. & He, Z. (1997). Spatio-Time Interaction with Disease Mapping. Submitted.

Sun, D., Tsutakawa, R.K. and Speckman, P.L. (1997). Bayesian inference for CAR (1) models with noninformative priors. *Biometrika*, in press.

Tsutakawa, R.K. (1988). Mixed model for analysing geographic variability in mortality rates. *Journal of the American Statistical Association*, **83**, 117–130.

Waller, L.A., Carlin, B.P., Xia, H., & Gelfand, A.E. (1997). Hierarchical spatio-temporal mapping of disease rates. *Journal of the American Statistical Association*, **92**, 607–617.

West, M. and Aguilar, O. (1997). Studies of Quality Monitor Time Series: The VA Hospital System, Discussion Paper 97-20a, ISDS, Duke University.

3

Prior Elicitation and Variable Selection for Generalized Linear Mixed Models

Joseph G. Ibrahim
Ming-Hui Chen

ABSTRACT Generalized linear models serve as a useful class of regression models for discrete and continuous data. In applications such as longitudinal studies, observations are typically correlated. The correlation structure in the data is induced by introducing a random effect, leading to the generalized linear mixed model (GLMM). In this chapter, we propose a class of informative prior distributions for the class of GLMM's and investigate their theoretical as well as their computational properties. Specifically, we investigate conditions for propriety of the proposed priors for the class of GLMM's and show that they are proper under some very general conditions. In addition, we examine recent computational methods such as hierarchical centering and semi-hierarchical centering for doing Gibbs sampling for this class of models. One of the main applications of the proposed priors is variable subset selection. Novel computational tools are developed for sampling from the posterior distributions and computing the posterior model probabilities. We demonstrate our methodology with a real longitudinal dataset.

1. Introduction

Generalized linear models (McCullagh and Nelder 1989; Nelder and Wedderburn 1972) are a unified approach to regression methods. They can be applied to a wide array of discrete, continuous, and censored outcomes, and are most commonly used when the outcomes are independent. However, in many applications, this independence is not a reasonable assumption. This is particularly obvious in longitudinal studies, where multiple measurements made on the same individual are likely to be correlated. One recent technique for the analysis of such general correlated data is the generalized estimating equation approach introduced by Liang and Zeger (1986) and Zeger and Liang (1986). This approach has the desirable quality that allows for independence between subjects while introducing a correlation structure within subjects. A drawback to this approach, however, is that it assumes all subjects have the same covariance structure. For continuous outcomes with normal errors, Laird and Ware (1982) present the random effects model. In this model, a subject-specific covariance structure is generated by assuming that each individual has a unique set of regression coefficients, the random effects, which are distributed around the mean regression coefficients for the population, also known as the fixed effects. There may also be regression coefficients which are equal for all individuals. Conditional on the random effects, repeated observations on a subject are considered independent, while marginalizing over the random effects, a unique covariance structure for the observations within each subject is obtained.

Zeger and Karim (1991) present a generalization of the normal random effects model to the class of generalized linear models, generating a generalized linear mixed model (GLMM). They frame the model from the Bayesian perspective and fit it using a Gibbs sampler. They point out that attempts to fit this model using classical (frequentist) techniques are limited by the need for multidimensional numerical integrations, except in special cases. As a result of these analytically intractable integrations, classical analysis of GLMMs has relied on approximations to maximum likelihood techniques (see Breslow and Clayton 1993).

In this chapter, we address the problem of informative prior elicitation and Bayesian variable selection for the class of generalized linear mixed models. Specifically, we discuss a class of informative prior distributions for the regression parameters that are extensions of the priors proposed in Ibrahim, Ryan, and Chen (1998) and Chen, Ibrahim, and Yiannoutsos (1999). We also give several attractive theoretical and computational properties of the priors, as well as give efficient computational algorithms for computing posterior model probabilities in the variable selection problem.

The construction of the prior distributions is based on the notion of the availability of historical data from a similar previous study or studies. As is well known, historical data are often available in applied research settings where the investigator has access to previous studies measuring the same response and covariates as the current study. For example, in many cancer and AIDS clinical trials, current studies often use treatments that are very similar or slight modifications of treatments used in previous studies. In carcinogenicity studies, large historical databases exist for the control animals from previous experiments. In all of these situations, it is natural to incorporate the historical data into the current study by quantifying it with a suitable prior distribution on the model parameters. Our proposed methodology can be applied to each of these situations as well as in other applications that involve historical data.

The prior specification is based on the notion of specifying a vector of prior predictions, y_0, for the response vector of the current study, along with a covariate matrix X_0 corresponding to y_0. Then y_0 and X_0 are used to specify an automated parametric informative prior for the regression coefficients. The quantity y_0 can be taken as the raw response vector from the historical data, a vector of fitted values based on the historical data, a vector obtained from a theoretical prediction model, or a vector specified from expert opinion or case-specific information. Thus y_0 is viewed as a prior prediction for y, the actual data in the current study. Similarly, X_0 can be taken as the raw covariate matrix based on the historical data or it can be specified in other ways. In any case, taking y_0 and X_0 to be the raw historical data results in a more natural, interpretable, and automated specification. Throughout the remainder of this chapter, we will refer to y_0 as a prior prediction, though it need not be a prediction in any formal sense.

In Section 2.1, we present notation for the GLMM and give the likelihood. In Sections 2.2-2.4, we discuss the priors in detail, giving several theoretical properties. In Section 3, we discuss Bayesian variable selection and discuss novel computational methods for implementation. In Section 4, we demonstrate the proposed priors on a longitudinal dataset. We conclude this chapter with a brief discussion.

2. Generalized Linear Mixed Models

2.1 Models

First, we define the normal linear random effects model, then introduce random effects into generalized linear models. For individual i, with n_i repeated measurements, the normal linear random effects model for outcome vector y_i is given by

$$y_i = X_i\beta + Z_i b_i + e_i, \quad i = 1, \ldots, N,$$

where y_i is $n_i \times 1$, X_i is an $n_i \times p$ matrix of fixed covariates, β is a $p \times 1$ parameter vector of regression coefficients, commonly referred to as fixed effects in these models, Z_i is an $n_i \times q$ matrix of covariates for the $q \times 1$ vector of random effects b_i, and e_i is an $n_i \times 1$ vector of errors. We let y_{it} denote the tth component of y_i, $t = 1, \ldots, n_i$. It is standard in implementations of this model to assume e_i and b_i are independent and that both are distributed Normal, with $e_i \sim N_{n_i}(0, \sigma^2 I_{n_i})$ and

$$b_i \sim N_q(0, V) \,, \tag{1}$$

where I_s is the $s \times s$ identity matrix and $N_s(\mu, V)$ denotes the s-dimensional multivariate Normal distribution with mean μ and covariance matrix V. Throughout the chapter, it will be more convenient to work with $T = V^{-1}$. Under these assumptions,

$$(y_i|\beta, b_i, \sigma^2) \sim N_{n_i}(X_i\beta + Z_i b_i, \sigma^2 I_{n_i}). \tag{2}$$

Notice that marginally,

$$(y_i|\beta, \sigma^2, T) \sim N_{n_i}(X_i\beta, Z_i T Z_i^T + \sigma^2 I_{n_i}), \tag{3}$$

which shows the unique covariance structure for subject i. For convenience, we call model (2) the normal random effects model.

Now suppose the sampling distribution of y_{it}, $t = 1, \ldots, n_i$ is from the exponential family, so that

$$p(y_{it}|\theta_{it}, \tau) = \exp\left\{\tau\left[y_{it}\theta_{it} - g(\theta_{it})\right] + c(y_{it}, \tau)\right\} \,,$$

where

$$\mu_{it} = E(y_{it}|\theta_{it}, \tau) = \frac{dg(\theta_{it})}{d\theta_{it}},$$

$$v_{it} = \text{var}(y_{it}|\theta_{it}, \tau) = \tau^{-1}\frac{d^2 g(\theta_{it})}{d\theta_{it}^2} \,,$$

and τ is a scalar dispersion parameter.

In the generalized linear mixed model, the canonical parameter θ_{it} is related to the covariates by

$$\theta_{it} = \theta(\eta_{it}) \,,$$

where $\eta_{it} = x_{it}'\beta + z_{it}' b_i$, and x_{it}' and z_{it}' are rows of the X_i and Z_i matrices, $\theta(\cdot)$ is a monotonic differentiable function, often referred to as the θ-link, and η_{it} is called the linear predictor. Throughout, we write

$$p(y_{it}|\theta_{it}, \tau) \equiv p(y_{it}|\beta, b_i, \tau) \,,$$

where

$$p(y_{it}|\beta, b_i, \tau) = \exp\{\tau[y_{it}\theta(\eta_{it}) - g(\theta(\eta_{it}))] + c(y_{it}, \tau)\}. \qquad (4)$$

When $\theta_{it} = \eta_{it}$, then the link is said to be the canonical link. For example, for the logistic regression model, we have

$$p(y_{it}|\beta, b_i, \tau) = \exp\{y_{it}(x_{it}'\beta + z_{it}'b_i) - \log(1 + \exp(x_{it}'\beta + z_{it}'b_i))\},$$

so that $\tau = 1$.

Note that the GLMM imitates the normal random effects model in that we assume that conditional on the random effect b_i, the repeated observations on subject i are independent. For ease of exposition, we assume that $\tau = \tau_0$, where τ_0 is known as $\tau_0 = 1$ in logistic and Poisson regression, and denote $c(y) = c(y, \tau_0)$ and $p(y_{it}|\beta, b_i) = p(y_{it}|\beta, b_i, \tau_0)$ in (4). Letting $b = (b_1, \ldots, b_N)'$, $y = (y_{11}, \ldots, y_{Nn_N})'$, $X = (X_1, \ldots, X_N)'$, $Z = (Z_1, \ldots, Z_N)$, $\eta = X\beta + Zb$, the joint density of (y, b) based on N subjects for the GLMM is

$$p(y, b \mid \beta, T) = \prod_{i=1}^{N}\prod_{t=1}^{n_i} p(y_{it}|\beta, b_i)\, p(b_i \mid T), \qquad (5)$$

where $p(b_i \mid T)$ is the distribution of b_i in (1), with $T = V^{-1}$. Letting $\eta = X\beta + Zb$, it will be convenient to write (2) in vector notation as

$$\begin{aligned} p(y, b \mid \beta, T) &= p(y \mid b, \beta)\, p(b \mid T) \\ &= \exp\{\tau_0[y'\theta(\eta) - J'g(\theta(\eta))] + J'c(y)\}\, p(b \mid T), \end{aligned} \qquad (6)$$

where J is a vector of ones, $\theta(\eta)$ are elementwise vectorized versions of those in (2), and

$$p(b \mid T) = \prod_{i=1}^{N}(2\pi)^{-q/2} \mid T \mid^{1/2} \exp\{b_i'Tb_i\}.$$

There are several attractive properties of (6). First, it takes within-subject correlation into account while allowing each individual to have a unique correlation structure and maintaining independence between subjects. Second, the model accommodates unbalanced data, in that response vectors need not be of the same length. Similarly, irregularly timed measurements can be fit within this model without any adjustment. Finally, posterior distributions or estimates of the random effects have interpretive value when the trend of the mean function of individuals is of interest. To induce the correlation structure on the responses, we integrate out the random effect, which leads to the likelihood

$$p(y \mid \beta, T) = \int_{R^{Nq}} p(y, b \mid \beta, T)\, db, \qquad (7)$$

where R^{Nq} denotes the Nq dimensional Euclidean space.

2.2 The Prior Distributions

Informative prior elicitation is a very important part of a Bayesian analysis. We propose a class of informative priors for the regression coefficients β, since these parameters are typically of primary inferential interest in these problems. The proposed prior distributions are useful in situations such as variable selection and prediction. Likelihood-based frequentist methods for inference with these models is virtually impossible, due to the high dimensional integrations required for integrating out the random effects (see Zeger and Karim, 1991). Our prior construction for β is based on the notion of the existence of similar previous studies, i.e., historical data, as was motivated in Section 1. For ease of exposition, we will assume one previous study as the generalization of the prior to multiple previous studies will become immediately clear. Suppose there exist historical data with N_0 subjects that yield the $n_{0i} \times 1$ response vector y_{0i} for subject i. Let X_{0i} be an $n_{0i} \times p$ matrix of fixed covariates, and Z_{0i} be an $n_{0i} \times q$ matrix of covariates for the $q \times 1$ vector of random effects b_{0i} for subject i, $i = 1, 2, \ldots, N_0$ for the historical data. Also let $b_0 = (b_{01}, \ldots, b_{0N_0})'$, $y_0 = (y_{011}, \ldots, y_{0N_0 n_{0N_0}})'$, $X_0 = (X_{01}, \ldots, X_{0N_0})'$, and $Z_0 = (Z_{01}, \ldots, Z_{0N_0})$. Finally let $D_0 = (N_0, X_0, y_0, Z_0)$ denote the historical data.

We propose a prior distribution for β taking the form

$$\pi(\beta \mid D_0, T, a_0) \propto \prod_{i=1}^{N_0} \left(\int_{R^q} \prod_{t=1}^{n_{0i}} [p(y_{0it} \mid \beta, b_{0i})]^{a_0} \; p(b_{0i} \mid T) \; db_{0i} \right), \tag{8}$$

where $p(y_{0it} \mid \beta, b_{0i}, \tau)$ is (4) with $(y_{0it}, b_{0i}, \tau_0)$ in place of (y_{it}, b_i, τ). That is, $p(y_{0it} \mid \beta, b_{0i})$ is the GLMM based on the prior data y_{0it}. The quantity a_0 is a prior parameter which weights the historical data relative to the likelihood function of the current study. It is reasonable to restrict a_0 to $0 \leq a_0 \leq 1$. The parameter a_0 can also be interpreted as a dispersion parameter which takes into account between and within study variability in the historical data. The prior in (6) does not have a closed form but it has several attractive theoretical and computational properties as shown in Sections 2.3 and 3. In vector notation, the prior can be written as

$$\begin{aligned}\pi(\beta \mid D_0, T, a_0) &= \int_{R^{N_0 q}} [p(y_0 \mid \beta, b_0)]^{a_0} \; p(b_0 \mid T) db_0 \\ &= \int_{R^{N_0 q}} \exp\left\{a_0 \left\{\tau_0 [y_0' \theta(\eta_0) - J_0' g(\theta(\eta_0))] + J_0' c(y_0)\right\}\right\} \; p(b_0 \mid T) \; db_0,\end{aligned} \tag{9}$$

where J_0 is a vector of ones and $\eta_0 = X_0\beta + Z_0 b_0$ are elementwise vectorized versions of those in (6).

The single most common choice for the structure of $V = T^{-1}$ is a scalar with Z_i a column of ones. In general, it is usual to take V to be unstructured but by choice of parameterizations of V one can model rather different covariance structures. For example, taking Z_i to be the identity matrix and V to be an autoregressive (AR-1) structure, one gets a common time series model with independent measurement error e_i. We can incorporate both of these examples in a single model. We take $V = \sigma_b^2 \Sigma_b$, where the $(j, j^*)^{th}$ element of Σ_b has the form $\sigma_{jj^*} = \rho^{|j-j^*|}$, where $\rho^{|j-j^*|}$ is the correlation between (b_{ij}, b_{ij^*}), and $-1 \leq \rho \leq 1$. This AR-1 structure for V is quite general and appears to be quite suitable in practice for many longitudinal datasets.

The prior specification is thus completed by specifying priors for (a_0, σ_b^2, ρ). We take these parameters independent a priori. We specify a beta prior for a_0, an inverse gamma prior for σ_b^2, denoted $IG(\delta_0, \gamma_0)$, and a scaled beta prior for ρ, denoted scbeta(ν_0, ψ_0). Thus, we propose a joint prior distribution of the form

$$\begin{aligned}&\pi(\beta, a_0, \sigma_b^2, \rho \mid D_0) \\ &\propto \prod_{i=1}^{N_0} \left(\int_{R^q} \prod_{t=1}^{n_{0i}} [p(y_{0it} \mid \beta, b_{0i})]^{a_0} \; p(b_{0i} \mid T) \; db_{0i} \right)\end{aligned}$$

$$\times a_0^{\alpha_0-1}(1-a_0)^{\lambda_0-1} \times (\sigma_b^2)^{-(\delta_0+1)}\exp(-\sigma_b^{-2}\gamma_0)$$
$$\times(1+\rho)^{\nu_0-1}(1-\rho)^{\psi_0-1}, \quad (10)$$

where $\theta_0 \equiv (\delta_0, \gamma_0, \alpha_0, \lambda_0, \nu_0, \psi_0)$ are known prior parameters. In our analyses, we take vague choices of the prior hyperparameters θ_0.

2.3 Propriety of the Prior Distribution

It is critical to examine the conditions under which the joint prior in (9) is proper. This issue is crucial in Bayesian variable selection (see, for example, Ibrahim, Ryan, and Chen (1998) and Chen, Ibrahim, and Yiannoutsos (1999)), as it is well known that Bayesian variable selection requires a proper prior distribution. It is also an important issue in Bayesian hypothesis testing problems, and in particular, in the calculation of Bayes factors and related quantities (see, for example, Berger, 1985, pp. 145-157). Here, we establish some very general results concerning the propriety of the joint prior distribution of $(\beta, a_0, \sigma_b^2, \rho)$ for generalized linear mixed models. We have the following theorem.

Theorem 2..1 *Assume that*

$$\exp\left\{\tau_0\left[y_{0it}\theta(\eta_{0it}) - g(\theta(\eta_{0it}))\right] + c(y_{0it})\right\} \le M_0 \quad (11)$$

for $1 \le t \le n_{0i}$, $1 \le i \le N_0$, *where* M_0 *is some finite constant. Suppose that there exist* $y_{0i_1t_1}$, $y_{0i_2t_2}, \ldots, y_{0i_pt_p}$ $(1 \le i_1 \le i_2 \le \cdots \le i_p)$ *such that*

$$\int_{-\infty}^{\infty} e^{t_0|\eta|}\exp\left\{\tau_0\left[y_{0i_jt_j}\theta(\eta) - g(\theta(\eta))\right] + c(y_{0i_jt_j})\right\}d\eta < \infty \quad (12)$$

or

$$\int_{-\infty}^{\infty} e^{t_0|\eta|^2}\exp\left\{\tau_0\left[y_{0i_jt_j}\theta(\eta) - g(\theta(\eta))\right] + c(y_{0i_jt_j})\right\}d\eta < \infty \quad (13)$$

for some $t_0 > 0$ *and* $j = 1, 2, \ldots, p$ *and the corresponding design matrix* $(x_{0i_1t_1}, x_{0i_2t_2}, \ldots, x_{0i_pt_p})'$ *has full rank. Then the joint prior distribution* $\pi(\beta, a_0, \sigma_b^2, \rho|D_0)$ *is proper, i.e.,*

$$\int_{-1}^{1}\int_{0}^{\infty}\int_{0}^{1}\int_{R^p}\prod_{i=1}^{N_0}\left(\int_{R^q}\prod_{t=1}^{n_{0i}}\left[p(y_{0it}\mid\beta, b_{0i})\right]^{a_0}\ \pi(b_{0i}\mid T)\ db_{0i}\right)$$
$$\times a_0^{\alpha_0-1}(1-a_0)^{\lambda_0-1} \times (\sigma_b^2)^{-(\delta_0+1)}\exp(-\sigma_b^{-2}\gamma_0)$$
$$\times(1+\rho)^{\nu_0-1}(1-\rho)^{\psi_0-1}\ d\beta\ da_0\ d\sigma_b^2\ d\rho < \infty \quad (14)$$

if one of the following conditions is satisfied:

(i) $\alpha_0 > p, \lambda_0 > 0$, *and* (12) *holds;*

(ii) $\alpha_0 > p/2, \lambda_0 > 0$, *and* (13) *holds.*

The proof of the theorem is omitted for brevity. The conditions stated in Theorem 2..1 are sufficient and they hold for many generalized linear mixed models such as the normal, Poisson, and binomial models. However, when the y_{0i} are binary responses, i.e., $y_{0it} = 0$ or 1, we have

$$\pi(\beta, a_0, \sigma_b^2, \rho \mid D_0)$$
$$\propto \prod_{i=1}^{N_0} \left(\int \prod_{t=1}^{n_{0i}} \left[F(x_{0it}'\beta + z_{0it}'b_{0i})^{y_{0it}} (1 - F(x_{0it}'\beta + z_{0it}'b_{0i}))^{1-y_{0it}} \right]^{a_0} \right.$$
$$\left. p(b_{0i} \mid T)\, db_{0i} \right) \times a_0^{\alpha_0 - 1}(1 - a_0)^{\lambda_0 - 1} \times (\sigma_b^2)^{-(\delta_0+1)} \exp(-\sigma_b^{-2}\gamma_0)$$
$$\times (1+\rho)^{\nu_0 - 1}(1-\rho)^{\psi_0 - 1},$$

where F is a cumulative distribution function with $0 < F(\eta_i) < 1$, and F^{-1} is called a link function. In this binary response case, neither (12) nor (13) will be satisfied. But, under some additional regularity conditions on the fixed covariates x_{0it}, the propriety of the prior distribution $\pi(\beta, a_0, \sigma_b^2, \rho \mid D_0)$ can still be established. The main result is stated in the next theorem.

Theorem 2..2 *Let* $c_{0it} = 1$ *if* $y_{0it} = 0$ *and* $= -1$*if* $y_{0it} = 1$ *and define* $I_0 = \{(i,t) : 1 \le t \le n_{0i}, 1 \le i \le N_0,\ c_{0it}\ x_{0it1} > 0\}$ *and* $J_0 = \{(i,t) :\ 1 \le t \le n_{0i}, 1 \le i \le N_0,\ c_{0it}x_{0it1} < 0\}$. *Assume that the following conditions are satisfied*

$(C1)$ I_0 *and* J_0 *are non-empty sets;*

$(C2)$ $\forall\ (\beta_2, \cdots, \beta_p)' \neq 0,$

$$\min_{(j,t)\in J_0} \left(\frac{\sum_{l=2}^p x_{0jtl}\beta_l}{x_{0jt1}} \right) < \max_{(i,t)\in I_0} \left(\frac{\sum_{l=2}^p x_{0itl}\beta_l}{x_{0it1}} \right);$$

$(C3)$ $\int_{-\infty}^{\infty} e^{t_0|u|} dF(u) < \infty$ *for some* $t_0 > 0$;

$(C4)$ $\alpha_0 > p,\ \lambda > 0,\ \delta_0 > p/2,$

then for the binary responses y_{0it}*'s,* (14) *holds.*

The proof of this theorem is omitted for brevity. Note that in Theorem 2..2, x_{0it1} may be 1, which corresponds to the inclusion of an intercept in the model.

2.4 Specifying the Hyperparameters

In the context of model selection and estimation of β in GLMM, (ρ, σ^2) are viewed as nuisance parameters, and therefore we take vague choices for their prior hyperparameters. In particular, it is reasonable to take $\nu_0 = \psi_0 = 1$ which yields a uniform prior for ρ on $[-1, 1]$. Also, we take $\delta_0 \to p/2$ and $\gamma_0 \to 0$ for σ_b^2 to ensure a proper joint prior $\pi(\beta, a_0, \sigma_b^2, \rho | D_0)$.

For a_0, we recommend that several values of the hyperparameters be chosen and sensitivity analyses conducted. For elicitation purposes, it is easier to work with the prior mean and variance of a_0, given by $\mu_{a_0} = \alpha_0/(\alpha_0 + \lambda_0)$, and $\sigma_{a_0}^2 = \mu_{a_0}(1 - \mu_{a_0})(\alpha_0 + \lambda_0 + 1)^{-1}$. From Theorems 2..1 and 2..2, a sufficient condition for the propriety of the prior distribution is that $\alpha_0 > p$ for the full model. Therefore, a reasonable starting point for the analysis is to choose $\alpha_0 = \lambda_0 = p+1$, which gives $\mu_{a_0} = 1/2$. Then we conduct several sensitivity analyses within a suitable range of the uniform prior, using various values of $(\mu_{a_0}, \sigma^2_{a_0})$. Small and large values of $(\mu_{a_0}, \sigma^2_{a_0})$ should be considered. We do not recommend doing an analysis based on one set of proposed values of $(\mu_{a_0}, \sigma^2_{a_0})$, nor do we propose specifying $(\mu_{a_0}, \sigma^2_{a_0})$ by a one-time automated procedure. The choice of $(\mu_{a_0}, \sigma^2_{a_0})$ depends on the context of the problem and the structure of the historical data. In Section 4, we demonstrate several choices of $(\mu_{a_0}, \sigma^2_{a_0})$ and conduct sensitivity analyses and examine their impact on variable selection and estimation of β.

2.5 The Posterior Distribution and its Computation

Let $D = (N, X, y, Z)$ denote the data for the current study. Then, the joint posterior distribution of $(\beta, a_0, \sigma_b^2, \rho)$ is given by

$$p(\beta, a_0, \sigma_b^2, \rho \mid D, D_0) \quad \propto \quad p(y \mid \beta, T)\pi(\beta, a_0, \sigma_b^2, \rho \mid D_0), \tag{15}$$

where $p(y \mid \beta, T)$ is given by (3) and $\pi(\beta, a_0, \sigma_b^2, \rho|D_0)$ is given by (9). Clearly, when the prior is proper, then so is the posterior. The posterior in (15) does not have a closed form in general, but it has attractive computational properties due to the recent development of the Gibbs sampler for generalized linear mixed models; see, for example, Gelfand, Sahu, and Carlin (1996), and Gelfand and Sahu (1996). Instead of directly sampling $(\beta, a_0, \sigma_b^2, \rho)$ from the posterior distribution $p(\beta, a_0, \sigma_b^2, \rho \mid D, D_0)$, we sample $(\beta, a_0, \sigma_b^2, \rho, b, b_0)$ from the joint posterior distribution $p(\beta, a_0, \sigma_b^2, \rho, b, b_0 \mid D, D_0)$, which is proportional to

$$\begin{aligned} & p^*(\beta, a_0, \sigma_b^2, \rho, b, b_0 \mid D, D_0) \\ &= p(y, b \mid \beta, T)p(y_0 \mid \beta, b_0)]^{a_0} p(b_0 \mid T) \times a_0^{\alpha_0 - 1}(1 - a_0)^{\lambda_0 - 1} \\ & \quad \times (\sigma_b^2)^{-(\delta_0+1)} \exp(-\sigma_b^{-2}\gamma_0) \ \times (1+\rho)^{\nu_0 - 1}(1-\rho)^{\psi_0 - 1}, \end{aligned} \tag{16}$$

where $p(y, b \mid \beta, T)$, $p(y_0 \mid \beta, b_0)$, and $p(b_0 \mid T)$ are given by (6) and (9), respectively. The complete or semi hierarchical centering reparameterization technique of Gelfand, Sahu, and Carlin (1996) or Gelfand and Sahu (1996) is particularly suitable for the implementation of the Gibbs sampler for our problem. Here, we note that if all the n_i's, the n_{0i}'s and q are the same, we can directly apply the complete hierarchical centering reparameterization of Gelfand, Sahu, and Carlin (1996). Otherwise, we can use the semi-hierarchical centering reparameterization of Gelfand and Sahu (1996). After the (semi) hierarchical centering reparameterization, efficient algorithms can be adopted to sample the reparametrized parameters, which are functions of b, b_0, and β, and the other parameters including β, a_0, σ_b^2, and ρ. To preserve space, we omit the details of these algorithms here.

3. Bayesian Variable Selection

In this section, we consider Bayesian variable selection procedures for generalized linear mixed models. We develop an efficient Monte Carlo approach for computing posterior model probabilities.

We first introduce some necessary notation used for variable subset selection. Let $\mathcal{M}$ denote the model space. We enumerate the models in $\mathcal{M}$ by $m = 1, 2, \ldots, \mathcal{K}$, where $\mathcal{K}$ is the dimension of $\mathcal{M}$ and model $\mathcal{K}$ denotes the full model. The full model is defined here as the model containing all of the available covariates in the study, and thus $\mathcal{K} = 2^p$. Also, let $\beta^{(\mathcal{K})} = (\beta_1, \ldots, \beta_p)'$ denote the regression coefficients for the full model, and let $\beta^{(m)}$ denote a $k_m \times 1$ vector of regression coefficients for model m with a specific choice of k_m covariates. We write $\beta^{(\mathcal{K})} = (\beta^{(m)'}, \beta^{(-m)'})'$, where $\beta^{(-m)}$ is $\beta^{(\mathcal{K})}$ with $\beta^{(m)}$ deleted. In addition, we let $D^{(m)}$ and $D_0^{(m)}$ denote the current data and the historical data under model m.

To elicit the prior distribution on the model space $\mathcal{M}$, we let

$$\begin{aligned} & p_0^*(\beta^{(m)}|D_0^{(m)}) \\ &= \int_{-1}^1 \int_0^\infty \int_0^1 \left\{ \prod_{i=1}^{N_0} \left(\int_{R^q} \prod_{t=1}^{n_{0i}} \left[p(y_{0it} \mid \beta^{(m)}, b_{0i}) \right]^{a_0} p(b_{0i} \mid T)\, db_{0i} \right) \right. \\ & \quad \times a_0^{\alpha_0 - 1} (1 - a_0)^{\lambda_0 - 1} \times (\sigma_b^2)^{-(\delta_0 + 1)} \exp(-\sigma_b^{-2} \gamma_0) \\ & \quad \left. \times (1 + \rho)^{\nu_0 - 1} (1 - \rho)^{\psi_0 - 1} \right\} da_0\, d\sigma_b^2\, d\rho. \end{aligned} \tag{17}$$

We see that $p_0^*(\beta^{(m)}|D_0^{(m)})$ is proportional to the marginal prior of $\beta^{(m)}$. We propose to take the prior probability of model m as

$$p(m) = \frac{\int p_0^*(\beta^{(m)}|D_0^{(m)})\, d\beta^{(m)}}{\sum_{j=1}^{\mathcal{K}} \int p_0^*(\beta^{(j)}|D_0^{(j)})\, d\beta^{(j)}}. \tag{18}$$

This choice for $p(m)$ in (11) is a natural one since the numerator is just the normalizing constant of the joint prior of $(\beta^{(m)}, a_0, \sigma_b^2, \rho)$ under model m. The prior model probabilities in (11) are based on coherent Bayesian updating. It can be shown that $p(m)$ in (11) corresponds to the posterior probability of model m based on the data $D_0^{(m)}$ using a uniform prior for the previous study, $p_0(m) = 2^{-p}$ for $m \in \mathcal{M}$ as $\alpha_0 \to \infty$. That is, $p(m) \propto p(m \mid D_0^{(m)})$, and thus $p(m)$ corresponds to the usual Bayesian update of $p_0(m)$ using $D_0^{(m)}$ as the data.

Using the prior model probability $p(m)$ given in (11) and Bayes theorem, the posterior probability of model m is given by

$$p(m|D^{(m)}) = \frac{p(D^{(m)}|m)\, p(m)}{\sum_{j=1}^{\mathcal{K}} p(D^{(j)}|j)\, p(j)}, \tag{19}$$

where $p(D^{(m)}|m)$ denotes the marginal distribution of the data $D^{(m)}$ for the current study under model m, which has the following expression:

$$\begin{aligned} & p(D^{(m)}|m) = \\ & \int_{-1}^1 \int_0^\infty \int_0^1 \int_{R^{k_m}} p(y \mid \beta^{(m)}, T) \pi(\beta^{(m)}, a_0, \sigma_b^2, \rho \mid D_0) d\beta^{(m)} da_0 d\sigma_b^2 d\rho, \end{aligned}$$

where $p(y \mid \beta^{(m)}, T)$ and $\pi(\beta^{(m)}, a_0, \sigma_b^2, \rho \mid D_0)$ are defined by (3) and (9) under model m. It is easy to see that $p(D^{(m)}|m)$ is indeed the normalizing constant of the joint posterior distribution $p(\beta^{(m)}, a_0, \sigma_b^2, \rho \mid D, D_0)$ given in (16) under model m. Now it can be shown that the posterior probability $p(m|D^{(m)})$ in (12) of model m is given by

$$p(m|D^{(m)}) = \frac{p(\beta^{(-m)} = 0|D^{(\mathcal{K})}, D_0^{(\mathcal{K})})}{\sum_{j=1}^{\mathcal{K}} p(\beta^{(-j)} = 0|D^{(\mathcal{K})}, D_0^{(\mathcal{K})})}, \tag{20}$$

$m = 1, \ldots, \mathcal{K}$, where $p(\beta^{(-m)} = 0|D^{(\mathcal{K})}, D_0^{(\mathcal{K})})$ is the marginal posterior density of $\beta^{(-m)}$ for the full model evaluated at $\beta^{(-m)} = 0$. In (16), for notational convenience we assume that $p(\beta^{(-\mathcal{K})} = 0|D^{(\mathcal{K})}, D_0^{(\mathcal{K})}) = 1$. The result in (16) is very attractive since it shows that the posterior model probability $p(m|D^{(m)})$ is simply a function of the marginal posterior density functions of $\beta^{(-m)}$ for the full model evaluated at $\beta^{(-m)} = 0$. This formula does not algebraically depend on the prior model probability $p(m)$ since it cancels out in the derivation due to the structure of $p(m)$. This is an important feature since it allows us to compute the posterior model probabilities directly *without* numerically computing the prior model probabilities. This has a clear computational advantage and as a result, allows us to compute posterior model probabilities very efficiently. We note that this computational device works best if all of the covariates are standardized to have mean 0 and variance 1. This is not restrictive since this is a typical transformation taken quite often in practice to numerically stabilize the Gibbs sampler.

Due to the complexity of our model, the analytical evaluation of $p(\beta^{(-m)} = 0\ |D^{(\mathcal{K})}, D_0^{(\mathcal{K})})$ does not appear possible. However, we can adopt the importance weighted marginal posterior density estimation (IWMDE) method of Chen (1994) to estimate these marginal posterior densities. The IWMDE is a Monte Carlo method developed by Chen (1994), which is particularly suitable for estimating marginal posterior densities when the joint posterior density is known up to a normalizing constant. The IWMDE method requires using only one MCMC sample from the posterior distribution for the full model, making the computation of complicated posterior model probabilities feasible. It directly follows from the IWMDE that a simulation consistent estimator of $p(\beta^{(-m)} = 0|D^{(\mathcal{K})})$ is given by

$$\begin{aligned}
&\hat{p}(\beta^{(-m)} = 0|D^{(\mathcal{K})}) \\
&= \frac{1}{B}\sum_{l=1}^{B} w\left(\beta_{(l)}^{(-m)} \mid \beta_{(l)}^{(m)}, a_{0(l)}, \sigma_{b(l)}^2, \rho_{(l)}, b_{(l)}, b_{0(l)}\right) \\
&\times \frac{p^*(\beta_{(l)}^{(m)}, \beta^{(-m)} = 0, a_{0(l)}, \sigma_{b(l)}^2, \rho_{(l)}, b_{(l)}, b_{0(l)}|D^{(\mathcal{K})}, D_0^{(\mathcal{K})})}{p^*(\beta_{(l)}^{(\mathcal{K})}, a_{0(l)}, \sigma_{b(l)}^2, \rho_{(l)}, b_{(l)}, b_{0(l)}|D^{(\mathcal{K})}, D_0^{(\mathcal{K})})},
\end{aligned}$$

where $w(\beta^{(-m)} \mid \beta^{(m)}, a_0, \sigma_b^2, \rho, b, b_0)$ is a completely known conditional density of $\beta^{(-m)}$ given $\beta^{(m)}$, a_0, σ_b^2, ρ, b, and b_0, $p^*(\beta, a_0, \sigma_b^2, \rho, b, b_0 \mid D, D_0)$ is given in (16) for the full model, and $\{(\beta_{(l)}^{(\mathcal{K})}, a_{0(l)}, \sigma_{b(l)}^2, \rho_{(l)}, b_{(l)}, b_{0(l)}), l = 1, 2, \ldots, B\}$ is a MCMC sample from the joint posterior distribution $p(\beta, a_0, \sigma_b^2, \rho, b, b_0 \mid D, D_0)$. Note that the choice of the weight density function $w(\beta^{(-m)} \mid \beta^{(m)}, a_0, \sigma_b^2, \rho, b, b_0)$ is somehow arbitrary. However, Chen (1994) showed that the best choice of w is the conditional posterior density of $\beta^{(-m)}$ given $\beta^{(m)}$, a_0, σ_b^2, ρ, b, and b_0. For the complete hierarchical centering case, after reparameterization the conditional posterior distribution of $\beta^{(\mathcal{K})}$ given the other parameters is normal, and thus the closed form of the best w is available. For the other cases, we can follow an empirical procedure provided by Chen (1994) to select w. As demonstrated in Ibrahim and Chen (1998), it is sufficient to choose w to be conditional density of the p dimensional normal distribution with its mean and covariance constructed by the MCMC sample,

$$\{(\beta_{(l)}^{(\mathcal{K})}, a_{0(l)}, \sigma_{b(l)}^2, \rho_{(l)}, b_{(l)}, b_{0(l)}), l = 1, 2, \ldots, B\},$$

for our variable selection problem.

4. Pediatric Pain Data

We illustrate our methodology on a repeated measures data set from Pediatric Pain. The response for each of the 58 children with complete data is a two dimensional vector of the total number of nurse visits taken each year over the two-year period. The covariance structure is a random intercept model for all models with $q \equiv 2$ and each Z a 2×2 identity matrix. The children are classified into one of two groups, attenders or distracters (A or D), depending on their style of coping (CS) with the pain of the cold. Attenders pay attention to their arm in the water and the experimental apparatus during the trial; distracters think about topics unrelated to the trial. Prior to the fourth trial, an intervention occurs. The intervention (treatment or TMT) is a short counseling session. Three types of counseling are given: counseling to attend (A); counseling to distract (D); or a null counseling without instructions (N). If TMT has any effect, then a priori, CS and TMT were presumed to interact. Interest lies in whether the two CS groups have different baseline response times, and given that CS has an effect, whether treatment affects the response time. Thus, the full model contains seven covariates and an intercept term. The seven covariates are age (x_1), the two treatment indicator variables (x_2 and x_3), coping style (x_4), tolerance (x_5), rating (x_6), and a coping style by rating interaction (x_7). The response variable y is the total number of nurse visits, which we model as a Poisson GLMM. The dataset contains 33 girls and 18 boys. For the purposes of illustration, we use the boys as the historical data, from which we will elicit our prior, and use the data for the girls as the current data. Thus, for the Pediatric Pain data, we have $N = 33$, $N_0 = 18$, and all n_i's and n_{0i} are equal to 2. Since all the n_i's, the n_{0i}'s and q are the same, we can directly apply the complete hierarchical centering reparameterization of Gelfand, Sahu, and Carlin (1996). We used 1,000 iterations to "burn in" the Gibbs sampler and then generate 20,000 iterations to obtain the estimates of all posterior model probabilities in all the following computations.

Table 3.1 below gives results for the top three models with $\alpha_0 = 10$, $\lambda_0 = 10$. i.e., $\mu_{a_0} = .5$ and $\sigma_{a_0} = .11$. In addition, we take vague priors for σ_b^2 and ρ. Specifically, for σ_b^2, we take $(\delta_0, \gamma_0) = (.005, .005)$, and for ρ, we take a uniform prior on $[-1, 1]$, i.e., $\nu_0 = \psi_0 = 1$. From Table 3.1, we see that there is not a dominant top model. The table does indicate that treatment, rating, coping style, and rating by coping style interaction are important covariates for explaining the number of nurse visits. To examine the sensitivity of model selection to the choices of $(\mu_{a_0}, \sigma_{a_0})$, we computed posterior model probabilities for several choices of $(\mu_{a_0}, \sigma_{a_0})$. The results for the top model are given in Table 3.2.

Table 3.1: Posterior Model Probabilities for $(\mu_{a_0}, \sigma_{a_0}) = (.5, .11)$

m	$p(m\|D)$
(x_2, x_4, x_6, x_7)	.119
$(x_2, x_3, x_4, x_6, x_7)$	.111
$(x_2, x_4, x_5, x_6, x_7)$	.059

Table 3.2: Posterior Model Probabilities for Several Values of $(\mu_{a_0}, \sigma_{a_0})$

$(\mu_{a_0}, \sigma_{a_0})$	m	$p(m\|D)$
(.5, .078)	(x_2, x_4, x_6, x_7)	.100
(.5, .064)	(x_2, x_4, x_6, x_7)	.085
(.5, .050)	(x_2, x_4, x_6, x_7)	.073
(.91, .027)	$(x_2, x_3, x_4, x_5, x_6, x_7)$	.046

From Table 3.2, we see that for several choices of $(\mu_{a_0}, \sigma_{a_0})$, the ($x_2$, x_4, x_6, x_7) model obtains the largest posterior probability. The pattern of the posterior probability structure for the other models for these choices of prior parameters is similar to that of Table 3.1. However, model selection does become sensitive to the choice of $(\mu_{a_0}, \sigma_{a_0})$ when we give large weight to the historical data as demonstrated in the last line of Table 3.2. Here, we see that the top model is $(x_2, x_3, x_4, x_5, x_6, x_7)$. In all of the above choices of $(\mu_{a_0}, \sigma_{a_0})$, there was not a dominant top model which obtained a large posterior probability. Thus, it appears for this dataset that there is no clear cut top model, but perhaps two or three adequate models, which all contain the covariates treatment, rating, coping style, and rating by coping style interaction.

5. Discussion

We have developed a general class of informative prior distributions for the generalized linear mixed model and have also implemented some novel computational tools. One of the main applications for our methods is variable selection, but the priors, as well as the computational methods, can be used for other applications, such as in the analysis of developmental toxicology data, or in general situations when historical data is available. To date, we do not know of any Bayesian or frequentist methods for doing variable selection for this class of models. We have proposed novel methods which are computationally feasible. Our prior distributions are quite general in that they can be used in any application for which there is historical data available. The priors are semi-automatic in the sense that they take the form of a likelihood function based on the historical data, and they require very few hyperparameters, thus making them especially attractive for variable selection. In addition, the proposed priors are proper under some very general conditions.

References

Berger, J. O. (1985), *Statistical Decision Theory and Bayesian Analysis*, Second Edition, New York: Springer-Verlag.

Breslow N.E., and Clayton, D. G. (1993). Approximate Inference in Generalized Linear Mixed Models. *Journal of the American Statistical Association, 88*, 9-25.

Chen, M.-H. (1994). Importance-weighted Marginal Bayesian Posterior Density Estimation. *Journal of the American Statistical Association, 89*, 818-824.

Chen, M.-H., Ibrahim, J. G., and Yiannoutsos, C. (1999). Prior Elicitation, Variable Selection, and and Bayesian Computation for Logistic Regression Models. *Journal of the Royal Statistical Society, Series B*, 61, 223-242.

Gelfand, A. E. and Sahu, S. K. (1996). Identifiability, Propriety and Parameterization with regard to Simulation-Based Fitting of Generalized Linear Mixed Models. Technical Report #9636, Department of Statistics, University of Connecticut.

Gelfand, A. E., Sahu, S. K., and Carlin, B. P. (1996). Efficient Parametrisations for Generalized Linear Mixed Models (with discussion) in *Bayesian Statistics 5*, eds. J.M. Bernardo, J.O. Berger, A.P. Dawid and A.F.M. Smith, Oxford: Oxford University Press, 165–180.

Ibrahim, J. G. and Chen, M.-H. (1998). Prior Distributions and Bayesian Computation For Proportional Hazards Models. *Sankhya, Series B, 60*, 48-64.

Ibrahim, J. G., Ryan, L. M., and Chen, M.-H. (1998). Use of Historical Controls to Adjust for Covariates in Trend Tests for Binary Data. *Journal of the American Statistical Association, 93*, 1282-1293.

Laird, N. M. and Ware, J. H. (1982). Random-effects models for longitudinal data. *Biometrics, 38*, 963-974.

Liang, K.-Y. and Zeger, S.L. (1986). Longitudinal Data Analysis Using Generalized Linear Models. *Biometrika, 73*, 13-22.

McCullagh, P. and Nelder, J. (1989). *Generalized Linear Models*, 2nd edition, New York: Chapman & Hall.

Nelder, J. A. and Wedderburn, R. W. M. (1972). Generalized Linear Models. *Journal of the Royal Statistical Society, Series A, 135*, 370-384.

Zeger, S. L., and Karim, M. R. (1991). Generalized Linear Models With Random Effects; A Gibbs Sampling Approach. *Journal of the American Statistical Association, 86*, 79-86.

Zeger, S. L. and Liang, K.-Y. (1986). Longitudinal Data Analysis for Discrete and Continuous Outcomes. *Biometrics, 42*, 121-130.

Part II

Extending the GLMs

4

Dynamic Generalized Linear Models

Marco A. R. Ferreira
Dani Gamerman

ABSTRACT Dynamic Generalized Linear Models are generalizations of the Generalized Linear Models when the observations are time series and the parameters are allowed to vary through the time. They have been used increasingly in different areas such as epidemiology, econometrics and marketing. Here we make an overview of the different statistical methodologies that have been proposed to deal with these models from the Bayesian viewpoint. Also, we present some of the challenges involved in the estimation process. Finally, two applications in epidemiology are presented showing the power of MCMC-based methodologies.

1. Introduction

Real world often leads to the necessity of nonnormal data analysis. This issue was highly enlightened with the introduction of generalized linear models (GLM), clever extensions of linear regressions, by Nelder and Wedderburn (1972), and the Bayesian point of view on this subject can be found in chapter 1. As pointed out there, the observations are distributed in the exponential family. Hence, if we denote the observations by y_t, $t = 1, \ldots, T$ then their distribution can be represented through the density (or probability function)

$$p(y_t|\theta_t) \quad \propto \quad \exp\left\{\frac{y_t\theta_t - b(\theta_t)}{\phi_t}\right\} \tag{1}$$

In addition, a suitable link function is introduced relating the mean $\mu_t = E(y_t|\theta_t) = b'(\theta_t)$ to the regressor vector F_t through $g(\mu_t) = \nu_t = F_t'\beta$. Also, it is supposed that $y_1, \ldots, y_T$ are independent, conditionally on β. Many nonnormalities can be accommodated in this framework, including for example the Binomial and Poisson models for counting data and Gamma distribution for positive continuous data.

This class of models cannot be used directly in other situations in practice when the data consist of a time series, hence dependent observations. For example, in epidemiology the analysis of the number of cases of a particular disease through the time is clearly nonnormal. The study is important for the definition of control policies by showing the trend and seasonal patterns, identifying epidemics, measuring the impact of mass vaccination and so on. Another example in medicine is the study of treatments to prevent malaria crisis. Some people submitted to different treatments are accompanied during some time, and each individual has his status (0 = no crisis, 1 = crisis) recorded each day. Additionally, values of some regressors are known for each individual, as for example age, sex, education and living area. This problem can be thought as a longitudinal study with Bernoulli observations.

One option often used in practice to analyze time series is the class of dynamic linear models, where the observation depends on a set of components and the components evolve independently through the time. Generally they assume normal observations, but there is a generalization to the exponential family formalized by West, Harrison and Migon (1985), called dynamic generalized linear models. This class can also be seen as a generalization of the generalized linear models, with the parameters changing through the time. In this chapter we present a review of the literature on dynamic generalized linear models and associated inference and we illustrate with two applications.

The organization of the chapter is as follows: In section 2, the dynamic linear models are revised; Section 3. contains the definition of the dynamic generalized linear models, and presents initial approaches for inference about these models; section 4 presents Markov Chain Monte Carlo based approaches to do inference; in section 5 applications illustrating some aspects of the models are shown; Discussion and possible extensions are presented in section 6.

2. Dynamic linear models

Dynamic linear models (DLM), also known as state space models, have been widely used to analyze time series. They provide a very flexible framework that permits smooth and abrupt changes in the time series generating process and permits the natural accommodation of subjective information. A good reference on dynamic linear models from a Bayesian point of view is West and Harrison (1997). The DLM is usually formed by an observation equation describing the relationship between the observation and regressors, trend, seasonallity and other components that takes the form of a multivariate regression

$$y_t = F_t \beta_t + v_t \tag{2}$$

and a system equation describing the evolution of the vector of regression coefficients or state parameters β_t through time

$$\beta_t = G_t \beta_{t-1} + u_t \tag{3}$$

The set of disturbances v_t and u_t are independent and $v_t \sim N(0, \Sigma_t)$, $u_t \sim N(0, W_t)$ and the model is completed with a prior $\beta_1 \sim N(a_1, R_1)$.

The DLM can be seen as a generalization of regression models that allows changes in parameter values through time. The components of the model are usually defined in blocks, so we have a block describing the trend, another describing the seasonallity and so on.

Let D_t denote the information until time t. Then D_0 is the prior information and $D_t = D_{t-1} \bigcup \{y_t\}$ if there is no information out of the sample at time t. Traditionally, inference for dynamic models is made sequentially by obtaining for each time t the prior, predictive and updated distributions for the system parameter β_t. The first two distributions are respectively obtained by

$$p(\beta_t|D_{t-1}) = \int p(\beta_t|\beta_{t-1}) p(\beta_{t-1}|D_{t-1}) d\beta_{t-1} \tag{4}$$

$$p(y_t|D_{t-1}) = \int p(y_t|\beta_t)p(\beta_t|D_{t-1})d\beta_t \tag{5}$$

and the last one is obtained via Bayes' theorem as

$$p(\beta_t|D_t) \propto p(y_t|\beta_t)p(\beta_t|D_{t-1}) \tag{6}$$

The last equation involves implicitly an integration to find the normalizing constant. If F_t, G_t, Σ_t and W_t are known matrices then all the integrals in (4), (5) and (6) can be obtained exactly. The resulting algorithm is the so called Kalman Filter. But usually Σ_t, W_t and in some cases elements of F_t and G_t are not known, implying in problems to calculate analytically the integrals. These unknown quantities are called hyperparameters and denoted by Ψ here.

A Bayesian attempt to solve the problem of unknown W_t is the use of discount factors (see West and Harrison, 1997), widely used in practice. If $\delta \in (0, 1]$ is the known discount factor and C_{t-1} is the variance of $\beta_{t-1}|D_{t-1}$, then the variance of $\beta_t|D_{t-1}$ is defined by $R_t = G_tC_{t-1}G_t'/\delta$. Thus, the discount factor is related with the increase of uncertainty. It is easy to show that $W_t = G_tC_{t-1}G_t'(1-\delta)/\delta$ thus solving the problem of specification of the unknown value of W_t.

A classical procedure presented in Harvey (1989) is to integrate expressions with respect to β_t. The accumulation of expression (5) through time leads to the predictive likelihood depending only on Ψ. He then maximizes this likelihood to estimate Ψ. Inference for state parameters proceeds as before with Ψ replaced by the estimate.

The extension of DLM to nonlinear state space models has been widely used in control engineering. In these models the observational equation is replaced by $y_t = F_t(\beta_t) + v_t$ or by $y_t = F_t(\beta_t, v_t)$, or the evolution equation is replaced by $\beta_t = G_t(\beta_{t-1}) + u_t$ or by $\beta_t = G_t(\beta_{t-1}, u_t)$ with v_t and u_t still supposed normal. The nonlinear state space models are in some sense related to the DGLM's. For example, one of the options to do inference in nonlinear state space models is the extended Kalman filter and smoother, in which the non-linearities are linearized by Taylor expansions. A modification of this approach was proposed by Farhmeir in order to draw inference for DGLM's, as we explain in subsection 3.3. More information about nonlinear state space models can be found in Anderson and Moore (1979).

Extensions of DLM to nonnormal data were proposed by West (1981), Meinhold and Singpurwalla (1989) and Carlin, Polson and Stoffer (1992), but these extensions were to distributions that are in some way linked to the normal distribution, such as scale mixtures of normals.

3. Definition and first approaches to inference

Perhaps a first attempt to analyze counting time series was through the use of data transformation, as described in Stevens (1974). This was used to improve the normal approximation for the transformed observations and to make the variance independent of the location. The disadvantages are the difficulty in interpreting the results and the inadequacy of the transformations when the observations are counts close to zero.

The modeling of variance laws, as in Harrison and Stevens (1976), has been widely used in applications. In this approach, it is supposed that the variances Σ_t can be approximated by a function $h(F_t\beta_t)$, thus incorporating the dependence on the mean. For example, if the data are univariate Poisson then $h(x) = x$ and if the data are multinomial then $h(x) = N_t[diag(p_t) - p_t p_t']$ where p_t is the vector of multinomial probabilities. As they use the Kalman filter, they substitute β_t in $h(F_t\beta_t)$ by its best estimate. The main drawbacks are that it still supposes normality and does not account for the uncertainty about the mean in the variance.

West, Harrison and Migon (1985) formalized the extension of dynamic models to allow observations in the exponential family, thus defining the Dynamic Generalized Linear Models (DGLM). The extension is based on the generalized linear models proposed in Nelder and Wedderburn (1972), which are well covered in McCullagh and Nelder (1989). The observation equation (2) is replaced by (1) and the link function $g(\mu_t) = \nu_t = F_t'\beta_t$ relates the successive means $\mu_t = E(y_t|\theta_t) = b'(\theta_t)$ to the state parameters, usually mapping the range of μ_t into the real line. The model generalizes GLM by allowing a different vector of regression coefficients for each time and is completed with (3), relating these coefficients through time.

In the case of DGLM the integrals in (4), (5) and (6) cannot be obtained exactly, and so the inference cannot be made exactly. Many proposals to solve this problem have been presented in literature. In the following subsections we present some of them.

3.1 Linear Bayes Approach

West, Harrison and Migon (1985) proposed an approximation based on linear Bayes. This idea was also described in Migon (1984) within the context of dynamic normal nonlinear models.

Basically, the distribution of the system errors u_t and the distribution of $\beta_{t-1}|D_{t-1}$ are specified only by the first and second order moments, namely $u_t \sim WS(0, W_t)$ and $\beta_{t-1}|D_{t-1} \sim WS(m_{t-1}, C_{t-1})$, with WS standing for wide sense. Then, the prior distribution for β_t is $\beta_t|D_{t-1} \sim WS(a_t, R_t)$ where $a_t = G_t m_{t-1}$ and $R_t = G_t C_{t-1} G_t' + W_t$. Here again they suggest the use of discount factors to bypass the difficult problem of specification or estimation of W_t. Hence, the prior distribution for $\nu_t = g(\mu_t) = F_t'\beta_t$ is $\nu_t|D_{t-1} \sim WS(f_t, q_t)$ where $f_t = F_t' a_t$ and $q_t = F_t' R_t F_t$.

In order to calculate the posterior for θ_t the following conjugate prior for θ_t is assumed:

$$p(\theta_t|D_{t-1}) \propto \exp\left\{\frac{r_t\theta_t - b(\theta_t)}{s_t}\right\}$$

where r_t and s_t are such that $E[g(b'(\theta_t))|D_{t-1}] = f_t$ and $V[g(b'(\theta_t))|D_{t-1}] = q_t$. Hence, the posterior for θ_t is

$$p(\theta_t|D_t) \propto \exp\left\{\frac{\theta_t(\phi_t r_t + s_t y_t)/(\phi_t + s_t) - b(\theta_t)}{\phi_t s_t/(\phi_t + s_t)}\right\}$$

It implies that the posterior distribution to ν_t is $\nu_t|D_t \sim WS(f_t^*, q_t^*)$ where $f_t^* = E[g(b'(\theta))|D_t]$ and $q_t^* = V[g(b'(\theta))|D_t]$. The next step is to obtain $E[\beta_t|\nu_t, D_{t-1}]$ and $V[\beta_t|\nu_t, D_{t-1}]$. This cannot be done exactly, and West, Harrison and Migon (1985) estimated these moments using linear Bayes, leading to

$$\hat{E}[\beta_t|\nu_t, D_{t-1}] = a_t + R_t F_t(\nu_t - f_t)/q_t$$

and

$$\hat{V}[\beta_t|\nu_t, D_{t-1}] = R_t - R_t F_t F_t' R_t/q_t$$

Finally, moments of $\beta_t|D_t$ are calculated using properties $E[\beta_t|D_t = E\{E[\beta_t|\upsilon_t, D_{t-1}]|D_t\}$ and $V[\beta_t|D_t] = V\{E[\beta_t|\nu_t, D_{t-1}]|D_t\} + E\{V[\beta_t|\nu_t, D_{t-1}]|D_t\}$, and replacing the conditional moments by their linear Bayes estimators. Then, $\beta_t|D_t \sim WS(m_t, C_t)$ with

$$m_t = a_t + R_t F_t (f_t^* - f_t)/q_t$$

and

$$C_t = R_t - R_t F_t F_t' R_t (1 - q_t^*/q_t)/q_t$$

Hence, the above development allows sequential analysis of DGLM. The approach of West, Harrison and Migon (1985) was the first to treat DGLM's in general, in a time when MCMC techniques were not well known, and in many cases the approximation is quite good.

3.2 *Piecewise Linear Approximation*

Kitagawa (1987) presented a method to analyze non-Gaussian and nonlinear state-space models using piecewise linear functions. He suggested that the densities $p(\beta_t|D_{t-1})$, $p(\beta_t|D_t)$, $p(\beta_t|D_T)$ and the density of u_t are approximated by piecewise linear functions. Then, the value of each function in a given grid of points define the approximated function. Kitagawa (1987) showed that under the assumption of piecewise linearity the integrals in expressions (4), (5) and (6) are just sums with the same order as the number of points on the grid. It is generally necessary to adapt the grid of points through the time to include the region where the densities are concentrated, which is not an easy task. In addition, as the dimension of β_t increases the number of points rises exponentially and it is very hard to locate the grid, thus the procedure becomes prohibitive in terms of computational cost and time.

3.3 *Posterior Mode Estimation*

Fahrmeir (1992) proposed a generalization of the extended Kalman filter and smoother applying it to multivariate DGLM to estimate the state parameters via their posterior mode. The algorithm is applied sequentially and provides an approximation of the posterior mode. Indeed, it can be seen as a simplified Fisher scoring algorithm. The use of the posterior mode as estimator is just to avoid integration. In order to estimate the hyperparameters a procedure based on an EM-type algorithm is proposed.

In order to estimate the state space parameters, Singh and Roberts (1992) proposed the iterative application of the linear Kalman smoothing to the DGLM with modified observation equation

$$\tilde{y}_t = F_t' \beta_t + \tilde{v}_t, \quad \tilde{v}_t \sim N[0, \tilde{V}_t] \tag{7}$$

with adjusted observations $\tilde{y}_t = \tilde{y}_t(\beta_t) = \nu_t + (y_t - \mu_t) g'(\mu_t)$ and associated variances $\tilde{V}_t = \tilde{V}_t(\beta_t) = b''(\theta_t)[g'(\mu_t)]^2$ for $t = 1, \ldots, T$. These modified observations and variances are defined at each iteration by using the value of β_t estimated at the previous iteration, by the Kalman filter and smoother. Singh and Roberts (1992) estimate $W_t = W$ using a moment-based approach.

Fahrmeir and Wagenpfeil (1997) also worked on obtaining the posterior mode of state parameters to multivariate DGLM. They showed that the algorithm proposed by Singh and Roberts (1992) leads to the posterior mode of the state parameters conditional on a fixed value of W. They also showed that the Fahrmeir's (1992) generalization of the extended Kalman filter is a particular case of this algorithm with just one iteration and a convenient choice of the initial values. They propose the use of a procedure based on the generalized cross-validation criterion to estimate the hyperparameters Ψ.

3.4 Other Approaches and Models

Frühwirth-Schnatter (1992) worked with finite mixture approximations for the predictive density $p(y_t|D_s)$, $t > s$ of DGLM. The fully exponential method of Laplace (Kass, Tierney and Kadane, 1989) and the Gauss-Hermite integration (Naylor and Smith, 1982) are within this class of approximations.

Harvey and Fernandes (1989) proposed models with conjugate property with the level varying through the time and the covariate effects constant. For example, to model Poisson observations with covariates they assumed $y_t|\mu_t, \beta \sim Poisson[\mu_t \exp(F_t'\beta)]$ and $\mu_{t-1}|D_{t-1} \sim Ga(a_{t-1}, b_{t-1})$, and by analogy with the Kalman Filter they proposed a implicit system equation such that $\mu_t|D_{t-1} \sim Ga(a_{t|t-1}, b_{t|t-1})$ with $a_{t|t-1} = \delta a_{t-1}$ and $b_{t|t-1} = \delta b_{t-1}$, δ being the discount factor. Then, the conjugacy implies $\mu_t|D_t \sim Ga(a_t, b_t)$ with $a_t = \delta a_{t-1} + y_t$ and $b_t = \delta b_{t-1} + \exp(F_t'\beta)$. In their approach, δ and β are regarded as hyperparameters and are estimated by the maximization of the predictive likelihood.

4. MCMC-based Approaches

Markov Chain Monte Carlo (MCMC) techniques have been widely used to solve complex Bayesian inference problems in the 90s. They allow great flexibility in defining the model and drawing inference about parameters or functions of the parameters and predictions. Basically, a Markov chain is defined with the equilibrium distribution given by the posterior distribution of the model parameters. A realization of this chain is generated until convergence is reached. After convergence, the following iterations of the chain are in the equilibrium and can be used to form a sample from the posterior distribution. The main advantage of this approach is the possibility of doing full Bayesian analysis of the problem, which means that the uncertainty due to the fact that Ψ is unknown is considered, or in other words, it is possible to integrate Ψ out in order to draw inference on $(\beta_1, \ldots, \beta_T)$. In addition, point and interval estimation of Ψ can be done based on the posterior distribution. References about this subject include Gilks et al. (1996) and Gamerman (1997).

For the DGLM's the posterior distribution is of the form

$$p(\beta_1, \ldots, \beta_T, W|D_T) \propto \left[\prod_{t=1}^{T} p(y_t|\theta_t)\right] \left[\prod_{t=2}^{T} p(\beta_t|\beta_{t-1}, W)\right] p(\beta_1)p(W)$$

where for expository purposes we assume that $W_t = W$ and F_t and G_t are known, $\forall t$. The analysis is no longer sequential, and (4), (5) and (6) are not calculated at all. The density calculated is the posterior distribution $p(\beta_1, \ldots, \beta_T, \Psi|D_T)$ and the necessary integrals are performed numerically.

The simplest MCMC technique is the Gibbs sampling. In this approach the transitions are defined by the full conditional distributions for each parameter, that is, the distribution of the parameter conditional on all other parameters. If we denote $\{\beta_{\neq t}\} = \{\beta_1, \ldots, \beta_{t-1}, \beta_{t+1}, \ldots, \beta_T\}$ then the full conditional distribution of

β_t is $p(\beta_t|\{\beta_{\neq t}\}, D_T, W)$. Assume also an inverted Wishart prior distribution for W, denoted by $IW(n_W/2, n_W S_W/2)$. Then the full conditional for W is obtained as

$$\begin{aligned} p(W|\{\beta\}, D_T) &\propto \left[\prod_{t=2}^{T} p(\beta_t|\beta_{t-1}, W)\right] p(W) \\ &\propto \left[\prod_{t=2}^{T} |W|^{-1/2} \exp[-\frac{1}{2}(\beta_t - \beta_{t-1})' W^{-1}(\beta_t - \beta_{t-1})]\right] \\ &\quad \times |W|^{-(n_W+p+1)/2} \exp[-\frac{1}{2} tr(W^{-1} S_W^{-1})] \\ &= |W|^{-(n_W+T-1+p+1)/2} \exp\{-\frac{1}{2} tr[W^{-1}(S_W^{-1} \\ &\quad + \sum_{t=2}^{T} (\beta_t - \beta_{t-1})(\beta_t - \beta_{t-1})')]\} \end{aligned}$$

Hence, the full conditional for W is also inverted Wishart from which it is easy to sample (see for example Devroye, 1986), and so it is said that the distribution is available for sampling. In DGLM's, the full conditional for β_t is not available for sampling in general. There are some proposals of implementation of Gibbs sampling to particular cases, as we show in subsection 4.1. For applications in general, the most indicated approaches seem to be those based on the Metropolis-Hastings algorithm, as explained in subsection 4.2.

4.1 Gibbs Sampling

Carlin, Polson and Stoffer (1992) introduced the use of Gibbs Sampler to perform inference for nonnormal and nonlinear state-space models, with observational and system disturbances distributed in the class of scale mixtures of normals (Andrews and Mallows, 1974), which includes for example exponential power, logistic and t densities. Although Carlin, Polson and Stoffer (1992) did not work with DGLM, the idea motivated a few MCMC-based solutions for DGLM.

Fahrmeir, Hennevogl and Klemme (1992) proposed Gibbs sampler to analyze DGLM using rejection sampling to generate from the full conditionals of β_t. In general, a value β_t^* is generated from a density $g(.)$ and it is accepted with probability $f(\beta_t^*)/K_t[g(\beta_t^*)]$, where $f(\beta_t)$ is proportional to $p(\beta_t|\{\beta_{\neq t}\}, D_T, W)$ and K_t is such that $K_t g(\beta_t) \geq f(\beta_t)$, $\forall \beta_t$. In the DGLM framework the system equation has normal disturbances, and so there exist m_t and C_t such that $p(\beta_t|\{\beta_{\neq t}\}, D_T, W) \propto p(y_t|\beta_t)N(m_t, C_t) = f(\beta_t)$, $\forall t$. Fahrmeir, Hennevogl and Klemme (1992) then proposed $g(\beta_t) = N(m_t, C_t)$ which sounds natural, but can lead to low acceptance rates.

In the context of counting time series, Carlin and Polson (1992) proposed the use of the Gibbs Sampler with latent continuous variables, which could evolve following the standard linear and normal observational equations. These latent variables were related to the discrete observations through a suitable transformation. For example, if y was Bernoulli and y^* was the related latent variable then the suitable transformation could be $y = 1$ if $y^* > 0$ or $y = 0$ if $y^* \leq 0$. These latent variables can be seen as nuisance parameters and their full conditionals are easy to sample. In addition, their inclusion makes the sampling from the full conditionals of β_t very easy. The drawbacks are that this approach is not applicable to exponential family in general but only to counting data, and it is difficult to find a suitable transformation mapping the latent variables to the discrete observations in most of the applications.

4.2 Metropolis-Hasting Algorithm

To overcome the difficulty in sampling of the full conditional for β_t, some authors have been presenting approaches based on the Metropolis-Hastings algorithm, which is also a MCMC technique. The idea is to use the Gibbs sampler with a Metropolis-Hasting step to do the transition for the state parameters. When sampling state parameters one at a time, this Metropolis-Hasting step is made up of two steps: first a value β_t^* is generated from a proposal transition density $q_t(\beta_t^{(old)}, \beta_t)$; then this value is accepted with probability

$$\min\left\{1, \frac{\pi_t(\beta_t^*)/q_t(\beta_t^{(old)}, \beta_t^*)}{\pi_t(\beta_t^{(old)})/q_t(\beta_t^*, \beta_t^{(old)})}\right\}$$

where π_t is the full conditional density of β_t for $t = 1, \ldots, T$. Then, one of the differences between the approaches presented by different authors is the choice of the proposal transition density.

Another important issue is that the convergence can be very slow when using MCMC to analyze state space models as pointed out by Carter and Kohn (1994) and Shephard (1994). The approaches to analyze DGLM using the Metropolis-Hastings algorithm differ also by the ways used to accelerate the convergence of the chain. A very related topic is how to update the state parameters. It is possible to use a single move step updating each β_t, a multimove step updating $\beta_1, \ldots, \beta_T$ jointly, or a block move step updating blocks of β_t jointly. As shown in Carter and Kohn (1994) the speed of convergence in state space models when using single move step is slow because of the high correlation between the state parameters, generating highly correlated chains. Carter and Kohn (1994) have shown that the multimove step uses the information of that correlation and optimally updates the state parameters through Gibbs sampler iterations, raising the speed of convergence.

In the DGLM context the introduction of the multimove Metropolis-Hastings step is not generally optimal because it can result in very low acceptance rates, leading the chain to stay in the same point for many iterations thus slowing down the convergence. It motivated the introduction of block move step Metropolis-Hastings updating in DGLM context by Shephard and Pitt (1997) and by Knorr-Held (1997).

Knorr-Held (1997) proposes the use of conditional prior proposals with fixed blocks. Basically, he uses the full conditional prior distribution $p(\beta_{r,s} \mid \beta_{\neq r,s}, W)$ to generate a proposal value $\beta_{r,s}^*$ for $\beta_{r,s} = (\beta_r, \ldots, \beta_s)$, $1 \leq r < s \leq T$. As the system equation disturbances are normal, this proposal is multivariate normal and it is easy to sample from. Indeed, since $p(\beta_{r,s}|\beta_{\neq r,s}, W, D_t) \propto \prod_{t=r}^{s} p(y_t|\beta_t) \times N(m_{r,s}, C_{r,s})$ for some value of $m_{r,s}$ and $C_{r,s}$, the acceptance probability simplifies to the likelihood ratio

$$\min\left\{1, \frac{\prod_{t=r}^{s} p(y_t|\beta_t^*)}{\prod_{t=r}^{s} p(y_t|\beta_t^{(old)})}\right\}$$

It is relatively simple to obtain the proposal transition density and to calculate the acceptance probability, thus reducing considerably the computational complexity. The problem with this approach is that it can lead to low acceptance rates, depending on the structure of the model and the time series under study.

Shephard and Pitt (1997) propose sampling random blocks of the disturbances $u_{r,s} = (u_r, \ldots, u_s)$ using as proposal transition density a second order Taylor expansion of $h_t(\nu_t) = \log\ p(y_t|\theta_t(\nu_t))$ in the full conditional of $u_{r,s}$. The proposal transition density is then multivariate normal, and is calculated via the definition of artificial observations $\hat{y}_t = F_t\hat{\beta}_t + \hat{V}_t h_t'(\nu_t)$, $t = r+1, \ldots, s+2$ where usually $\hat{V}_t^{-1} = -h_t''(\nu_t)$ and the replacement of the equation (1) by

$$\hat{y}_t = F_t'\beta_t + \hat{v}_t, \quad \hat{v}_t \sim N[0, \hat{V}_t] \tag{8}$$

The De Jong and Shephard (1995) simulation smoother is then used on model defined by equations (8) and (3) in order to generate the proposal for $u_{r,s}$. Shephard and Pitt (1997) also provide empirical evidence of gains in computational efficiency of randomly defined blocks over deterministic block definition.

Gamerman (1998) suggests the use of a proposal transition density very similar to that of Shephard and Pitt (1997) but instead of block movement he proposes single movement with reparametrization of the model in terms of the system disturbances and sampling from these disturbances. The proposal transition density is the full conditional distribution of u_t in the model with the modified observational equation (7), used by Singh and Roberts (1992) and Fahrmeir and Wagenpfeil (1997). The reparametrization rewrites the link function in terms of the system disturbances as $g(\mu_t) = \nu_t = F_t' \sum_{j=1}^{t} G^{t-j} u_j$, with $u_t \sim N(0, W)$, $t = 2, \ldots, T$ and $u_1 \sim N(a_1, R_1)$, if $G_t = G$, $\forall t$.

5. Applications

We present here two applications of this methodology to epidemiological data. The first is concerned with infectious disease control, when the number of cases of meningitis is observed through the time and the main purposes are to verify if the risk is below an acceptable level and if there is evidence of an epidemy. We present as example an analysis of the monthly series of meningococic meningitis in Rio de Janeiro, Brazil, from January 1976 to December 1992. The data was obtained from Rio de Janeiro State Health Secretary.

The other application is the study of relationships between the number of cases of a disease, or deaths due to a disease, and possible explanatory variables. This can suggest ways to reduce the risk of the disease. We present here a preliminary analysis of the relationship between the number of child deaths due to respiratory diseases in São Paulo, Brazil, from 1st January 1991 to 31st December 1991, obtained from São Paulo City Health Secretary, and the levels of pollutants, obtained from São Paulo State Environment Secretary.

We suppose here that the observations in both examples follow a Poisson distribution. The models include level, seasonallity and random effects for the first example, and level and regression variables for the second example. The random effects account for overdispersion, implying that the data dispersion may not be adequately explained by the Poisson model.

We denote y_t the number of disease cases in time t, $t = 1, \ldots, T$, and p the periodicity of the series. As usual, we use the logarithmic link function. The logarithm of the process mean evolve through the time decomposed into level, seasonallity, random effect and regression components as

$$y_t \sim Poisson(\lambda_t)$$

$$\begin{aligned}\log \lambda_t &= \nu_t = \mu_t + s_t + \phi_t + X_t'\beta \\ \mu_t &= \mu_{t-1} + u_{1t}, \qquad u_{1t} \sim N(0, W_1) \\ s_t &= -(s_{t-1} + \cdots + s_{t-p+1}) + u_{2t}, \quad u_{2t} \sim N(0, W_2)\end{aligned}$$

We complete the model with the following independent prior distributions: $\mu_1 \sim N(a_1, R_1)$, $s_t \sim N(a_2, R_2), t = 1, \ldots, p-1$, $\phi_t \sim N(0, \sigma_\phi^2)$, $\beta \sim N(m_\beta, C_\beta)$, $W_1 \sim IG(n_{W1}/2, n_{W1}S_{W1}/2)$ and $W_2 \sim IG(n_{W2}/2, n_{W2}S_{W2}/2)$.

The parameters $\{\mu_t\}$, $\{s_t\}$, $\{\phi_t\}$ and β do not have full conditionals available for sampling. We use the Metropolis-Hasting step within the Gibbs sampler to make the transitions for these parameters, as proposed by Gamerman (1998) and explained in subsection 4.2. Empirical evidence from studies with simulated data has shown that the methodology works well in estimating the parameters of the model. In addition, in the present applications we assessed the convergence by graphical inspection. Formal methods of convergence diagnosis, as those proposed by Geweke (1992) and Raftery and Lewis (1992) can also be used.

5.1 *Application 1: Meningococcic Meningitis*

We analyze here the monthly series of meningococic meningitis cases recorded in Rio de Janeiro city from January 1976 to December 1992, which is shown in figure 4.1a. The main concerns are:

- Is there a trend, or in other words, is it necessary to do something to reverse a rising movement in the number of cases?
- Is there seasonal movement, implying need for more stringent surveillance in certain months?
- Is it possible to tell something about the efficiency of the government surveillance?
- How can one ascertain whether the disease is under control?

Figure 4.1b shows the posterior mean and 95% credibility intervals for the level of the series. It can be seen that the level fell from January 1976 to January 1982 and then rose until January 1990. The movement in the end is not clear, but it seems that there is a tendency for stabilization or decay.

The graph of seasonallity shown in figure 4.1c presents a clear pattern, with less number of cases in January, February and March, and more cases in July and August. In addition, the seasonal pattern is very stable through the years.

The efficiency of the government surveillance defines the distribution of the number of cases. If the government succeeds in isolating the new cases then the occurrence of a case in a region does not increase significantly the risk in that region, and so the Poisson model can be a good approximation to the process. On the other hand, if the control is not efficient then the cases tend to appear in clusters implying in an increase of the variance of the number of cases and so in a departure from the Poisson process. Moreover, the random effects accommodate this type of departure: the module of the random effects are larger as the observations have greater variability than that expected in a Poisson process. The estimates of the random effects are shown in figure 4.1d and it seems that the module of the random effects

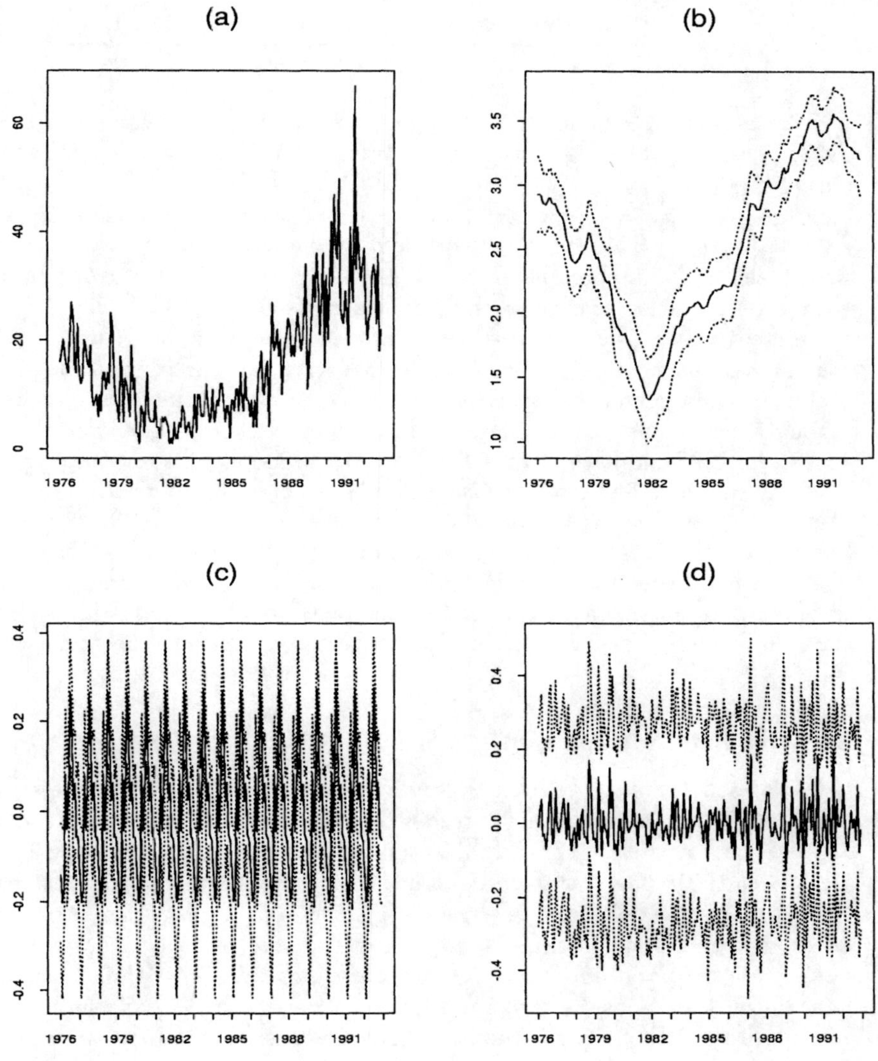

FIGURE 4.1. Summary of posterior inference for application 1: a. original series; b. estimates of the level of the series; c. estimates of the seasonal pattern; d. estimates of the random effects. In graphs b, c and d, solid lines represent mean and dashed lines represent the 95% credibility limits.

TABLE 4.1. Logarithm of the cross validation predictive densities

	With seasonality	Without seasonality
With overdispersion	6143.19	6141.21
Without overdispersion	6135.79	6136.22

were greater from September 1978 to January 1980 and from January 1987 to July 1991. Therefore, it seems that the government surveillance was less efficient in these periods.

In order to get some quantitative indication of the need for seasonality and random effects in the model, models with and without seasonality and overdispersion were fitted and compared using the cross-validation predictive densities, whose use with Monte Carlo techniques is explained in Gelfand (1996). The logarithms of these cross validation predictive densities are shown in table 4.1 up to an arbitrary constant. Based on this criterion, the model with seasonality and overdispersion is prefered. The understanding of the state of control of the disease must be obtained through the analysis of all these graphs. In our example, the random effects are smaller in 1992 compared to the five previous years and the level seems to stabilize or decay in 1992. Therefore the surveillance seems to be more efficient in this year than in the previous years. In addition, we can easily obtain the predictive distribution of future observations for each month and compare it with the observation itself when it becomes available. This can be used to build a monitoring system to trigger an alarm for mass vaccination. Finally, it would be interesting to incorporate the information about the vaccination programs to measure their impact.

5.2 Application 2: Respiratory Diseases and Level of Pollutants

We present here a preliminary analysis of the relationship between the daily number of children deaths due to respiratory diseases and the levels of the pollutants NO_2 (g/m^3) and CO (ppm), in São Paulo, Brazil, from 1st January 1991 to 31st December 1991. There are records of other pollutants, but the correlations between the levels of the pollutants are very high and it is very difficult to measure the effect of each pollutant individually. Therefore, we removed from the model the less significant pollutants using an exploratory analysis. We plan to report a fuller analysis on this dataset elsewhere. Figure 4.2a presents the daily number of deaths ranging from 0 to 4. In our model we do not consider the medium and long term effects of the pollutants, and therefore we are measuring the instantaneous impact of the pollutants on the number of deaths. We supposed that the vector of regressors coefficient in the model is constant, as in the models of Harvey and Fernandes (1989), but this assumption can be easily relaxed.

Figures 4.2c and 4.2d show the histograms of the coefficients of NO_2 and CO, respectively. They are located predominantly in the positive part of the real line. Indeed, $\hat{Pr}(\beta_{NO_2} > 0|D_t) = 0.98062$ and $\hat{Pr}(\beta_{CO} > 0|D_t) = 0.91424$ indicating high probability of relation between level of pollutants and number of children deaths due to respiratory diseases. In addition, the estimated posterior means are $\hat{E}(\beta_{NO_2}|D_t) = 0.00222$ and $\hat{E}(\beta_{CO}|D_t) = 0.06530$.

Figure 4.2b shows the histogram of W_1, the variance of the disturbances in the system equation for the level. As it would be expected, the distribution is skewed and

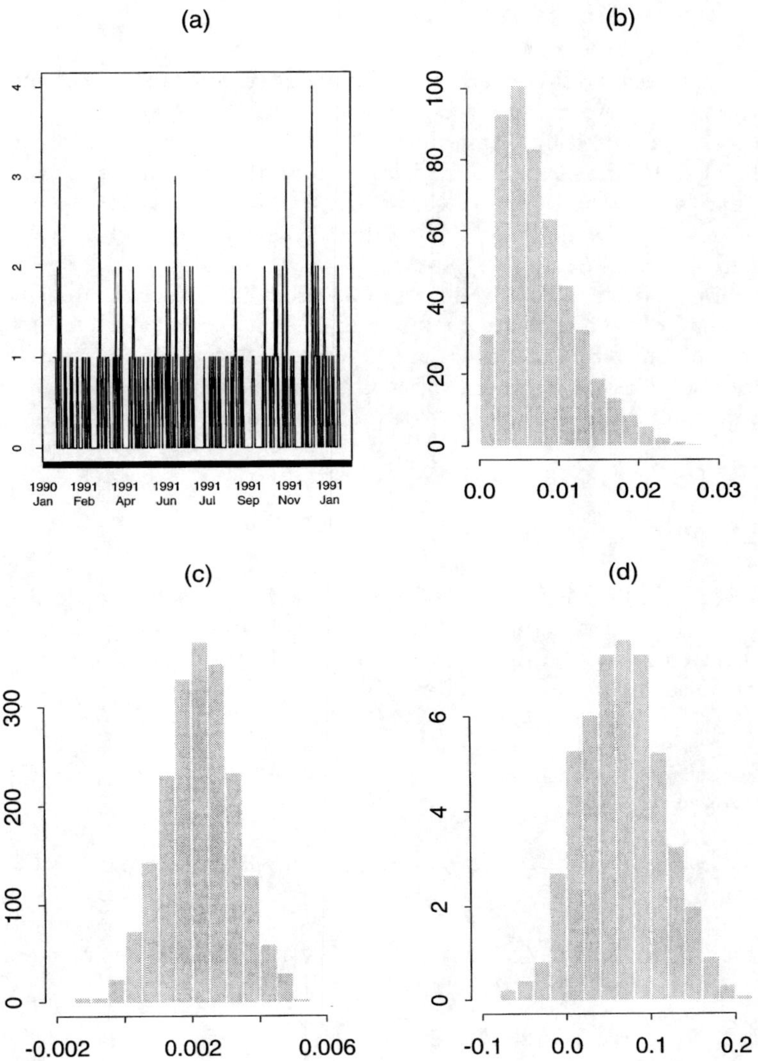

FIGURE 4.2. Summary of posterior inference for application 2: a. original series; b. histogram of W_1; c. histogram of β_{NO_2}; d. histogram of β_{CO}.

has just one mode. The estimated posterior mean for W_1 is $\hat{E}(W_1|D_t) = 0.00758$. It illustrates that in this approach there is no difficult in estimating the hyperparameters.

6. Discussions and Extensions

We presented in this chapter the dynamic generalized linear models, some alternatives of inference and two applications using the Monte Carlo Markov chain alternative proposed by Gamerman (1998). This approach allows great flexibility in modeling and inference. The models can be stated through a study of the problem and there are few mathematical limitations, which means that the modeler is almost free to set the best model without thinking about the complexity of inference.

There are many extensions that can be thought mainly with respect to the models. For example, in the second application the model measures the instantaneous impact of the level of pollutants on the number of deaths. Transfer functions can be used in order to measure the impact of the regressors in medium term. There are many functional forms for transfer functions that can be used and that have meaningful interpretation in epidemiology. We expect to address this question in a future paper.

Acknowledgements

Research was partially supported by grants from CNPq and PRONEX. We thank Getúlio Silveira for allowing us the use of the dataset of meningococic meningitis (application 1) and CEA-IME/USP for allowing us the use of the dataset of respiratory diseases and level of pollutants (application 2).

References

Anderson, B. D. O. and Moore, J. B. (1979), *Optimal Filtering*, Prentice-Hall, New Jersey.

Andrews, D. F. and Mallows, C. L. (1974). Scale mixtures of normality. *Journal of the Royal Statistical Society* (Ser. B) 36, pp. 99-102.

Besag, J., York, J. and Mollié, A. (1991). Bayesian image restoration with two applications in spatial statistics. *Annals of the Institute of Statistical Mathematics* 43, pp. 1-59.

Carlin, P. B., Polson, N. G. and Stoffer, D. S. (1992). A Monte Carlo approach to nonnormal and nonlinear state-space modeling. *Journal of the American Statistical Association* 87, 493-500.

Carter, C. K. and Kohn, R. (1994). On Gibbs sampling for state space models. *Biometrika* 81, 541-53.

De Jong, P. and Shephard, N. (1995). The simulation smoother for time series models. *Biometrika* 82, 339-50.

Devroye, L. (1986). *Non-uniform random variate generation.* Springer-Verlag, New York.

Fahrmeir, L. (1992). Posterior mode estimation by extended Kalman filtering for multivariate dynamic linear models. *Journal of the American Statistical Association* 87, pp. 501-9.

Fahrmeir, L., Hennevogl, W. and Klemme, K. (1992). Smoothing in dynamic generalized linear models by Gibbs sampling. In *Advances in GLIM and Statistical Modelling*, Eds. L. Fahrmeir, B. Francis, R. Gilchrist and G. Tutz, Lecture Notes in Statistics 78, pp. 85-90. Springer, New York.

Fahrmeir, L. and Wagenpfeil, S. (1997). Penalized likelihood estimation and iterative Kalman filtering for non-Gaussian dynamic regression models. *Comp. Stat. Data Anal.* 24, 295-320.

Fruhwirth-Schnatter, S. (1992). Approximate predictive integrals for dynamic generalized linear models. In *Advances in GLIM and Statistical Modelling*, Eds. L. Fahrmeir, B. Francis, R. Gilchrist and G. Tutz, Lecture Notes in Statistics 78, pp. 101-6. New York: Springer.

Gamerman, D. (1997). *Markov Chain Monte Carlo: Stochastic Simulation for Bayesian Inference.* Chapman and Hall, London.

Gamerman, D. (1998). Markov chain Monte Carlo for dynamic generalized linear models. *Biometrika* 85, pp. 215-227.

Gelfand, A. E. (1996). Model determination using sampling-based methods, in *Markov Chain Monte Carlo in Practice.* Eds. W. R. Gilks, S. Richardson and D.J. Spiegelhalter, Chapman and Hall, London, pp. 145-61.

Geweke, J. (1992). Evaluating the accuracy of sampling-based approaches to the calculation of posterior moments (with discussion). In *Bayesian Statistics 4* (eds. J. M. Bernardo et al.), Oxford University Press, Oxford, pp. 169-93.

Gilks, W. R., Richardson, S. and Spiegelhalter, D.J. (1996) (eds.). *Markov Chain Monte Carlo in Practice.* Chapman and Hall, London.

Harrison, P. J. and Stevens, C.F. (1976). Bayes forecasting in action: case studies. *Research Report 14*, Department of Statistics, University of Warwick.

Harvey, A. C. (1989). *Forecasting, Structural Time Series Models and the Kalman Filter.* Cambridge University Press, Cambridge.

Harvey, A. C. and Fernandes, C. (1989). Time series models for count data or qualitative observations. *Journal of Business and Economic Statistics* 7, 407-17.

Kass, R. E., Tierney, L. and Kadane, J. B. (1989). Fully exponential Laplace approximations to expectations and variances of non-positive functions. *Journal of the American Statistical Association* 84, 710-16.

Kitagawa, G. (1987). Non-Gaussian state-space modeling of nonstationary time series. *Journal of the American Statistical Association* 82, 1032-63.

Knorr-Held, L. (1997). *Hierarchical Modelling of Discrete Longitudinal Data - Applications of Markov Chain Monte Carlo.* Herbert Utz Verlag Wissenschaff, München.

McCullagh, P. and Nelder, J. A. (1989). *Generalized Linear Models.* 2nd ed., Chapman and Hall, London.

Meinhold, R. J. and Singpurwalla, N. D. (1989). Robustification of Kalman filter models. *Journal of the American Statistical Association* 84, 479-86.

Migon, H. S. (1984). *An approach to non-linear Bayesian forecasting problems with applications.* Unpublished Ph.D. thesis, Department of Statistics, University of Warwick.

Naylor, J. C. and Smith, A. F. M. (1982). Application of a method for the efficient computation of posterior distributions. *Applied Statistics* 31, 214-25.

Nelder, J. A. and Wedderburn, R. W. M. (1972). Generalized Linear Models. *Journal of the Royal Statistical Society* (Ser. A) 135, 370-84.

Raftery, A. E. and Lewis, S. (1996). How many iterations in the Gibbs sampler?. In *Bayesian Statistics 4* (eds. J. M. Bernardo et al.), Oxford University Press, Oxford, pp. 763-73.

Shephard, N. (1994). Partial non-Gaussian state space. *Biometrika* 81, 115-31.

Shephard, N. and Pitt, M. K. (1997). Likelihood analysis of non-Gaussian measurement time series. *Biometrika* 84, 653-67.

Singh, A. C. and Roberts, G. R. (1992). State space modelling of cross-classified time series of counts. *International Statistical Review* 60, 321-36.

Stevens, C. F. (1974). On the variability of demand for families of items. *Oper. Res. Quart.* 25, 411-20.

West, M (1981). Robust sequential approximate Bayesian estimation. *Journal of the Royal Statistical Society* (Ser. B) 43, 157-66.

West, M. and Harrison, J.(1997). *Bayesian Forecasting and Dynamics Models.* 2nd ed., Springer-Verlag, New York.

West, M., Harrison, P.J. and Migon, H.S. (1985). Dynamic generalized linear models and Bayesian forecasting (with discussion). *Journal of the American Statistical Association* 80, 73-96

5

Bayesian Approaches for Overdispersion in Generalized Linear Models

Dipak K. Dey
Nalini Ravishanker

ABSTRACT Generalized linear models (GLM's) have been routinely used in statistical data analysis. The evolution of these models as well as details regarding model fitting, model checking and inference have been thoroughly documented in McCullagh and Nelder (1989). However, in many applications, heterogeneity in the observed samples is too large to be explained by the simple variance function which is implicit in GLM's. To overcome this, several parametric and nonparametric approaches for creating overdispersed generalized linear models (OGLM's) were developed. In this article, we summarize recent approaches to OGLM's, with special emphasis given to the Bayesian framework. We also discuss computational aspects of Bayesian model fitting, model determination and inference through examples.

1. Introduction

Generalized linear models (GLM) are a standard class of models in contemporary statistical data analysis (McCullagh and Nelder 1989). The widely available GLIM software as well as SPlus facilitate computation under these models. Bayesian fitting of GLM's via Gibbs sampling is discussed in Dellaportas and Smith (1993). In GLM's, the underlying distribution of responses is assumed to be of the exponential family form, and a link function transformation of its expectation is modeled as a linear function of observed covariates, assuming that the variance of the response is a specified function of its mean. This allows modeling in various nonnormal situations such as the binomial, Poisson, negative binomial etc. However, in many applications, such a simple functional relationship is inadequate to handle the heterogeneity in the data; a common problem is the so-called overdispersion problem, where the variance of the response exceeds the nominal variance (Cox 1983). For instance, regression analysis of count data in biomedical research areas such as toxicology, epidemiology etc., must handle extra-Poisson variation. Count data analyzed under a Poisson assumption (Breslow 1984; Lawless 1987a,b; McCullagh and Nelder 1989, Sec. 6.2) often exhibit overdispersion. In modeling categorical or ordered categorical data in longitudinal studies, toxicity studies, or in biological experimental research, the use of binomial or multinomial distributions with overdispersion is encountered.

Crowder (1985) and Williams (1982) analyzed proportions using overdispersed binomial distributions. Overdispersion often results from latent heterogeneity, i.e., the sample of responses is drawn from a population consisting of many subpopulations.

Historically, the presence of overdispersion prompted the need to define a wider class of models than the GLM. The different approaches include finite mixture models (Everitt and Hand 1981; Titterington, Makov and Smith 1985), exponential dispersion models, EDM (Jorgensen 1987), Efron's (1986) double exponential family models, EDM's embedded within Efron's double exponential family (Ganio and Schafer 1992), and generalized linear mixed models, GLMM (Breslow and Clayton 1993). These classes are described in Section 2.

Numerous methods have been proposed in the literature for estimation in overdispersed GLM's with particular emphasis given to binomial or Poisson models. The quasi-likelihood (QL) approach requires specification of the mean-variance relationship rather than a full likelihood function. If y is a random variable with mean μ and variance $V(\mu)$ (a known function), the QL for y is defined as (Wedderburn 1974)

$$Q(\mu; y) = \int_y^\mu \frac{y-t}{V(t)} dt.$$

For a sequence of independent observations $\tilde{y} = (y_1, \cdots, y_n)$, the QL is defined as $Q(\tilde{\mu}; \tilde{y}) = \sum_{i=1}^n Q(\mu_i, y_i)$, where $\tilde{\mu} = (\mu_1, \cdots, \mu_n)$. In general, if $\tilde{\mu}$ is a function of p parameters $\tilde{\theta} = (\theta_1, \cdots, \theta_p)$, the QL estimates of $\tilde{\theta}$ may be obtained by maximizing $Q(\tilde{\mu}; \tilde{y})$ under mild conditions (see McCullagh and Nelder 1989). For the exponential family, the QL coincides with the log-likelihood, and therefore the estimation of $\tilde{\theta}$ retains full asymptotic efficiency (Nelder and Lee 1992). The QL approach for a model with a single dispersion parameter in the variance function was discussed by Finney (1971), McCullagh and Nelder (1989) and Wedderburn (1974). Moore and Tsiatis (1989) and Williams (1988) showed that the QL approach generally gave consistent estimates of the regression coefficients even under a mispecified variance function. Wang (1996) presented a QL approach for ordered categorical data with overdispersion and illustrated using fish mortality data from quantal response experiments. The extended quasi-likelihood was defined in Nelder and Pregibon (1987), using saddlepoint arguments.

There has been some concern over the use of overly simple models for overdispersion in GLMs, especially in the context of the widely used binomial and Poisson models. To overcome this, it has been proposed that explanatory variables be included into the model for dispersion (see Carroll and Ruppert 1982; Efron 1986; Jorgensen 1987; McCullagh and Nelder 1989; Nelder and Pregibon 1987; Smyth 1989).

The QL/M approach (also called the pseudolikelihood approach) uses only the mean and variance structure implied by a mixture model; the regression parameters are estimated by quasi-likelihood while the variance parameter is estimated by the method of moments. For Poisson counts, use of a gamma mixing distribution results in a negative binomial distribution for the observed data. The QL approach yields asymptotically efficient estimates of the regression coefficients for the negative binomial model with fixed shape parameter, although the variance estimation by the method of moments is less efficient (see Breslow 1984; Lawless 1987a; Williams 1982). Carroll and Ruppert (1982) discuss the QL/M approach for general heteroscedastic models. The penalized quasi-likelihood (PQL) was proposed as an approximate Bayes procedure for GLMM's (Laird 1978; Stiratelli, Laird and Ware 1984).

The marginal quasi-likelihood (MQL) procedure, which is similar to Goldstein's

(1991) apparoach for GLMM's with nested random effects, is appropriate when interest is focused on the marginal relationship between covariates and response (Liang, Zeger and Qaqish 1992). Both PQL and MQL may be implemented by standard software for variance components analysis of normally distributed responses and provide shrinkage estimates of random error terms and for estimates of the variance components. Breslow and Clayton (1993) used PQL and MQL to analyze data on seed germination (Crowder 1978) under an overdispersed binomial assumption and a GLMM with a linear predictor including a random effect. They observed that there was little difference between PQL and MQL from a practical viewpoint; a major distinction between the two methods is that whereas the regression estimates of the former depend strongly on the estimated variance components under a non-identity link, those of the latter do not. The PQL also has a limitation in that inferences on variance components are not very accurate and it fails to account for the contribution of estimated variance components when assessing uncertainty in both fixed and random effects (the latter problem exists in empirical Bayes methods as well). They also point out that use of PQL and MQL is more natural for simple overdispersion problems involving a log-linear, conditionally Poisson model than a competing approach in Liang and Waclawiw (1990). Zeger (1988) considered a time series model for the random effects in a simple overdispersion model for count data. Recent Bayesian procedures for GLLM models have used importance sampling ideas (Raghunathan 1994) or Gibbs sampling techniques (Besag, York and Mollié 1991; Zeger and Karim 1991). A general estimating equation approach (Diggle, Liang and Zeger 1994) provides nearly efficient estimation relative to maximum likelihood estimates in overdispersion problems. Efron (1992) discussed *asymetric maximum likelihood* (AML) estimation of overdispersed generalized linear regression models, focusing on Poisson regression with application to an archeological data set. The AML approach is easy to implement, does not require explicit specification of a model for overdispersion and can be a good starting point for a more complete parametric analysis based on quasi-likelihood models, double exponential families or the overdispersion models (Gelfand and Dalal 1990; Dey, Gelfand and Peng 1997).

Several tests for overdispersion have been suggested in the literature. A test for Poisson overdispersion is given by Böhning (1994). Dean (1992) discussed tests for overdispersion with respect to a natural exponential family, which are powerful against arbitrary alternative mixture models with just the first two moments of the mixture distributions specified. Breslow (1990) derived tests of hypotheses in overdispersed Poisson regression and other quasi-likelihood models.

The format of the rest of this paper is as follows. We review various models for handling overdispersion in Section 2, parametric approches for Bayesian model fitting using MCMC techniques in Section 3 and nonparametric approaches in Section 4. Section 5 discusses extension to multistage modeling.

2. Classes of Overdispersed General Linear Models

Unfortunately, finite parametric mixtures are awkward to work with and in many cases it is unclear how many components should be used. Further, even for a given number of components, inconvenient constraints on the model are necessary to insure identifiability, when the mixing weights and the parameters of the component densities in the mixture are unknown. Although continuous mixtures alleviate these problems to a certain extent, once again the choice of a mixing distribution is a problem. Mixture models, especially finite mixture models have been most frequently used for creating a larger class of models for overdispersion. For instance, the one parameter exponential family defining the GLM is mixed with a two parameter exponential family for the canonical parameter θ (or equivalently the mean parameter μ), resulting in a two parameter marginal mixture family for the data. For the Poisson case, the gamma mixing distribution has been widely used, resulting in a negative binomial distribution for the observed data (Manton, Woodbury, and Stallard 1981 and Margolin, Kaplan and Zeiger 1981) while Hinde (1982) proposed a log-normal mixing distribution. Shaked (1980) showed that such mixing necessarily inflates the model variance, while Gelfand and Dalal (1990) argued that taking additional observations within a population does not provide more information about heterogeneity across populations. Finite parametric mixture models have another drawback in that the resulting overdispersed family of mixture models will no longer be an exponential family and consequently would be awkward to work with. In many cases, it is unclear how many components should be used. Further, even for a given number of components, inconvenient constraints on the model are necessary to insure identifiability when the mixing weights and the parameters of the component densities in the mixture are unknown. Although continuous mixtures alleviate these problems to a certain extent, once again the choice of a mixing distribution is a problem.

For a given one parameter exponential family, Gelfand and Dalal (1990) considered a class of two parameter exponential family of models

$$f(y \mid \theta, \tau) = b(y)e^{\theta y + \tau T(y) - \rho(\theta,\tau)} \tag{1}$$

where depending on whether y is continuous or discrete, f is assumed to be a density with respect to Lebesgue measure or counting measure respectively. They showed that under the assumption that (1) is integrable over $y \epsilon \mathcal{Y}$, and $T(y)$ is convex, then for a common mean, var(y) increases in τ. It is presumed that the natural parameter space contains a two dimensional rectangle which, by translation, can be taken to contain the line $\tau = 0$. The associated one parameter exponential family which is obtained at $\tau = 0$ has the form

$$f(y \mid \theta) = b(y)e^{\theta y - \chi(\theta)} \tag{2}$$

with $\chi(\theta) = \rho(\theta, 0)$. As τ increases from 0, var(y) increases relative to the variance under the associated one parameter exponential family so that τ describes the overdispersion.

A GLM is usually developed from the form of the one parameter exponential family (2). In particular $\mu \equiv E(y) = \chi^{'}(\theta)$, var$(y) = \chi^{''}(\theta) = V(\mu)$, the variance function. Here $\chi^{'}(\theta)$ is strictly increasing in θ so that μ and θ are one-to-one ($\theta = (\chi^{'})^{-1}(\mu)$). A link function g is defined, which is a strictly increasing differentiable transformation from μ to $\eta \epsilon R^1$, so that $g(\mu) = \eta = \tilde{x}^T\tilde{\beta}$, where $\tilde{x}$ and $\tilde{\beta}$ are, respectively, a $p \times 1$ vector of known explanatory variables and an unknown vector of model parameters.

Efron (1986) presented an alternative approach through so-called double exponential families. Such families are derived as the saddle point approximation to the density of an average of n^* random variables from a one parameter exponential family for large n^* . The parameter n^* written suggestively by Efron as $n\rho$, $0 < \rho < 1$ for actual sample size n, introduces ρ as a second parameter in the model along with the canonical parameter θ. The density corresponding to the double exponential family is

$$\bar{f}(y \mid \theta, \rho, n) = c(\theta, \rho, n)\rho^{\frac{1}{2}} e^{n\rho(\theta y - \chi(\theta)) + n(1-\rho)(\theta(y)y - \chi(\theta(y)))} \tag{3}$$

where $\theta(y) = (\chi')^{-1}(y)$, y may be viewed as an average of n i.i.d. random variables, θ is the canonical parameter and ρ is a dispersion parameter. For regression problems, Efron assumed that $\theta = \tilde{x}^T\tilde{\beta}$ and that $\rho = h(\tilde{z}^T\tilde{\alpha})$ for a suitable h, where $\tilde{z}$ is a known $q \times 1$ vector and $\tilde{\alpha}$ is an unknown parameter vector. Using various expansions, he showed that (3) permits attractive approximation as n increases. Most notably, $\bar{f}$ behaves like (2) with $w = n$ and $\phi = \rho^{-1}$.

Model (2) is extended to an exponential dispersion model, EDM (Jorgensen 1987) by incorporating a dispersion parameter ϕ, to give

$$f(y \mid \theta, \phi) = b(y, \phi)e^{w(\theta y - \chi(\theta))/\phi}, \tag{4}$$

where w is a known "sample size"; here $\text{var}(y) = \phi V(\mu)/n$. Whereas (1) is a customary two parameter exponential family, (4) is a one parameter family for each fixed ϕ. The ensuing problem was circumvented in an approximate fashion by Ganio and Schafer (1992) who considered the EDM as embedded in Efron's double exponential family and appealed to the associated asymptotic inference. Unlike the mixture case, these asymptotics result in overdispersion relative to the original exponential family which approaches a constant as $n \to \infty$ (Efron 1986; Gelfand and Dalal 1990). Specifically, Ganio and Schafer directly set $\phi = h(\tilde{z}^T\tilde{\alpha})$ in (3). Assuming that $\tilde{z}^T\tilde{\alpha}$ includes an intercept, they fit double exponential family models by the method of maximum likelihood and using an approximation to (3). Note that the Normal or Gamma families have scale parameters and hence can be written as EDM's.

Gelfand and Dalal (1990) argued that an appeal to asymptotics is not necessary to justify these models. Specifically, (1) not only includes Efron's model as a special case, but also a family discussed by Lindsay (1986). This is attractive because retaining the exponential family structure simplifies inference and the relative overdispersion behaves as in Efron's model. Note that regardless of n, (3) is of the form (1) with $T(y) = \theta(y)y - \chi(\theta(y))$, $\tau = n(1-\rho)$ and $\theta = n\rho\theta$. Straightforward calculation shows this $T(y)$ is convex so that, in fact, (3) is a special case of (1).

Although Gelfand and Dalal suggested that with the specification of link functions, these two parameters could each be given the usual GLM structure, they did not pursue the matter. These models, which are referred to as overdispersed generalized linear models (OGLM's), were fully examined by Dey *et al.* (1997). They assumed independent responses y_i with associated covariates $\tilde{x}_i$, and $\tilde{z}_i$, $i = 1, 2, .., n$, where the components of $\tilde{x}$ and $\tilde{z}$ need not be exclusive. Let $\tilde{y} = (y_1, \cdots, y_n)$ and define $\theta_i = g(\tilde{x}_i^T\tilde{\beta})$ and $\tau_i = h(\tilde{z}_i^T\tilde{\alpha})$ where g and h are strictly increasing. The corresponding joint density is from (1)

$$f(\tilde{y}|\tilde{\beta}, \tilde{\alpha}) = \prod_{i=1}^{n} e^{\theta_i y_i + \tau_i T(y_i) - \rho(\theta_i, \tau_i)}. \tag{5}$$

The monotonicity of g and h is natural and insures that θ_i is monotonic in x_{il} and that τ_i is monotonic in z_{il}, facilitating interpretation. Of course such monotonicity does not imply that, e.g., θ_i is monotone in each covariate. The form $\tilde{x}_i^T\tilde{\beta}$ allows a covariate to enter as a polynomial. If an explanatory variable appears in $\tilde{x}_i$ but not in $\tilde{z}_i$, say x_{il}, then $\frac{\partial \mu_i}{\partial x_{il}} = \rho^{(2,0)}(\theta_i, \tau_i)g'(\tilde{x}_i^T\tilde{\beta})\beta_l$. But since $\rho^{(2,0)}$ and g' are strictly positive, μ_i is strictly monotone in x_{il} with the sign of β_l determining the direction. If the variable appears in $\tilde{z}_i$ as well, its influence on μ_i is less clear, since now $\frac{\partial \mu_i}{\partial x_{il}}$ involves $\rho^{(1,1)}$, the covariance between y and $T(y)$. A practical drawback to working with (1) is that $\rho(\theta, \tau)$ is not available explicitly. While $\chi(\theta)$ in (2) is usually an explicit function of θ, $\rho(\theta, \tau) = \log \int b(y) e^{\theta y + \tau T(y)} dy$ usually requires a univariate numerical integration or summation. Dey *et al.* (1997) mentioned that this was not a problem in the practical examples they investigated. Gelfand and Dalal (1990) argued that the performance of (1) is not sensitive to the choice of $T(y)$. Attention should focus on the specification of θ_i and τ_i rather than on the stochastic mechanism (1).

Dey, Peng and Larose (1995) modeled heterogeneity and overdispersion through the notion of a parametrized weighted distribution. Suppose the random variable of interest Y is distributed over a population of interest with probability density $f(y|\theta)$, where θ is the underlying parameter of interest. The observed data is now a random sample from the following weighted distribution with density function

$$f^w(y|\theta, \tau) = \frac{w(y, \tau) f(y|\theta)}{E_f[w(y, \tau)]}, \tag{6}$$

where the expectation in the denominator is the normalizing constant. The overdispersed model is viewed as a perturbation of the original model, so that the latter can be formed as a weighted distribution of the form (6). These models are regular and parsimonious compared to a GLM and enable the capture of overdispersion within an exponential family framework. Section 3 describes fitting (5) in a parametric setting while Section 4 describes a nonparametric framework using MCMC techniques.

3. Fitting OGLM in the Parametric Bayesian Framework

Dey *et al.* (1997) examined inference for OGLM's using Markov chain Monte Carlo methods. The advantages of their approach include the familiarity of exponential families, the ready interpretation of model parameters, the unification of modeling by absorbing earlier cases and exact inference rather than inference based on asymptotic approximations. They adopted a Bayesian perspective in fitting these models but used noninformative prior specifications since primary concern lies in the modeling incorporated in the likelihood in (5). Hence inference will be close to that arising from maximum likelihood though estimates of variability will be exact rather than asymptotic and entire posteriors for model parameters would result. Required Bayesian computation is handled through Markov chain Monte Carlo using a Metropolis algorithm resulting in samples essentially from the joint posterior distribution which may be summarized to provide any desired inference. Such samples may also be used as the starting point for sampling from predictive distributions to investigate questions of model adequacy and model choice.

3.1 Model Fitting

The Fisher information matrix associated with (5) is of interest in Bayesian model fitting since the square root of the determinant of this matrix, as a function of $\tilde{\beta}$ and $\tilde{\alpha}$, is known as Jeffreys' prior and is commonly used as a "noninformative" specification. Denoting the right side of (5) as $L(\tilde{\beta}, \tilde{\alpha}, \tilde{y})$, Dey *et al.* (1997) showed that

$$E\left(\frac{\partial^2 logL(\tilde{\beta},\tilde{\alpha},\tilde{y})}{\partial\tilde{\beta}_i\partial\tilde{\beta}_j}\right) = -\sum_i \rho^{(2,0)}(\theta_i,\tau_i)x_{ij}x_{ik}(g^{'}(\tilde{x}_i^T\tilde{\beta}))^2,$$

$$E\left(\frac{\partial^2 logL(\tilde{\beta},\tilde{\alpha},\tilde{y})}{\partial\tilde{\alpha}_i\partial\tilde{\alpha}_j}\right) = -\sum_i \rho^{(0,2)}(\theta_i,\tau_i)z_{ij}z_{ik}(h^{'}(\tilde{z}_i^T\tilde{\alpha}))^2,$$

$$E\left(\frac{\partial^2 logL(\tilde{\beta},\tilde{\alpha},\tilde{y})}{\partial\tilde{\beta}_j\partial\tilde{\alpha}_k}\right) = -\sum_i \rho^{(1,1)}(\theta_i,\tau_i)x_{ij}z_{ik}(g^{'}(\tilde{x}_i^T\tilde{\beta}))(h^{'}(\tilde{z}_i^T\tilde{\alpha})).$$

Let $\tilde{X}$ denote the $n \times p$ design matrix arising from the $\tilde{x}_i^{'}$s, $\tilde{Z}$ the $n \times q$ design matrix arising from the $\tilde{z}_i^{'}$s, $\tilde{M}_\theta$ an $n \times n$ diagonal matrix with $(\tilde{M}_\theta)_{ii} = \rho^{(2,0)}(\theta_i,\tau_i)(g^{'}(\tilde{x}_i^T\tilde{\beta}))^2$, $\tilde{M}_\tau$ an $n{\times}n$ diagonal matrix with $(\tilde{M}_\tau)_{ii} = \rho^{(0,2)}(\theta_i,\tau_i)(h^{'}(\tilde{z}_i^T\tilde{\alpha}))^2$ and $\tilde{M}_{\theta,\tau}$ an $n \times n$ diagonal matrix with $(\tilde{M}_{\theta,\tau})_{ii} = \rho^{(1,1)}(\theta_i,\tau_i)(g^{'}(\tilde{x}_i^T\tilde{\beta}))(h^{'}(\tilde{z}_i^T\tilde{\alpha}))$. Then

$$I(\tilde{\beta},\tilde{\alpha}) = \begin{pmatrix} \tilde{X}^T\tilde{M}_\theta\tilde{X} & \tilde{X}^T\tilde{M}_{\theta,\tau}\tilde{Z} \\ \tilde{Z}^T\tilde{M}_{\theta,\tau}\tilde{X} & \tilde{Z}^T\tilde{M}_\tau\tilde{Z} \end{pmatrix} \tag{7}$$

and Jeffreys' prior is $|I(\tilde{\beta},\tilde{\alpha})|^{\frac{1}{2}}$. Computation of (7) requires calculation of ρ, $\rho^{(2,0)}$, $\rho^{(0,2)}$ and $\rho^{(1,1)}$ which in turn requires numerical integration or summation of the form $\int y^c T(y)^d b(y) e^{\theta y + \tau T(y)} dy$ for various c's and d's.

Dey *et al.* showed that the posterior is proper under Jeffreys' prior or a flat prior for $\tilde{\beta}$ and $\tilde{\alpha}$, provided $L(\tilde{\beta},\tilde{\alpha};\tilde{y})$ is log concave. Log concavity of $L(\tilde{\beta},\tilde{0};\tilde{Y})$ is discussed in Wedderburn (1976) in the context of properties of MLE's and in Dellaportas and Smith (1993) with respect to simplifying Monte Carlo sampling of $\tilde{\beta}$. In particular, for OGLM Dey *et al* established log concavity by verifying a simple nonnegative definiteness condition. Letting $\theta = g(\eta)$, $\tau = h(\gamma)$ they considered the function $a(\eta,\gamma) = \theta(\eta)y + \tau(\gamma)T(y) - \rho(\theta(\eta),\tau(\gamma))$. If $-\frac{\partial^2 a}{\partial\eta^2} \geq 0$, $-\frac{\partial^2 a}{\partial\gamma^2} \geq 0$ and $\begin{vmatrix} -\frac{\partial^2 a}{\partial\eta^2}, & -\frac{\partial^2 a}{\partial\eta\partial\gamma} \\ -\frac{\partial^2 a}{\partial\gamma\partial\eta}, & -\frac{\partial^2 a}{\partial\gamma^2} \end{vmatrix} \geq 0$, (5) is log concave. For the canonical case $\theta = \eta$, $\tau = \gamma$, $-\frac{\partial^2 a}{\partial\eta^2} = \text{var}(y)$, $-\frac{\partial^2 a}{\partial\gamma^2} = \text{var}(T(y))$ and the determinant becomes $\text{var}(y)\text{var}(T(y))$ - $cov^2(y,T(y))$ which is nonnegative whence log concavity holds.

Sampling based fitting of the OGLM in the Bayesian framework was carried out using standard tools. Dey *et al.* (1997) used a Gaussian proposal to implement a Metropolis algorithm (Hastings 1970) although adaptive rejection sampling within the Gibbs sampler as in Gilks and Wild (1992) is another possibility. They used multiple starts in the vicinity of the MLE for $\tilde{\beta}$ (under $\tau = 0$) and at $\tilde{\alpha} = \tilde{0}$. Evaluation of $L(\tilde{\beta},\tilde{\alpha};\tilde{Y})$ requires repeated calculation of the function $\rho(\theta_i,\tau_i)$, which fortunately is a routine univariate numerical integration.

3.2 Example: Overdispersed Poisson Regression Model

Dey *et al.* (1997) used as a motivating example data involving damage incidents to cargo ships (McCullagh and Nelder 1989, p204). For each of 34 ships the aggregate months in service were recorded as well as the number of damage incidents over that period. Explanatory factors are ship type having 5 levels (A, B, C, D, E), year of construction having 4 levels (CP1, CP2, CP3, CP4) and period of operation having two levels (SP1, SP2). Since the response is a count, McCullagh and Nelder proposed a Poisson regression presuming that the expected number of damage incidents is directly proportional to the aggregate months in service, i.e., the total period of risk. Using a canonical link the GLM sets

$$\begin{aligned} \theta &= \log(\text{aggregate months service}) + \beta_0 + \text{effect due to ship type} \\ &+ \text{effect due to year of construction} + \text{effect due to service period}. \end{aligned} \tag{8}$$

The log(aggregate months service) term is called an offset, its coefficient fixed at 1 as a result of the proportionality assumption. They also incorporated a dispersion parameter ϕ in the Poisson model, whose estimate was $\hat{\phi} = 1.69$, indicating overdispersion relative to the standard Poisson density which, intrinsically, has $\phi = 1$.

On examining the nonstandardized residuals $y_i - \hat{\mu}_i$ under the GLM in (8), Dey *et al.* noted that some observations did not show a good model fit, and this situation is not improved through an EDM. The use of OGLM's to fit this data is illustrated in Dey *et al.*, a brief summary of which follows.

Taking (1) as the model for $\theta = \tilde{x}^T \tilde{\beta}$ in (7), Dey *et al.* considered three specifications for τ. In the simple case (model 1) they set $\tau = 0$, which resulted in a 9 parameter Bayesian GLM for (1). Model 2 incorporated a constant dispersion parameter $\tau = \alpha_0$ with the convex function, $T(y) = (y+1)\log(y+1)$, yielding a 10 parameter model. Finally, anticipating that overdispersion might increase with exposure, model 3 set $\tau = \alpha_0 + \alpha_1 \log(\text{aggregate months service})$, using the same $T(y)$, a 11 parameter model.

Results from these model fits provide evidence of overdispersion ($\tau > 0$) and, in addition, evidence that $\alpha_1 > 0$, supporting the hypothesis that overdispersion increases with exposure to risk. It was also seen that model 3 better gave a better fit for larger values of y, which are associated with greater exposure.

3.3 Model Determination for Parametric OGLM's

Model determination includes model adequacy and model choice. Formal Bayesian model determination proceeds from the marginal predictive distribution of the data $f(\tilde{y})$ evaluated at the observed data, $\tilde{y}_{obs}$. Since this high dimensional density is hard to estimate well and its value is hard to calibrate it may be preferable, as argued in Gelfand (1995), to look at alternative predictive distributions, in particular univariate ones such as the posterior predictive density at a new y_0, $f(y_0|\tilde{y}_{obs})$ or the cross–validated predictive density of say y_r, $f(y_r|\tilde{y}_{(r),obs})$ (which can be compared with $y_{r,obs}$). Although the model determination diagnostics are admittedly informal, they are in the spirit of widely used classical EDA approaches and are attractive since they permit examination of model performance at the level of the individual observation. Required calculations may be carried out by sampling $f(y_r|\mathbf{y}_{(r),obs})$ which may be achieved using the posterior sampler as described in Gelfand and Dey (1994).

For the cargo data analysis, Dey *et al.* (1997) adopted a cross-validation approach, paralleling widely used classical regression strategy. In particular, they considered the proper densities $f(y_r \mid \tilde{y}_{(r)})$, $r = 1, 2, ..., 34$, where $\tilde{y}_{(r)}$ denotes $\tilde{y}$ with y_r removed. In fact, they conditioned on the actual observations $\tilde{y}_{(r),obs}$ creating the predictive distribution for y_r under the model and all the data except y_r. For model determination this would require comparison, in some fashion, of $f(y_r \mid \tilde{y}_{(r),obs})$ with the r_{th} observation, $y_{r,obs}$. Such cross validation is discussed in Gelfand, Dey and Chang (1992) and in further references provided therein.

A natural approach for model adequacy is to draw, for each r, a sample from $f(y_r \mid \tilde{y}_{(r),obs})$, and compare this sample with $y_{r,obs}$. In particular, based on this sample, it is possible to obtain the .025 and .975 quantiles of $f(y_r \mid \tilde{y}_{(r),obs})$ say $\underline{y}_r$ and $\overline{y}_r$ and see how many of the $y_{r,obs}$ belongs to $[\underline{y}_r, \overline{y}_r]$. Under each of the three models fit to the cargo ship data, at least 27 of the 34 intervals contained the corresponding $y_{r,obs}$. They also obtained the lower and upper quartiles of $f(y_r \mid \tilde{y}_{(r),obs})$ to see how many $y_{r,obs}$ belonged in their interquartile ranges; they found 13, 15 and 18 under model 1, model 2 and model 3 respectively. Since we would expect half, i.e., 17 under the true model, both model 2 and model 3 perform close to expectation though all three models seem adequate.

A well established tool for model choice is the conditional predictive ordinate (CPO), $f(y_{r,obs} \mid \tilde{y}_{(r),obs})$, a large value of which implies agreement between the observation and the model. For comparing models, the ratio $d_r = \frac{f(y_{r,obs}|y_{(r),obs},M_i)}{f(y_{r,obs}|y_{(r),obs},M_{i'})}$ (or perhaps $\log d_r$) indicates support by the rth point for one model versus the other (see Pettit and Young, 1990). For the cargo data, model 3 provided the best fit. A simple diagnostic with a frequentist flavor, $\sum_{r=1}^{34} (y_{r,obs} - E(\mu_r|\boldsymbol{y}))^2/34$ yielded values 6.74, 7.11 and 8.61 respectively for model 1, model 2 and model 3; compared with the EDM result of 6.70, all three models are adequate. All of the foregoing calculations are carried out by sampling $f(y_r|\boldsymbol{y}_{(r),obs})$ which may be achieved using the posterior sampler as described in Gelfand and Dey (1994).

4. Modeling Overdispersion in the Nonparametric Bayesian Framework

In some problems, samples exhibit extra heterogeneity because observations are drawn from an overall population which would be more effectively modeled as a mixture of subpopulations. Such behavior cannot be captured by univariate exponential family models and even OGLM's cannot explain all types of heterogeneity adequately. To alleviate the problems in finite parametric mixture models, Mukhopadhyay and Gelfand (1997) captured the specification of overdispersion nonparametrically through Dirichlet Process (DP) mixing. They extended GLM's to DPMGLM's (Dirichlet process mixed GLM's) and OGLM's to DPMOGLM's, with specific emphasis on binomial and Poisson regression models. DPMOGLM's afford the possibility of capturing a very broad range of heterogeneity resulting in the most flexible GLM's yet proposed and are an alternative to GLMM's (Breslow and Clayton 1993). Note that they intentionally retained the GLM aspect with regard to the mean. In theory, although one could DP mix over the coefficients in the presumed linear mean structure on a translated scale, these coefficients would vanish from the likelihood resulting in a mixture model which would no longer be a linear model in any sense. Their structure retains these coefficients and the associated linear structure to permit the appealing interpretation they provide with regard to the relationship between a response variable and a collection of explanatory variables.

4.1 Fitting DP Mixed GLM and OGLM

The basics of DP mixing as described in Blackwell and McQueen (1973), Ferguson (1973) and Sethuraman and Tiwari (1982) are extended to mixture modeling as follows (for details, see Mukhopadhyay and Gelfand, 1997). Consider $\{f(\cdot|\tilde{\theta}) : \tilde{\theta} \in \Theta \subset \Re^d\}$ to be a parametric family of densities with respect to a dominating measure μ and consider also the family of probability distributions $\mathcal{F} = \{F_G : G \in \mathcal{P}\}$ with densities

$$f(y|G) = \int f(y|\tilde{\theta})dG(\tilde{\theta}) \tag{9}$$

with respect to μ, so that $\mathcal{F}$ becomes a nonparametric family of mixture distributions; it is assumed that the mixing distribution G comes from a DP on $\mathcal{P}$, i.e., $G \sim$ DP(νG_0), where ν is a precision parameter and G_0 is a proper base probability distribution in $\mathcal{P}$. A semiparametric class of models may be obtained by choosing to DP mix with respect to a subset $\tilde{\omega}$ of $\tilde{\theta}$ instead of $\tilde{\theta}$ itself. Such DP mixture models have become increasingly popular for modeling when conventional parametric models are either hard to fit or impose unreasonably strict constraints on the class of distributions (see for e.g. Escobar and West (1993); MacEachern 1992; Bush and MacEachern 1993 in the context of hierarchical modeling).

Mukhopadhyay and Gelfand (1997) extended the GLM's which arises from (2) by introducing DP mixed GLM's (DPMGLM's) as follows. In the form of the usual GLM's they write $\tilde{x}'\tilde{\beta}$ as $\alpha + \tilde{x}'\tilde{\beta}$, so that the intercept term is written separately; under a canonical link $\tilde{\theta} = \alpha + \tilde{x}'\tilde{\beta}$ and they mix over α. DP mixing over α extends the basic GLM model to capture heterogeneity in the population with regard to the *location* of the mean, presuming that the covariate relationship is unaffected by such centering. The resulting density for an observation $\tilde{y} = (y_1, \cdots, y_n)$ with associated covariate $\tilde{x}$ under this model is, following (9),

$$f(\tilde{y}|\tilde{x}, \tilde{\beta}, G) = \int f(\tilde{y}|\tilde{x}, \tilde{\beta}, \alpha)dG(\alpha), \tag{10}$$

where $f(\tilde{y}|\tilde{x}, \tilde{\beta}, \alpha)$ is the model in (2) under the canonical link. They used a vague normal prior specification for $\tilde{\beta}$ with $G \sim DP(\nu G_0)$ and carried out Gibbs sampling for models created under (10). The complete conditional densities for the β_i's are log concave (Dellaportas and Smith 1992) and are sampled using adaptive rejection sampling (Gilks and Wild 1992). Alternatively, Metropolis steps may be used (Tierney, 1994). The latent α_i are sampled following the approach of MacEachern and Müller (1994). This approach is applicable when a general link function is used and two possible forms for $\tilde{\theta}$ are considered, viz. $\tilde{\theta} = h(\alpha + \tilde{x}'\tilde{\beta})$ or $\tilde{\theta} = \alpha + h(\tilde{x}'\tilde{\beta})$.

Mukhopadhyay and Gelfand discussed mixing of OGLM's to introduce further flexibility into the family of models. Imitating the earlier discussion, they replaced $\tilde{\theta}$ by $\alpha + \tilde{x}'\tilde{\beta}$ in (1), writing the resultant density as $f(\tilde{y}|\tilde{x}, \tilde{\beta}, \alpha, \tau)$. A Bayesian model requires specification of a prior for $(\tilde{\beta}, \alpha, \tau)$. Three possibilities exist:

$$f(\tilde{y}|\tilde{x}, \tilde{\beta}, G_\alpha, \tau) = \int f(\tilde{y}|\tilde{x}, \tilde{\beta}, \alpha, \tau)d\,G_\alpha(\alpha), \tag{11}$$

$$f(\tilde{y}|\tilde{x}, \tilde{\beta}, \alpha, G_\tau) = \int f(\tilde{y}|\tilde{x}, \tilde{\beta}, \alpha, \tau)d\,G_\tau(\tau), \tag{12}$$

and

$$f(\tilde{y}|\tilde{x}, \tilde{\beta}, G_{\alpha,\tau}) = \int f(\tilde{y}|\tilde{x}, \tilde{\beta}, \alpha, \tau)d\,G_{\alpha,\tau}(\alpha, \tau). \tag{13}$$

Estimation under all these forms may be handled in the Bayesian framework and details are given in Mukhopadhyay and Gelfand (1997).

4.2 Example: Overdispersed Binomial Regression Model

Lindsey (1993) analyzed data from an egg hatching experiment conducted using 72 tanks. Only two covariates are available, temperature (x_1) and salinity (x_2). The responses are the number of eggs hatched (y_i) out of n_i eggs in the ith tank ($i = 1, \cdots, 72$). Lindsey observed that, for this data, heterogeneity is too great to be explained by a standard binomial regression using these covariates. Mukhopadhyay and Gelfand investigated the binomial GLM and their proposed extensions. They assumed a flat prior for each coefficient (β_1, β_2). With $T(y) = y^2$, a uniform prior on the interval $[-1, 1]$ was chosen for τ in each of the OGLM and the DPMOGLM. Other prior specifications are the same as described in the general case.

Using the log CPO ratios, they compared the performance of the fitted models. This showed that approximately 96% of the data are better explained by the OGLM than the GLM. Again, comparing the DP mixed models with the OGLM, they found that approximately 76% and 85% of the data support the DPMGLM and the DPMOGLM respectively. In each of these comparisons the average of the log CPO's was at least 33.253, lending strong support to their conclusions. Lastly, in comparing the DPMOGLM to the DPMGLM, the CPO's were roughly split in half and the average of the log CPO's was only 0.795, suggesting little improvement by the former model and adoption of the latter in the interest of parsimony.

To compare common parameters across different models they plotted the posterior densities; the posterior density of τ for the OGLM was almost degenerate at 1, indicating the inadequacy of the OGLM in explaining the heterogeneity present in the data. The posterior density of τ under the DPMOGLM appeared to be unimodal with mode near 0.7.

4.3 Model Determination for Dirichlet Process Mixed Models

Mukhopadhyay and Gelfand showed explicit computation of $f(y_0 \mid \tilde{y}_{obs})$ and $f(y_{r,obs} \mid \tilde{y}_{(r),obs})$ for a generic DP mixed model in a regression setting with covariate values $\tilde{x}_i$ associated with responses, y_i, $i = 1, \cdots, n$. Under the assumption that the density of y_i depends upon a vector of parameters $\tilde{\theta}$ which is partitioned as $\tilde{\theta} = (\tilde{\eta}, \tilde{\omega})$, they DP mixed over $\tilde{\omega}$, assuming that ν is specified. The case of unknown ν can be handled in a fully Bayesian way using Gibbs sampling as well. They assumed a specific parametric prior on $\tilde{\eta}$, $\pi(\tilde{\eta})$ which could also arise hierarchically say as $\pi(\tilde{\eta}|\tilde{\gamma}) \cdot \pi(\tilde{\gamma})$ where $\tilde{\gamma}$ is a vector of hyperparameters. Note that although $\pi(\tilde{\gamma})$ need not be proper to consider posterior or cross–validated predictive densities, it must be proper in order that the marginal predictive density be proper and hence interpretable.

With the above assumptions the nonparametric Bayesian regression model takes the form

$$\prod_{i=1}^{n} f(y_i|\tilde{x}_i, \tilde{\eta}, G) \cdot f(G|G_0, \nu) \cdot \pi(\tilde{\eta}) = \prod_{i=1}^{n} \int f(y_i|\tilde{x}_i, \tilde{\eta}, \tilde{\omega}_i)\, dG(\tilde{\omega}_i) \cdot f(G|G_0, \nu) \cdot \pi(\tilde{\eta}) \qquad (14)$$

Marginalizing the posterior predictive density for y_0 at $\tilde{x}_0$ over G they obtained

$$f(y_0|\tilde{x}_0, \tilde{y}_{obs}) = \int_{\tilde{\eta}} \int_{\tilde{\omega}} \int_{\tilde{\omega}_0} f(y_0|\tilde{x}_0, \tilde{\eta}, \tilde{\omega}_0) \cdot f(\tilde{\omega}_0|\tilde{\omega}) \cdot f(\tilde{\omega}, \tilde{\eta}|\tilde{y}_{obs}). \qquad (15)$$

Assuming the sampling based fitting of (14) has yielded draws $(\tilde{\omega}_l^*, \tilde{\eta}_l^*)$, $l = 1, \cdots, B$ from $f(\tilde{\omega}, \tilde{\eta}|\tilde{y}_{obs})$, Monte Carlo integration for (15) is done in two ways. If, given $(\tilde{\omega}_l^\star, \tilde{\eta}_l^\star)$ we draw $\tilde{\omega}_{0l}^\star$ from $f(\tilde{\omega}_0|\tilde{\omega})$, we obtain

$$\hat{f}(y_0|\tilde{x}_0, \tilde{y}_{obs}) = B^{-1} \sum_{l=1}^{B} f(y_0|\tilde{x}_0, \tilde{\eta}_l^\star, \tilde{\omega}_{0l}^\star). \tag{16}$$

Alternatively, if we do the innermost integration in (15) we need not sample the $\tilde{\omega}_{0l}^\star$, obtaining

$$\begin{aligned} \tilde{f}(y_0|\tilde{x}_0, \tilde{y}_{obs}) &= B^{-1}(\nu+n)^{-1} \\ & \sum_{l=1}^{B}\{\sum_{i=1}^{n} f(y_0|\tilde{x}_0, \tilde{\eta}_l^\star, \tilde{\omega}_{il}^\star) + \nu \int f(y_0|\tilde{x}_0, \tilde{\eta}_l^\star, \tilde{\omega}) dG_0(\tilde{\omega})\}, \end{aligned} \tag{17}$$

where $\tilde{f}$ is a mixture of distributions. Though (17) would generally be preferable to (16), since exact integration replaces sampling, the unpleasant integral in (17) must be computed B times. Also, should we wish to obtain samples from $f(y_0|\tilde{x}_0, \tilde{y}_{obs})$ we can do so by drawing $y_{0l}^\star$ from $f(y_0|\tilde{x}_0, \tilde{\eta}_l^\star, \tilde{\omega}_{0l}^\star)$ again for $l = 1 \cdots, B$. They showed that computations for the cross validation predictive density through $f(y_r|\tilde{x}_r, \tilde{y}_{(r),obs})$ was similar. Denoting the new y_r and the corresponding w_r by y_0 and w_0 respectively, they approximated $f(y_0|\tilde{x}_r, \tilde{y}_{(r),obs})$ through a ratio of Monte Carlo integrations. Since the Monte Carlo integration for the denominator in this ratio can be unstable, they suggest preference for the posterior predictive density.

For the egg hatching data, use of the log CPO ratios showed that approximately 96% of the data are better explained by the OGLM than the GLM. Again, comparing the DP mixed models with the OGLM, approximately 76% and 85% of the data support the DPMGLM and the DPMOGLM respectively. In each of these comparisons the average of the log CPO's was at least 33.253, lending strong support to these conclusions. Lastly, in comparing the DPMOGLM to the DPMGLM, the CPO's were roughly split in half and the average of the log CPO's was only 0.795, suggesting little improvement by the former model and adoption of the latter in the interest of parsimony. To compare common parameters across different models they plotted the posterior densities; the posterior density of τ for the OGLM was almost degenerate at 1, indicating the inadequacy of the OGLM in explaining the heterogeneity present in the data. The posterior density of τ under the DPMOGLM appeared to be unimodal with mode near 0.7. Although the posterior densities of β_1 and of β_2 for all the four models were shifted to the left and were less concentrated, they still strongly supported nonzero coefficients.

5. Overdispersion in Multistage GLM

Models are sometimes formulated as hierarchical or multistage GLM's in which case we may incorporate overdispersion or DP mixing at one or more of the stages. For illustration, consider two examples: semiparametric nested random effects models and semiparametric errors in variables models.

The case of nested normal random effect models was discussed in Goldstein (1986). More generally, suppose response Y_{ijk} is modeled as a GLM with canonical parameter $\mu + \alpha_i + \beta_{ij}$. For instance, Y_{ijk} might be a count recorded for the kth child in the

jth class in the ith school with α_i being a random school effect and β_{ij} being a random class effect nested within school. They defined $\gamma_i = \mu + \alpha_i$, $\rho_{ij} = \mu + \alpha_i + \beta_{ij}$ and the hierarchical model $f(y_{ijk}|\rho_{ij}) \cdot f(\rho_{ij}|\gamma_i) \cdot f(\gamma_i|\mu)$. Here $f(\gamma_i|\mu)$ would be extended to a GLM involving school level covariates say $\tilde{x}_i$, i.e., the canonical parameter associated with γ_i would be $\mu + \tilde{x}_i'\tilde{\beta}^{(\gamma)}$. Similarly, $f(\rho_{ij}|\gamma_i)$ would be extended to a GLM involving class within school covariates say $\tilde{\omega}_{ij}$ and having canonical parameter $\gamma_i + \tilde{\omega}_{ij}'\tilde{\beta}^{(\rho)}$. One can introduce DP mixing at the third stage, i.e., mixing on μ, based on which the model for γ_i becomes $f(\gamma_i|\tilde{x}_i, \tilde{\beta}^{(\gamma)}, G)$. The Bayesian model is completely specified with priors on $\tilde{\beta}^{(\gamma)}$, $\tilde{\beta}^{(\rho)}$ and G. OGLM's could be added, perhaps most naturally at the first stage and Gibbs sampling proceeds straightforwardly.

In the errors–in–variables problem, the class of models described in Carroll (1992) may be extended. Mukhopadhyay and Gelfand (1997) modeled the response Y_i as $f(y_i|\tilde{x}_i, \tilde{\beta}^{(y)}, G_y) = \int f(y_i|\tilde{x}_i, \tilde{\beta}^{(y)}, \alpha_i^{(y)})\, d\,G_y(\alpha_i^{(y)})$. That is, conditional on $\alpha_i^{(y)}$, Y_i follows a GLM with canonical parameter $\alpha_i^{(y)} + \tilde{x}_i'\tilde{\beta}^{(y)}$ and they DP mix over $\alpha_i^{(y)}$. Suppose a component of $\tilde{x}_i$ say x_{1i} is not directly observable. Its actual level may only be known to arise with error around some nominal level or, more generally, in place of x_{1i}, a vector $\tilde{\omega}_i$ of surrogate variables may be observed. Then they model x_{1i} as $f(x_{1i}|\tilde{\omega}_i, \tilde{\beta}^{(x)}, G_x) = \int f(x_{1i}|\tilde{\omega}_i, \tilde{\beta}^{(x)}, \alpha_i^{(x)})\, d\,G_x(\alpha_i^{(x)})$. That is, conditional on $\alpha_i^{(x)}$, x_{1i} follows a GLM with canonical parameter $\alpha_i^{(x)} + \tilde{\omega}_i'\tilde{\beta}^{(x)}$ and they DP mix over the $\alpha_i^{(x)}$. This introduces DP mixing at both modeling stages. The Bayesian model is completely specified with priors on $\tilde{\beta}^{(y)}$, $\tilde{\beta}^{(x)}$, G_y and G_x. They fit this model using Gibbs sampling, the details of which may be obtained from their paper. Additionally, OGLM's could replace the GLM's in the above discussion, leading to an extension of the DPMOGLM to a multistage setup.

References

Barndorff-Nielsen, O.E. (1978). *Information and Exponential Families in Statistical Theory.* John Wiley & Sons, New York.

Besag, J., York, J. and Mollié, A. (1991). Bayesian Image Restoration, with Two Applications in Spatial Statistics (with discussion). *Annals of the Institute of Statistical Mathematics,* 43, 1-59.

Böhning, D. (1994). A note on a test for Poisson overdispersion. *Biometrika,* 81, 418-419.

Box, G.E.P. and Tiao, G.C. (1992). *Bayesian Inference in Statistical Analysis.* John Wiley & Sons, New York.

Breslow, N. (1984). Extra-Poisson Variation in Log-Linear Models. *Applied Statistics,* 33, 38-44.

Breslow, N. (1990). Tests of Hypotheses in Overdispersed Poisson Regression and Other Quasi-Likelihood Models. *Journal of the American Statistical Association,* 85, 565-571.

Breslow, N. and Clayton, D. (1993). Approximate inference in Generalized Linear Mixed Models. *Journal of the American Statistical Association,* 88, 9-25.

Carroll, R.J. and Ruppert, D. (1982). Robust Estimation in Heteroscedastic Linear Models. *Annals of Statistics,* 10, 429-441.

Cox, D.R. (1983). Some remarks on overdispersion. *Biometrika,* 70, 269-274.

Cox, D.R. and Reid, N. (1987). Parameter Orthogonality and Approximate Conditional Inference (with discussion). *Journal of the Royal Statistical Society,* Ser. B, 49, 1-39.

Crowder, M.J. (1978). Beta-Binomial ANOVA for Proportions. *Applied Statistics,* 27, 34-37.

Crowder, M.J. (1985). Gaussian Estimation for Correlated Binomial data. *Journal of the Royal Statistical Society, Ser.* B, 47, 229-237.

Dean, C.B. (1992). Testing for Overdispersion in Poisson and Binomial Regression Models. *Journal of the American Statistical Association,* 87, 451-457.

Dellaportas, P. and Smith, A.F.M. (1993). Bayesian Inference for Generalized Linear and Proportional Hazards Models via Gibbs Sampling. *Applied Statistics.,* 42, 443-460.

Dey, D.K., Gelfand, A.E. and Peng, F. (1997). Overdispersed Generalized Linear Models. *Journal of Statistical Planning and Inference,* 64, 93-108.

Dey, D.K., Peng, F. and Larose, D. (1995). Modeling Heterogeneity and Extraneous Variation using Weighted Distributions. In *Model Oriented Data Analysis,* C. P. Kitsos and W.G. Müller, eds., Physica Verlag, Heidelberg, 241-249.

Diggle, P.J., Liang, K-Y, and Zeger, S.L. (1994). *Analysis of Longitudinal Data.* Oxford University Press, Oxford.

Efron, B. (1986). Double Exponential Families and Their Use in Generalized Linear Regression. *Journal of the American Statistical Association,* 81, 709-721.

Efron, B. (1992). Poisson Overdispersion Estimates Based on the Method of Asymmetric Maximum Likelihood. *Journal of the American Statistical Association,* 87, 98–107.

Everitt, B.S. and Hand, D.J. (1981). *Finite Mixture Distributions.* Chapman and Hall, London.

Finney, D.J. (1971). *Probit Analysis* (3rd ed.). Cambridge, U.K.: Cambridge University Press.

Ganio, L.M. and Schafer, D.W. (1992). Diagnostics for Overdispersion. *Journal of the American Statistical Association,* 87, 795-804.

Gelfand, A.E. (1995). Model determination using sampling-based methods. In *Markov Chain Monte Carlo in Practice,* eds. W. Gilks, S. Richardson and D. Spiegelhalter. Chapman and Hall, London, 145-161.

Gelfand, A.E. and Dalal, S.R. (1990). A Note on Overdispersed Exponential Families. *Biometrika,* 77, 55-64.

Gelfand, A.E. and Dey, D.K. (1994). Bayesian Model Choice: Asymptotics and Exact Calculations. *Journal of the Royal Statistical Society,* Ser. B, 56, 501-514.

Gelfand, A.E., Dey, D.K. and Chang H. (1992). Model Determination Using Predictive Distributions with Implementation Via Sampling-Based Methods. In *Bayesian Statistics* 4, (J. Bernardo et al. eds.), Oxford University Press, Oxford, 147-167.

Gilks, W.R. and Wild, P. (1992). Adaptive Rejection Sampling for Gibbs Sampling. *Journal of the Royal Statistical Society,* Ser. C, 41, 337-348

Goldstein, H. (1991). Nonlinear Multilevel Models, With an Application to Discrete Response Data. *Biometrika,* 78, 45-51.

Hinde, J. (1982). Compound Poisson Regression Models. in GLIM 82: *Proceedings of the International Conference on Generalized Linear Models*, ed. R. Gilchrist, Berlin: Springer-Verlag, 109-121.

Jorgensen, B. (1987). Exponential Dispersion Models (with discussion). *Journal of the Royal Statistical Society,* Ser. B, 49, 127-162.

Laird, N.M. (1978). Empirical Bayes Methods for Two-Way Contingency Tables. *Biometrika,* 65, 581-590.

Lawless, J.F. (1987a). Negative Binomial and Mixed Poisson Regression. *Canadian Journal of Statistics,* 15, 209-225.

Lawless, J.F. (1987b). Regression Methods for Poisson Process Data. *Journal of the American Statistical Association,* 82, 808-815.

Liang, K.Y. and Waclawiw, M.A. (1990). Extension of the Stein estimating procedure Through the use of Estimating Functions. *Journal of the American Statistical Association,* 85, 435-440.

Liang, K.Y, Zeger, S.L. and Qaqish, B. (1992). Multivariate Regression Analysis for Categorical Data (with discussion). *Journal of the Royal Statistical Society, Ser.*B, 54, 3-40.

Lindsay, B. (1986). Exponential Family Mixture Models (with least squares estimators). *The Annals of Statistics*, 14,124-37.

MacEachern, S.N. and Müller, P. (1994). Estimating Mixture of Dirichlet Process Models. Technical Report No. 94-11, Duke University, ISDS.

Manton, K.G., Woodbury, M.A., and Stallard, E. (1981). A Variance Components approach to Categorical Data Models with Heterogeneous Cell Populations: Analysis of Spatial Gradients in Lung Cancer Mortality Rates in North Carolina counties. *Biometrics,* 37, 259-269.

Margolin, B.H., Kaplan, N. and Zeiger, E. (1981). Statistical Analysis of the Ames *Salmonella* Microsome Test. *Proceedings of the National Academy of Sciences,* 76, 3779-3783.

McCullagh, P., and Nelder, J.A. (1989). *Generalized Linear Models.* Chapman & Hall, London.

Moore, D.F. and Tsiatis, A. (1989). Robust Estimation of the Standard Error in Moment Methods for Extra-Binomial and Extra-Poisson Variation. Unpublished manuscript.

Mukhopadhyay, S. and Gelfand, A.E. (1997). Dirichlet Process Mixed Generalized Linear Models. *Journal of the American Statistical Association*, 92, 633-639.

Nelder, J.A. and Lee, Y. (1992). Likelihood, quasi-likelihood and pseudolikelihood: Some comparisons. *Journal of the Royal Statistical Society, Series B,* 54, 273-284.

Nelder, J.A. and Pregibon, D. (1987). An extended quasi-likelihood function. *Biometrika,* 74, 221-232.

Pettit, L.I. and Young, K.D.S. (1990). Measuring the Effect of Observations on Bayes Factors. *Biometrika,* 77, 455-466.

Raghunathan, T.E. (1994). Monte Carlo Methods for Exploring Sensitivity to Distributional Assumptions in a Bayesian Analysis of a Series of 2×2 Tables. *Statistics in Medicine*, 1525-1538.

Shaked, M. (1980). On Mixtures from Exponential Families. *Journal of the Royal Statistical Society,* Ser. B, 42, 192-198.

Smyth, G.K. (1989). Generalized Linear Models with Varying Dispersion. *Journal of the Royal Statistical Society, Ser.* B, 51, 47-60.

Stiratelli, R., Laird, N.M. and Ware, J.H. (1984). Random Effects Models for Serial Observatiosn with Binary Response. *Biometrics,* 40, 961-971.

Titterington, D.M., Makov, U.E. and Smith, A.F.M. (1985). *Statistical Analysis of Finite Mixture Distributions.* J. Wiley & sons, Chichester.

Wang, Y. (1996). A Quasi-Likelihood Approach for Ordered Categorical Data with Overdispersion. *Biometrics,* 52, 1252-1258.

Wedderburn, R. W.M. (1974). Quasilikelihood functions, generalized linear models and the Gauss-Newton method. *Biometrika,* 61, 439-47.

Wedderburn, R. (1976). On the Existence and Uniqueness of the Maximum Likelihood Estimates for Certain Generalized Linear Models. *Biometrika,* 63, 27-32.

Williams, D.A. (1982). Extra-Binomial Variation in Logistic Linear Models. *Applied Statistics,* 31, 144-148.

Williams, D.A. (1988). Extra-Binomial Variation in toxicology. In *Proceedings of the Fourteenth International Biometrics Conference,* Namur, Belgium: Biometric Society, 301-313.

Zeger, S.L. (1988). A Regression Model for Time Series of Counts. *Biometrika,* 75, 621-629.

Zeger, S.L. and Karim, M.R. (1991). Generalized Linear Models with Random Effects: A Gibbs Sampling Approach. *Journal of the American Statistical Association,* 86, 79-86.

6

Bayesian Generalized Linear Models for Inference About Small Areas

Balgobin Nandram

ABSTRACT Small area estimation is concerned with the estimation of parameters corresponding to small geographical areas or subpopulations when the underlying theme is to pool the data from other areas to estimate the parameters for a particular area. The purpose of this work is to present a review of the use of Bayesian generalized linear models in small area estimation. In particular, we review recent research that use hierarchical logistic and Poisson regression models in small area estimation or are potentially useful for small area estimation. Also, we discuss computational issues in small area estimation. We present an example of Poisson regression in which mortality rates are estimated for 798 U.S. health service areas (small areas) for all cancer for white males. We also discuss challenges that confront statisticians using small area methodology via generalized linear models. Finally, some remarks are made with respect to future research in this area.

1. Introduction

Small area estimation is concerned with the estimation of parameters corresponding to small geographical areas or subpopulations when the underlying theme is to pool the data from other areas to estimate the parameters for a particular area. Interest in small area estimation has grown tremendously in recent years, more so after the elegant review paper of Ghosh and Rao (1994). More sophisticated models are being constructed to take care of many sources of variation, and these models can include both discrete data and continuous data. As can be envisioned there is a fairly large literature on continuous data models while the literature on discrete data models is very scanty. The literature on generalized linear models is relatively large, but the literature on Bayesian generalized linear models for small area estimation is limited.

Sample survey designs are usually constructed to provide accuracy at a high level of aggregation. But interest is sometimes on very small areas which are not well represented by the survey data (i.e., the sample sizes of these small areas are inherently small). Most of these small areas tend to be areas for which the survey is not designed (or intended) to answer questions about. Thus, direct estimators such as design-based estimators (e.g., Cochran 1970) are not available or would have unacceptably large standard errors. Some countries (e.g., Canada) are beginning to design large-scale surveys to include small areas directly. But, as is expected, a complete enumeration of all small areas would be prohibitively expensive. Thus, statisticians must rely on methods of estimation not common in survey sampling.

Hierarchical Bayes (HB) and empirical Bayes (EB) approaches have been extensively used in recent years for small area estimation. These models are particularly suitable for a systematic connection of small areas. Therefore, the underlying theme in small area estimation is to "borrow strength from the ensemble" and this leads to improved precision. An attractive feature of this borrowing of strength is that shrinkage towards the grand mean is done adaptively because the estimates for areas with large sample sizes are shrunk less than those based on smaller sample sizes.

General theories, methods and applications of the EB and HB approaches in small area estimation are presented by Fay and Herriot (1979), Ghosh and Meeden (1986), Ghosh and Lahiri (1987, 1989), Battese, Harter and Fuller (1988), Prasad and Rao (1990), Datta and Ghosh (1991), Ghosh and Lahiri (1992), Nandram and Sedransk (1993 a,b,c), Stroud (1987, 1991), Malec and Sedransk (1985), Nandram (1994), Arora et al. (1997), Nandram (1999), and others. Some of these papers are not directly related to small area estimation, the objective there being Bayesian predictive inference to estimate an overall finite population quantity such as the finite population mean (e.g., Nandram and Sedransk 1993a and Malec and Sedransk 1985). However, the proposed methodology is useful for inference about small areas as well.

For disease mapping, Marshall (1991) followed an EB approach to provide a local shrinkage estimator in which the crude rate is shrunk towards a local, neighborhood, rate. He applied this estimator for the analysis of infant mortality rates in Auckland, New Zealand for the period of 1977-1985. It is worth noting that in disease mapping data exist for all areas, and this is different from survey sampling where not all geographical areas are sampled.

In passing we may note that small area estimation is not limited to survey sampling. Essentially, whenever the underlying theme is a borrowing of strength, one is performing small area estimation. See, for example, Hulting and Harville (1991) for a more general notion than survey sampling. They pointed out that the small areas under consideration need not be geographical regions, and they presented an example in which the areas are batches of raw material in an industrial application. Another example is estimation and prediction for many time series with most of these series being very short. For example, Nandram and Petruccelli (1997) showed that there are gains in precision for estimation and forecasting when similar series are pooled. For an approach using hierarchical Bayesian multivariate time series method directly applicable to small area estimation, see Ghosh et al. (1996).

Until recently much of the work in small area estimation was restricted to continuous variates. Survey data are often categorical and methods used for continuous data are then inappropriate. There is a quick introductory review of the hierarchical Bayesian generalized linear models in Gelman et al. (1995, ch. 14) who discussed binomial, Poisson, multinomial models, and overdispersed models. See also Zeger and Karim (1991) who used Gibbs sampling to incorporate random effects in generalized linear models.

Ghosh et al. (1998) provides a unified approach to the analysis of both continuous and categorical data through hierarchical Bayesian generalized linear models which include logistic regression and Poisson regression. However, their models do not include random regression coefficients. Recently, Malec et al. (1997) described a hierarchical Bayesian logistic model including both area-specific and element-specific auxiliary variables and Christiansen and Morris (1997) and Waller et al. (1997) described hierarchical models for Poisson regression including only area-specific auxiliary data. (It may be more challenging to include element-specific auxiliary data for Poisson regression because it is really each area's rate that is modeled in Poisson regression.)

The rest of the paper is organized as follows. In Section 2 we review hierarchical logistic regression models with emphasis on the work of Malec et al. (1997). In Section 3 we review hierarchical Poisson regression models with emphasis on the work of Christiansen and Morris (1997). In Section 4 we describe computational issues, highlighting a theorem of Ghosh et al. (1998). In Section 5 an example on the estimation of mortality rates for U.S. health service areas is presented. In Section 6 we describe challenges for Bayesian statisticians using generalized linear models in small area estimation. Section 7 has concluding remarks. Pertinent references are listed in Section 8.

2. Logistic Regression Models

Dempster and Tomberlin (1980) first proposed empirical and hierarchical Bayes methods using logistic regression to incorporate random effects model-based inference for small area binary data. The random effects model permits the data to determine a compromise between the classical unbiased estimates which depend on the data only in the specific local area, and the fixed effects estimates which pool information across areas. They applied this method to census undercount from a post enumeration survey. MacGibbon and Tomberlin (1989) showed how to use logistic regression models to estimate (empirical Bayes) proportions for small areas for a multi-stage survey.

Malec et al. (1993, 1997) describe how to use a predictive approach to estimate finite population proportions for small areas using models motivated by Wong and Mason (1985), Dempster and Tomberlin (1980) and MacGibbon and Tomberlin (1989). Malec et al. (1997) made inferences for finite population proportions such as the probability of at least one visit to a doctor within the past 12 months. They used data from the National Health Interview Survey (NHIS) for the 50 states and the District of Columbia for many subpopulations within these 51 areas. Three major difficulties are the very large size of the NHIS sample, the use of models with many parameters, and the need to make predictions for many small areas. In fact, they made inferences for 72 age/race/sex categories for each U.S. county, approximately 216,000 subpopulations.

Malec et al. (1997) assume that each individual in the population is assigned to one of K mutually exclusive and exhaustive classes which are based on the individual's socioeconomic/demographic status. Let Y_{ikj} denote a binary random variable for individual j in class k, cluster i where $i = 1, \ldots, L, k = 1, \ldots, B$, and $j = 1, \ldots, N_{ik}$, and let p_{ik} denote the probability that an individual has the characteristic in cluster i and class k. Note that these probabilities are not allowed to vary with the individuals although this could have been done if there are covariates specific to each individual. Within cluster i and class k, and conditional on p_{ik}, the Y_{ikj} are assumed to be independent Bernoulli random variables with $Pr(Y_{ikj} = 1 \mid p_{ik}) = p_{ik}$. A column vector of M covariates, $\mathbf{x}_k = (x_{k1}, \ldots, x_{kM})^t$, is assumed to be the same for each individual in class k and cluster i. Given $\mathbf{x}_k$ and a column vector of regression coefficients, $\beta_i = (\beta_{i1}, \ldots, \beta_{iM})'$, they assume that

$$\ln\{p_{ik}/(1-p_{ik})\} \equiv \text{logit}(p_{ik}) = \mathbf{x}_k'\beta_i. \tag{1}$$

Thus, (1) is a linear regression with "dependent variable" logit(p_{ik}) and independent variable $\mathbf{x}_k$ which does not depend on i. Then, they assume that, conditional on $\boldsymbol{\eta}$ and $\boldsymbol{\Gamma}$, the β_i are independently distributed with

$$\beta_i \sim N(\mathbf{G}_i\boldsymbol{\eta}, \boldsymbol{\Gamma}) \tag{2}$$

where each row of $\mathbf{G}_i$ is a subset of the cluster-level covariates $(Z_{i1}, \ldots, Z_{ic})$, not necessarily related to $\mathbf{x}_k$, η is a vector of regression coefficients, and $\boldsymbol{\Gamma}$ is an $M \times M$ positive definite matrix. The regression in (2) is especially important, because it permits correlation between individuals in a cluster, and provides the opportunity for increased precision.

Finally, reference prior distributions are assigned to $\boldsymbol{\eta}$ and $\boldsymbol{\Gamma}$ as well, such that

$$p(\boldsymbol{\eta}, \boldsymbol{\Gamma}) \propto \text{ constant.} \tag{3}$$

There is no additional computational complexity if one replaces the reference prior in (3) with normal and inverse Wishart distributions for $\boldsymbol{\eta}$ and $\boldsymbol{\Gamma}$. However, they have used reference prior distributions, because as they stated "published estimates are used by many secondary data analysts, and thus there is a need to minimize subjectivity." One caveat is that one might need to use proper (or proper diffuse) priors at least for $\boldsymbol{\Gamma}$ because it is not clear that the joint posterior distributions of the parameters are proper with reference prior distributions on $\boldsymbol{\eta}$ and $\boldsymbol{\Gamma}$.

In their application (1) is a piecewise linear model, linear in age; that is,

$$\text{logit}(p_{ik}) = \alpha + \beta_{i1}x_{0k} +$$

$$\beta_{i2}x_{15,k} + \beta_{i3}x_{25,k} + \beta_{i4}x_{55,k} + \beta_{i5}Y_k x_{15.k} + \beta_{i6}Y_k x_{25,k} + \beta_{i7}Z_k$$

where Y_k and Z_k are binary variables with $Y_k = 1$ if class k corresponds to males, $Z_k = 1$ if class k corresponds to whites, and $x_{ak} = \max(0, k-a)$ with age k denoting the midpoint of the ages of the individuals in class k [e.g., if class k corresponds to black females ages $40 - 45$, $X_{15,k} = \max(0, 42.5 - 15)$]. They discuss the relationships between logit(p_{ik}) and age (for the four race/sex combinations). Formula (2) is a second regression model with the vector β_i as the dependent variable, and $E(\beta_{il}) = g_{il1}\eta_{l1} + \cdots + g_{ilc_l}\eta_{lc_l}$, where $\{g_{il1}, \ldots, g_{ilc_l}\}$ is a subset of $\{Z_{i1}, \ldots, Z_{ic}\}$. The elements of $\mathbf{G}_i$ are county-level covariates, such as county per capita income, education level, and so on.

The objective is to make inference about a finite population proportion, P, for a specified small area and subpopulation. In general,

$$P = \sum_{i\in I}\sum_{k\in K}\sum_{j=1}^{N_{ik}} Y_{ikj} / \sum_{i\in I}\sum_{k\in K} N_{ik} \tag{4}$$

where I is the collection of clusters that define the small area, K is the collection of classes that define the subpopulation, and N_{ik} is the total number of individuals in cluster i, class k. If P is the proportion of male Iowans who have made at least one visit to a doctor, then I is the 99 counties in Iowa and K is the collection defined by the cross-classification of race, age (in 5-year groups), and males. Throughout, they assume that $\sum_{i\in I}\sum_{k\in K} N_{ik}$ is known, and for convenience they define

$$\Theta = \left(\sum_{i\in I}\sum_{k\in K} N_{ik}\right) P = \sum_{i\in I}\sum_{k\in K}\sum_{j=1}^{N_{ik}} Y_{ikj}. \tag{5}$$

For example, Θ is the total number of male Iowans who have made at least one visit to a doctor.

Let s_{ik} denote the set of sampled individuals in class k, cluster i that has size n_{ik}. Then

$$\Theta = \sum_{i\in I}\sum_{k\in K}\sum_{j\in s_{ik}} y_{ikj} + \sum_{i\in I}\sum_{k\in K}\sum_{j\notin s_{ik}} Y_{ikj}.$$

Let y_s denote the vector of sample observations. Because $E(Y_{ikj} \mid p_{ik}) = p_{ik}$, the posterior expected value of Θ is

$$E(\Theta \mid y_s) = \sum_{i\in I}\sum_{k\in K}\sum_{j\in s_{ik}} y_{ikj} + \sum_{i\in I}\sum_{k\in K}\sum_{j\notin s_{ik}} E(p_{ik} \mid y_s) =$$

$$\sum_{i\in I}\sum_{k\in K}\sum_{j\in s_{ik}} y_{ikj} + \sum_{i\in I}\sum_{k\in K}(N_{ik} - n_{ik})E(p_{ik} \mid y_s) \tag{6}$$

where

$$p_{ik} = \exp\{\mathbf{x}_k'\beta_i\}/\{1 + \exp(\mathbf{x}_k'\beta_i)\}.$$

Formula (6) defines the hierarchical Bayes point estimator of Θ. The empirical Bayes estimator is the special case of (6) obtained by using only (1) and (2) and replacing (η, Γ) with a point estimate, $(\hat{\eta}, \hat{\Gamma})$. The synthetic estimator is also a special case of (6) obtained by taking $\Gamma = 0$ in (2).

A measure of variability is obtained by using the posterior variance of Θ. In fact,

$$\mathrm{Var}(\Theta \mid y_s) = \sum_{i\in I}\sum_{k\in K}(N_{ik} - n_{ik})E\{p_{ik}(1 - p_{ik}) \mid y_s\} +$$

$$\mathrm{Var}\{\sum_{i\in I}\sum_{k\in K}(N_{ik} - n_{ik})p_{ik} \mid y_s\}. \tag{7}$$

Malec et al. (1997) investigate the quality of inferences about P in two studies. First, they used a Bayesian cross-validation deleting an individual or a county at a time. The second is a study that compares their estimates for small geographical areas or subpopulations to the true values of the parameters. Both studies permit them to validate their models and methods. (For details see Sec. 5.3 of their paper.)

Malec et al. (1997) compare hierarchical Bayes, empirical Bayes, synthetic and randomization based estimates. They found that the hierarchical Bayes estimates are very versatile. For some subpopulations, the hierarchical Bayes estimates are more variable than the synthetic estimates and less variable than the randomization estimates, and for other subpopulations, the hierarchical Bayes estimates are similar to the synthetic estimates. However, for large subpopulations, as expected, the hierarchical Bayes estimates are close to the randomization-based estimates, and the synthetic estimates are farther from the randomization-based estimates than are the hierarchical Bayes estimates.

3. Poisson Regression Models

Hierarchical Poisson models have been used for the analysis of different kinds of data. Much of the work on disease mapping starts with a Poisson sampling process. Clayton and Kaldor (1987) described empirical Bayes approaches that account for spatial similarities among neighboring rates. Bernardinelli and Montomoli (1992) compared empirical Bayes and hierarchical Bayes methods, the latter being implemented by Markov chain Monte Carlo (MCMC) methods. Breslow and Clayton (1993) used generalized linear mixed models to study the disease mapping problem, providing approximation schemes for inference; approximation of the marginal quasi-likelihood using Laplace's method leads eventually to estimating equations based on penalized quasi-likelihood for the mean parameters and pseudo-likelihood for the variances. Waller et al. (1997) presented spatio-temporal hierarchical Bayesian models to model regional disease rates over space and time including space-time interactions.

Christiansen and Morris (1997) proposed a hierarchical Poisson regression model not in connection with survey sampling. But this method can be used to analyze mortality rates for small areas when a two-level model is used. The elegance is that, unlike current approaches to Bayesian analysis, it does not use MCMC methods, and simple but approximate closed form expressions are obtained for credible intervals of small area effects.

Christiansen and Morris (1997) noted that their approach, Poisson regression interactive multilevel modeling (PRIMM), has several advantages over earlier methods. Unlike some of the other methods, PRIMM provides interval estimates for all of the parameters, it has better nominal operating characteristics than most commonly used alternatives, and it is much faster than BUGS which is a software normally used for hierarchical Bayesian analysis. This makes the procedure versatile because it can be used for many statistical applications.

Let z_i be the observed number of deaths for area i with exposure n_i and mortality rate $\lambda_i, i = 1, \ldots, k$. Then they assume

$$z_i \mid \lambda_i \overset{ind}{\sim} \text{Poisson}(n_i \lambda_i) \tag{8}$$

$$\lambda_i \mid \tau, \boldsymbol{\beta} \overset{iid}{\sim} \text{Gamma}(e^{\tau}, e^{\tau - \mathbf{x}'\boldsymbol{\beta}}) \tag{9}$$

where $\mathbf{x}_i = (x_{i0}, x_{i1}, \ldots, x_{i,r-1})'$ and $\boldsymbol{\beta} = (\beta_0, \beta_1, \ldots, \beta_{r-1})'$. In (9), the random variable $X \sim \text{Gamma}(a, b)$ if $f(x) = b^a x^{a-1} \exp(-bx)/\Gamma(a), x > 0$. (A transformation made later in their paper leads to (9).) Essentially, this is a negative binomial regression model, used to address the issue of overdispersion in Poisson models (e.g., see Lawless 1987).

Observe that (8) and (9) are conjugate leading to exact posterior inference about λ_i conditional on $\boldsymbol{\beta}$ and τ because, letting $\mathbf{z}$ be the vector of the z_i,

$$\lambda_i \mid \boldsymbol{\beta}, \tau, \mathbf{z} \sim \text{Gamma}(z_i + e^{\tau}, n_i + e^{\tau - \mathbf{x}_i'\boldsymbol{\beta}}).$$

For the hyperparameters $\boldsymbol{\beta}$ and τ, they use the prior (after the transformation)

$$\pi(\boldsymbol{\beta}, \tau) \propto \frac{z_0 e^{\tau}}{(z_0 + e^{\tau})^2} \quad (\tau, \boldsymbol{\beta})' \epsilon R^{r+1}$$

where one can take $z_0 = n_0 m_0$ with $n_0 = \min_i n_i$ and $m_0 = \sum_{i=1}^{k} z_i / \sum_{i=1}^{k} n_i$. The prior on τ is proper and is chosen to ensure that the maximum likelihood estimate of τ

is finite. This also ensures a proper posterior distribution for τ. Nevertheless, the prior for $\boldsymbol{\beta}$ is improper. However, the joint posterior density for $\boldsymbol{\beta}$ and τ can be shown to be proper provided that the number of cases (c_o) for which $z_i > 0$ is at least r and that the $c_o \times r$ submatrix of $\mathbf{X} = (\mathbf{x}_1, \ldots, \mathbf{x}_k)'$ for these cases is full rank (see Christiansen and Morris 1997).

We describe briefly how Christiansen and Morris (1997) obtained their main results (see their paper for details). There are three steps.

First, after integrating out the λ_i from (8) and (9), the likelihood function is

$$L(\boldsymbol{\beta}, \tau) = \prod_{i=1}^{k} \frac{\Gamma(e^\tau + z_i)}{\Gamma(e^\tau) z_i!} (1 - B_i)^{z_i} B_i^{e^\tau} \tag{10}$$

where, letting $\mu_i = e^{\mathbf{x}_i'\boldsymbol{\beta}}$, $B_i = e^\tau/(e^\tau + n_i\mu_i)$ are the shrinkage factors. Thus, conditional on τ, the likelihood function in $\boldsymbol{\beta}$ is concave and there is a unique mode which can be obtained easily. Let $\hat{\boldsymbol{\beta}}_\tau$ and $\hat{H}_\tau$ denote the modal estimate and the negative Hessian matrix of the loglikelihood function $\mathcal{L}(\boldsymbol{\beta}, \tau) = \log(L(\boldsymbol{\beta}, \tau))$.

Second, a restricted maximum likelihood (REML) type correction is made which takes into account the presence of the nuisance factor $\boldsymbol{\beta}$ in computing a modal estimate of τ. The approximate marginal density,

$$p_2(\tau) \propto |\hat{H}_\tau|^{-1/2} L(\hat{\boldsymbol{\beta}}_\tau, \tau \mid \mathbf{z}) e^\tau z_o/(e^\tau + z_0)^2, \tag{11}$$

is maximized with respect to τ.

Third, they made a REML type adjustment to the log density, with respect to τ in (11), of $(\boldsymbol{\beta}, \tau)$, denoted by $l_R(\boldsymbol{\beta}, \tau)$ where

$$l_R(\boldsymbol{\beta}, \tau) = \mathcal{L}(\boldsymbol{\beta}, \tau) + (1 - \frac{r}{2})\tau - 2\log(\exp(\tau) + z_o) + \frac{r}{2k}\sum_{i=1}^{k} \log(\exp(\tau) + e_i m_o). \tag{12}$$

Then, the adjusted loglikelihood function in (12) is maximized to obtain the modal estimates $\hat{\boldsymbol{\beta}}$ and $\hat{\tau}$.

Christiansen and Morris (1997) argued that asymptotically in k

$$\begin{pmatrix} \boldsymbol{\beta} \\ \tau \end{pmatrix} \mid \mathbf{z} \sim N_{r+1}\left\{ \begin{pmatrix} \hat{\boldsymbol{\beta}} \\ \hat{\tau} \end{pmatrix}, \Sigma \right\} \tag{13}$$

with

$$\Sigma = \hat{\sigma}_\tau^2 \begin{bmatrix} \hat{\sigma}_\tau^{-2} \hat{H}_{\hat{\tau}}^{-1} - \hat{\boldsymbol{\nu}}\hat{\boldsymbol{\nu}}' & \hat{\boldsymbol{\nu}} \\ \hat{\boldsymbol{\nu}}' & 1 \end{bmatrix}$$

where $\hat{\boldsymbol{\nu}}$ is obtained by differentiating $\mathcal{L}(\boldsymbol{\beta}, \tau)$ with respect to τ and solving the resulting equation for $\partial\hat{\boldsymbol{\beta}}/\partial\tau$ setting $\partial\hat{\boldsymbol{\beta}}/\partial\tau \equiv \hat{\boldsymbol{\nu}}$ and $\hat{\sigma}_\tau^2 \equiv \text{var}(\tau \mid \mathbf{z}) \approx -(\partial^2 l_R(\boldsymbol{\beta}, \tau)/\partial\tau^2 + \hat{\boldsymbol{\nu}}'\hat{H}_{\hat{\tau}}^{-1}\hat{\boldsymbol{\nu}})^{-1}$, all quantities in (13) being evaluated at $(\hat{\boldsymbol{\beta}}, \hat{\tau})$.

Recalling $\mu_i = e^{\mathbf{X}_i'\boldsymbol{\beta}}$, it follows immediately that

$$E(\mu_i \mid \mathbf{z}) = \hat{\mu}_i = e^{\mathbf{x}_i'\hat{\boldsymbol{\beta}} + \mathbf{x}_i'\Sigma_{11}\mathbf{x}_i/2}$$

and

$$\text{Var}(\mu_i \mid \mathbf{z}) = \hat{\sigma}_{\mu_i}^2 = \hat{\mu}_i^2\{\exp(\mathbf{x}_i'\Sigma_{11}\mathbf{x}_i) - 1\}$$

where $\Sigma_{11} = \hat{\sigma}_\tau^2 \left(\hat{\sigma}_\tau^{-2}\hat{H}_{\hat{\tau}}^{-1} - \hat{\boldsymbol{\nu}}\hat{\boldsymbol{\nu}}'\right)$. With further approximations on integrals based on the adjusted density method (ADM), they establish their main theorem which we describe next.

First, we need some notation. Let $w_i^2 \equiv \text{var}(\tau - x_i'\beta)$,

$$E_0 B_i = \hat{B}_i = e^{\hat{\tau}}/(e^{\hat{\tau}} + n_i e^{\mathbf{x}_i'\hat{\beta}}) = a_{i1}/(a_{i1} + a_{i2}).$$

$$a_{i1} = w_i^{-2}/(1 - \hat{B}_i), \quad a_{i2} = w_i^{-2}/\hat{B}_i,$$

$b_i = \text{cov}(\mathbf{x}_i'\beta, \tau - \mathbf{x}_i'\boldsymbol{\beta})/w_i^2$, $\hat{\mu}_i$, $E\mu_i^2 = \hat{\sigma}_{\mu_i}^2 + \hat{\mu}_i^2$, $E_1(B_i) = (a_{i1} + b_i)/(a_{i1} + a_{i2})$, and for $s = 0, 1, 2$,

$$E_s(B_i^2) = (a_{i1} + sb_i)(a_{i1} + sb_i + 1)/((a_{i1} + a_{12})(a_{i1} + a_{i2} + 1)).$$

Then assume the distribution (13) for β and τ, given the data. For $i = 1, \ldots, k$, the ADM approximations to the first two moments of λ_i, given the data, are

$$\hat{\lambda}_i \equiv E(\lambda_i \mid \mathbf{z}) = (1 - \hat{B}_i)z_i + \hat{\mu}_i E_1(B_i) \tag{14}$$

and

$$\hat{\sigma}_{\lambda_1}^2 \equiv \text{var}(\lambda_i \mid \mathbf{z}) = \frac{1}{n_i} z_i E_0(1 - B_i)^2 + \frac{1}{n_i}\hat{\mu}_i E_1 B_i(1 - B_i) + (E\mu_i^2)(E_2 B_i^2)$$

$$-2z_i\hat{\mu}_i(E_1 B_i^2) + z_i^2 E_0 B_i^2 - (z_i\hat{B}_i - \hat{\mu}_i E_1 B_i)^2. \tag{15}$$

It follows from the theorem that given the data, approximately

$$\frac{2\lambda_i\hat{\lambda}_i}{\hat{\sigma}_{\lambda_i}^2} \sim \chi_{2\hat{\nu}}$$

where $\hat{\nu} = \hat{\lambda}_i^2/\hat{\sigma}_{\lambda_i}^2$. Thus, credible intervals for λ_i are obtained.

Christiansen and Morris (1997) study the operating characteristics of the credible interval. They compare their method with six alternative methods. One of these uses a MCMC method through the BUGS software. The PRIMM method obtains noncoverage probabilities closest to the nominal value of .05. In addition, PRIMM has relatively fast computing time, and the computer program (S-PLUS environment) is publicly available to practitioners through Statlib. Your data set is simply read into a standard S-PLUS program which anyone can do quickly without any prior knowledge about S-PLUS.

PRIMM is potentially useful for small area statisticians interested in overdispersed Poisson models with covariates because answers can be obtained very quickly. The simplicity of the model permits quick data analysis within small regions (e.g., strata) which contain small areas. While PRIMM assumes a very simple model, one may be able to extend it to more complex small area models, a job not to be underestimated.

4. Computational Issues

It is inherently difficult to compute quantities of interest in nonlinear parametric problems, thus to simplify the computations approximations are usually used. This is true even within the framework of MCMC methods.

It is worth noting that Albert (1998) discussed computational methods for a Bayesian hierarchical generalized linear model. A Bayesian two-stage prior distribution is used, and the posterior distributions of the two hyper-parameters are intractable. The focus of his article is on tractable accurate approximations to these posterior distributions. In particular, he discussed the Laplace method, a quasi likelihood method, and the Brooks method, and found that the Laplace method performs best for the binomial-logit hierarchical model.

However, the current approach is to use a sampling based method. Usually the conditional posterior distributions do not exist in closed forms, making the Gibbs sampler (Gelfand and Smith 1990) difficult to use. Then one needs to use a version of the Metropolis-Hastings algorithm (Chib and Greenberg 1995 and Tierney 1994). However, one needs to note that if the nonstandard posterior conditional distributions are log concave, the Gibbs sampler can be used with the Gilks-Wild algorithm (Gilks and Wild 1992).

With many parameters in the model it becomes necessary to accelerate the MCMC. For example, see Nandram and Chen (1996) for a method to accelerate the Gibbs sampler for the probit model when latent variables (Albert and Chib 1993) are used.

It is always important to demonstrate that the joint posterior distributions of the parameters are proper for any model. Ghosh et al. (1998) demonstrate how to do this for the generalized linear model applied to small area estimation.

They started with m strata or local areas. Let Y_{ik} denote the minimal sufficient statistic (discrete or continuous) corresponding to the kth unit within the ith stratum ($k = 1, \ldots, n_i; i = 1, \ldots, m$). The Y_{ik} are assumed to be conditionally independent with pdf

$$f(y_{ik} \mid \theta_{ik}, \phi_{ik}) = \exp[\phi_{ik}^{-1}(y_{ik}\theta_{ik} - \psi(\theta_{ik})) + \rho(y_{ik}; \phi_{ik})] \tag{16}$$

where $k = 1, \ldots, n_i, \quad i = 1, \ldots, m$. The density (16) is parameterized with respect to the canonical parameters θ_{ik} and the scale parameters $\phi_{ik}(> 0)$. It is assumed that the scale parameters ϕ_{ik} are known.

The natural parameters θ_{ik} are first modeled as

$$h(\theta_{ik}) = \mathbf{x}_{ik}'\boldsymbol{\beta} + u_i + \epsilon_{ik} \quad (k = 1, \ldots, n_i; i = 1, \ldots, m) \tag{17}$$

where h is a strictly increasing function; the $\mathbf{x}_{ik}(p \times 1)$ are known design vectors, $\boldsymbol{\beta}(p \times 1)$ is the unknown regression coefficient, the u_i are the random effects, and the ϵ_{ik} are the errors. It is assumed that the u_i and the ϵ_{ik} are mutually independent with $u_i \stackrel{iid}{\sim} N(0, \sigma_u^2)$ and $\epsilon_{ik} \stackrel{iid}{\sim} N(0, \sigma^2)$.

It appears that (16) and (17) do not form a hierarchical Bayesian model. But it is now standard to represent such a model in a hierarchical framework as was done by Ghosh et al. (1998). Let $R_u = \sigma_u^{-2}$ and $R = \sigma^{-2}$. Also, let $\boldsymbol{\Theta} = (\theta_{11}, \ldots, \theta_{1n_1}, \ldots, \theta_{m1}, \ldots, \theta_{mn_m})'$ and $\mathbf{u} = (u_1, \ldots, u_m)'$. Then the hierarchical model is given by

(I) conditional on $\boldsymbol{\Theta}, \boldsymbol{\beta}, \mathbf{u}, \mathbf{R}_u = r_u$ and $R = r$, the Y_{ik} are independent with densities given in (16);

(II) conditional on $\boldsymbol{\beta}, \mathbf{u}, R_u = r_u$ and $R = r, h(\theta_{ik}) \stackrel{iid}{\sim} N(\mathbf{x}_{ik}'\boldsymbol{\beta} + u_i, r^{-1})$;

(III) conditional on $\boldsymbol{\beta}, R_u = r_u$ and $R = r, u_i \stackrel{iid}{\sim} N(0, r_u^{-1})$.

To complete the hierarchical model, Ghosh et al. (1998) assign the following prior to $\boldsymbol{\beta}, R_u = r_u$ and $R = r$:

(IV) $\boldsymbol{\beta}, R_u = r_u$ and $R = r$ are mutually independent with $\boldsymbol{\beta} \sim$ uniform $(\mathbf{R}^p)$, $(p < m)$, $R_u \sim \text{Gamma}\left(\frac{1}{2}a, \frac{1}{2}b\right)$ and $R \sim \text{Gamma}\left(\frac{1}{2}c, \frac{1}{2}d\right)$.

In (IV) a random variable $Z \sim \text{Gamma}(\alpha, \beta)$ if Z has pdf

$$f(z) = \beta^\alpha \exp(-\beta z) z^{\alpha-1}/\Gamma(\alpha), \quad z > 0.$$

The model in (I)–(IV) is very similar the one considered by MacGibbon and Tomberlin (1989) and Breslow and Clayton (1993). As pointed out by Ghosh et al. (1998), this model is not strictly contained in the one considered by Zeger and Karim (1991). This is true because Zeger and Karim (1991) consider $h(\theta_{ik}) = \mathbf{x}'_{ik}\boldsymbol{\beta} + u_i$, where $h(\cdot)$ is a strictly increasing function, but this formulation does not include possible error in misspecifying this model. In fact, the uncertainty in specifying the model in (I)–(IV) consists of two components: (i) the effect of the local area and (ii) the error component, permitting the possibility to account for overdispersion by introducing an extra variance component.

Interest is on finding the joint posterior distribution of the $g(\theta_{ik})$, given the data $\mathbf{y} = (y_{11}, \ldots, y_{1n_1}, \ldots, y_{m1}, \ldots, y_{mn_m})'$, where g is a strictly increasing function, and in particular in finding the posterior means, variances and covariances of these parameters. In typical applications, $g(\theta_{ik}) = \psi'(\theta_{ik}) = E(Y_{ik} \mid \theta_{ik})$.

First, however, one needs to ensure that the joint posterior distribution of the Θ_{ik} given $\mathbf{y}$ is proper. The most important result in the paper is the theorem that establishes this result. Let the support of θ_{ik} be the open interval $(\underline{\Theta}_{ik}, \bar{\Theta}_{ik})$, where the lower endpoint of the interval can be $-\infty$, the upper endpoint can be $+\infty$, or both.

The theorem follows: assume $a > 0, c > 0, \sum_i n_i - p + d > 0$, and $m + b > 0$. Then, if

$$\int_{\underline{\Theta}_{ik}}^{\bar{\Theta}_{ik}} \exp\{[\theta y_{ik} - \phi(\Theta)]/\phi_{ik}\} h'(\theta) d\theta < \infty \tag{18}$$

for all y_{ik} and ϕ_{ik} (> 0), the joint posterior pdf of the θ_{ik} given $\mathbf{y}$ is proper.

The theorem covers the two important special cases of logistic regression and Poisson regression. For the logistic case,

$$Y_{ik} \mid \theta_{ik} \sim \text{Binomial}(n_{ik}, \exp(\theta_{ik})/\{1 + \exp(\theta_{ik})\})$$

and h is the identity function (i.e., the link is canonical). Also, let $g(\theta_{ik}) = \psi'(\theta_{ik})/n_{ik} = \exp(\theta_{ik})/[1 + \exp(\theta_{ik})]$. Then writing $p_{ik} = \exp(\theta_{ik})/[1 + \exp(\theta_{ik})]$, (18) reduces to $\int_0^1 p_{ik}^{y_{ik}-1}(1 - p_{ik})^{n - y_{ik} - 1} dp_{ik} < \infty$ which requires $1 \le y_{ik} \le (n_{ik} - 1)$, i.e., excludes cases of all failures or all successes. For the Poisson case,

$$Y_{ik} \mid \theta_{ik} \sim \text{Poisson}(\exp(\theta_{ik})).$$

Then, if h is the canonical link, and $g(\theta_{ik}) = \psi'(\theta_{ik}) = \exp(\theta_{ik})$, the condition (18) reduces to $\int_0^\infty \zeta_{ik}^{y_{ik}-1} \exp(-\zeta_{ik}) d\zeta_{ik} < \infty$ which holds for $y_{ik} = 1, 2, \ldots$.

As is apparent, direct evaluation of the joint posterior distribution of the $g(\theta_{ik})$ given $\mathbf{y}$ involves high-dimensional numerical integration, and is not computationally feasible except, of course, by using MCMC methods (e.g., Gibbs sampler, Gelfand and Smith 1990). The implementation of the Gibbs sampler requires generating samples from the conditional posterior distributions. Let

$$\mathbf{h}(\boldsymbol{\Theta}) = (h(\theta_{11}), \ldots, h(\theta_{1n_1}), \ldots, h(\theta_{m1}), \ldots, h(\theta_{m}n_m))',$$

$$\mathbf{X} = (\mathbf{x}_{11}, \ldots, \mathbf{x}_{1n_1}, \ldots, \mathbf{x}_{m1}, \ldots, \mathbf{x}_{mn_m})'$$

and $\mathbf{X}'\mathbf{X}$ be nonsingular. Then the necessary conditional posterior distributions based on the hierarchical Bayesian model given in (I)–(IV) are:

(i) $\boldsymbol{\beta} \mid \boldsymbol{\Theta}, \mathbf{u}, r_u, r, \mathbf{y} \sim N((\mathbf{X}'\mathbf{X})^{-1}(\mathbf{X}'\mathbf{h}(\boldsymbol{\Theta}) - \sum_i u_i \sum_k \mathbf{x}_{ik}), r^{-1}(\mathbf{X}'\mathbf{X})^{-1})$;

(ii) $u_i \mid \boldsymbol{\Theta}, \boldsymbol{\beta}, r_u, r, \mathbf{y} \overset{iid}{\sim} N((rn_i + r_u)^{-1} \sum_k (h(\theta_{ik}) - \mathbf{x}'_{ik}\boldsymbol{\beta}), (rn_i + r_u)^{-1})$;

(iii) $R \mid \boldsymbol{\Theta}, \boldsymbol{\beta}, \mathbf{u}, r_u, \mathbf{y} \sim \text{Gamma}\left(\frac{1}{2}(c + \sum_i \sum_k (h(\theta_{ik}) - \mathbf{x}'_{ik}\boldsymbol{\beta} - u_i)^2), \frac{1}{2}(d + \sum_1^m n_i)\right)$;

(iv) $R_u \mid \boldsymbol{\Theta}, \boldsymbol{\beta}, \mathbf{u}, r, \mathbf{y} \sim \text{Gamma}\left(\frac{1}{2}(a + \sum_i u_i^2), \frac{1}{2}(b + \sum_1^m n_i)\right)$;

(v) $\theta_{ik} \mid \boldsymbol{\beta}, \mathbf{u}, r_u, r, \mathbf{y} \overset{iid}{\sim} \pi(\theta_{ik} \mid \boldsymbol{\beta}, \mathbf{u}, r_u, r, \mathbf{y}) \propto$
$\exp\left[(y_{ik}\theta_{ik} - \psi(\theta_{ik}))\phi_{ik}^{-1} - \frac{r}{2}(h(\theta_{ik}) - \mathbf{x}'_{ik}\boldsymbol{\beta} - u_i)^2\right] h'(\theta_{ik})$.

Samples can be generated easily from the normal and gamma distributions given in (i)–(iv). However, the conditional posterior distribution in (v), θ_{ik} given $\boldsymbol{\beta}, \mathbf{u}, r_u, r$ and $\mathbf{y}$, is known only up to a multiplicative constant, making it difficult to draw samples from this conditional posterior distribution. In the special case where $h(z) = z$ for all z, Ghosh et al. (1998) noted that it is straightforward to show that $\log \pi(\theta_{ik} \mid \boldsymbol{\beta}, \mathbf{u}, r, r_u, \mathbf{y})$ is a concave function of Θ_{ik}. In such cases, one can use the adaptive rejection sampling (ARS) scheme of Gilks and Wild (1992).

It is worth noting here, however, that although the conditional posterior distribution is theoretically log concave, it is possible for the ARS to fail (see Gilks and Wild 1992). Instead, one can use a Metropolis step to obtain a Metropolis-Hastings algorithm (Chib and Greenberg 1995). The main issue then is how to construct efficient proposal densities; see Nandram et al. (1998) who obtained proposal densities for many generalized linear models and Nandram (1998) who obtained proposal densities for the three-stage hierarchical multinomial-Dirichlet model.

Inference about $\boldsymbol{\Theta}$, based on (i)–(v), can now be obtained in a straightforward manner by performing an output analysis from the Gibbs sampler. That is, $E(\theta_{ik} \mid \mathbf{y})$, $V(\theta_{ik} \mid \mathbf{y})$ and $\text{Cov}(\theta_{ik}, \theta_{i'k'} \mid \mathbf{y})$ $(i,k) \neq (i',k')$ can be easily obtained from formulas for iterated conditional expectations and variances. (These are Rao-Blackwellized estimates as described by Gelfand and Smith 1990.)

Ghosh et al. (1998) extend the model in (I)–(IV) in two important directions. The first extension covers the analysis of multi-category data. Again they considered m strata, and within each stratum, they assumed that several units are selected and the responses of individuals within each selected unit are independent, and can be classified into J categories. The second extension considers a spatial hierarchical Bayesian generalized linear model. In this case the u_i in (17) represent variables which display spatial structure. In particular, they model the u_i so that a pair of contiguous zones would have stronger positive correlation than non-contiguous ones. They use the pairwise difference prior (e.g., Besag et al. 1995) on the u_i. Finally, they proved a theorem about the propriety of the θ_{ik} in this case.

Also Ghosh and Natarajan (1998) have generalized the theorem by relaxing the conditions. Consider $k = 1$ and drop this subscript. They assume that $f(y_i \mid \theta_i)$ is bounded for all i and $h'(\theta) = 1$. Let I_i denote (18), $S = \{i : I_i < \infty\}$, and s be the cardinality of S. Then $s + b > p$ and $a > 0$ are the conditions needed for the propriety of the posterior distribution.

Models with random regression coefficients are not covered by the theory of Ghosh et al. (1998), and the computations are more difficult. Other techniques must be considered. Gelfand et al. (1995) describe centering, an important technique which provides some useful tricks for computations. In addition, many of the conditional posterior distributions do not exist in closed forms, and they may not be log concave. Then, the Metropolis-Hastings algorithm is the obvious choice to perform computations, creating a possible need to construct a reasonably accurate proposal density for the conditional posterior distributions. For good examples of how these approximations can be obtained see the details in Christiansen and Morris (1997), Nandram et al. (1998).

5. Models for the U.S. Mortality Data

Pickle et al. (1996, 1997) discussed mixed effects models for estimating mortality rates for the eighteen leading causes of death in the U.S. Nandram et al. (1999) discussed alternative Bayesian models and methods for producing age specific and age adjusted mortality rates for "all cancer" for white males using data from 1988-92.

Nandram et al. (1999) use the same geographical units, health service areas (HSAs), as used in the Atlas (Pickle et al. 1996). The U.S. is divided up into 805 HSAs, and excluding those in Alaska and Hawaii, there are 798 HSAs. Each HSA is a group of counties, and the numbers of HSAs per state varies quite a bit ranging from 1 to 58 with median 16. There are 1 to 20 counties within a HSA with a median of about 2. With the exception of NYC, each HSA is at least 250 square miles in size. The states are too heterogeneous and counties are too small for any meaningful analysis. For the statistical analysis, there are twelve 'regions'; three of the nine Census Divisions were split to achieve greater homogeneity (see discussion on pg. 5 and Appendix 1 of Pickle et al. 1996). Numbers of deaths by age, race, sex, place of residence, and cause of death are based on original death certificates reported to the National Center for Health Statistics (NCHS) from which the mortality data are obtained.

Let d_{ij} and n_{ij} denote, respectively, the number of deaths and exposure for age class j in HSA i ($i = 1, \ldots, 798; j = 1, \ldots, 10$). The age classes are 0-4, 5-14, 15-24,...,75-84, 85 and up, coded as .25, 1,..., 9. Assume for fixed λ_{ij} that

$$d_{ij}|n_{ij}, \lambda_{ij} \sim \text{Poisson}(n_{ij}\lambda_{ij}).$$

Inference is desired for the age specific mortality rate, λ_{ij}, and the age adjusted rate $R_i = \sum_{j=1}^{10} a_j \lambda_{ij}$ where the a_j are the proportions of people in a standard million U.S. population. (It is surprising that for the Atlas the standard million is based on the 1940 U.S. population.)

The basis for the analysis in the Atlas is the first order Taylor series approximation of $\log r_{ij}$ where $r_{ij} = d_{ij}/n_{ij}$, the observed age specific mortality rate; i.e.,

$$\log\ r_{ij} \sim N\big(\log\ \lambda_{ij}, (n_{ij}\lambda_{ij})^{-1}\big) \tag{19}$$

and

$$\log\ \lambda_{ij} = \mathbf{x}_j' \boldsymbol{\beta}_i \tag{20}$$

where

$$\mathbf{x}_j' = \Big(1, \text{decade } j, (\text{decade } j)^2, (\text{decade } j)^3, max\ \big\{0, (\text{decade } j - \text{knot})^3\big\}\Big)$$

with decade $1 = .25$, decade $j = j - 1$ for $j = 2, \ldots, 10$ and the value of the knot is 6.

Let $r_{ij}^* = r_{ij}$ if $r_{ij} > 0$ and 10^{-6} if $r_{ij} = 0$, and define $y_{ij} = \log(r_{ij}^*)$. Denote by $r_{[k]j}$ the observed mortality rate for region $k, k = 1, \ldots, 12$, and let $w_{ij} = d_{ij}$ if $d_{ij} \geq 3$ and $n_{ij} r_{[k]j}$ if $d_{ij} < 3$.

Nandram et al. (1999) fitted several models. One of them is a Bayesian version of the mixed effects model used for constructing the Atlas. They have investigated three alternative versions of (19) and (20). First, $\log(r_{ij})$ is replaced by y_{ij} and the variance term, $(n_{ij}\lambda_{ij})^{-1}$ is replaced by $\phi_{[k]}/w_{ij}$. Observing that d_{ij} is a sample based estimator of $n_{ij}\lambda_{ij}$, Pickle et al. (1997) showed that replacing d_{ij} with w_{ij} provides better estimates of the λ_{ij}. The parameter $\phi_{[k]}$ is added to try to capture (regional) dispersion that is different from the Poisson distribution assumed in deriving (19) (see Efron 1986). Then, the model closest to the one in the Atlas is

$$y_{ij} \sim N(\mathbf{x}_j' \gamma_i, \phi_{[k]}/w_{ij}),$$

$$\boldsymbol{\gamma}_i = \boldsymbol{\beta} + \begin{pmatrix} \mathbf{b}_{i1} \\ \cdots \\ \mathbf{b}_{i2} \end{pmatrix} \tag{21}$$

with independence over i and j where $\mathbf{b}_{i1}$ and $\mathbf{b}_{i2}$ are independent with $\mathbf{b}_{il} \overset{iid}{\sim} N(0, D_l), D_l = \text{diag}(\delta_l^2)$. There are locally uniform prior distributions for $\boldsymbol{\beta}$ and the $\{\phi_{[k]}\}$, and proper, *diffuse* (i.e., proper with large variance) prior distributions for the components of D_l. The model used for the Atlas is, essentially, the model in (21) with a *separate analysis* in each of the 12 regions.

Nandram et al. (1999) constructed several other models without the approximation on the Poisson sampling process. Each model uses the same sampling distribution

$$d_{ij}|n_{ij}, \lambda_{ij} \overset{ind}{\sim} \textit{Poisson } (n_{ij}\lambda_{ij}). \tag{22}$$

We describe three of these models. The first model is

$$\log \lambda_{ij} = \mathbf{x}_j' \boldsymbol{\beta} + \nu_i, \tag{23}$$

$$\nu_i \mid \sigma^2 \overset{iid}{\sim} N(0, \sigma^2), i = 2, \ldots, 798.$$

There is a locally uniform prior distribution on $\boldsymbol{\beta}$, and a proper diffuse prior on σ^2.

The second model is

$$\log \lambda_{ij} = \mathbf{x}_j' \boldsymbol{\beta}_i \tag{24}$$

$$\beta_i | \theta, \Delta \overset{iid}{\sim} N(\theta, \Delta)$$

where the prior distribution on θ is locally uniform, and the prior on Δ is proper but diffuse. Note that Δ is not diagonal. The third model is the one just described with (24) replaced by

$$\log \lambda_{ij} = \mathbf{x}_j' \beta_i + \delta_j \tag{25}$$

where

$$\delta_j \overset{iid}{\sim} N(0, \sigma^2)$$

with a proper, diffuse prior on σ^2.

Nandram et al. (1999) analyzed the models in (21), (23), (24) and (25) fit separately within regions, and to the entire population of HSAs. The model in (21) was fitted using the simple Gibbs sampler and the models in (23), (24), and (25) were fitted using the Metropolis-Hastings sampler. For computational details such as construction of proposal densities and centering see Nandram et al. (1998).

Nandram et al. (1999) used three different measures to assess the fit of the models. The first is the posterior expected predicted deviance (EPD),

$$E\{P(\mathbf{d}^{obs}, \mathbf{d}^{new})|\mathbf{d}^{obs}\} \tag{26}$$

where $\mathbf{d}^{new}$ is a random vector with distribution

$$f(\mathbf{d}^{new}|\mathbf{d}^{obs}) = \int g(\mathbf{d}^{new}|\lambda)h(\lambda|\mathbf{d}^{obs})d\lambda.$$

This is a measure of discrepancy between $\mathbf{d}^{obs}$, the *observed* vector of the d_{ij}, and $\mathbf{d}^{new}$, a set of "new" observations selected from the posterior predictive distribution of $\mathbf{d}^{new}$ in (22). If the model and data are concordant (26) should be small. One choice of $P(\cdot,\cdot)$, based on the Chi-squared statistic, is

$$P(\mathbf{d}^{obs}, \mathbf{d}^{new}) = \sum_i \sum_j (d_{ij}^{obs} - d_{ij}^{new})^2/(d_{ij}^{new} + 0.5).$$

Nandram et al. (1999) give two other choices; see also Waller et al. (1997).

The second measure that they used to assess the fit of the models is the posterior predicted p-value; i.e.,

$$Pr\{T(\mathbf{d}^{new}, \lambda) \geq T(\mathbf{d}^{obs}, \lambda)|\mathbf{d}^{obs}\}.$$

One choice of checking function $T(\mathbf{d}^{new}, \lambda)$, analogous to the Chi-squared discrepancy measure, is

$$\sum_{i,j} (d_{ij} - n_{ij}\lambda_{ij})^2/n_{ij}\lambda_{ij}.$$

Again they have considered two more checking functions. While the EPD is used to rank the models, the p–value is primarily used for goodness of fit.

The third measure of evaluating the alternative models is to use standardized residuals. Let $d_{(ij)}$ denote the set of all d's *except* the $(ij)^{th}$ component itself. Then define the standardized residual as

$$DRES_{ij} = \frac{r_{ij} - E(r_{ij}|\mathbf{d}_{(ij)})}{SD(r_{ij}|\mathbf{d}_{(ij)})}.$$

That is, the $(ij)^{th}$ observed r_{ij} is "held out" and compared with its point estimator, $E(r_{(ij)}|d_{(ij)})$, which is evaluated *without* using the observed d_{ij}.

Of the four models that they considered (i.e., those given by (21), (23), (24) and (25), fit to all 798 HSAs), the values of the posterior predicted p-value were acceptable only for the model (25). Using the expected predicted deviances as the criteria, model (25) was best, followed by (24), (23) and (21). They also fit model (25) separately in each region. The EPD values for this latter case were almost the same as those when (25) was fit to all 798 HSAs; however, the p-values provide greater support for (25) fit to all 798 HSAs.

6. Challenges in Small Area Estimation

First, the problem of model diagnostic is an important one. The most promising diagnostic procedures are based on cross-validation (Gelfand et al. 1992 and Gelfand

and Dey 1994) and expected predictive deviances (Ibrahim and Laud 1995). One would need to assess distribution of residuals and tail area probabilities to find threshold values not based on normality. The problem is exaggerated in small area estimation because of the obvious difficulty with small sample sizes and the complexity of the likelihood function for generalized linear models.

Second, overshrinkage is a serious issue in small area estimation. Louis (1984) and Ghosh (1992) used constrained Bayes estimates to avoid overshrinkage. Shen and Louis (1998) describe triple-goal estimates in two-stage hierarchical models in which the estimation method is linked to an inferential goal via a loss function. Triple-goal estimates are necessary because it may be desirable to have a set of estimates that produce good ranks, a good parameter histogram and good co-ordinate-specific estimates. Motivated by the comments of Thomsen in Ghosh and Rao (1994), Rashid and Nandram (1998) used rank-based methods in an attempt to fix the overshrinkage problem for an error components model (see Battese, Harter and Fuller 1988), but it is not clear how to apply it to a Bayesian generalized linear model.

One obvious way to overcome some overshrinkage is to use several parameters in the modeling. Fitting several regressions may be better than fitting a single regression when the coefficients are permitted to share effects (i.e., random regression model). This is a difficult problem (both theoretically and computationally) not covered by the general theory of Ghosh et al. (1998). But such models are desirable because they are potentially useful to reduce the effects of over shrinkage. See the comments of Holt in Ghosh and Rao (1994).

Another method that can help with overshrinkage is the concept of "uncertain borrowing" suggested by Malec and Sedransk (1992) for pooling the results from several experiments. Instead of assuming all the small area effects are exchangeable, it is assumed that subsets of small areas are exchangeable and the composition of the partition sets is uncertain. Consoni and Veronese (1995) applied the ideas of Malec and Sedransk (1992) to a set of binomial experiments. Evans and Sedransk (1998) applied this methodology for pooling subpopulation regressions. However, when there are many small areas, this method is difficult to perform because there are too many partitions, but one might tolerate larger partition sets. It is possible to perform this method for the simple Poisson model of Christiansen and Morris (1997) but it may not be so easy for a disease mapping application.

This uncertain borrowing method is really to include information of clustering about the area effects in the model. It is necessary to incorporate other sources of information especially when data are sparse. For example, the area effects may be restricted in some set or there may be an order restriction on the effects. One example of this order restricted inference is on an application to the age composition of a fish population (see Nandram et al. 1997).

It is possible to provide models that fit better by introducing a scale parameter. This is particularly important for binomial or Poisson models in which the mean and variance are functionally related making such distribution assumptions doubtful. One would like to expand these models to include under-dispersion and especially over-dispersion. There is a method that can be used for doing so; see West (1985) and Efron (1986). Essentially these authors used Hoeffding's representation of the exponential family through the deviance function. One of the main problems is that the normalization constant is not unity which increases the computation problem. This method can help indirectly to correct for overshrinkage.

A problem that everyone seems to overlook in small area estimation is that the sample sizes from the small areas are random because the survey is usually not designed to collect data from the small areas. Not incorporating this randomness can potentially lead to optimistic precision especially if the sample sizes are informative about the parameters of interest. This randomness should be incorporated into the model. Thus, there are further difficulties when the generalized linear model is used in small area estimations. In addition, if these small areas are finite populations, then the finite population sizes of these areas may also be unknown. This creates a difficult problem for Bayesian predictive inference. One way to overcome this difficulty is to construct a survey design which incorporates the small areas, but with many small areas this will be prohibitively expensive. Otherwise, approximations are inevitable.

The multivariate nature of small area effects is also an important consideration. Inference about small area effects is sensitive to the prior specification of these small area effects. This is particularly true when a subgroup analysis (whose purpose is to identify the areas that have values above or below some cutoff point) or a ranking of the small area effects is done. Correlation among the small area effects can markedly change the subgroups and the rankings. Thus, in a Bayesian analysis, effort needs to be placed on the construction of robust priors for the small area effects. This is expected to be more difficult within the generalized linear model framework.

Finally, we must agree that it is desirable to obtain simple approximate closed form expressions for distributions of parameters of interest. The MCMC method can be used to check the accuracy of these approximations. One example is Christiansen and Morris (1997) on Poisson regression. While the MCMC method provides interval estimators in an output analysis, approximate closed form intervals are desirable. The difficulty here is that, unless common features can be found, an extensive amount of work has to be done for each model.

7. Concluding Remarks

We have reviewed how generalized linear models are currently used in small area estimation. Among many problems we summarize three that exist in small area estimation.

First, the problem of overshrinkage will always exist. Methods now available to cope with overshrinkage can work for simple models. Other ideas are needed for more complex models. A more applied Bayesian approach rather than a decision-oriented approach can be useful.

Second, computation using MCMC methods is perhaps the only way to proceed to fit and study the fit of these models. Theorems like those proved by Ghosh et al. (1998) are needed for models that do not fall in their framework. This adds credence to the use of the Bayesian methodology.

Third, in the area of diagnostics there should be a search for other checking functions and deviance measures. For example, checking functions depending on sufficient statistics are no good (see Gelman et al. 1995). Checking functions based on ranks seem to be a reasonable alternative. Also residual plots for generalized linear models ought to be calibrated by, for example, using training samples.

It is sensible to fit many models to each small area data set. We need to study all these models carefully to understand their strengths and weaknesses. An alternative approach is Bayesian model averaging (BMA); see Hoeting et al. (1998) for an elegant tutorial on BMA. It is computationally more difficult to implement BMA; the present approach is by using the reversible jump Markov chain (Green 1995).

Faster computers will be helpful for these problems characterized by complicated generalized linear models, large data sets and the desire to produce estimates for small areas and subpopulations. Also, it is very important to understand the mechanism that generates the data before any sensible model can be obtained. The data analyst must work with the scientists, and the statisticians working in small area estimation with generalized linear models must ride on technology.

Acknowledgment

The author thanks Professors Joseph Sedransk and Malay Ghosh for their assistance. The research was supported by a research contract with the National Center for Health Statistics, U.S. Department of Health and Human Services.

References

Albert, J. H. (1988). Computational methods using a Bayesian hierarchical generalized linear model. *Journal of the American Statistical Association*, 83, 1037-1044.

Albert, J. H. and Chib, S. (1993). Bayesian analysis of binary and polychotomous response data. *Journal of the American Statistical Association*, 88, 669-679.

Arora, V., Lahiri, P. and Mukherjee, K. (1997). Empirical Bayes estimation of finite population means from complex surveys. *Journal of the American Statistical Association*, 92, 1555-1562.

Battese, G. E., Harter, R. M. and Fuller, W.A. (1988). An error components model for prediction of county crop areas using survey and satellite data. *Journal of the American Statistical Association*, 83, 28-36.

Bernardinelli, L. and Montomoli, C. (1992). Empirical Bayes versus fully Bayesian analysis of geographical variation in disease risk. *Statistics in Medicine*, 11, 983-1007.

Besag, J., Green, P., Higdon, D. and Mengersen, K. (1995) Bayesian computation and stochastic systems (with discussion). *Statistical Science*, 10, 3-66.

Chib, S. and Greenberg, E. (1995). Understanding the Metropolis-Hastings algorithm. *The American Statistician*, 49, 327–335.

Christiansen, C. L. and Morris, C. N. (1997). Hierarchical Poisson regression modeling. *Journal of the American Statistical Association*, 92, 618-632.

Clayton, D. and Kaldor, J. (1987). Empirical Bayes estimates of age-standardized relative risks for use in disease mapping. *Biometrics*, 43, 671-681.

Cochran, W.G. (1977). *Sampling Techniques.* 3rd edition, New York: Wiley.

Consoni, G. and Veronese, P. (1995). A Bayesian method for combining results from several binomial experiments. *Journal of the American Statistical Association*, 90, 935-944.

Datta, G. and Ghosh, M. (1991). Bayesian prediction in linear models: Applications to small area estimation. *Annals of Statistics*, 19, 1748-1770.

Dempster, A.P., and Tomberlin, T. J. (1980). The analysis of census undercount from a postenumeration Survey. In: *Proceedings of the Conference on Census Undercount, Arlington, VA*, pp. 88-94.

Efron, B. (1986). Double exponential families and their use in generalized linear regression. *Journal of the American Statistical Association*, 81, 709-721.

Evans, R. and Sedransk, J. (1998). Methodology for pooling subpopulation regressions when the sample sizes are small and there is uncertainty about which subpopulations are similar. *Technical Report*, Department of Statistics, Case Western Reserve University.

Gelfand, A.E. and Dey, D.K. (1994). Bayesian model choice: Asymptotics and exact calculations. *Journal of the Royal Statistical Society, B, 56*, 501-514.

Gelfand, A.E., Dey, D.K. and Chang, H. (1992). Model determination using predictive distributions with implementation via sampling-based methods. In: *Bayesian Statistics, 3, (J. Bernardo, et al., eds.)*, Oxford University Press, Oxford, 147-158.

Gelfand, A. E., Sahu, S. K., and Carlin, B. P. (1995). Efficient reparametrizations for normal linear models. *Biometrika*, 82, 479-488.

Gelman, A., Carlin, J., Stern, H., and Rubin, D. (1995). *Bayesian Data Analysis.* London: Chapman and Hall.

Ghosh, M. (1992), Constrained Bayes estimation with applications. *Journal of the American Statistical Association, 87.* 533-540.

Ghosh, M. and Lahiri, P. (1987). Robust empirical Bayes estimation of means from stratified samples. *Journal of the American Statistical Association*, 82, 1153-1162.

Ghosh, M. and Meeden, G. (1986). Empirical Bayes estimation in finite population sampling. *Journal of the American Statistical Association*, 81, 739-757.

Ghosh, M. and Natarajan, K. (1998). Small area estimation: a Bayesian perspective. In: *Multivariate Analysis, Design of Experiments and Sample Surveys*, Ed. Subir Ghosh, New York: Marcel Dekker, to appear.

Ghosh, M., Nangia, N., and Kim, D. H. (1996). Estimation of median income of four-person families: A Bayesian time series approach. *Journal of the American Statistical Association*, 91, 1423-1431.

Ghosh, M., Natarajan, K., Stroud, T. W. F., and Carlin, B. P. (1998). Generalized linear models for small area estimation. *Journal of the American Statistical Association*, 93, 273-282.

Ghosh, M. and Rao, J. N. K. (1994). Small area estimation: An appraisal. *Statistical Sciences*, 9, 55-93.

Gilks, W. R. and Wild, P. (1992). Adaptive rejection sampling for Gibbs sampling. *Journal of the Royal Statistical Society*, Ser. C, 41, 337-348.

Green, P. J. (1995). Reversible jump Markov chain Monte Carlo computation and Bayesian model determination. *Biometrika*, 82, 711-732.

Hoeting, A. J., Madigan, D., Raftery, A. E., and Volinsky, C. (1998). Bayesian model averaging," *Technical Report*, Department of Statistics, Colorado State University.

Hulting, F. L. and Harville, D. A. (1991) Some Bayesian and non-Bayesian procedures for the analysis of comparative experiments and for small-area estimation: computational aspects, frequentists properties and relationships. *Journal of the American Statistical Association*, 86, 557-568.

Laud, P. W. and Ibrahim, J.G. (1995). Predictive model selection. *Journal of the Royal Statistical Society, B 57*, 247-262.

Lawless J. F. (1987). Negative binomial and mixed Poisson regression. *Canadian Journal of Statistics*, 15, 209-225.

MacGibbon, B. and Tomberlin, T. J. (1989). Small area estimates of proportions via empirical Bayes techniques. *Survey Methodology*, 15, 237-252.

Malec, D. and Sedransk, J. (1992). Bayesian methodology for combining the results from different experiments when the specifications for pooling are uncertain. *Biometrika*, 79, 593-601.

Malec, D. and Sedransk, J. (1985). Bayesian methodology for predictive inference for finite population parameters in multistage cluster sampling. *Journal of the American Statistical Association*, 80, 897-902.

Malec, D., Sedransk, J., Moriarity, C. L., and LeClere, F. B. (1997). Small area inference for binary variables in the National Health Interview Survey. *Journal of the American Statistical Association*, 92, 815-826.

Malec, D., Sedransk, J., and Tompkins, L. (1993).Bayesian predictive inference for small areas for binary variables in the National Health Interview Survey. In: *Case Studies in Bayesian Statistics*, eds. C. Gatsonis, J.S. Hodges, R.E. Kass, and N.D. Singpurwalla, New York: Springer-Verlag, pp. 377-389.

Marshall, R. J. (1991). Mapping disease and mortality rates using empirical Bayes estimators. *Journal of the Royal Statistical Society*, C, 40, 283-294.

Nandram, B. (1994). Bayesian predictive inference for multivariate sample surveys. *Journal of Official Statistics*, 167-179.

Nandram, B. (1998). A Bayesian analysis of the three-stage hierarchical multinomial model. *Journal of Statistical Computation and Simulation*, 61, 97-126.

Nandram, B. (1999). Empirical Bayes interval estimation for the finite population mean of a small area. *Statistica Sinica*, to appear.

Nandram, B. and Chen, M-H. (1996). Reparameterizing the generalized linear model to accelerate Gibbs sampler convergence. *Journal of Statistical Computation and Simulation*, 54, 129-144.

Nandram, B. and Petruccelli, J. D. (1997). A Bayesian analysis of autoregressive time series panel data. *Journal of Business and Economic Statistics*, 15, 328-334.

Nandram, B. and Sedransk, J. (1993a). Bayesian predictive inference for longitudinal sample surveys. *Biometrics*, 49, 1045-1055.

Nandram, B. and Sedransk, J. (1993b). Empirical Bayes estimation for the finite population mean on the current occasion. *Journal of the American Statistical Association*, 88, 994 -1000.

Nandram, B. and Sedransk J. (1993c). Bayesian predictive inference for a finite population proportion: Two-stage cluster sampling. *Journal of the Royal Statistical Society*, Ser. B, 55, 399-408.

Nandram, B., Sedransk, J. and Pickle, L.W. (1998). Bayesian analysis of mortality rates for U.S. Health Service Areas. *Technical Report*, Mathematical Sciences, Worcester Polytechnic Institute.

Nandram, B., Sedransk, J. and Pickle, L.W. (1999). Bayesian analysis of mortality rates for U.S. Health Service Areas. *Sankhya*, Ser. B, to appear.

Nandram, B., Sedransk, J. and Smith, S. J. (1997). Order restricted Bayesian estimation of the age composition of a population of Atlantic cod. *Journal of the American Statistical Association*, 92, 33-40.

Pickle, L. W., Mungiole, M., Jones, G. K., and White, A. A. (1996). *Atlas of United States Mortality*. Hyattsville, Maryland: National Center for Health Statistics.

Pickle, L. W., Mungiole, M., Jones, G. K., and White, A. A. (1997). Analysis of mapped mortality data by mixed effects models. Technical Report, National Center for Health Statistics.

Prasad, N. G. N. and Rao, J. N. K. (1990). On the estimation of mean squared error of small area predictors. *Journal of the American Statistical Association, 85*, 163 -171.

Rashid, M. M. and Nandram, B. (1998). A rank-based predictor for the finite population mean of a small area: An application to crop production. *Journal of Agricultural, Biological, and Environmental Statistics*, 3, 201-222 (1998).

Shen, W. and Louis, T.A. (1998). Triple-goal estimates in two-stage hierarchical models. *Journal of the Royal Statistical Society*, B, 60, 455-472.

Stroud, T. W. F. (1987). Bayes and empirical Bayes approaches to small area estimation. In: *Small Area Statistics*, eds. R. Platek, J.N.K. Rao, C.E. Sarndal and M.P. Singh, New York: Wiley, pp. 124-137.

Tierney, L. (1994). Markov chains for exploring posterior distributions (with discussion). *Annals of Statistics*, 22, 1701-1762.

Waller, L., Carlin, B., Xia, H., and Gelfand, A. (1997). Hierarchical spatio-temporal mapping of disease rates. *Journal of the American Statistical Association*, 92, 607-617.

West, M. (1985), "Generalized linear models: scale parameters, outlier accommodation and prior distributions. In *Bayesian Statistics 2*, eds., J. M. Bernardo, M.H. Degroot, D.V. Lindley and A. F. M. Smith, North-Holland: Elsevier Science Publishers, pp. 531-558.

Wong, G. Y. and Mason, W. M. (1985). The hierarchical logistic regression model for multilevel analysis. *Journal of the American Statistical Association*, 80, 513-524.

Zeger, S. L. and Karim, M. R. (1991). Generalized linear models with random effects: A Gibbs sampling approach. *Journal of the American Statistical Association*, 86, 79-86.

Part III

Categorical and Longitudinal Data

7

Bayesian Methods for Correlated Binary Data

Siddhartha Chib

ABSTRACT This paper reviews some of the recent developments in the fitting and comparison of Bayesian models for correlated binary data. The importance of the multivariate probit model is highlighted and Markov chain simulation algorithms for the fitting of the multivariate probit, probit normal and hierarchical probit models (the last two for longitudinal data) are discussed. The fitting of each model is illustrated with the Six Cities Data on the health effects of pollution. The computations are conducted using the new Windows^{TM} software package BAYESTAT that has been developed by the author for the fitting of these and many other Bayesian models. The paper also shows how alternative Bayesian models for correlated binary data can be compared. Marginal likelihoods from Chib's (1995) method are computed for each of five models that reflect different assumptions about the correlation structure and the extent of heterogeneity in the sample. Details of the fitting algorithms are reported in the Appendix.

1. Introduction

Statistical methods for correlated, continuous response data have been studied for a long time and many sophisticated models and methods are now available. In the last ten years or so, increasing attention has been devoted to questions for which the standard methods are no longer applicable. For example, in a biostatistical context, one might be interested in the joint probability distribution of several binary variables, each representing a particular disease outcome [Betensky and Whittemore (1996)]. In an economics problem, one might be interested in the outcomes from several related binary financial decisions made by a subject. Correlated binary data can also arise from the analysis of a single binary variable. In a longitudinal problem, for example, one might observe a time series of binary responses (e.g., an indicator of whether the subject visited a physician in a given year) that are likely to be correlated.

The analysis of correlated binary data is an exciting and important area of statistics in which both frequentist and modern Bayesian methods have had much to offer. For reasons of tractability, some of the first developments occurred in the analysis of longitudinal binary data where subject-specific random effects provided a simple means to model intra-cluster correlation [Stiratelli, Laird and Ware (1984)]. Three recent surveys covering this material are those of Agresti (1989), Fitzmaurice,

Laird and Rotnitzky (1993) and Pendergast, Gange, Newton, Lindstrom, Palta and Fisher (1996). More recently, significant methods for the analysis of general multivariate correlated data have begun to appear. Leading papers in this field are those by Carey, Zeger and Diggle (1993) and Glonek and McCullagh (1995) on the multivariate logit, and by Chib and Greenberg (1998) on the multivariate probit.

One purpose of this paper is to show case the multivariate probit model, along with some of its variants, as the canonical model for correlated binary data. This model was introduced more than twenty five years ago by Ashford and Sowden (1970) in the context of bivariate binary responses and was analyzed under simplifying assumptions on the correlation structure by Amemiya (1972), Ochi and Prentice (1984) and Lesaffre and Kaufman (1992). The general version of the model was considered to be intractable. The problems, however, have now disappeared with the recent work of Chib and Greenberg (1998), which builds on the framework of Albert and Chib (1993). An efficient Markov chain Monte Carlo approach for both classical and Bayesian estimation and model comparison is now available. In addition, software that implements the Chib-Greenberg methods is also available.

The MVP model provides a relatively straightforward way of modeling correlated binary data. In this model, the marginal probability of each response is given by a probit function that depends on covariates and response specific parameters. Associations between the binary variables are incorporated by assuming that the vector of binary outcomes are a function of correlated Gaussian variables, taking the value one if the corresponding Gaussian component is positive and the value zero otherwise. This connection with a latent Gaussian random vector means that the regression coefficients can be interpreted independently of the correlation parameters (unlike the case of log-linear models). The link with Gaussian data is also helpful in estimation. By contrast, models based on marginal odds ratios [Connolly and Liang (1992)] tend to proliferate nuisance parameters as the number of variables increase and they become difficult to interpret and estimate. Finally, the Gaussian connection enables generalizations of the model to ordinal outcomes [see Chen and Dey (this volume)], mixed continuous and binary outcomes, and also to spatial data [Oliveira (1997)].

The rest of this chapter is organized as follows. In Section 2 we consider the multivariate probit model and some of its variants. In Section 3 we consider models for clustered binary data and discuss random effect models and Markov-type transition models. Section 4 will take up the issue of model comparison and show how Bayes factors for competing correlated binary data models can be computed. All the methods are illustrated using the new Bayesian software package BAYESTAT (developed by the author) that provides a user-friendly Windows^{TM} based environment for the fitting of these and many other Bayesian models.

2. The Multivariate Probit Model

Let Y_{ij} denote a binary response on the ith observation unit and jth variable, and let $\mathbf{Y}_i = (Y_{i1}, \ldots, Y_{iJ})'$, $1 \leq i \leq n$, denote the collection of responses on all J variables. Also let $\mathbf{x}_{ij}$ denote the set of covariates for the jth response and $\beta_j \in R^{k_j}$ the conformable vector of covariate coefficients. Let $\beta' = (\beta_1', \ldots, \beta_J') \in R^k$, $k = \sum k_j$ denote the complete set of coefficients (one for each response) and let

$\mathbf{\Sigma} = \{\sigma_{jk}\}$ denote a $J \times J$ correlation matrix. Finally let

$$\mathbf{X}_i = \begin{pmatrix} \mathbf{x}_{i1}' & 0' & \cdots & 0' \\ 0' & \mathbf{x}_{i2}' & 0' & 0' \\ \vdots & \vdots & \vdots & \vdots \\ 0' & 0' & \cdots & \mathbf{x}_{iJ}' \end{pmatrix}$$

denote the $J \times k$ covariate matrix on the ith subject. In the case that the covariate effects are not response specific and the same number of covariates influence the response y_{ij}, the covariate matrix is $\mathbf{X}_i = (\mathbf{x}_{i1}, ..., \mathbf{x}_{iJ})'$.

Then, according to the multivariate probit model, the marginal probability that $Y_{ij} = 1$ is given by the probit form

$$\text{pr}(Y_{ij} = 1|\beta) = \Phi(\mathbf{x}_{ij}'\beta_j)$$

and the probability that $Y_i = y_i$, conditioned on parameters $\beta, \mathbf{\Sigma}$, and covariates $\mathbf{x}_{ij}$, is given by

$$\text{pr}(Y_i = y_i|\beta, \mathbf{\Sigma}) \equiv \text{pr}(y_i|\beta, \mathbf{\Sigma}) = \int_{A_{iJ}} \cdots \int_{A_{i1}} \phi_J(t|0, \mathbf{\Sigma})\, dt, \tag{1}$$

where $\phi_J(t|0, \mathbf{\Sigma})$ is the density of a J-variate normal distribution with mean vector 0 and correlation matrix $\mathbf{\Sigma}$ and A_{ij} is the interval

$$A_{ij} = \begin{cases} (-\infty, \mathbf{x}_{ij}'\beta_j) & \text{if} \quad y_{ij} = 1 \\ [\mathbf{x}_{ij}'\beta_j, \infty) & \text{if} \quad y_{ij} = 0. \end{cases} \tag{2}$$

Note that each outcome is determined by its own set of k_j covariates $\mathbf{x}_{ij}$ and covariate effects β_j.

The multivariate discrete mass function presented in (1) can be specified in terms of latent Gaussian random variables. This alternative formulation also forms the basis of the computational scheme that is described below. Let $\mathbf{Z}_i = (z_{i1}, \ldots, z_{iJ})$ denote a J-variate normal vector and let

$$\mathbf{Z}_i \sim N_J(\mathbf{X}_i\beta, \mathbf{\Sigma}). \tag{3}$$

Now let Y_{ij} be 1 or 0 according to the sign of z_{ij}:

$$y_{ij} = I(z_{ij} > 0), \quad j = 1, \ldots, J, \tag{4}$$

where $I(A)$ is the indicator function of the event A. The probability in (1) may be expressed as

$$\int_{B_{iJ}} \cdots \int_{B_{i1}} \phi_J(\mathbf{Z}_i|\mathbf{X}_i\beta, \mathbf{\Sigma})\, d\mathbf{Z}_i, \tag{5}$$

where B_{ij} is the interval $(0, \infty)$ if $y_{ij} = 1$ and the interval $(-\infty, 0]$ if $y_{ij} = 0$. It is easy to confirm that this integral reduces to the form given above. It should also be noted that due to the threshold specification in (4), the scale of Z_{ij} cannot be identified. As a consequence, the matrix $\mathbf{\Sigma}$ must be in correlation form (with units on the main diagonal).

2.1 Dependence Structures

One basic question in the analysis of correlated binary data is the following: How should correlation between binary outcomes be defined and measured? The point of view behind the MVP model is that the correlation is modeled at the level of the latent data which then induces correlation amongst the binary outcomes. This modeling perspective is both flexible and general. In contrast, attempts to model correlation directly (as in the classical literature using marginal odds ratios) invariably lead to difficulties, partly because it is difficult to specify pair-wise correlations in general, and partly because the binary data scale is not natural for thinking about dependence.

Within the context of the MVP model, alternative dependence structures are easily specified and conceived, due to the connection with Gaussian latent data. Some of the possibilities are enumerated below.

- Unrestricted form. Here $\mathbf{\Sigma}$ is fully unrestricted except for the unit constraints on the diagonal. The unrestricted $\mathbf{\Sigma}$ matrix has $p^* = p(p-1)/2$ unknown correlation parameters that must be estimated.

- Equicorrelated form. In this case, the correlations are all equal and described by a single parameter ρ. This form can be a starting point for the analysis when one is dealing with outcomes where are all the pair-wise correlations are believed to have the same sign. The equicorrelated model may be also be seen as arising from a random effect formulation.

- Toeplitz form. Under this case, the correlations depend on a single parameter ρ but under the restriction that $\mathrm{Corr}(Z_{ik}, Z_{il}) = \rho^{|k-l|}$. This version can be useful when the binary outcomes are collected from a longitudinal study where it is plausible that the correlation between outcomes at different dates will diminish with the lag. In fact, in the context of longitudinal data, the $\mathbf{\Sigma}$ matrix can be specified in many other forms with analogy with the correlation structures that arise in standard time series ARMA modeling.

2.2 Student-t Specification

Now suppose that the distribution on the latent $\mathbf{Z}_i$ is multivariate-t with specified degrees of freedom ν. This gives rise to a model that may be called the multivariate-t link model. Albert and Chib (1993) extended the probit link to the t-link in the binary response case and provided a simple approach for estimating the resulting model. Under the multivariate-t assumption, $\mathbf{Z}_i|\beta, \mathbf{\Sigma} \sim \mathrm{MVT}_J(\mathbf{X}_i\beta, \mathbf{\Sigma}, \nu)$ with density

$$f(\mathbf{Z}_i|\beta, \mathbf{\Sigma}) \propto |\mathbf{\Sigma}|^{-1/2}\left\{1 + \frac{1}{\nu}(\mathbf{Z}_i - \mathbf{X}_i\beta)'\mathbf{\Sigma}^{-1}(\mathbf{Z}_i - \mathbf{X}_i\beta)\right\}^{-(\nu+J)/2}. \tag{6}$$

As before, the matrix $\mathbf{\Sigma}$ is in correlation form and the observed outcomes are defined by (4). The model for the latent $\mathbf{Z}_i$ may be expressed as a scale mixture of normals by introducing a random variable $\lambda_i \sim \mathrm{Gamma}(\frac{\nu}{2}, \frac{\nu}{2})$ and letting

$$\mathbf{Z}_i|\beta, \mathbf{\Sigma}, \lambda_i \sim N(\mathbf{X}_i\beta, \lambda_i^{-1}\mathbf{\Sigma}). \tag{7}$$

Conditionally on λ_i, this model is equivalent to the MVP model.

Chen and Dey (1998) have further extended this idea by letting $\mathbf{Z}_i$ follow a general scale mixture of normal distributions.

2.3 *Estimation of the MVP Model*

Consider now the question of inference in the MVP model. We are given a set of data on n subjects with outcomes $\mathbf{Y} = \{\mathbf{Y}_i\}_{i=1}^n$ and interest centers on the parameters of the model $\theta = (\beta, \mathbf{\Sigma})$ and the posterior distribution $\pi(\beta, \mathbf{\Sigma}|Y)$, given some prior distribution $\pi(\beta, \mathbf{\Sigma})$ on the parameters.

To begin with, note that the likelihood function of the model is composed of the probabilities

$$\text{pr}(y_i|\beta, \mathbf{\Sigma}) = \int_{A_{iJ}} \cdots \int_{A_{i1}} \phi_J(t|0, \mathbf{\Sigma})\, dt, \ i \leq n\,.$$

As has been known for some time, these integrals are difficult to evaluate by non-simulation based methods if Σ is unrestricted and J is large. This problem, however, has been bypassed entirely by Chib and Greenberg (1998).

Let $\sigma = (\sigma_{12}, \sigma_{31}, \sigma_{32}, ..., \sigma_{JJ})$ denote the $J(J-1)/2$ distinct elements of $\mathbf{\Sigma}$. It can be shown that the admissible values of σ (that lead to a positive definite $\mathbf{\Sigma}$ matrix) form a convex solid body in the hypercube $[-1, 1]^p$. Denote this set by C. Now let $\mathbf{Z} = (\mathbf{Z}_1, \ldots, \mathbf{Z}_n)$ denote the latent values corresponding to the observed data $\mathbf{Y} = \{\mathbf{Y}_i\}_{i=1}^n$. Consider the simulation of the augmented posterior density

$$\begin{aligned}\pi(\beta, \sigma, \mathbf{Z}|y) &\propto \pi(\beta, \sigma) f(\mathbf{Z}|\beta, \mathbf{\Sigma}) \text{pr}(\mathbf{y}|\mathbf{Z}, \beta, \mathbf{\Sigma}) \\ &\propto \pi(\beta, \sigma) \prod_{i=1}^{n} \left(\phi_J(\mathbf{Z}_i|\beta, \mathbf{\Sigma}) \text{pr}(y_i|\mathbf{Z}_i, \beta, \mathbf{\Sigma})\right), \beta \in \Re^k, \sigma \in C \qquad (8)\end{aligned}$$

where

$$\text{pr}(y_i|\mathbf{Z}_i, \beta, \mathbf{\Sigma}) = \prod_{j=1}^{J} \{I(z_{ij} > 0) I(y_{ij} = 1) + I(z_{ij} \leq 0) I(y_{ij} = 0)\}$$

due to the fact that $\text{pr}(y_i|Z_i, \beta, \mathbf{\Sigma})$ is one if the z_{ij} respect the constraint imposed by the observed value of y_{ij}, indicated by the functions in curly braces.

To sample this posterior density, the following Markov chain Monte Carlo scheme is available. Fuller details are in the appendix.

Algorithm 1 (Chib-Greenberg (1998)):

1. Sample z_{ij} from the distribution $z_{ij}|\mathbf{Z}_{i(-j)}, \mathbf{y}_i, \beta, \mathbf{\Sigma}$ for $j \leq J$ and $i \leq n$, where $\mathbf{Z}_{i(-j)}$ are the latent values on the ith subject excluding z_{ij};
2. Sample β from the distribution $\beta|\mathbf{Z}, \mathbf{y}, \mathbf{\Sigma}$;
3. Sample σ from the conditional distribution $\sigma|\mathbf{Z}, \mathbf{y}, \beta$ using the Metropolis-Hastings algorithm;
4. Repeat Steps 1-3 using the most recent values of the conditioning variables.

This algorithm achieves the intended purpose in that it allows one to sample the posterior distribution without requiring the computation of the likelihood function. In Step 3, the Metropolis-Hastings algorithm [see Chib and Greenberg (1995)] is used to sample the distinct elements of $\mathbf{\Sigma}$. This method of sampling σ is quite general and can be used to sample unrestricted covariance matrices under non-Wishart prior assumptions.

In the source paper, the algorithm above is tested in a problem where $J = 7$ and the dimension of σ is twenty one. The method is found to work quickly and efficiently. The paper also provides some additional discussion about sampling σ in blocks if the dimension of σ is excessively large, say over thirty. It is mentioned that estimation of the MVP model when Σ is restricted to be in equicorrelated form (say) requires no new analysis because Step 3 is based on the distinct elements of Σ. Additional restrictions, therefore, simplify the simulation.

Example: MVP model applied to longitudinal data. Consider a data set on the health effects of pollution on young children. The response variable is an indicator of wheezing status. The data are collected on 537 children in Stuebenville, Ohio, each observed at ages 7, 8, 9 and 10 years (see Chib and Greenberg, 1998 for the data). One assumes that the marginal probability of wheeze status of the ith child at the jth time point is

$$\Pr(y_{ij} = 1|\beta) = \Phi(\beta_0 + \beta_1 x_{1ij} + \beta_2 x_{2ij} + \beta_3 x_{3ij}) \, , i \leq 537, \, j \leq 4 \, ,$$

where x_1 is the age of the child centered at nine years, x_2 is a binary indicator variable representing the mother's smoking habit during the first year of the study, and $x_3 = x_1 x_2$. We are assuming that β is constant across categories j. Let the association between the binary responses on the ith child be modeled by an unrestricted correlation matrix. The prior on $\beta = (\beta_1, \beta_2, \beta_3, \beta_4)$ is independent Gaussian with a mean of zero and a variance of ten. Also let the prior on the correlations σ, a six component vector, be proportional to a normal distribution with mean zero and covariance equal to the identity matrix. From 10,000 runs of algorithm 1 (coded in the software package BAYESTAT), one obtains the following covariate effects and posterior distributions of the correlations.

	Prior		Posterior				
	Mean	Std dev	Mean	NSE	Std dev	Lower	Upper
β_1	0.000	3.162	-1.108	0.001	0.062	-1.231	-0.985
β_2	0.000	3.162	-0.077	0.001	0.030	-0.136	-0.017
β_3	0.000	3.162	0.155	0.002	0.101	-0.043	0.352
β_4	0.000	3.162	0.036	0.001	0.049	-0.058	0.131

TABLE 7.1. Covariate effects in the Six Cities example: MVP model with unrestricted correlations. In the table, NSE denotes the numerical standard error, lower is the 2.5th percentile and upper is the 97.5th percentile of the simulated draws. The results are based on 10000 draws from Algorithm 1.

These posterior distributions suggest that an equicorrelated correlation structure might be appropriate for these data. A formal comparison of the two correlation structures is provided below using marginal likelihoods.

2.4 Fitting of the Multivariate t-link Model

Algorithm 1 is easily modified for the fitting of the multivariate-t link version of the MVP model. One simple possibility is to include the $\{\lambda_i\}$ into the sampling since conditioned on the value of λ_i, the t-link binary model reduces to the MVP model. With this augmentation, one implements Steps 1-3 conditioned on the value of $\{\lambda_i\}$. The sampling is completed with an additional step involving the simulation of $\{\lambda_i\}$.

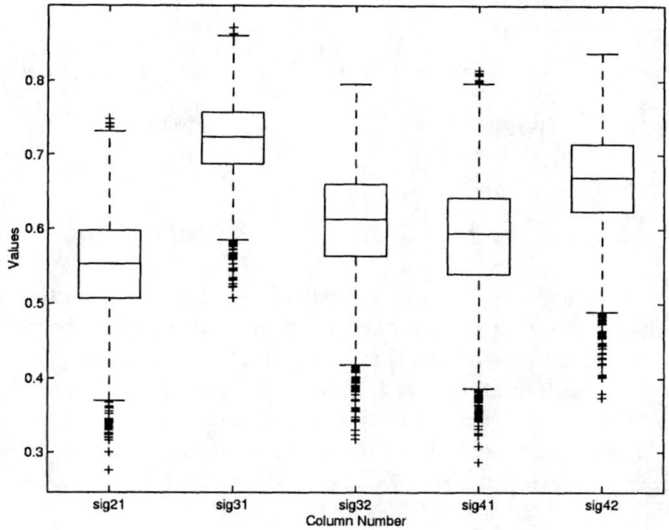

FIGURE 7.1. Posterior boxplots of the correlations in the Six Cities example: MVP model

A straightforward calculation shows that the updated full conditional distribution of λ_i is

$$\lambda_i|\mathbf{Z}_i,\beta,\mathbf{\Sigma} \sim \text{Gamma}\left(\frac{\nu+J}{2}, \frac{\nu+(\mathbf{Z}_i-\mathbf{X}_i\beta)'\mathbf{\Sigma}^{-1}(\mathbf{Z}_i-\mathbf{X}_i\beta)}{2}\right), \ i \leq n,$$

which is easily sampled. The modified sampler thus requires little extra coding.

3. Longitudinal Binary Data

Correlated binary data can arise from a longitudinal study, as in the Six Cities example above. For such data, one can use the MVP model or one of the other special models that have been designed for longitudinal data. In this section we discuss the important classes of models that are available.

3.1 Probit (or logit) Normal Model

Consider a sequence of binary measurements $\mathbf{Y}_i = (y_{i1}, ..., y_{in_i})'$, $y_{ij} \in \{0,1\}$ on the ith unit taken at n_i specific time points (assumed here to be equally spaced). Let x_{ij} denote a set of covariates on the ith subject at the jth time point ($j \leq n_i$). In modeling a sequence of such binary random variables the main issue is about a probability model for $\mathbf{Y}_i$ that incorporates intra-cluster correlation.

A quite popular means of modeling intra-cluster correlation is via a random effects formulation [Stiratelli, Laird and Ware (1984)]. Under this model, the probability of a positive response is

$$\begin{aligned} \Pr(y_{ij}=1|\mathbf{b}_i) &= F(\mathbf{x}_{ij}'\beta + \mathbf{w}_{ij}'\mathbf{b}_i) \\ \mathbf{b}_i &\sim N(\mathbf{0},\mathbf{D}), \end{aligned} \tag{9}$$

where $\mathbf{w}_{ij}$ is a subset of $\mathbf{x}_{ij}$ and $\mathbf{b}_i$ denotes a vector of random effects. The parameters β and $\mathbf{D}$ are unknown.

3.2 Inference

For the n_i outcomes $\mathbf{Y}_i$ on the ith subject, the joint probability mass function is

$$\int \left\{ \prod_{j=1}^{n_i} \left[F(\mathbf{x}'_{ij}\beta + \mathbf{w}'_{ij}\mathbf{b}_i)\right]^{y_{ij}} \left[1 - F(\mathbf{x}'_{ij}\beta + \mathbf{w}_{ij}\mathbf{b}_i)\right]^{1-y_{ij}} \right\} \phi(\mathbf{b}_i|\mathbf{0}, \mathbf{D})\, db_i \quad (10)$$

which, although quite different from the MVP formulation, is also difficult to evaluate.

In existing work, both in the classical and Bayesian contexts, F is taken to be cdf of either the logit or standard normal distributions. Zeger and Karim (1991), in early work, provide a Bayesian analysis of the logit case, while the probit case is discussed by Albert and Chib (1996) and Chib and Carlin (1997). The latter two papers demonstrate that the probit model is considerably easier to handle than the logit. All full conditional distributions are tractable and the MCMC sampling does not require a Metropolis step.

3.3 Computations for the Probit-Normal Model

The most effective strategy for dealing with the probit normal model is to express the model in terms of latent Gaussian variables. Let z_{it} be distributed as normal with

$$z_{ij}|\mathbf{b}_i \sim \mathcal{N}(\mathbf{x}'_{ij}\beta + \mathbf{w}'_{ij}\mathbf{b}_i, 1), \quad 1 \le j \le n_i; \ 1 \le i \le n, \quad (11)$$

and write the model for the ith cluster as $\mathbf{Z}_i \sim \mathcal{N}(\mathbf{X}_i\beta + \mathbf{W}_i\mathbf{b}_i, \mathbf{I})$, where $\mathbf{W}_i$ is a $n_i \times q$ matrix (under obvious notation). As in the MVP model, let the observed response y_{ij} be given by

$$y_{ij} = \begin{cases} 1 & \text{if} \quad z_{ij} > 0 \\ 0 & \text{if} \quad z_{ij} \le 0 \end{cases} .$$

Then, it can be seen that the joint probability mass function of $\mathbf{Y}_i$ from this specification reduces to (10). With the introduction of the latent data, one simple MCMC strategy, in line with that proposed by Gelfand and Smith (1990) for Gaussian panel models with fully observed outcomes, is discussed by Albert and Chib (1996). This algorithm was recently revised by Chib and Carlin (1997) with a view to improving its mixing properties. The idea behind the modification is to sample β and $\{\mathbf{Z}_i\}$ in one block, marginalized over $\{\mathbf{b}_i\}$. That this is possible is a consequence of the fact that $\mathbf{b}_i$ can be analytically integrated out of the model. Then, for the ith cluster, it follows that

$$\mathbf{Z}_i|\beta, \mathbf{D} \sim \mathcal{N}(\mathbf{X}_i\beta, \mathbf{V}_i)$$

where $\mathbf{V}_i = \mathbf{W}_i\mathbf{D}\mathbf{W}_i + \mathbf{I}$. This model now looks identical to the MVP model with a particular form for the covariance matrix. Therefore, conditioned on $(\mathbf{y}, \mathbf{D})$ (but marginalized over $\{\mathbf{b}_i\}$), one can sample β and $\{\mathbf{Z}_i\}$ exactly as in Steps 1 and 2 of Algorithm 1. Given sampled values of $\{\mathbf{Z}_i\}$ and β, one then samples $\{\mathbf{b}_i\}$ from the model

$$\begin{aligned} (\mathbf{Z}_i - \mathbf{X}_i\beta) &\sim \mathcal{N}(\mathbf{W}_i\mathbf{b}_i, \mathbf{V}_i) \\ \mathbf{b}_i &\sim \mathcal{N}_q(\mathbf{0}, \mathbf{D}). \end{aligned}$$

In specifying the prior on $\mathbf{D}^{-1}$ in this model, a useful approach is to assume that $\mathbf{D}$ follows the inverse Wishart distribution with mean $\mathbf{D}_0$ on a_0 degrees of freedom.

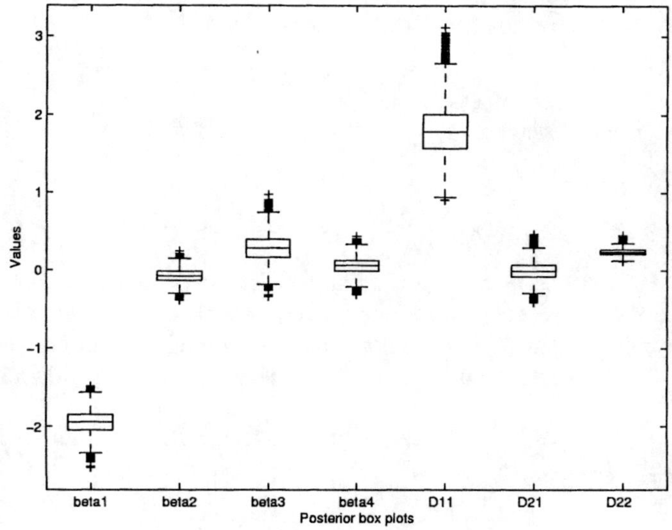

FIGURE 7.2. Posterior boxplots in the Six Cities example: Probit normal model

Then, the prior on $\mathbf{D}^{-1}$ is Wishart with $\nu_0 = a_0 - q - 1$ degrees of freedom and scale matrix $\mathbf{R}_0 = \mathbf{D}_0^{-1}/(\nu_0 - 2q - 2)$. This leads to the following algorithm for sampling the probit-normal model.

Algorithm 2 (Chib-Carlin (1997)):

1. Sample z_{ij} from the distribution $z_{ij}|\mathbf{Z}_{i(-j)}, y_i, \beta, \mathbf{D}$ for $j \leq J$ and $i \leq n$, where $\mathbf{Z}_{i(-j)}$ are the latent values on the ith subject excluding z_{ij};
2. Sample β from the distribution $\beta|\mathbf{Z}, \mathbf{y}, \mathbf{D}$;
3. Sample $\mathbf{b}_i$ from the conditional distribution $\mathbf{b}_i|\mathbf{Z}_i, \beta, \mathbf{D}$;
4. Sample $\mathbf{D}^{-1}$ from the distribution $\mathbf{D}^{-1}|\{\mathbf{b}_i\}$;
5. Repeat Steps 1-4 using the most recent values of the conditioning variables.

Example (contd.). Probit normal model applied to the Six Cities data. Consider the application of the probit normal model to the Six Cities data analyzed earlier through the multivariate probit model. Assume now that the marginal probability of wheeze status of the ith child at the jth time point is

$$\Pr(y_{ij} = 1|\beta) = \Phi(\beta_0 + \beta_1 x_{1ij} + \beta_2 x_{2ij} + \beta_3 x_{3ij})\,, i \leq 537,\, j \leq 4\,,$$

with $\mathbf{w}_{ij} = (1, x_{2ij})$, implying that both the intercept and the age effects are children-specific. The parameter vector β is given the same Gaussian prior as before and $\mathbf{D} : 2 \times 2$ is assumed apriori to have an inverse Wishart distribution with mean of $\mathbf{I}_2$ on 15 degrees of freedom. From 10,000 runs of algorithm 2 (coded in the software package BAYESTAT), one obtains the following posterior distribution of the covariate effects and elements of the matrix $\mathbf{D}$. The posterior distributions indicate that the age effect is not significant but that there is considerable heterogeneity in the effect across the children in the sample.

3.4 Binary Response Hierarchical Model

The random effect model for clustered binary data can be extended to the case where all covariate effects are subject-specific. Then, $\mathbf{w}_{ij} = \mathbf{x}_{ij}$ and the model can be written as

$$\begin{aligned} \Pr(y_{ij} = 1|\mathbf{b}_i) &= F(\mathbf{x}_{ij}'\mathbf{b}_i) \\ \mathbf{b}_i &\sim \mathcal{N}(\mathbf{A}_i\beta, \mathbf{D}) \end{aligned} \tag{12}$$

where $\mathbf{A}_i$ denotes a $k \times m$ matrix consisting of cluster-specific covariates. This model may be called a two stage binary response hierarchical model. It is a non-linear version of the hierarchical models considered by Lindley and Smith (1972). In this model, it is not required that the dimension of β be smaller than the dimension of $\mathbf{b}_i$.

Algorithm 2 has been easily adapted to deal with this model since one can define the latent data model as $\mathbf{Z}_i \sim \mathcal{N}(\mathbf{X}_i\mathbf{b}_i, \mathbf{I})$ and integrate out $\mathbf{b}_i$ to produce the distribution

$$\mathbf{Z}_i|\beta, \mathbf{D} \sim \mathcal{N}(\mathbf{X}_i\mathbf{A}_i\beta, \mathbf{V}_i^* = \mathbf{I} + \mathbf{X}_i\mathbf{D}\mathbf{X}_i')$$

Once again $\mathbf{Z}_i$ can be sampled from its full conditional distribution as in Step 1 of the MVP algorithm and β can be sampled as in Step 2 of the MVP algorithm (both using $\mathbf{X}_i\mathbf{A}_i$ as the new covariate matrix). The remaining steps are similar to those in Algorithm 2. The resulting algorithm is given as follows.

Algorithm 3 (Chib and Carlin (1997)):

1. Sample z_{ij} from the distribution $z_{ij}|\mathbf{Z}_{i(-j)}, \mathbf{y}_i, \beta, \mathbf{D}$ for $j \le J$ and $i \le n$, where $\mathbf{Z}_{i(-j)}$ are the latent values on the ith subject excluding z_{ij};
2. Sample β from the distribution $\beta|\mathbf{Z}, \mathbf{y}, \mathbf{D}$;
3. Sample $\mathbf{b}_i$ from the conditional distribution $\mathbf{b}_i|\mathbf{Z}_i, \beta, \mathbf{D}$;
4. Sample $\mathbf{D}^{-1}$ from the distribution $\mathbf{D}^{-1}|\{\mathbf{b}_i\}$;
5. Repeat Steps 1-4 using the most recent values of the conditioning variables.

Example (contd.). To illustrate the fitting of the binary response hierarchical model, consider again the Six Cities data set. Now suppose that

$$\Pr(y_{ij} = 1|\mathbf{b}_i) = \Phi(b_{i0} + b_{i1}x_{1ij} + b_{i2}x_{3ij})$$

where, as before, x_1 is the age covariate and x_3 is the interaction between age and the mother's smoking indicator. Treat x_{2i} as a covariate that influences the subject-specific covariate effects $\mathbf{b}_i$ and write the second stage mean function as

$$\mathbf{A}_i\beta = \begin{pmatrix} 1 & x_{2i} & 0 & 0 & 0 & 0 \\ 0 & 0 & 1 & x_{2i} & 0 & 0 \\ 0 & 0 & 0 & 0 & 1 & x_{2i} \end{pmatrix} \beta$$

where $\beta = (\beta_1, ..., \beta_6)$ is a six dimensional fixed effect and the matrix $\mathbf{D}$ is a full six dimensional matrix of variances and covariances. Assume that the prior information about β can be described by a $N(0, 10\mathbf{I})$ distribution and that of $\mathbf{D}^{-1}$ by a Wishart with 25 degrees of freedom with a mean equal to the identity matrix. Algorithm 3, which is coded in the software package BAYESTAT, produces the following

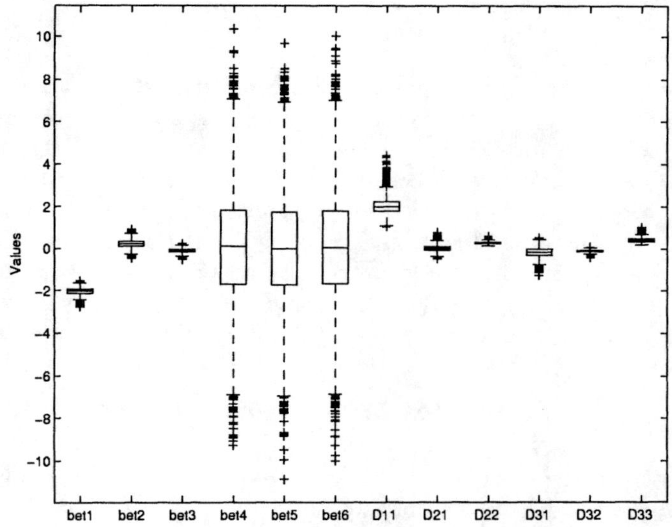

FIGURE 7.3. Posterior boxplots in the Six Cities example: Probit hierarchical model

posterior distributions of β and $\mathbf{D}$ from 10,000 iterations, beyond a burn-in of 500 cycles. Because the posterior distributions of β_4 to β_6 are quite dispersed, one can conjecture that the random effects b_{i1} and b_{i2} do not follow a hierarchical model with mean depending on x_{2i}. There is, however, clear indication of heterogeneity in the coefficients across the subjects.

3.5 Other Models

Other models for clustered binary outcomes have been proposed in the literature. One class of models is the Markov transition model in which the probability of a positive outcome at observation j, conditioned on the past outcomes, depends only on the last outcome. The simplest such model is given by

$$\Pr(y_{ij} = 1 | y_{il}\ (l \leq j-1), \beta) = F(\mathbf{x}'_{it}\beta + \phi_1 y_{ij-1}),$$

where y_{ij-1} is the value of the response at time point $j-1$ and F is the cdf of some distribution, usually that of the standard normal or the standard logit. Under this model dependence is of the first-order Markov kind. Higher order Markov dependence can be modeled by letting

$$\Pr(y_{ij} = 1 | y_{il}\ (l \leq j-1), \beta) = F(\mathbf{x}'_{it}\beta + \phi_1 y_{ij-1} + \phi_2 y_{ij-2}).$$

Other variations on this theme are possible [see Muenz and Rubinstein (1992)]. These models are easily fit using latent data as discussed above. There are no real problems in fitting these models except that one must specify the start-up values of the response before the first time point. One possibility is to condition the analysis on the first few observations in the observed data. This effectively reduces the sample size on each cluster but the cost of doing this is minimal provided the time series is sufficiently long.

4. Comparison of Alternative Models

In the discussion above, we have considered the MCMC based fitting of a variety of different correlated binary data models. We now turn to the question of how these alternative models can be compared on the basis of a given set of data and priors. One Bayesian approach is to compute the model marginal likelihood and to compare competing models using the ratio of marginal likelihoods (called Bayes factors). The model marginal likelihood is defined as the integral of the likelihood function with respect to the prior density on the parameters. Specifically, if we let $\mathcal{M}_j$ denote model j, and $f(y|\mathcal{M}_j, \theta)$ and $\pi(\theta|\mathcal{M}_j)$ denote the likelihood function and prior density, respectively, then the marginal likelihood is defined as

$$m(\mathbf{y}|\mathcal{M}_j) = \int f(\mathbf{y}|\mathcal{M}_j, \theta)\pi(\theta|\mathcal{M}_j)d\theta\,.$$

Alternatively, the marginal likelihood, by virtue of being the normalizing constant of the posterior density, can be expressed as

$$m(\mathbf{y}|\mathcal{M}_j) = \frac{f(\mathbf{y}|\mathcal{M}_j, \theta)\pi(\theta|\mathcal{M}_j)}{\pi(\theta|\mathbf{y}, \mathcal{M}_j)} \tag{13}$$

This expression is an identity in θ. Chib (1995) has called this expression the *basic marginal likelihood identity* and has developed a technique for computing the marginal likelihood by a two step method: first selecting a high posterior density point θ^* (such as the posterior mean) and second computing the ordinate $\pi(\theta^* | \mathbf{y}, \mathcal{M}_j)$ by a conditional /marginal decomposition, where each conditional/marginal ordinate is estimated from the MCMC output. Given this estimate of the posterior ordinate (denoted $\hat{\pi}(\theta^*|\mathbf{y}, \mathcal{M}_j)$), the marginal likelihood estimate on the log scale is given by

$$\log m(\mathbf{y}|\mathcal{M}_j) = \log f(\mathbf{y}|\mathcal{M}_j, \theta^*) + \log \pi(\theta^*|\mathcal{M}_j) - \log \hat{\pi}(\theta^*|\mathbf{y}, \mathcal{M}_j)\,. \tag{14}$$

4.1 Likelihood Ordinate

Estimation of the various correlated binary data models does not require the computation of the likelihood function. In the expression of the marginal likelihood, however, one must compute the likelihood ordinate at the single point θ^*. As it turns out, the computation of this ordinate for each of the models presented above involves an integral that is similar to that which arises in the context of the multivariate probit model. In particular, the likelihood contribution $f(\mathbf{y}_i|\mathcal{M}_j, \theta^*)$ of the ith observation in each model is of the type

$$\int_{B_{iJ}} \cdots \int_{B_{i1}} \phi_J(\mathbf{Z}_i|\mu_i, \Sigma_i)\, d\mathbf{Z}_i, \tag{15}$$

where B_{ij} is the interval $(0, \infty)$ if $y_{ij} = 1$ and the interval $(-\infty, 0]$ if $y_{ij} = 0$. In the case of the MVP model, $\mu_i = \mathbf{X}_i\beta$ and $\Sigma_i = \Sigma$; in the case of the probit normal model, $\mu_i = \mathbf{X}_i\beta$ and $\Sigma_i = \mathbf{I} + \mathbf{W}_i\mathbf{D}\mathbf{W}_i'$; while in the case of hierarchical model, $\mu_i = \mathbf{X}_i\mathbf{A}_i\beta$ and $\Sigma_i = \mathbf{I} + \mathbf{X}_i\mathbf{D}\mathbf{X}_i'$.

One simple way to compute the likelihood contribution is through an application of Chib's method. One can note that the likelihood contribution is the normalizing constant of the truncated normal density $f(\mathbf{Z}_i|\mathbf{y}_i, \mu_i, \mathbf{\Sigma}_i)$ and hence the quantity of interest is given by

$$f(\mathbf{y}_i|\mu_i, \mathbf{\Sigma}_i) = \frac{f(\mathbf{Z}_i^*|\mu_i, \mathbf{\Sigma}_i)}{f(\mathbf{Z}_i^*|\mathbf{y}_i, \mu_i, \mathbf{\Sigma}_i)}$$

where $\mathbf{Z}_i^*$ is some arbitrarily chosen vector. In this expression the numerator is multivariate normal and hence is easily available while the denominator can be computed by the approach of Ritter and Tanner (1992) as follows. Write the denominator (by extending the argument) as

$$f(\mathbf{Z}_i^*|\mathbf{y}_i, \mu_i, \mathbf{\Sigma}_i) = \int f(\mathbf{Z}_i^*|\mathbf{y}_i, \mu_i, \mathbf{\Sigma}_i, \mathbf{Z}_i)\pi(\mathbf{Z}_i|\mathbf{y}_i, \mu_i, \mathbf{\Sigma}_i)d\mathbf{Z}_i \tag{16}$$

and note that the integrand is the Markov transition density of moving from $\mathbf{Z}_i$ to $\mathbf{Z}_i^*$. Hence it can be written as

$$f(\mathbf{Z}_i^*|\mathbf{y}_i, \mu_i, \mathbf{\Sigma}_i, \mathbf{Z}_i) = \prod_{j=1}^{J} f(z_{ij}^*|z_{ik}^*\,(k<j), z_{il}\,(l>j), \mathbf{y}_i, \mu_i, \mathbf{\Sigma}_i)$$

where each of the terms in the product is a univariate truncated normal density, truncated to the interval $(0, \infty)$ if $y_{ij} = 1$ or to the region $(-\infty, 0)$ if $y_{ij} = 0$. Now the integral in (16) can be computed by taking a large number of draws of $\mathbf{Z}_i$ from $\pi(\mathbf{Z}_i|\mathbf{y}_i, \mu_i, \mathbf{\Sigma}_i)$, computing $f(\mathbf{Z}_i^*|\mathbf{y}_i, \mu_i, \mathbf{\Sigma}_i, \mathbf{Z}_i)$ for each such simulated $\mathbf{Z}_i$, and averaging the results.

4.2 Posterior Ordinate

The starting point in this computation is a decomposition of the posterior ordinate into marginal and conditional ordinates followed by a Monte Carlo estimate of each ordinate. In the case of the MVP model, $\theta = (\beta, \sigma)$ and the posterior ordinate is written as

$$\pi(\theta^*|y) = \pi(\sigma^*|\mathbf{y})\pi(\beta^*|\mathbf{y}, \mathbf{\Sigma}^*)$$

where

$$\pi(\beta^*|\mathbf{y}, \mathbf{\Sigma}^*) = \int \pi(\beta^*|\mathbf{Z}, \mathbf{y}, \mathbf{\Sigma}^*)\pi(\mathbf{Z}|\mathbf{y}, \mathbf{\Sigma}^*)d\mathbf{Z}\,.$$

The latter ordinate can be computed by taking draws of $\mathbf{Z}$ from $\pi(\mathbf{Z}|\mathbf{y}, \mathbf{\Sigma}^*)$ and averaging the Gaussian density $\pi(\beta^*|\mathbf{Z}, \mathbf{y}, \mathbf{\Sigma}^*)$ over these draws. Specifically, as shown in the Appendix under Algorithm 1,

$$\pi(\beta^*|\mathbf{Z}, \mathbf{y}, \mathbf{\Sigma}^*) = \phi_J(\beta^*|\hat{\beta}, \mathbf{B}^{-1})\,,$$

where $\hat{\beta} = \mathbf{B}^{-1}(\mathbf{B}_0\beta_0 + \sum_{i=1}^{n} \mathbf{X}_i'\mathbf{\Sigma}^{*-1}\mathbf{Z}_i)$, $\mathbf{B} = \mathbf{B}_0 + \sum_{i=1}^{n} \mathbf{X}_i'\mathbf{\Sigma}^{*-1}\mathbf{X}_i$ and β_0 and $\mathbf{B}_0$ are the prior hyperparameter values. This ordinate can be averaged over the different values of $\mathbf{Z}$ that are drawn from $\pi(\mathbf{Z}|\mathbf{y}, \mathbf{\Sigma}^*)$. How should these draws of $\mathbf{Z}$ be obtained? A simple approach suggested by Chib is to employ the idea of a reduced MCMC run. The reduced run involves running Algorithm 1 with $\mathbf{\Sigma}$ fixed at $\mathbf{\Sigma}^*$, sampling β from $\beta|\mathbf{Z}, \mathbf{y}, \mathbf{\Sigma}^*$ and $\mathbf{Z}$ from $\mathbf{Z}|\mathbf{y}, \beta, \mathbf{\Sigma}^*$. The draws of $\mathbf{Z}$ from this run follow the desired distribution. The estimation of the posterior ordinate is completed by estimating the posterior ordinate $\pi(\sigma^*|\mathbf{y})$ by kernel smoothing applied to the draws on σ from the full run of Algorithm 1.

This same approach can be applied to the remaining models. For example, in the probit normal model, $\theta = (\beta, \mathbf{D}^{-1})$ and the posterior ordinate is written as

$$\pi(\theta^* | \mathbf{y}) = \pi(\mathbf{D}^{-1*} | \mathbf{y}) \pi(\beta^* | \mathbf{y}, \mathbf{D}^*)$$

where $\pi(\mathbf{D}^{-1*}|\mathbf{y})$ is obtained by averaging $\pi(\mathbf{D}^{*-1}|\{\mathbf{b}_i\})$ (a Wishart density - see Algorithm 2 in the Appendix) over the draws of $\{\mathbf{b}_i\}$, and $\pi(\beta^*|\mathbf{y}, \mathbf{D}^*)$ is obtained by averaging the Gaussian density $\pi(\beta^*|\mathbf{y}, \mathbf{Z}, \mathbf{D}^*)$ over draws of $\mathbf{Z}$ from the reduced run consisting of the distributions $\beta|\mathbf{y}, \mathbf{Z}, \mathbf{D}^*$; $\mathbf{Z}|\mathbf{y}, \beta, \mathbf{D}^*$. Precisely the same approach works in the case of the binary response hierarchical model.

Example: Consider the comparison of five alternative models for correlated binary data. Three of these models are the ones that were considered above. Two other models are in the MVP class with restricted correlation matrices. One of the very interesting features of marginal likelihoods is that they allow for these comparisons amongst non-nested models. In BAYESTAT, the marginal likelihood is computed using (3) for each model that is simulated. The results for the five models under comparison are reported in Table 2. On the basis of these marginal likelihoods we

	$\mathcal{M}_1$	$\mathcal{M}_2$	$\mathcal{M}_3$	$\mathcal{M}_4$	$\mathcal{M}_5$
$\ln f(Y_n\|\theta^*)$	-795.1869	-798.5567	-804.4102	-805.0249	-812.0036
$\ln m(Y_n)$	-823.9188	-818.009	-824.0001	-829.9736	-840.6648

TABLE 7.2. Maximized log-likelihood and log marginal likelihood of five models fit in the examples: $\mathcal{M}_1$ is MVP with unrestricted correlations; $\mathcal{M}_2$ is MVP with an equicorrelated correlation; $\mathcal{M}_3$ is the MVP with Toeplitz correlation structure; $\mathcal{M}_4$ is the Probit normal; and $\mathcal{M}_5$ is the hierarchical probit.

conclude that the data tend to support the MVP model with equicorrelated correlations. For completeness, the prior-posterior summary for the best fitting model is reproduced in Table 3.

	Prior		Posterior				
	Mean	Std dev	Mean	NSE	Std dev	Lower	Upper
β_1	0.000	3.162	-1.103	0.001	0.064	-1.227	-0.978
β_2	0.000	3.162	-0.077	0.001	0.030	-0.137	-0.021
β_3	0.000	3.162	0.158	0.002	0.103	-0.046	0.361
β_4	0.000	3.162	0.039	0.001	0.047	-0.051	0.133
β_1	0.500	1.000	0.651	0.002	0.039	0.571	0.727
Log Marginal likelihood					-818.009		

TABLE 7.3. Prior-posterior summary for the equicorrelated MVP model applied to the Six Cities data. In the table, NSE denotes the numerical standard error, lower is the 2.5th percentile and upper is the 97.5th percentile of the simulated draws. The results are based on 10000 draws from Algorithm 1.

5. Concluding Remarks

This chapter has summarized some of the recent developments in the fitting and comparison of Bayesian models for correlated binary data. The discussion has emphasized the multivariate probit model, which may be applied to clustered and non-clustered responses, and the probit-normal and binary response hierarchical models (the latter two for use with clustered binary data). We discussed the fitting of these models and the issue of model comparisons. The discussion did not explore residual diagnostics and model fit issues. Early developments on some relevant ideas are contained in Albert and Chib (1995) and Chen and Dey (1998). Another problem that is not covered in this paper relates to the question of drop-outs in longitudinal data analysis [see Cowles, Carlin and Connett (1996)].

The comparison of alternative models for correlated binary data is an important achievement of the last few years. We have shown how the marginal likelihood of each model can be computed based on the output of the MCMC simulations. At this time, other approaches, for example, the method of Carlin and Chib (1995), have not been applied to compare alternative models for correlated binary data. Further developments in these areas will likely occur in the near future.

6. Appendix

This appendix reports full details of each algorithm discussed above. These algorithms are coded in the software package BAYESTAT that has been developed by the author. This software program includes a spreadsheet and graphics environment for data entry and analysis and pull-down menus for the specification of a large variety of Bayesian models. This program automatically includes the computation of the marginal likelihood by Chib's method for every model that is included in the program.

6.1 Algorithm 1

1. Sample z_{ij} from the distribution $z_{ij}|\mathbf{Z}_{i(-j)}, \mathbf{y}_i, \beta, \mathbf{\Sigma}$ for $j \leq J$ and $i \leq n$, where $\mathbf{Z}_{i(-j)}$ are the latent values on the ith subject excluding z_{ij} and

$$z_{ij}|\mathbf{Z}_{i(-j)}, \mathbf{y}_i, \beta, \mathbf{\Sigma} \sim \begin{cases} TN_{(0,\infty)}(a_{ij}, b_{ij}) & \text{if} \quad y_{ij} = 1 \\ TN_{(-\infty,0])}(a_{ij}, b_{ij}) & \text{if} \quad y_{ij} = 0 \end{cases}$$

   In this expression TN denotes the truncated normal distribution and a_{ij} and b_{ij} are the parameters of the conditional distributions $z_{ij}|\mathbf{Z}_{i(-j)}, \beta, \mathbf{\Sigma}$ (ignoring truncation). The truncated normal distributions are sampled by the inverse cdf method, as described in Devroye (1987).

2. Sample β from the distribution $\beta|\mathbf{Z}, \mathbf{y}, \mathbf{\Sigma}$, where, under the prior $\beta \sim N(\beta_0, \mathbf{B}_0^{-1})$,

$$\beta|Z, \mathbf{y}, \mathbf{\Sigma} \sim N_k(\hat{\beta}, \mathbf{B}^{-1}),$$

   with $\hat{\beta} = \mathbf{B}^{-1}(\mathbf{B}_0\beta_0 + \sum_{i=1}^{n} \mathbf{X}_i'\mathbf{\Sigma}^{-1}\mathbf{Z}_i)$ and $\mathbf{B} = \mathbf{B}_0 + \sum_{i=1}^{n} \mathbf{X}_i'\mathbf{\Sigma}^{-1}\mathbf{X}_i$.

3. Sample σ from the conditional distribution $\sigma|\mathbf{Z},\mathbf{y},\beta$, where the density of the latter distribution (up to an irrelevant norming constant) is

$$g(\sigma) = \pi(\sigma) \prod_{i=1}^{n} \phi(\mathbf{Z}_i|\mathbf{X}_i\beta, \Sigma) I[\sigma \in \mathbf{C}]$$

and $\pi(\sigma)$ is the assumed prior on σ (say multivariate normal truncated to C). To sample $g(\sigma)$, let $q(\sigma|\mu, \mathbf{V})$ denote a multivariate-t density with parameters μ and $\mathbf{V}$ defined as the mode and inverse of the negative Hessian, respectively, of $\log g(\sigma)$. Then:

   (a) Sample a proposal value σ' from the density $q(\sigma'|\mu, \mathbf{V})$

   (b) Move to σ' given the current point σ with probability of move

$$\alpha(\sigma, \sigma') = \min\left(\frac{\pi(\sigma')\, f(\mathbf{Z}|\beta, \Sigma') I[\sigma' \in \mathbf{C}]}{\pi(\sigma)\, f(\mathbf{Z}|\beta, \Sigma) I[\sigma \in \mathbf{C}]} \frac{q(\sigma|\mu, \mathbf{V})}{q(\sigma'|\mu, \mathbf{V})}, 1\right),$$

   and stay at σ with probability $1 - \alpha(\sigma, \sigma')$.

4. Repeat Steps 1-3 using the most recent values of the conditioning variables.

6.2 *Algorithm 2*

1. Sample z_{ij} from the distribution $z_{ij}|\mathbf{Z}_{i(-j)}, \mathbf{y}_i, \beta, \mathbf{D}$ for $j \le J$ and $i \le n$, where $\mathbf{Z}_{i(-j)}$ are the latent values on the ith subject excluding z_{ij} and

$$z_{ij}|\mathbf{Z}_{i(-j)}, \mathbf{y}_i, \beta, \mathbf{D} \sim \begin{cases} TN_{(0,\infty)}(a_{ij}, b_{ij}) & \text{if} \quad y_{ij} = 1 \\ TN_{(-\infty,0])}(a_{ij}, b_{ij}) & \text{if} \quad y_{ij} = 0 \end{cases}$$

In this expression TN denotes the truncated normal distribution and a_{ij} and b_{ij} are the parameters of the conditional distributions $z_{ij}|\mathbf{Z}_{i(-j)}, \beta, \mathbf{V}_i$ (ignoring truncation).

2. Sample β from the distribution $\beta|\mathbf{Z}, \mathbf{y}, \mathbf{D}$, where, under the prior $\beta \sim N(\beta_0, \mathbf{B}_0^{-1})$,

$$\beta|\mathbf{Z}, \mathbf{y}, \Sigma \sim N_k(\hat{\beta}, \mathbf{B}^{-1}),$$

with $\hat{\beta} = \mathbf{B}^{-1}(\mathbf{B}_0\beta_0 + \sum_{i=1}^{n} \mathbf{X}_i'\mathbf{V}_i^{-1}\mathbf{Z}_i)$ and $\mathbf{B} = \mathbf{B}_0 + \sum_{i=1}^{n} \mathbf{X}_i'\mathbf{V}_i^{-1}\mathbf{X}_i$.

3. Sample $\mathbf{b}_i$ from the conditional distribution $\mathbf{b}_i|\mathbf{Z}_i, \beta, \mathbf{D}$ where

$$\mathbf{b}_i|\mathbf{Z}_i, \beta, \mathbf{D} \sim N_k(\hat{\mathbf{b}}_i, \mathbf{D}_i^{-1}),$$

$\hat{\mathbf{b}}_i = \mathbf{D}_i^{-1}(\mathbf{W}_i\mathbf{V}_i^{-1}(\mathbf{Z}_i - \mathbf{X}_i\beta)$ and $\mathbf{D}_i = (\mathbf{D} + \mathbf{W}_i'\mathbf{V}_i^{-1}\mathbf{W}_i)^{-1}$.

4. Sample $\mathbf{D}^{-1}$ from the distribution $\mathbf{D}^{-1}|\{\mathbf{b}_i\}$ where

$$\mathbf{D}^{-1}|\{\mathbf{b}_i\} \sim \text{Wishart}\left(\nu_0 + n, \left(\mathbf{R}_0^{-1} + \sum_{i=1}^{n} \mathbf{b}_i\mathbf{b}_i'\right)^{-1}\right),$$

under the assumption that the prior on $\mathbf{D}^{-1}$ is Wishart$(\nu_0, \mathbf{R}_0)$.

5. Repeat Steps 1-4 using the most recent values of the conditioning variables.

6.3 *Algorithm 3*

1. Sample z_{ij} from the distribution $z_{ij}|\mathbf{Z}_{i(-j)}, \mathbf{y}_i, \beta, \mathbf{D}$ for $j \leq J$ and $i \leq n$, where $\mathbf{Z}_{i(-j)}$ are the latent values on the ith subject excluding z_{ij} and

$$z_{ij}|\mathbf{Z}_{i(-j)}, \mathbf{y}_i, \beta, \mathbf{D} \sim \begin{cases} TN_{(0,\infty)}(a_{ij}, b_{ij}) & \text{if} \quad y_{ij} = 1 \\ TN_{(-\infty,0])}(a_{ij}, b_{ij}) & \text{if} \quad y_{ij} = 0 \end{cases}$$

In this expression TN denotes the truncated normal distribution and a_{ij} and b_{ij} are the parameters of the conditional distributions $z_{ij}|\mathbf{Z}_{i(-j)}, \beta, \mathbf{V}_i^*$ (ignoring truncation).

2. Sample β from the distribution $\beta|\mathbf{Z}, \mathbf{y}, \mathbf{D}$, where, under the prior $\beta \sim N(\beta_0, \mathbf{B}_0^{-1})$,

$$\beta|\mathbf{Z}, \mathbf{y}, \mathbf{\Sigma} \sim N_k(\hat{\beta}, \mathbf{B}^{-1}),$$

with $\hat{\beta} = \mathbf{B}^{-1}(\mathbf{B}_0\beta_0 + \sum_{i=1}^n \mathbf{A}_i'\mathbf{X}_i'\mathbf{V}_i^{*-1}\mathbf{Z}_i)$ and $\mathbf{B} = \mathbf{B}_0 + \sum_{i=1}^n \mathbf{A}_i'\mathbf{X}_i'\mathbf{V}_i^{*-1}\mathbf{A}_i\mathbf{X}_i$.

3. Sample $\mathbf{b}_i$ from the conditional distribution $\mathbf{b}_i|\mathbf{Z}_i, \beta, \mathbf{D}$ where

$$\mathbf{b}_i|\mathbf{Z}_i, \beta, \mathbf{D} \sim N_k(\hat{\mathbf{b}}_i, \mathbf{D}_i^{-1}),$$

$\hat{\mathbf{b}}_i = \mathbf{D}_i^{-1}(\mathbf{D}^{-1}\mathbf{A}_i\beta + \mathbf{X}_i\mathbf{V}_i^{*-1}\mathbf{Z}_i)$ and $\mathbf{D}_i = (\mathbf{D} + \mathbf{X}_i'\mathbf{V}_i^{*-1}\mathbf{X}_i)^{-1}$.

4. Sample $\mathbf{D}^{-1}$ from the distribution $\mathbf{D}^{-1}|\{\mathbf{b}_i\}$ where

$$\mathbf{D}^{-1}|\beta, \{\mathbf{b}_i\} \sim Wishart\left(\nu_0 + n, \left(\mathbf{D}^{-1} + \sum_{i=1}^n (\mathbf{b}_i - \mathbf{A}_i\beta)(\mathbf{b}_i - \mathbf{A}_i\beta)'\right)^{-1}\right).$$

5. Repeat Steps 1-4 using the most recent values of the conditioning variables.

References

Agresti, A. (1989). A survey of models for repeated ordered categorical response data. Statistics in Medicine, **8**: 1209-1224.

Albert, J. and Chib, S. (1993). Bayesian analysis of binary and polychotomous response data. *Journal of the American Statistical Association*, 88, 669–679.

Albert, J. and Chib, S. (1995). Bayesian Residual Analysis for Binary Response Regression Models. *Biometrika,* (1995), 82, 747-759.

Albert, J. and Chib, S. (1996). Bayesian Probit Modeling of Binary Repeated Measures Data with an Application to a Cross-Over Trial. In *Bayesian Biostatistics* (eds. D. A. Berry and D. K. Stangl), 577-599, New York: Marcel Dekker.

Amemiya, T. (1972). Bivariate probit analysis: minimum chi-square methods. *Journal of American statistical association* **69**, 940-44.

Ashford, J. R. & Sowden, R. R. (1970). Multivariate probit analysis. *Biometrics* **26**, 535–46.

Betensky, R A., Whittemore, A. S. (1996). An analysis of correlated multivariate binary data: Application to familial cancers of the ovary and breast. *Applied Statistics*, **45**: 411-429.

Carey, V, Zeger, S. L., Diggle, P. (1993). Modelling multivariate binary data with alternating logistic regressions. *Biometrika*, **80**: 517-526.

Carlin, B and Chib, S. (1995). Bayesian Model Choice via Markov Chain Monte Carlo. *Journal of the Royal Statistical Society*, Ser B, 57, 473-484.

Chen, M-H. and Dey, D.K. (1998). Bayesian analysis of correlated binary responses via scale mixture of multivariate normal link functions. Technical report, Department of Statistics, University of Connecticut.

Chib, S. (1995). Marginal Likelihood From the Gibbs Output. *Journal of the American Statistical Association,* 90, 1313-1321.

Chib, S. and Greenberg, E (1995). Understanding the Metropolis-Hastings algorithm. *American Statistician*, 49, 327–335.

Chib, S. and Greenberg, E (1998). Analysis of multivariate probit models. *Biometrika*, 85, 347-361.

Chib, S. and Carlin, B (1997). On MCMC Sampling in Hierarchical Longitudinal Models. *Statistics and Computing*, in press.

Connolly, M. A., Liang, Kung-Yee (1988). Conditional logistic regression models for correlated binary data. *Biometrika*, **75**: 501-506.

Cowles, M. K., Carlin, B. P., and Connett, J. E. (1996). Bayesian Tobit modeling of longitudinal ordinal clinical tral compliance data with nonignorable missingness. *Journal of the American Statistical Association*, 91, 86-98.

Fitzmaurice, Garrett M., Laird, NanM., Rotnitzky, A. G. (1993). Regression models for discrete longitudinal responses (Disc: p300-309). *Statistical Science*, **8**: 284-299.

Gelfand, A.E. and Smith, A.F.M. (1990). Sampling-Based Approaches to Calculating Marginal Densities. *Journal of the American Statistical Association*, 85, 398-409.

Glonek, G. F. V. and McCullagh, P. (1995). Multivariate logistic models. *J. R. Statist. Soc.* B **57**, 533–46.

Lesaffre, E. and Kaufmann, H. K. (1992). Existence and uniqueness of the maximum likelihood estimator for a multivariate probit model. *Journal of the American Statistical Association*, **87**: 805-811.

Lindley, D.V., and Smith, A.F.M. (1972). Bayes estimates for the linear model (with discussion). *J. Roy. Statist. Soc., Ser. B*, **34**, 1-41.

Muenz, L. R and Rubinstein, L. V. (1985). Markov models for covariate dependence of binary sequences. *Biometrics*, **41**: 91-101.

Ochi, Y., and Prentice, R. L. (1984). Likelihood inference in a correlated probit regression model. *Biometrika*, **71**: 531-543.

Oliveira, V. (1997). Bayesian prediction of clipped Gaussian random fields. Technical report.

Pendergast, J. F., Gange, S. J., Newton, M. A., Lindstrom, M. J., Palta, M. and Fisher, M. R. (1996). A survey of methods for analyzing clustered binary response data. International Statistical Review, **64**: 89-118.

Ritter, C. and Tanner, M.A. Facilitating the Gibbs stopper and the Griddy-Gibbs sampler. *Journal of the American Statistical Association*, 87, 861-868.

Stiratelli, R, Laird, N., and Ware, J. H (1984). Random-effects models for serial observations with binary response. *Biometrics*, **40**: 961-971.

Zeger, S. L. and Karim, M.R. (1991), Generalized linear models with random effects: A Gibbs sampling approach. *Journal of the American Statistical Association* 86, 79–86.

8

Bayesian Analysis for Correlated Ordinal Data Models

Ming-Hui Chen
Dipak K. Dey

ABSTRACT Bayesian methods are considered for analyzing correlated ordinal data. We present a new approach to Bayesian hierarchical generalized linear models using a rich class of scale mixture of multivariate normal (SMMVN) link functions. Fully parametric classical approaches to these are intractable and thus Bayesian methods are pursued using a Markov chain Monte Carlo sampling-based approach. Marginal likelihood approaches are used to compare the proposed models and simulation-based diagnostic procedures are developed to assess model adequacy. Novel computational algorithms to perform the Bayesian analysis are further developed and a real data example is used to illustrate the methodologies.

1. Introduction

There is a growing interest in the statistical literature concerning the modeling and analysis of correlated binary data. Prentice (1988) provided a comprehensive review of various modeling strategies using generalized linear regression analysis of correlated binary data with covariates associated at each binary response. Following Liang and Zeger (1986) and Zeger and Liang (1986), Prentice used the generalized estimating equation (GEE) approach to obtain consistent and asymptotically normal estimators of regression coefficients. In a Bayesian framework, Chib and Greenberg (1998) used the multivariate probit (MVP) model for correlated binary data, while Chen and Dey (1998) considered general scale mixture of multivariate normal (SMMVN) link functions for longitudinal binary responses. However, the literature is still sparse in modeling and analyzing correlated or repeated ordinal data. A correlated ordinal data problem is not a simple generalization of the one for the correlated binary data. More importantly, correlated ordinal data problems are encountered in many practical applications. Data obtained from surveys are inherently categorical, items in a questionnaire usually consist of two-to-five options (e.g., "disagree," "neutral" and "agree"), and two or more responses are typically taken from the same individuals. Similarly, cholesterol level (low, medium, high) and blood sugar level (low, medium, high) on the same individual are measured over the time in a longitudinal fashion.

In this chapter, we assume that two or more ordinal responses are taken one at time for the same subjects or repeated ordinal measurements are taken over time such as in longitudinal studies. Generalized linear regression methods are considered for such correlated ordinal data to study the relation between various covariates and the polychotomous outcome measure.

Cowles, Carlin and Connett (1996) used multivariate tobit models for analyzing longitudinal ordinal data which include correlations among the latent variables. However, they considered only the three-levels ordinal responses. Here we extend work of Ashford and Sowden (1970) in a very novel way to correlated ordinal models based on multivariate link functions using a very rich class of scale mixtures of normal distributions. Our models are very flexible and include, as a special case, multivariate probit (MVP), t-link (MVT), logit (MVL), stable distribution family links (MVS), and exponential power distribution family links (MVEP). These models are more attractive than random effects models since the exchangeability of correlation structure is not required. Our approach produces results with desirable features; furthermore, it addresses issues which have not previously been considered.

The first objective of this chapter is to explore different modeling strategies for the analysis of correlated ordinal responses from a Bayesian perspective by incorporating scale mixture of multivariate normal distributions as link functions. By considering a rich class of link functions in such models (within a parametric framework) this approach unifies all the previous methods as well as allowing enough flexibility. As one entertains a collection of such models for a given data set one needs to address the problem of model determination, i.e., model comparisons and model diagnostics. Specifically, we adopt the Markov chain Monte Carlo (MCMC) framework (e.g., Gelfand and Smith, 1990 and Tierney, 1994) to simulate the posterior distribution for proposed models. Due to the nature of correlated ordinal data, we use marginal likelihoods (Chib, 1995) to compare proposed models and we consider several simulation-based diagnostic procedures to assess model adequacy.

In this chapter, we also present various efficient computational algorithms for the complex simulation problems. For example, we use a reparameterization technique of Nandram and Chen (1996) to obtain SMMVN-link reparameterized models. Such a reparameterization technique greatly eases the computational burden. We use the collapsed Gibbs sampler of Liu (1994) to derive an efficient algorithm for simulating from the posterior distribution. One of the nice features in our algorithm is that we provide an efficient and nearly automatic Hastings scheme (Hastings, 1970) to generate the cutpoints needed to classify the ordinal response data, which is considered to be a very challenging problem in analyzing the Bayesian generalized linear models (see, for example, Cowles, 1996 and Nandram and Chen, 1996). Furthermore, in order to compare different SMMVN-link models, we provide an elegant implementation scheme of the data-augmentation-based method of Chib (1995) for computing marginal likelihoods. The computational algorithms presented in this chapter make the use of SMMVN-link models for analyzing the correlated ordinal data very attractive and convenient.

The rest of the chapter is organized as follows. In Section 2, we discuss the general structure of the scale mixture of multivariate normal link models and several special cases of this general setting. Section 3 is devoted to the development of the prior distributions as well as the distribution theory involved in the posterior calculations. In Section 4, we discuss different methods involved in model comparisons and model diagnostics. Here we consider some novel graphical approaches to implement different types of newly developed discrepancy measures and standardized latent residuals. In Section 5, we use an item response data example to illustrate the methodologies. Finally, Section 6 gives brief concluding remarks.

2. Models

We first introduce some notation which will be used throughout the paper. Suppose that we observe an ordinal (1 through L) response Y_{ij} on the ith observations and the jth variable and let $x_{ij} = (x_{ij1}, x_{ij2}, \ldots, x_{ijp_j})$ be the corresponding p_j-dimensional row regression vector for $i = 1, 2, \ldots, n$ and $j = 1, 2, \ldots, J$. (Note that x_{ij1} may be 1, which corresponds to an intercept.) Denote $Y_i = (Y_{i1}, Y_{i2}, \ldots, Y_{iJ})'$ and assume that Y_{i1}, Y_{i2}, ..., Y_{iJ} are dependent whereas Y_1, Y_2, ..., Y_n are independent. Let $y_i = (y_{i1}, y_{i2}, \ldots, y_{iJ})'$ and $y = (y_1, y_2, \ldots, y_n)$ be the observed data. Also let $\beta_j = (\beta_{j1}, \beta_{j2}, \ldots, \beta_{jp_j})'$ be a p_j-dimensional column vector of regression coefficients and $\beta = (\beta_1', \beta_2', \ldots, \beta_J')'$.

In order to set up the scale mixture of multivariate normal (SMMVN) link models for the correlated ordinal response data, we introduce a J-dimensional (latent) random vector $w_i^* = (w_{i1}^*, w_{i2}^*, \ldots, w_{iJ}^*)'$ such that

$$Y_{ij} = l, \quad \text{if} \quad \gamma_{j,l-1}^* \leq w_{ij}^* < \gamma_{jl}^*, \tag{1}$$

where $-\infty = \gamma_{j0}^* \leq \gamma_{j1}^* \leq \gamma_{j2}^* \leq \gamma_{j,L-1}^* \leq \gamma_{jL}^* = \infty$ are cutpoints for the jth ordinal response, which divide the real line into L intervals. As explained by Nandram and Chen (1996), we specify $\gamma_{j1}^* = 0$ to ensure the identifiability of the cutpoint parameters. Here, we introduce different sets of cutpoints for different ordinal responses since in many practical problems, each ordinal response may behave quite differently. We further assume that

$$w_i^* \sim N(x_i\beta^*, \kappa(\lambda)\Sigma^*), \tag{2}$$

and

$$\lambda \sim \pi(\lambda), \tag{3}$$

where $\kappa(\lambda)$ is a positive function of one-dimensional positive-valued scale mixing variable λ, $\pi(\lambda)$ is a mixing distribution which is either discrete or continuous, $x_i = diag(x_{i1}, x_{i2}, \ldots, x_{iJ})$, and $\beta^* = (\beta_1^{*\prime}, \beta_2^{*\prime}, \ldots, \beta_J^{*\prime})'$ is a $p = \sum_{j=1}^J p_j$ dimensional column vector of regression coefficients corresponding to the cutpoints $\gamma_j^* = (\gamma_{j2}^*, \gamma_{j3}^*, \ldots, \gamma_{j,L-1}^*)'$ for $j = 1, 2, \ldots, J$. In (2) we further take $\Sigma^* = \left(\rho_{jj^*}^*\right)_{J\times J}$ to be a correlation matrix such that $\rho_{jj}^* = 1$ to ensure the identifiability of the parameters. Such a w_i^* is sometimes called a tolerance variable since in a bioassay setting w_i^* can be a lethal dose of a drug.

For a special case where $\Sigma^* = I_J$, and I_J is the $J \times J$ identity matrix, Albert and Chib (1993) fitted the Bayesian independent ordinal probit model using the Gibbs sampler. Even for this simple independent ordinal probit model, the primary Gibbs sampler considered in Albert and Chib (1993) may present challenging problems in achieving convergence. In view of that, Cowles (1996) provided an algorithm which substantially improves convergence not only for the probit model but also for the cumulative logit and complementary log-log link models. Recently, Nandram and Chen (1996) proposed an algorithm using reparameterization technique which improves convergence even further. For the above general SMMVN-link models, the computation is even more challenging since we face two difficult sampling problems, i.e., (i) generating cutpoints and (ii) generating correlation matrix.

To ease the computational burden, we consider the following reparameterization:

$$\delta_j = 1/\gamma_{j,L-1}^*, \quad \gamma_{jl} = \delta_j\gamma_{jl}^*, \quad \beta_j = \delta_j\beta_j^*, \quad \text{and} \quad w_{ij} = \delta_j w_{ij}^* \tag{4}$$

for $j = 1, 2, \ldots, J$ and $i = 1, 2, \ldots, n$. With reparameterization (4), the SMMVN-link models given by (6) and (2) become

$$Y_{ij} = l, \text{ if } \gamma_{j,l-1} \leq w_{ij} < \gamma_{jl}, \tag{5}$$

and

$$w_i \sim N(x_i\beta, \kappa(\lambda)\Sigma), \tag{6}$$

where the reparameterized cutpoints are $-\infty = \gamma_{j0} \leq \gamma_{j1} = 0 \leq \gamma_{j2} \leq \cdots \leq \gamma_{j,L-1} = 1 \leq \gamma_{jL} = \infty$, $\Sigma = (\sigma_{jj^*})$, $\sigma_{jj} = \delta_j^2$, and $\sigma_{jj^*} = \delta_j\delta_{j^*}\rho^*_{jj^*}$ for $j \neq j^*$. The models given by (4) and (5) are thus called the SMMVN-link reparameterized models.

Notice that in (4), for each j, we have only $L - 3$ unknown cutpoints, and in (5), Σ is an unrestricted variance-covariance matrix, which has a great advantage in the implementation of MCMC sampling. We also note that reparameterization (4) does not affect the distribution of the scale mixing variable λ. That is, we still have the same mixing distribution $\pi(\lambda)$ for the mixing variable λ. The SMMVN-link reparameterized models have several attractive features. First, the number of unknown cutpoints is reduced by J. Second, all unknown cutpoints γ_{jl} are between 0 and 1, i.e., $0 \leq \gamma_{jl} \leq 1$ for $l = 2, 3, \ldots, L-2$ and $j = 1, 2, \ldots, J$. Third, the variance-covariance matrix Σ for w_i is unrestricted. Fourth, when $L = 3$, there are no unknown cutpoints. Due to these nice features, we use the SMMVN-link reparameterized models throughout the rest of this chapter.

Finally, we note that the distribution of w_i determines the joint distribution of Y_i through (4) and the variance-covariance matrix Σ captures the correlations among the Y_{ij}'s. More specifically, we have the joint distribution of the correlated ordinal responses given by

$$\begin{aligned} &P(Y_{i1} = y_{i1}, Y_{i2} = y_{i2}, \ldots, Y_{iJ} = y_{iJ} | \beta, \Sigma, \gamma, \lambda, x_i) \\ &= \int_{A_{i1}} \int_{A_{i2}} \cdots \int_{A_{iJ}} \frac{1}{(2\pi\kappa(\lambda))^{J/2} |\Sigma|^{\frac{1}{2}}} \\ &\times \exp\left\{ -\frac{[\kappa(\lambda)]^{-1}}{2} (w_i - x_i\beta)'\Sigma^{-1}(w_i - x_i\beta) \right\} dw_i, \end{aligned} \tag{7}$$

where $\gamma = (\gamma_1', \gamma_2', \ldots, \gamma_J')'$, $\gamma_j = (\gamma_{j2}, \gamma_{j3}, \ldots, \gamma_{j,L-2})'$, and

$$A_{ij} = (\gamma_{j,l-1}, \ \gamma_{jl}] \text{ if } y_{ij} = l, \text{ for } j = 1, 2, \ldots, J. \tag{8}$$

The class of SMMVN-link models is very rich and flexible, which include common MVP, MVT, MVL, MVS, and MVEP-link models. A brief explanation of such models is given as follows.

Taking $\kappa(\lambda) = 1$ and the mixing distribution $\pi(\{1\}) = 1$, the SMMVN models reduce to the MVP models. Similar to the MVP models, when we take $\kappa(\lambda) = 1/\lambda$ and $\lambda \sim \mathcal{G}(\nu/2, \nu/2)$, i.e.,

$$\pi(\lambda) = \frac{1}{\Gamma\left(\frac{\nu}{2}\right)} \left(\frac{\nu}{2}\right)^{\nu/2} \lambda^{\nu/2-1} \exp\left\{-\frac{\nu}{2}\lambda\right\},$$

the SMMVN-link models give the MVT models. As a special case, the MVT links reduce to the multivariate Cauchy (MVC) link when $\nu = 1$ and the MVP when $\nu \to \infty$.

Logit models are widely used to fit binary data (e.g., see Prentice 1988). As pointed out by Choy (1995), the SMMVN-link model leads to the MVL model when $\kappa(\lambda) = 4\lambda^2$ and λ follows an asymptotic Kolmogorov distribution with density

$$\pi(\lambda) = \pi_K(\lambda) = 8\sum_{k=1}^{\infty}(-1)^{k+1}k^2\lambda\exp\{-2k^2\lambda^2\}. \tag{9}$$

The MVL models are attractive since the exchangeability on the correlation structure is not required, which is advantageous compared to the random effects type of logistic regression models, for example, stratified and mixture models as given in Prentice (1988).

A multivariate stable distribution can be obtained as a scale mixture of multivariate normal distributions with $\kappa(\lambda) = 2\lambda$ and the mixing distribution $\pi(\lambda) = S^P(\alpha, 1)$ where the density of the positive stable distribution $S^P(\alpha, 1)$ in the polar form is given by

$$\pi_{S^P}(\lambda|\alpha, 1) = \frac{\alpha}{1-\alpha}\lambda^{-(\frac{\alpha}{1-\alpha}+1)}\int_0^1 s(u)\exp\left\{-\frac{s(u)}{\lambda^{\frac{\alpha}{1-\alpha}}}\right\}du, \text{ for } 0<\alpha<1, \tag{10}$$

with $s(u) = \left(\frac{\sin(\alpha\pi u)}{\sin(\pi u)}\right)^{\frac{\alpha}{1-\alpha}}\left(\frac{\sin[(1-\alpha)\pi u]}{\sin(\pi u)}\right)$. In our scenario, to obtain a heavy-tailed multivariate link model, we consider a symmetric multivariate stable (MVS) distribution $S_J(2\alpha, 0, x_i\beta, \Sigma)$ for w_i, where the characteristic function of $S_J(2\alpha, 0, x_i\beta, \Sigma)$ on the natural log scale is given by

$$\ln\psi(t) = \mathbf{i}\,(x_i\beta)'t - (t'\Sigma t)^{\alpha}, \quad \text{for } \alpha\in[1/2, 1),$$

with $t = (t_1, \ldots, t_J)'$ and $\mathbf{i}^2 = -1$. Note that when $\alpha = 1/2$, $S_J(1, 0, x_i\beta, \Sigma)$ is the multivariate Cauchy distribution, while

$$S_J(2, 0, x_i\beta, \Sigma) = \lim_{\alpha\to 1} S_J(2\alpha, 0, x_i\beta, \Sigma)$$

is a multivariate normal distribution. Therefore, MVC is a special case of MVT as well as a special case of MVS, and MVP is a limiting case of MVS.

Exponential power family distributions play an important role in Bayesian modeling, as indicated in Box and Tiao (1992), where they used a univariate exponential power family to model random effects in linear and nonlinear models. Formally, the density function of the multivariate exponential power family (MVEP) distribution has the form

$$\begin{aligned}&\pi_{EP}(w_i|x_i\beta, \Sigma, \alpha)\\ &= c_J\,|\Sigma|^{-1/2}\exp\left\{-\left[c_0(w_i - x_i\beta)'\Sigma^{-1}(w_i - x_i\beta)\right]^{\alpha}\right\},\end{aligned} \tag{11}$$

for $1/2 \le \alpha \le 1$, where α is called the kurtosis parameter and constants c_0 and c_J are defined as

$$c_0 = \frac{\Gamma\left(\frac{3}{2\alpha}\right)}{\Gamma\left(\frac{1}{2\alpha}\right)} \quad \text{and} \quad c_J = \frac{\alpha c_0^{J/2}\Gamma\left(\frac{J}{2}\right)}{\Gamma\left(\frac{J}{2\alpha}\right)\pi^{J/2}}. \tag{12}$$

We notice that a MVEP-link model is a special case of the SMMVN-link model with $\kappa(\lambda) = 1/(2c_0\lambda)$ and $\pi(\lambda) = \left(\frac{1}{\lambda}\right)^{J/2}\pi_{S^P}(\lambda|\alpha, 1)$, where $\pi_{S^P}(\lambda|\alpha, 1)$ is defined in (10). Further, there are two interesting special cases of the MVEP distributions, that is, the multivariate normal ($\alpha = 1$) and the multivariate double exponential distribution ($\alpha = 1/2$).

3. Prior Distributions and Posterior Computations

In this section we present prior distributions for SMMVN-link models and develop algorithms to perform posterior computations for such models.

3.1 Prior Distributions

First, we choose the same prior distribution for the regression coefficient vector β for all SMMVN-link models presented in Section 2. That is,

$$\pi(\beta|\beta_0, B_0) \propto \exp\left\{-\frac{1}{2}(\beta-\beta_0)'B_0(\beta-\beta_0)\right\}, \tag{13}$$

where B_0 is a precision matrix, β_0 is a location parameter vector, and both β_0 and B_0 are prespecified. Typically, we choose $\beta_0 = 0$ and $B_0 = diag\ (B_{11}, B_{12},\ldots, B_{1p}, B_{21}, B_{22},\ldots, B_{2p2},\ldots, B_{J1}, B_{J2},\ldots, B_{Jp_J})$ where B_{jl} is chosen to be small (e.g., B_{jl} = 0.01) so that a vague prior distribution for β is obtained, which ensures that the posterior is driven by the data.

Second, we choose

$$\Sigma^{-1} \sim W_J(n_0, Q_0), \tag{14}$$

where Q_0 is a $J \times J$ symmetric and positive definite matrix, $W_J(n_0, Q_0)$ denotes the Wishart distribution with degrees of freedom n_0 and mean matrix n_0Q_0, and n_0 and Q_0 are prespecified *a priori*. In our illustrative example, we take $n_0 = 11$ and $Q_0^{-1} = 0.001I_{10}$ where I_{10} is the 10-dimensional identity matrix, so that the prior is sufficiently diffuse.

Third, we take independent uniform priors on $\gamma_j = (\gamma_{j2}, \gamma_{j3}, \ldots, \gamma_{j,L-2})'$, i.e.,

$$\pi_g(\gamma_j) \propto 1, \text{ for } 0 \le \gamma_{j2} \le \gamma_{j3} \le \cdots \le \gamma_{j,L-2} \le 1, \tag{15}$$

for $j = 1, 2, \ldots, J$.

3.2 Posterior Computations

We use Gibbs sampling (e.g., Geman and Geman, 1984 and Gelfand and Smith, 1990) along with the Metropolis algorithms within Gibbs steps to perform the posterior computation. To sample from the posterior distribution, we need to generate β, Σ, γ_j, w_i and λ_i from their respective conditional distributions. The technical detail of computational implementation is given as follows.

Let $B = B_0 + \sum_{i=1}^n [\kappa(\lambda_i)]^{-1} x_i'\Sigma^{-1}x_i$ and

$$\hat{\beta} = B^{-1}\left(B_0\beta_0 + \sum_{i=1}^n [\kappa(\lambda_i)]^{-1} x_i'\Sigma^{-1}w_i\right).$$

Also let $w = (w_1', w_2', \ldots, w_n')'$ and $\Lambda = (\lambda_1, \lambda_2, \ldots, \lambda_n)'$. Then, given Σ, w, and Λ, we have

$$\beta \mid \Sigma, w, \Lambda, y \sim N\left(\hat{\beta}, B^{-1}\right). \tag{16}$$

From (5) and (13), we have that the conditional distribution of Σ^{-1} given β, w, and Λ is a Wishart distribution, that is,

$$\Sigma^{-1} \mid \beta, w, \Lambda, y$$

$$\sim \quad W_J \left(n + n_0, \left[Q_0^{-1} + \sum_{i=1}^{n} [\kappa(\lambda_i)]^{-1} (w_i - x_i\beta)(w_i - x_i\beta)' \right]^{-1} \right). \tag{17}$$

Therefore, generating β and Σ from (32) and (17) is straightforward. We notice that without reparameterization (4), we must draw the correlation matrix Σ^* from its conditional posterior distribution based on (2). From Chen and Dey (1998), it can be seen that generating a correlation matrix is much more difficult than drawing a variance-covariance matrix from a Wishart distribution.

To generate γ_j and w from their conditional distributions, the primary Gibbs sampler considered in Albert and Chib (1993) may present challenging problems in achieving convergence as discussed in Section 2. Therefore, we consider a more efficient MCMC sampling scheme as follows. Let $w_{(j)} = (w_{1j}, w_{2j}, \ldots, w_{nj})'$ and let $w_{(-j)}$ denote w with $w_{(j)}$ deleted for $j = 1, 2, \ldots, J$. Then, we use a cycle of J Gibbs steps to generate γ_j and $w_{(j)}$ jointly from their conditional distributions for $j = 1, 2, \ldots, J$ in turn. To draw γ_j and $w_{(j)}$ jointly from the conditional distribution $[\gamma_j, w_{(j)}|\beta, \Sigma, w_{(-j)}, \Lambda, y]$, we first draw γ_j from $[\gamma_j|\beta, \Sigma, w_{(-j)}, \Lambda, y]$, and then draw $w_{(j)}$ from $[w_{(j)}|\gamma_j, \beta, \Sigma, w_{(-j)}, y]$.

Given γ_j, β, Σ, $w_{(-j)}$, and Λ, the conditional distribution of w_{ij}, the ith component of $w_{(j)}$, is a truncated normal over interval A_{ij} given in (7) with the mean and variance given as follows:

$$\tilde{\mu}_{ij} = x_{ij}\beta_j + \Sigma_{jj}\Sigma_{(-j)}^{-1}(w_{i(-j)} - x_{i(-j)}\beta_{(-j)}) \tag{18}$$

and

$$\tilde{\sigma}_{ij}^2 = \kappa(\lambda_i)\left(\sigma_{jj} - \Sigma_{jj}\Sigma_{(-j)}^{-1}\Sigma_{jj}'\right). \tag{19}$$

In (18) and (19), $w_{i(-j)}$ is w_i with w_{ij} deleted, $x_{i(-j)}$ is x_i with the jth row deleted, $\beta_{(-j)}$ is β with β_j deleted, $\Sigma_{(-j)}$ is Σ with the jth row and jth column deleted, and $\Sigma_{jj} = (\sigma_{j1}, \ldots, \sigma_{j,j-1}, \sigma_{j,j+1}, \ldots, \sigma_{jJ})$. Therefore, we can use the algorithm of Geweke (1991) to generate w_{ij} from the above truncated normal distribution for $i = 1, 2, \ldots, n$.

It can be easily observed that given γ_j, β, and Σ, w_{1j}, w_{2j}, …, w_{nj} are independent. Therefore, the conditional distribution $[\gamma_j|\beta, \Sigma, w_{(-j)}, \Lambda, y]$ is

$$\begin{aligned}
&\pi(\gamma_j|\beta, \Sigma, w_{(-j)}, \Lambda, y) \\
\propto &\prod_{i:\ y_{ij}=2} \left\{ \Phi\left(\frac{\gamma_{j2} - \tilde{\mu}_{ij}}{\tilde{\sigma}_{ij}}\right) - \Phi\left(-\frac{\tilde{\mu}_{ij}}{\tilde{\sigma}_{ij}}\right) \right\} \\
\times &\prod_{i:\ y_{ij}=3} \left\{ \Phi\left(\frac{\gamma_{j3} - \tilde{\mu}_{ij}}{\tilde{\sigma}_{ij}}\right) - \Phi\left(\frac{\gamma_{j2} - \tilde{\mu}_{ij}}{\tilde{\sigma}_{ij}}\right) \right\} \\
\cdots &\prod_{i:\ y_{ij}=L-1} \left\{ \Phi\left(\frac{1 - \tilde{\mu}_{ij}}{\tilde{\sigma}_{ij}}\right) - \Phi\left(\frac{\gamma_{j,L-2} - \tilde{\mu}_{ij}}{\tilde{\sigma}_{ij}}\right) \right\},
\end{aligned} \tag{20}$$

where $\tilde{\mu}_{ij}$ and $\tilde{\sigma}_{ij}$ are defined by (18) and (19).

Generating γ_j from (20) is a challenging problem. Cowles (1996) proposed a Hastings scheme using a truncated normal distribution. By drawing γ_j simultaneously from its conditional distribution (20), Nandram and Chen (1996) developed an improved algorithm with a proposal density based on the Dirichlet distribution. However, the latter algorithm works well only when the cell counts, i.e.,

$n_{jl} = \sum_{i=1}^{n} 1_{\{l\}}(y_{ij})$, where the indicator function $1_{\{l\}}(y_{ij}) = 1$ if $y_{ij} = l$ and 0 otherwise, are relatively balanced. More recently, Chen and Schmeiser (1998) suggested to use a nearly automatic algorithm, a random-direction interior-point (RDIP) approach, to generate γ_j. The RDIP requires the minimum input from a user, but it may not be very efficient due to the nature of black-box algorithms.

In many correlated ordinal problems, the cell counts n_{jl}'s may be unbalanced and some of them are possibly missing; see Section 5 for an example. Therefore, it is difficult to apply any of the aforementioned existing algorithms. To remedy this problem, we use a Metropolis-Hastings algorithm to generate γ_j from (20) using a transformation technique. Let

$$\gamma_{jl} = \frac{\gamma_{j,l-1} + e^{\zeta_{jl}}}{1 + e^{\zeta_{jl}}}, \quad l = 2, \ldots, L-2 \tag{21}$$

and $\zeta_j = (\zeta_{j2}, \ldots, \zeta_{j,L-2})'$. Then, the conditional distribution $[\zeta_j | \beta, \Sigma, w_{(-j)}, \Lambda, y]$ is

$$\pi(\zeta_j | \beta, \Sigma, w_{(-j)}, \Lambda, y) \propto \pi(\gamma_j | \beta, \Sigma, w_{(-j)}, \Lambda, y) \prod_{l=2}^{L-2} \frac{(1 - \gamma_{j,l-1}) e^{\zeta_{jl}}}{(1 + e^{\zeta_{jl}})^2}, \tag{22}$$

where $\pi(\gamma_j | \beta, \Sigma, w_{(-j)}, \Lambda, y)$ is given by (20) and γ_j is evaluated at $\gamma_{jl} = (\gamma_{j,l-1} + e^{\zeta_{jl}})/(1 + e^{\zeta_{jl}})$ for $l = 2, 3, \ldots, L-2$. Instead of directly generating γ_j from (20), we first generate ζ_j from (22) and then use (21) to obtain γ_j.

To generate ζ_j, we use a multivariate normal proposal $N(\hat{\zeta}_j, \hat{\Sigma}_{\zeta_j})$, where $\hat{\zeta}_j$ is a maximizer of the logarithm of the right hand side of (22), which can be obtained by using the Nelder-Mead algorithm implemented by O'Neill (1971), and $\hat{\Sigma}_{\zeta_j}$ is the minus inverse of Hessian matrix of $\hat{\zeta}_j$, that is,

$$\hat{\Sigma}_{\zeta_j}^{-1} = - \left. \frac{\partial^2 \ln \pi(\zeta_j | \beta, \Sigma, w_{(-j)}, \Lambda, y)}{\partial \zeta_j \partial \zeta_j'} \right|_{\zeta_j = \hat{\zeta}_j}.$$

Then, following Hastings (1970), our algorithm to generate ζ_j operates as follows:

Step 1. Let ζ_j be the current value.

Step 2. Generate a proposal value ζ_j^* from $N(\hat{\zeta}_j, \hat{\Sigma}_{\hat{\zeta}_j})$.

Step 3. A move from ζ_j to ζ_j^* is made with probability

$$\min \left\{ \frac{\pi(\zeta_j^* | \beta, \Sigma, w_{(-j)}, \Lambda, y) \exp\left[-\frac{1}{2} (\zeta_j - \hat{\zeta}_j)' \hat{\Sigma}_{\zeta_j}^{-1} (\zeta_j - \hat{\zeta}_j) \right]}{\pi(\zeta_j | \beta, \Sigma, w_{(-j)}, \Lambda, y) \exp\left[-\frac{1}{2} (\zeta_j^* - \hat{\zeta}_j)' \hat{\Sigma}_{\zeta_j}^{-1} (\zeta_j^* - \hat{\zeta}_j) \right]}, 1 \right\}.$$

The novelty of this Metropolis-Hastings algorithm is that (a) unlike the other existing algorithms, the parameters in the multivariate normal proposal are specified within each Gibbs-step in an automatic fashion; (b) the proposal distribution roughly has the same shape as the true conditional distribution $\pi(\zeta_j | \beta, \Sigma, w_{(-j)}, \Lambda, y)$; (c) this new algorithm no longer requires the cell counts n_{jl} balanced. Therefore, our algorithm is more advantageous and efficient than the existing ones.

The conditional distribution $[\lambda_i|\beta, \Sigma, w_i, y]$ is

$$\pi(\lambda_i|\beta,\Sigma,w_i,y) \propto \left[\left(\frac{1}{2\pi\kappa(\lambda_i)}\right)^{J/2}|\Sigma|^{-1/2} \times \exp\left\{-\frac{[\kappa(\lambda_i)]^{-1}}{2}(w_i - x_i\beta)'\Sigma^{-1}(w_i - x_i\beta)\right\}\right]\pi(\lambda_i). \tag{23}$$

To generate λ_i from (23), we need to know the form of the mixing distribution $\pi(\lambda)$. Therefore, we present the random generation algorithms for a few special cases of SMMVN-link models.

For an MVP model, it does not require generating λ_i, since $\pi(\{\lambda_i = 1\}) = 1$. For an MVT model, $\pi(\lambda_i|\beta, \Sigma, w_i, y)$ in (23) reduces to

$$\mathcal{G}\left(\frac{\nu + J}{2}, \frac{1}{2}\left[\nu + (w_i - x_i\beta)'\Sigma^{-1}(w_i - x_i\beta)\right]\right),$$

where $\mathcal{G}(u, v)$ denotes a gamma distribution with density $\pi_{\mathcal{G}}(\lambda|u,v) \propto \lambda^{u-1}e^{-v\xi}$. Therefore, sampling λ_i from its conditional distribution is trivial.

For an MVL model, using an appropriate Student t approximation to the logistic distribution, Chen and Dey (1998) discovered that a good proposal density for $\pi_K(\lambda)$ given by (9) is

$$g_L(\lambda|\nu, b) = \frac{\left(\frac{\nu}{8b^2}\right)^{\nu/2}}{\Gamma\left(\frac{\nu}{2}\right)(\lambda^2)^{\nu/2+1}}\exp\left\{-\left(\frac{\nu}{8b^2}\right)\frac{1}{\lambda^2}\right\}2\lambda. \tag{24}$$

They further showed that the best choices of ν and b are $\nu = 5$ and $b = .712$ and they also provided an efficient way to evaluate the infinite series of $\pi_K(\lambda)$. It is interesting to mention that when we take

$$\lambda^2 \sim \mathcal{IG}\left(\frac{\nu}{2}, \frac{\nu}{8b^2}\right),$$

where $\mathcal{IG}(u, v)$ is an inverse gamma distribution with pdf $\pi_{\mathcal{IG}}(\lambda|u,v) = \frac{v^u}{\Gamma(u)\lambda^{u+1}}e^{-v/\lambda}$, $\lambda > 0$, then

$$\lambda \sim g_L(\lambda|\nu, b).$$

Therefore, to draw λ_i from (23), we can use the Metropolis sampling scheme. Let λ_i be the current value. Generate

$$\lambda^{*2}_i \sim \mathcal{IG}\left(\frac{J+\nu}{2}, \frac{1}{8}\left[(w_i - x_i\beta)'\Sigma^{-1}(w_i - x_i\beta) + \frac{\nu}{b^2}\right]\right). \tag{25}$$

Then, a move to the proposal point λ_i^* is made with probability

$$\min\left\{\frac{\pi_K(\lambda_i^*)/g_L(\lambda_i^*|\nu, b)}{\pi_K(\lambda_i)/g_L(\lambda_i|\nu, b)}, 1\right\}, \tag{26}$$

where $\pi_K(\lambda)$ and $g_L(\lambda_i|\nu, b)$ are given in (9) and (24) respectively.

For an MVS-link model, to draw λ_i from (23), Choy (1995) proposed the Sampling/Importance Resampling (SIR) method (Tanner, 1996), and the generalized Ratio-of-Uniform algorithm (Wakefield, Gelfand, and Smith, 1991), while Chen and

Dey (1998) developed the Metropolis algorithm (Metropolis *et al.*, 1953) with an inverse gamma proposal distribution. Here, we present only the Metropolis algorithm. Letting λ_i be the current value, we draw

$$\lambda_i^* \sim \mathcal{IG}\left(\frac{J+1}{2}, \frac{1}{4}\left[(w_i - x_i\beta)'\Sigma^{-1}(w_i - x_i\beta) + 1\right]\right). \tag{27}$$

Then, a move to the proposal point λ_i^* is made with probability

$$\min\left\{\frac{\pi_{SP}(\lambda_i^*|\alpha, 1)/\pi_{\mathcal{IG}}(\lambda_i^*|1/2, 1/4)}{\pi_{SP}(\lambda_i|\alpha, 1)/\pi_{\mathcal{IG}}(\lambda_i|1/2, 1/4)}, 1\right\}. \tag{28}$$

In the above Metropolis scheme, the full proposal density is proportional to

$$(4\pi\lambda_i)^{-\frac{J}{2}}|\Sigma|^{-\frac{1}{2}}\exp\left\{-\frac{1}{4\lambda_i}(w_i - x_i\beta)'\Sigma^{-1}(w_i - x_i\beta)\right\}\pi_{\mathcal{IG}}(\lambda_i|1/2, 1/4). \tag{29}$$

The proposal distribution given in (29) has heavier tails than the conditional distribution of λ_i and works well when α is not far away from 1/2.

Similar to the MVS-link model, we use a Metropolis algorithm with an inverse Gaussian proposal to generate λ_i for an MVEP-link model. Letting λ_i be the current value, we draw

$$\lambda_i^* \sim \mathcal{IN}(\mu_{EP}^*, \sigma_{EP}^*), \tag{30}$$

where

$$\mu_{EP}^* = \left(4c_0(w_i - x_i\beta)'\Sigma^{-1}(w_i - x_i\beta)\right)^{-1/2} \quad \text{and} \quad \sigma_{EP}^* = \frac{1}{2},$$

and the density of the inverse Gaussian distribution, $\mathcal{IN}(\mu^*, \sigma^*)$, is

$$\pi_{\mathcal{IN}}(\lambda|\mu^*, \sigma^*) = \sqrt{\frac{\sigma^*}{2\pi\lambda^3}}\exp\left\{-\frac{\sigma^*(\lambda - \mu^*)^2}{2\mu^{*2}\lambda}\right\}, \quad \text{for } \lambda > 0,$$

with parameters $\mu^* > 0$ and $\sigma^* > 0$. Then, a move to the proposal point λ_i^* is made with probability

$$\min\left\{\frac{\pi_{SP}(\lambda_i^*|\alpha, 1)/\pi_{\mathcal{IG}}(\lambda_i^*|1/2, 1/4)}{\pi_{SP}(\lambda_i|\alpha, 1)/\pi_{\mathcal{IG}}(\lambda_i|1/2, 1/4)}, 1\right\}. \tag{31}$$

In the Metropolis algorithm, the full proposal density is proportional to

$$\left(\frac{c_0\lambda_i}{\pi}\right)^{J/2}|\Sigma|^{-1/2}\exp\left\{-c_0\lambda_i(w_i - x_i\beta)'\Sigma^{-1}(w_i - x_i\beta)\right\}$$
$$\times\left(\frac{1}{\lambda_i}\right)^{J/2}\pi_{\mathcal{IG}}(\lambda_i|1/2, 1/4),$$

which exactly matches with an inverse Gaussian distribution. Finally, we mention that an elegant algorithm given in Devroye (1986, p 148) can be used to generate $\lambda \sim \mathcal{IN}(\mu_{EP}^*, \sigma_{EP}^*)$.

4. Model Determination

Once we have accomplished the first two steps of Bayesian analysis, i.e., constructing probability models and computing the posterior distributions of all parameters of interest, using a sampling-based approach, it is natural to compare several proposed models. It is also important to assess the fit of the selected models to the data and to own substantive knowledge. A good Bayesian analysis should always include at least some model comparisons along with some model diagnostics.

4.1 Model Comparisons

In this subsection we consider the problem of accounting for uncertainty about model form. Here we are faced with many models within a class of SMMVN-link models. Although we may wish to summarize our findings with a single model, there are usually many choices to be made. In this context, we consider marginal likelihood approach (Chib and Greenberg, 1998) for model comparisons since this approach is particularly suitable for the correlated ordinal data models.

Suppose that there are m models $\mathcal{M}_1$, $\mathcal{M}_2$, ..., $\mathcal{M}_m$ in our consideration. For model $\mathcal{M}_i$, we let $\pi(\beta, \Sigma^{-1}, \gamma \mid y, \mathcal{M}_i)$ denote the posterior distribution, which is

$$\pi(\beta, \Sigma^{-1}, \gamma \mid y, \mathcal{M}_i) = \frac{L(y \mid \beta, \Sigma, \gamma, \mathcal{M}_i)\pi(\beta, \Sigma^{-1}, \gamma)}{m(y|\mathcal{M}_i)}, \tag{32}$$

where $\gamma = (\gamma_1', \gamma_2', \ldots, \gamma_J')'$. In (32) the likelihood is

$$L(y \mid \beta, \Sigma, \gamma, \mathcal{M}_i) = \prod_{k=1}^{n} L(y_k \mid \beta, \Sigma, \gamma, \mathcal{M}_i)$$

and

$$L(y_k \mid \beta, \Sigma, \gamma, \mathcal{M}_i) = \int_0^\infty \int_{A_{k1}} \int_{A_{k2}} \cdots \int_{A_{kJ}} \frac{1}{(2\pi\kappa_i(\lambda_k))^{J/2} |\Sigma|^{\frac{1}{2}}}$$
$$\times \exp\left\{-\frac{[\kappa_i(\lambda_k)]^{-1}}{2}(w_k - x_k\beta)'\Sigma^{-1}(w_k - x_k\beta)\right\} \pi_i(\lambda_k) dw_k d\lambda_k,$$

where $\kappa_i(\lambda_k)$ and $\pi_i(\lambda_k)$ are the scale mixing function and the density function of the mixing variable λ_j associated with model $\mathcal{M}_i$, while the A_{kj}'s are defined in (7) based on the observed ordinal responses y_{kj}. Furthermore, in (32) the prior distribution, $\pi(\beta, \Sigma^{-1}, \gamma)$, is given by (11), (13), and (15), which is the same across all SMMVN-link models, and $m(y \mid \mathcal{M}_i)$ is the marginal likelihood.

To compare different SMMVN-link models, we calculate the marginal likelihoods for each of the models. In fact, it is in the same spirit to use the Bayes factor to compare two models (see, Kass and Raftery, 1995), which can be seen through the following identity for comparing models $\mathcal{M}_i$ and $\mathcal{M}_{i^*}$:

$$B_{ii^*} = \exp\{\ln(m(y \mid \mathcal{M}_i)) - \ln(m(y \mid \mathcal{M}_{i^*}))\},$$

where B_{ii^*} is the Bayes factor. We choose the model which yields the largest marginal likelihood $m(y \mid \mathcal{M}_i)$.

To estimate the marginal likelihood, we adopt a data-augmentation-based method of Chib (1995). Given some point $(\beta^*, \Sigma^*, \gamma^*)$ (typically the posterior means of β, Σ, and γ), we have an identity for the marginal likelihood on the natural log scale as

$$\begin{aligned} &\ln m(y \mid \mathcal{M}_i) \\ = &\sum_{k=1}^{n} \ln L(y_k \mid \beta^*, \Sigma^*, \gamma^*, \mathcal{M}_i) \\ &+ \ln \pi(\beta^*, {\Sigma^*}^{-1}, \gamma^*) - \ln \pi(\beta^*, {\Sigma^*}^{-1}, \gamma^* \mid y, \mathcal{M}_i). \end{aligned} \tag{33}$$

To estimate $\ln m(y \mid \mathcal{M}_i)$, we need to estimate the first term and the third term of (24). Following Chib and Greenberg (1998), we write

$$\ln \pi(\beta^*, {\Sigma^*}^{-1}, \gamma^* \mid y, \mathcal{M}_i)$$

$$= \ln \pi(\beta^* \mid \Sigma^*, \gamma^*, y, \mathcal{M}_i) + \ln \pi(\Sigma^{*-1} \mid \gamma^*, y, \mathcal{M}_i) + \ln \pi(\gamma^* \mid y, \mathcal{M}_i). \tag{34}$$

In (34),

$$\pi(\beta^* \mid \Sigma^*, \gamma^*, y, \mathcal{M}_i) = \int \pi(\beta^* \mid \Sigma^*, \gamma^*, w, \lambda, y, \mathcal{M}_i) \times \pi(w, \lambda \mid \Sigma^*, \gamma^*, y, \mathcal{M}_i) dw d\lambda, \tag{35}$$

$\pi(w, \lambda \mid \Sigma^*, \gamma^*, y, \mathcal{M}_i)$ is the conditional marginal posterior distribution of $w = (w_1, w_2, \ldots, w_n)'$ and $\lambda = (\lambda_1, \lambda_2, \ldots, \lambda_n)'$ given $\Sigma = \Sigma^*$ and $\gamma = \gamma^*$,

$$\pi(\Sigma^{*-1} \mid \gamma^*, y, \mathcal{M}_i) = \int \pi(\Sigma^{*-1} \mid \beta, w, \lambda, \gamma^*, y, \mathcal{M}_i) \times \pi(\beta, w, \lambda \mid \gamma^*, y, \mathcal{M}_i) d\beta dw d\lambda, \tag{36}$$

where $\pi(\beta, w, \lambda \mid \gamma^*, y, \mathcal{M}_i)$ is the conditional marginal posterior distribution of β, w, and λ given $\gamma = \gamma^*$, and

$$\pi(\gamma^* \mid y, \mathcal{M}_i) = \int \pi(\gamma^* \mid \beta, \Sigma, w, \lambda, y, \mathcal{M}_i) \times \pi(\beta, \Sigma^{-1}, w, \lambda \mid y, \mathcal{M}_i) \beta d\Sigma^{-1} dw d\lambda, \tag{37}$$

where $\pi(\beta, \Sigma^{-1}, w, \lambda \mid y, \mathcal{M}_i)$ is the marginal posterior distribution of β, Σ^{-1}, w, and λ.

To obtain simulation-consistent estimates of (35), (36), and (37), we independently generate $\{(w_1^{(r)}, \lambda_1^{(r)}),\ r = 1,2,\ldots,R\}$ from $\pi(w, \lambda \mid \Sigma^*, \gamma^*, y, \mathcal{M}_i)$, $\{(\beta_2^{(r)}, w_2^{(r)}, \lambda_2^{(r)}),\ r = 1, 2, \ldots, R\}$ from $\pi(\beta, w, \lambda \mid \gamma^*, y, \mathcal{M}_i)$, and $\{(\beta_3^{(r)}, \Sigma_3^{(r)}, w_3^{(r)}, \lambda_3^{(r)}),\ r = 1, 2, \ldots, R\}$ from $\pi(\beta, \Sigma^{-1}, w, \lambda \mid y, \mathcal{M}_i)$. Note that all three MCMC outputs are easy to obtain by adopting the MCMC sampling algorithms presented in Section 3.2. Then, a simulation-consistent estimate of (35) is

$$\hat{\pi}(\beta^* \mid \Sigma^*, \gamma^*, y, \mathcal{M}_i) = \frac{1}{R} \sum_{r=1}^{R} \pi(\beta^* \mid \Sigma^*, \gamma^*, w_1^{(r)}, \lambda_1^{(r)}, y, \mathcal{M}_i). \tag{38}$$

Note that in (38),

$$\pi(\beta^* \mid \Sigma^*, \gamma^*, w_1^{(r)}, \lambda_1^{(r)}, y, \mathcal{M}_i) = \left(\frac{1}{2\pi}\right)^{p/2} |B^{(r)}|^{1/2} \exp\left\{ -\frac{(\beta^* - \hat{\beta}^{(r)})' B^{(r)} (\beta^* - \hat{\beta}^{(r)})}{2} \right\},$$

where $p = \sum_{j=1}^{J} p_j$,

$$\hat{\beta}^{(r)} = \left(B^{(r)}\right)^{-1} \left(B_0 \beta_0 + \sum_{k=1}^{n} [\kappa_i(\lambda_{k1}^{(r)})]^{-1} x_k' (\Sigma^*)^{-1} w_{l1}^{(r)} \right),$$

and $B^{(r)} = B_0 + \sum_{k=1}^{n} [\kappa_i(\lambda_{k1}^{(r)})]^{-1} x_k' (\Sigma^*)^{-1} x_k$. A simulation-consistent estimate of (36) is

$$\hat{\pi}(\Sigma^{*-1} \mid \gamma^*, y, \mathcal{M}_i) = \frac{1}{R} \sum_{r=1}^{R} \pi(\Sigma^{*-1} \mid \beta_2^{(r)}, w_2^{(r)}, \lambda_2^{(r)}, \gamma^*, y, \mathcal{M}_i). \tag{39}$$

In (39),

$$\begin{aligned}&\pi(\Sigma^{*-1} \mid \beta_2^{(r)}, w_2^{(r)}, \lambda_2^{(r)}, \gamma^*, y, \mathcal{M}_i) \\ &= \frac{|V|^{-(n+n_0)/2} \exp\left\{-\frac{1}{2} tr\left(V^{-1}\Sigma^{*-1}\right)\right\} |\Sigma^*|^{-\frac{n+n_0-J-1}{2}}}{2^{(n+n_0)J/2}\pi^{J(J-1)/4}\prod_{j=1}^{J}\Gamma\left(\frac{n+n_0-j+1}{2}\right)},\end{aligned}$$

where

$$V = \left[Q_0^{-1} + \sum_{l=1}^{n}[\kappa_i(\lambda_{k2}^{(r)})]^{-1}(w_{k2}^{(r)} - x_k\beta_2^{(r)})(w_{k2}^{(r)} - x_k\beta_2^{(r)})'\right]^{-1}.$$

Assume that all sets $\{i^* : \ y_{i^*j} = l, \ i^* = 1, 2, \ldots, n\}$ for $l = 2, 3, \ldots, L-1$ and $j = 1, 2, \ldots, J$ are not empty. Under the above assumption, a simulation-consistent estimate of (37) is given by

$$\hat{\pi}(\gamma^* \mid y, \mathcal{M}_i) = \frac{1}{R}\sum_{r=1}^{R}\pi(\gamma^* \mid \beta_3^{(r)}, \Sigma_3^{(r)}, w_3^{(r)}, \lambda_3^{(r)}, y, \mathcal{M}_i), \tag{40}$$

where

$$\begin{aligned}&\pi(\gamma^* \mid \beta_3^{(r)}, \Sigma_3^{(r)}, w_3^{(r)}, \lambda_3^{(r)}, y, \mathcal{M}_i) \\ &= \prod_{j=1}^{J}\prod_{l=2}^{L-2}\frac{1}{\min\{w_{i^*j3}^{(r)} : y_{i^*j} = l+1\} - \max\{w_{i^*j3}^{(r)} : y_{i^*j} = l\}}\end{aligned} \tag{41}$$

for $\max\{w_{i^*j3}^{(r)} : y_{i^*j} = l\} < \gamma_{jl}^* \le \min\{w_{i^*j3}^{(r)} : y_{i^*j} = l+1\}$, $l = 2, 3, \ldots, L-2$, and $j = 1, 2, \ldots, J$. Note that the derivation of equation (41) follows from the fact that if the assumption,

$$max\{w_{i^*j3}^{(r)} : y_{i^*j} = l\} < \min\{w_{i^*j3}^{(r)} : y_{i^*j} = l+1\},$$

for $l = 2, 3, \ldots, L-2$ and $j = 1, 2, \ldots, J$, holds, the cutpoints γ_{jl}'s are independent. If the above assumption is violated, which is rare in practice, (41) still works with an obvious adjustment. However, if L or J is large, the above Monte Carlo approach may not be efficient because of high dimensionality of the problem. A more efficient Monte Carlo method can be obtained by using a sequence of $J-3$ dimensional conditional marginal distributions for γ. To explore this point, we let $\gamma_{(+j)}^* = (\gamma_1^{*\prime}, \gamma_2^{*\prime}, \ldots, \gamma_j^{*\prime})'$ for $j = 1, 2, \ldots, J$. Then, we have

$$\begin{aligned}&\pi(\gamma^* \mid y, \mathcal{M}_i) = \pi(\gamma_1^* \mid y, \mathcal{M}_i) \\ &\times\pi(\gamma_2^* \mid \gamma_{(+1)}^*, y, \mathcal{M}_i)\cdots\pi(\gamma_J^* \mid \gamma_{(+(J-1))}^*, y, \mathcal{M}_i),\end{aligned} \tag{42}$$

and

$$\begin{aligned}\pi(\gamma_j^* \mid \gamma_{(+(j-1))}^*, y, \mathcal{M}_i) &= \int \pi(\gamma_j^* | \beta, \Sigma, w, \lambda, \gamma_{(+(j-1))}^*, y, \mathcal{M}_i) \\ &\times\pi(\beta, \Sigma^{-1}, w, \lambda | \gamma_{(+(j-1))}^*, y, \mathcal{M}_i) d\beta d\Sigma^{-1} dw d\lambda,\end{aligned} \tag{43}$$

for $j = 1,2,\ldots,J$. Then, similar to (40) and (41), an efficient estimate of π (γj*|γ*(+(j-1)),*Mi*) can be obtained by using a random sample generated from π(β,Σ,w,λ | C*(*|γ*(+(j-1)),*Mi*).

Next, we discuss a Monte Carlo method to estimate the probability L(*yk* | β*,Σ*,γ*, *Mi*).

Note that Monte Carlo algorithms proposed by Chib and Greenberg (1998) and

Chen and Dey (1998) for correlated binary response data problems may not be applicable here because of simulation inefficiency of their algorithms in high dimension. Let $\Sigma_d^* = diag(\Sigma^*)$, $c_L = L(y_k \mid \beta^*, \Sigma^*, \gamma^*, \mathcal{M}_i)$, and $c_L^* = L(y_k \mid \beta^*, \Sigma_d^*, \gamma^*, \mathcal{M}_i)$. Then, c_L^* can be evaluated numerically, which may involve a one-dimensional integral. Letting

$$\pi^*(w_k, \lambda \mid \beta^*, \Sigma^*, y, \mathcal{M}_i) = \frac{|\Sigma^*|^{-\frac{1}{2}}}{(2\pi\kappa_i(\lambda_k))^{J/2}}$$
$$\times \exp\left\{-\frac{[\kappa_i(\lambda_k)]^{-1}}{2}(w_k - x_k\beta^*)'(\Sigma^*)^{-1}(w_k - x_k\beta^*)\right\}\pi_k(\lambda_k),$$

we have

$$\text{Ra} = \frac{c_L}{c_L^*} = E\left[\frac{\pi^*(w_k, \lambda \mid \beta^*, \Sigma^*, y, \mathcal{M}_i)}{\pi^*(w_k, \lambda \mid \beta^*, \Sigma_d^*, y, \mathcal{M}_i)}\right], \tag{44}$$

where the expectation is taken with respect to $\pi(w_k, \lambda \mid \beta^*, \Sigma_d^*, y, \mathcal{M}_i)$, which is proportional to $\pi^*(w_k, \lambda \mid \beta^*, \Sigma_d^*, y, \mathcal{M}_i)$. Then, we use the following steps to obtain an estimate of $L(y_j \mid \beta^*, \Sigma^*, \gamma^*, \mathcal{M}_i)$:

Step 1. Generate $(w_k^{(r)}, \lambda^{(r)})$ from $\pi(w_k, \lambda \mid \beta^*, \Sigma_d^*, y, \mathcal{M}_i)$ using the Gibbs sampler for $r = 1, \ldots, R$. The necessary steps required in Gibbs sampling are:

(i) generate $w_k^{(r)} \mid \lambda^{(r-1)} \sim N\left(x_k\beta^*, \kappa_i(\lambda_k^{(r-1)})\Sigma_d^*\right)$ over the constrained space $A_{k1}^* \times A_{k2}^* \times \cdots \times A_{kJ}^*$,

(ii) generate $\lambda^{(r)}$ from $[\lambda \mid w_k^{(r)}]$ using a procedure presented in Section 3.2.3.

Step 2. Calculate the average

$$\widehat{\text{Ra}} = \frac{1}{R}\sum_{r=1}^{R}\frac{\pi(w_k^{(r)}, \lambda^{(r)} \mid \beta^*, \Sigma^*, y, \mathcal{M}_i)}{\pi(w_k^{(r)}, \lambda^{(r)} \mid \beta^*, \Sigma_d^*, y, \mathcal{M}_i)}, \tag{45}$$

and compute $\ln \widehat{L}(y_k \mid \beta^*, \Sigma^*, \gamma^*, \mathcal{M}_i) = \ln c_L^* + \ln \widehat{\text{Ra}}$.

Note that the above approach is indeed the one for estimating ratios of two normalizing constants (see, e.g., Meng and Wong, 1996 and Chen and Shao, 1997).

Finally we comment that as we can see from the above, computing the marginal likelihoods for the correlated ordinal data problems is very different from the one for the correlated binary data problems. We also comment that as empirically shown by Chen and Dey (1998), the methods that use ratios of normalizing constants (see, e.g., Meng and Wong, 1996 and Chen and Shao, 1997) to estimate directly the Bayes factors are not efficient for the SMMVN-link models. This is partially due to the fact that the posterior distributions in the class of the SMMVN-link models are relatively far apart from each other. Therefore, the above proposed methods will greatly gain precision of the Monte Carlo estimation for computing the marginal likelihoods.

4.2 Model Diagnostics

In the earlier sections we presented a collection of models and proceeded with Bayesian inference. The issue at stake is which model is most adequate for the given correlated ordinal data as it is well-known that an inadequate model could lead to a misleading conclusion. In classical statistics, goodness of fit tests have been employed to check the plausibility of the model fit to the data. The classical goodness of fit test quantifies the extremeness of a particular discrepancy measure by calculating an accumulated tail probability that the model under a specified null hypothesis is true. In the classical setup the test statistic and hence the accumulated tail probability is a function of both the data and the unknown parameters which are specified only under the null hypothesis. As an alternative to the classical goodness-of-fit test, we develop two Bayesian model checking methods for the correlated ordinal data.

Since we use a class of SMMVN-link models for the correlated ordinal responses in a unified fashion, our first approach for model diagnostic is to check the appropriateness of the link function at the second stage of the hierarchical model. Recall that in the general structure of the SMMVN-link models, the prior distribution of λ_i for the ith observation forms the desired link. Thus the appropriateness of the link at each observation i, $i = 1, 2, \ldots, n$, can be checked by comparing the prior distribution versus the posterior distribution of λ_i's. Such comparison will detect outlying group for a given SMMVN-link model. For example, the MVT models correspond to a nuisance parameter λ_i from a gamma distribution with prior mean one. Thus, if the posterior mean of λ_i is quite smaller than one, it indicates variance inflation of the normal link. As mentioned in Vounatsou, Smith, and Choy (1996), the posterior means of λ_i may not exist. In this scenario, we use transformed variables $\psi_i = \ln \lambda_i$, $i = 1, 2, \ldots, n$, instead.

Since the Monte Carlo samples of λ_i's from their posteriors are readily available from the MCMC outputs, we can simply calculate ψ_i and compare the posterior and prior distributions of $\psi_i = \ln \lambda_i$ through several quantiles. In spirit, this is essentially an exploratory data analysis approach of model checking as indicated in Dey, Gelfand, Swartz, and Vlachos (1998). We denote the respective 5-quantiles of the prior and posterior distributions of ψ_i to be

$$q_{\lambda_i}^{\mathrm{pr}} = (q_{\lambda_i,0.05}^{\mathrm{pr}}, q_{\lambda_i,0.25}^{\mathrm{pr}}, q_{\lambda_i,0.5}^{\mathrm{pr}}, q_{\lambda_i,0.75}^{\mathrm{pr}}, q_{\lambda_i,0.95}^{\mathrm{pr}})'$$

and

$$q_{\lambda_i}^{\mathrm{po}} = (q_{\lambda_i,0.05}^{\mathrm{pr}}, q_{\lambda_i,0.25}^{\mathrm{po}}, q_{\lambda_i,0.5}^{\mathrm{po}}, q_{\lambda_i,0.75}^{\mathrm{po}}, q_{\lambda_i,0.95}^{\mathrm{po}})'.$$

Note that $q_{\lambda_i}^{\mathrm{pr}}$ can be obtained analytically or by using Monte Carlo samples from their corresponding distributions while $q_{\lambda_i}^{\mathrm{po}}$ can be computed by using readily available Monte Carlo samples from the Gibbs outputs. Now, if individual observations are of interest, we define the following observational level discrepancy measure:

$$D_i = \left|\left|q_{\lambda_i}^{\mathrm{po}} - q_{\lambda_i}^{\mathrm{pr}}\right|\right|^2, \quad \text{for } i = 1, 2, \ldots, n. \tag{46}$$

Due to the complexity of SMMVN-link models, it does not appear possible to analytically derive the distribution of d_i. Therefore, it is difficult to find a cutoff point for D_i that discriminates between "good" observation and aberrant value. To overcome such difficulty, we consider the standardized discrepancy measures of the D_i's. Let $\overline{D} = (1/n)\sum_{i=1}^n D_i$ and let $S(D)$ be the sample standard deviation of the D_i's. Then, we define the standardized observational level discrepancy measure:

$$d_i = \frac{D_i - \overline{D}}{S(D)} \tag{47}$$

for $i = 1, 2, \ldots, n$. Thus, when $|d_i| > k^*$ (typically, we choose $k^* = 3$), the ith observation will be viewed as aberrant. Finally, we notice that since the MVP-link model is embedded within MVT-link model, one can choose a large value of the degrees of freedom ν for the gamma distribution of λ_i and do similar comparisons. Similar argument works if one thinks that the MVP is also embedded within stable family or double exponential family.

For the purposes of checking the model adequacy and detecting outliers at the component level, we use Bayesian latent residuals. By generalizing the univariate Bayesian residuals of Albert and Chib (1995), we introduce our latent residuals as

$$\epsilon_{ij} = \frac{w_{ij} - \mu_{ij}}{\sigma_{ij}}, \tag{48}$$

where $\mu_{ij} = E(w_{ij}|y)$ and $\sigma_{ij}^2 = Var(w_{ij}|y)$, that is, μ_{ij} and σ_{ij}^2 are the posterior mean and posterior variance of w_{ij}, for $j = 1, 2, \ldots, J$, $i = 1, 2, \ldots, n$. Note that μ_{ij} and σ_{ij} can be simply calculated by using the readily available Monte Carlo samples of w_{ij}'s from the Gibbs outputs. Therefore, no additional MCMC samples are needed in order to obtain the latent residuals ϵ_{ij}'s.

Based on the above Bayesian latent residuals, we can use similar tools such as boxplots of the posterior distributions of the ϵ_{ij}'s to assess the model fitting and to detect componentwise outliers. Instead of overlaying posterior boxplots for each component considered in Albert and Chib (1995), we suggest to calculate $P(|\epsilon_{ij}| \geq K^* \mid y)$ for each component and plot $P(|\epsilon_{ij}| \geq K^* \mid y)$ versus $E(y_{new,ij}|y)$ where the expectation is taken with respect to the posterior predictive distribution $\pi(y_{new}|y)$. Note that to obtain an efficient Monte Carlo estimate of $E(y_{new,ij}|y)$, we can use the following identity

$$E(y_{new,ij}|y) = E\left[E(y_{new,ij}|x_{ij}, \beta_j, \Sigma, \gamma_j, y)\right],$$

where the first expectation is taken with respect to the posterior distribution $\pi(\beta_j, \Sigma, \gamma_j|y)$ and the second expectation is

$$\begin{aligned} &E(y_{new,ij}|x_{ij}, \beta_j, \Sigma, \gamma_j, y) \\ &= \int_0^\infty \sum_{l=1}^{L} l \left[\Phi\left(\frac{\gamma_{jl} - x_{ij}\beta_j}{\sqrt{\kappa(\lambda)}\sigma_{jj}}\right) - \Phi\left(\frac{\gamma_{j,l-1} - x_{ij}\beta_j}{\sqrt{\kappa(\lambda)}\sigma_{jj}}\right)\right] \pi(\lambda) d\lambda. \end{aligned} \tag{49}$$

Note that in (49), we denote $\Phi\left(\frac{\gamma_{jl} - x_{ij}\beta_j}{\sqrt{\kappa(\lambda)}\sigma_{jj}}\right)$ to be 1 when $l = L$ and 0 when $l = 0$. Several different values of K^*, e.g., K^* =1,2,3, can be used. We feel that our residual plots are more effective because of high dimensionality of the problem. Here, we do not consider a multivariate version of Bayesian residuals since the multivariate Bayesian residuals suffer loss of identification, making the component level interpretation difficult.

5. Item Response Data Example

The Department of Mathematical Sciences at Worcester Polytechnic Institute (WPI) recently conducted a survey. The results from the survey were to be used in the renovation of the Master's degree program for secondary teachers. One survey question contains ten features (items) of Master's degree programs for secondary mathematics teachers and the teachers were asked to identify which features are

TABLE 8.1. Summary of the Data

Group	Response	Items									
		1	2	3	4	5	6	7	8	9	10
I	1	1	2	4	4	1	1	-	1	1	2
	2	-	-	12	10	2	2	8	2	1	5
	3	9	3	36	34	13	7	16	13	9	21
	4	26	25	17	20	31	30	30	33	35	33
	5	40	46	7	8	29	36	22	27	30	15
II	1	-	-	3	12	2	-	5	5	4	4
	2	-	2	6	8	5	-	5	3	7	10
	3	-	9	21	13	7	6	13	10	7	16
	4	12	13	13	16	25	11	18	19	15	15
	5	39	27	8	2	12	34	10	14	18	6

important. Each individual responded in one of "not important," "somewhat important," "average important," "important," and "very important" for each item (feature). The subjects included teachers from two groups: I. Prospective Students, and II. Students who have been part of the WPI program. Prospective Students were defined to be one faculty member (usually the department head) from every high school mathematics department within a sixty-mile radius of WPI. The survey was sent to 315 secondary mathematics teachers in Massachusetts in November 1993 and completed surveys were received from 127 teachers.

A summary of the data is given in Table 8.1. Note that in Table 8.1, the values of the response 1, 2, 3, 4, and 5 correspond "not important," "somewhat important", "average important," "important," and "very important," respectively, and each entry represents the count. See Ganter, Paulauskas, and Chen (1995) for the complete information about this survey. We use this example to demonstrate the computational feasibility of the methodologies described in Sections 2 to 4 as well as to illustrate how SMMVN-link models can be applied to a real item response data problem.

Let y_{ij} be the response of the jth item from the ith individual and let β_{jk} to denote the intercept for the jth item within group k where $k = 1$ and $k = 2$ correspond to Group I and Group II, respectively. To assess the importance of each item, we consider the following mean-ranking measure

$$\mu_{jk} = \int_0^\infty \sum_{l=1}^{5} l \left[\Phi\left(\frac{\gamma_{jl} - \beta_{jk}}{\sqrt{\kappa(\lambda)}\sigma_{jj}} \right) - \Phi\left(\frac{\gamma_{j,l-1} - \beta_{jk}}{\sqrt{\kappa(\lambda)}\sigma_{jj}} \right) \right] \pi(\lambda) d\lambda \tag{50}$$

for $j = 1, 2, \ldots, 10$ and $k = 1, 2$.

For illustrative purposes, we consider three SMMVN-link models to fit the WPI survey data. These models are the MVP, MVL, and MVC (i.e., MVT with $\nu = 1$). These models capture different aspects and features of the SMMVN-link models. For example, the MVP and the MVC correspond to the lightest and the heaviest tails respectively, while the MVL model is roughly in the "halfway" between the MVP and MVC models. In the implementation of the Gibbs sampler, we use the Metropolis-Hastings algorithm to generate the cutpoints γ_j in the Metropolis step. Note that the NC algorithm cannot be applied in this example because the counts n_{jl}

TABLE 8.2. Bayesian Estimates of the Mean-Ranking Measures μ_{jk} (Group I): Part (a)

Item	Model	Posterior Mean	Posterior Std Dev	95% HPD Interval
1	MVP	4.348	0.084	(4.180, 4.508)
	MVL	4.343	0.083	(4.172, 4.500)
	MVC	4.369	0.091	(4.188, 4.535)
2	MVP	4.465	0.090	(4.283, 4.628)
	MVL	4.485	0.081	(4.319, 4.634)
	MVC	4.556	0.077	(4.406, 4.704)
3	MVP	3.146	0.116	(2.922, 3.378)
	MVL	3.192	0.109	(2.964, 3.395)
	MVC	3.257	0.089	(3.086, 3.432)
4	MVP	3.227	0.127	(2.984, 3.479)
	MVL	3.208	0.119	(2.970, 3.435)
	MVC	3.255	0.099	(3.072, 3.453)
5	MVP	4.098	0.104	(3.899, 4.304)
	MVL	4.127	0.095	(3.933, 4.301)
	MVC	4.200	0.090	(4.030, 4.375)

are very unbalanced and indeed we have few missing cells (see Table 8.1). Also note that this Metropolis-Hastings algorithm works quite well and this algorithm results in an acceptance probability of approximately 87% for the MVP model, 86% for the MVL model, and 85.5% for the MVC model. We check the convergence of the Gibbs sampler using several diagnostic procedures as recommended by Cowles and Carlin (1996). After convergence, we generate a large number of Gibbs iterates for various Bayesian calculations. It is noteworthy that after the convergence, we calculate the autocorrelations for all model parameters and we find that the autocorrelations for all β_{jk}'s and μ_{jk}'s disappear at lag 5 while the autocorrelations for all cutpoints γ_{jl}'s disappear at lag 10.

First, using 50,000 Gibbs iterates after convergence, we compute the posterior estimates and 95% highest posterior density (HPD) intervals for the mean-ranking measures μ_{jk}'s for all three models.

TABLE 8.3. Bayesian Estimates of the Mean-Ranking Measures μ_{jk} (Group I): Part (b)

Item	Model	Posterior Mean	Posterior Std Dev	95% HPD Interval
6	MVP	4.273	0.098	(4.079, 4.462)
	MVL	4.281	0.092	(4.096, 4.453)
	MVC	4.298	0.096	(4.112, 4.480)
7	MVP	3.852	0.118	(3.611, 4.074)
	MVL	3.807	0.113	(3.588, 4.023)
	MVC	3.813	0.094	(3.627, 3.995)
8	MVP	4.050	0.113	(3.825, 4.267)
	MVL	4.023	0.110	(3.809, 4.236)
	MVC	4.028	0.095	(3.839, 4.217)
9	MVP	4.159	0.112	(3.931, 4.366)
	MVL	4.156	0.107	(3.936, 4.352)
	MVC	4.168	0.097	(3.982, 4.360)
10	MVP	3.705	0.116	(3.479, 3.930)
	MVL	3.693	0.106	(3.486, 3.899)
	MVC	3.689	0.088	(3.520, 3.860)

TABLE 8.4. Bayesian Estimates of the Mean-Ranking Measures μ_{jk} (Group II): Part (a)

Item	Model	Posterior Mean	Posterior Std Dev	95% HPD Interval
1	MVP	4.748	0.068	(4.615, 4.876)
	MVL	4.741	0.070	(4.597, 4.866)
	MVC	4.699	0.074	(4.541, 4.820)
2	MVP	4.264	0.126	(4.013, 4.496)
	MVL	4.318	0.118	(4.084, 4.545)
	MVC	4.444	0.122	(4.203, 4.670)
3	MVP	3.331	0.142	(3.051, 3.607)
	MVL	3.341	0.142	(3.341, 3.625)
	MVC	3.380	0.147	(3.107, 3.669)
4	MVP	2.755	0.160	(2.437, 3.057)
	MVL	2.897	0.161	(2.582, 3.208)
	MVC	3.185	0.137	(2.930, 3.472)
5	MVP	3.774	0.141	(3.500, 4.056)
	MVL	3.863	0.132	(3.602, 4.123)
	MVC	4.005	0.103	(3.811, 4.214)

From the results are presented in Tables 8.2 to 8.5, it can be seen that the results are not sensitive to the choice of link functions. Based on the 95% HPD intervals of the μ_{jk}'s, the patterns that can be seen throughout the two different groups are that Items 1, 2, and 6 are always ranked as the top three features, although not necessarily in the same order. Note that these three features are *Broaden mathematical thinking* (Item 1), *Connect mathematics to other disciplines and real word problem solving* (Item 2), and *Increase mathematical proficiency* (Item 6). Similarly, Items 5, 7, 8 and 9 are ranked in the middle of the ordering, while Items 3, 4, and 10 are ranked at the bottom.

Second, for the model diagnostics, we plot posterior probabilities $P(|\epsilon_{ij}| \geq K^* \mid y)$ of the absolute values of standardized Bayesian residuals greater than or equal to K^* versus $E(y_{new,ij}|y)$ for all three models under consideration. Figure 8.1 gives these plots for $K^* = 2$. Several other values of K^* are also tried, but the corresponding plots are not presented here due to the nature of similarity.

From Figure 8.1, it can been seen that all three models fit the data fairly well and no aberrant features have been found.

Third, we calculate the posterior estimate of the covariance matrix Σ and find that the equicorrelation assumption on the correlation structure is questionable. For example, the correlations between Item 1 and the other items are not significant because all HPD intervals contain 0 for all three models. However, some other correlations are strongly significant. For example, the HPD intervals for the correlation between Item 8 and Item 9 are (.575, .810), (.586, .824), and (.585, .840) for the MVP, MVL, and MVC models, respectively. It is interesting to mention that this

TABLE 8.5. Bayesian Estimates of the Mean-Ranking Measures μ_{jk} (Group II): Part (b)

Item	Model	Posterior Mean	Posterior Std Dev	95% HPD Interval
6	MVP	4.538	0.098	(4.354, 4.731)
	MVL	4.453	0.099	(4.343, 4.722)
	MVC	4.531	0.108	(4.311, 4.723)
7	MVP	3.439	0.156	(3.146, 3.758)
	MVL	3.549	0.153	(3.254, 3.846)
	MVC	3.759	0.137	(3.482, 4.024)
8	MVP	3.683	0.160	(3.363, 3.988)
	MVL	3.796	0.154	(3.494, 4.095)
	MVC	4.025	0.127	(3.774, 4.272)
9	MVP	3.725	0.161	(3.409, 4.035)
	MVL	3.839	0.153	(3.528, 4.127)
	MVC	4.128	0.136	(3.869, 4.393)
10	MVP	3.190	0.153	(2.883, 3.480)
	MVL	3.272	0.143	(2.996, 3.550)
	MVC	3.401	0.134	(3.157, 3.669)

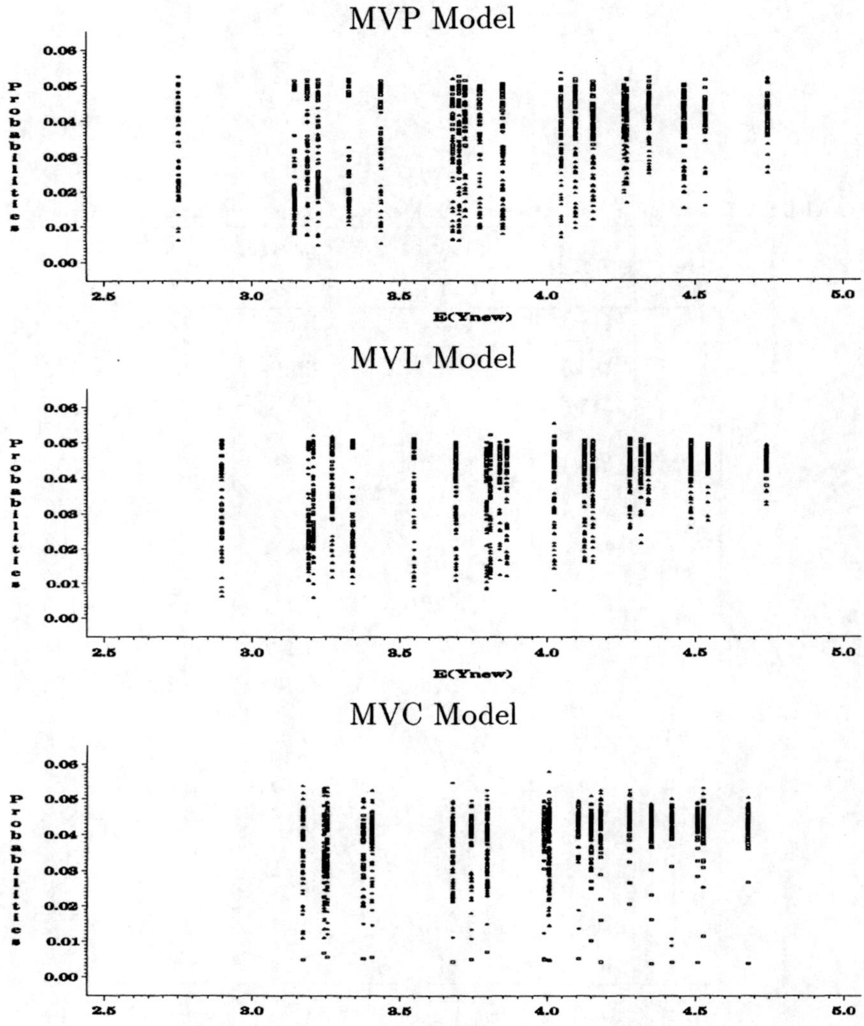

FIGURE 8.1. Plots of Posterior Probabilities $P(|\epsilon_{ij}| \geq K^*|y)$ versus $E(y_{new,ij}|y)$ for the MVP, MVL, and MVC models where the symbols • (dot), ⋆ (star), ∘ (circle), △ (triangle), and □ (square) correspond to ordinal responses $y_{ij} = 1$ to $y_{ij} = 5$

high correlation suggests a possible dimension reduction in survey sampling design. When we examine Item 8 and Item 9, we find that Item 8 is *Introduce modern teaching techniques and program assessment* while Item 9 is *Present ideas and applications for use in secondary school classrooms, e.g., hands-on and group learning activities.* Clearly, Items 8 and 9 are nested with each other, which is the main reason why we obtain such a high correlation between these two items. This finding also suggests that we either combine these two questions as one item or eliminate one of them.

Finally, we compute the marginal likelihoods for all three models. To obtain simulation-consistent estimates of the marginal likelihoods, the Monte Carlo sample sizes in (38), (39), (40), and (45) were taken to be $R = 10,000$. To obtain $\hat{\pi}(\gamma^* \mid y, \mathcal{M}_i)$, we use (42) instead of (41) since $J \times L = 50$ is relatively large. We find that the Monte Carlo method given in Section 4.1.1 works well in this example. Furthermore, we use a procedure provided by Chib (1995) to compute the simulation standard errors for marginal likelihood estimates. The estimated $\ln m(y|\mathcal{M}_i)$'s and the corresponding simulation standard errors in parentheses are -2055.6 (0.9), -1933.9 (0.6), and -1899.4 (0.6) for the MVC, MVL, and MVP models, respectively. Based on the marginal likelihoods, the MVP model is the best. However, this finding may not be important for this particular application because our primary interest is to compare the mean ranks among the items under study.

6. Concluding Remarks

Correlated ordinal data often arise in experiments when two or more measurements are taken at one time for the same subjects or when repeated measurements are taken over time. If such correlation is ignored in the model, overstatement or understatement of the precision of parameter estimates may result. We have considered a unified approach in this paper to incorporate the correlation structure, using the notion of multivariate generalized linear models.

Our suggested modeling approach is based on multivariate link functions using a very rich class of scale mixtures of normals. Such models are very flexible and include all the standard link functions in a generalized linear model scenario. Our data analyses include complete Bayesian model fitting along with model comparisons and model diagnostics, which are very difficult to implement in a classical setup. There are several effective graphical approaches presented in this chapter which give more insight to the data analysis.

There is another advantage of considering our approach over the usual random effects model which is based on the assumption of exchangeability. This is clearly reflected in our data analysis which shows that the equicorrelation assumption is not valid. In addition other advantages of Bayesian modeling over classical approach prevail in our studies. This includes more precise influence, exact small sample analysis, incorporation of the prior information, and inclusion of a large number of covariates.

Acknowledgement

Dr. Chen's research was supported by the National Science Foundation under Grant No. DMS-9702172.

References

Albert, J.H. and Chib, S. (1993). Bayesian Analysis of Binary and Polychotomous Response Data. *Journal of the American Statistical Association, 88*, 669-679.

Albert, J.H. and Chib, S. (1995). Bayesian Residual Analysis for Binary Response Regression Models. *Biometrika, 82*. 747-759.

Ashford, J.R. and Sowden, R.R. (1970). Multivariate Probit Analysis. *Biometrics, 26*, 535-546.

Box, G.E.P. and Tiao, G.C. (1992). *Bayesian Inference in Statistical Analysis.* Wiley: New York.

Chen, M.-H. and Dey, D.K. (1998). Bayesian Modeling of Correlated Binary Responses via Scale Mixture of Multivariate Normal Link Functions. *Sankhyā, Series A, 60*, 322-343.

Chen, M.-H. and Schmeiser, B.W. (1998). Towards Black-Box Sampling: A Random-Direction Interior-Point Markov Chain Approach. *Journal of Computational and Graphical Statistics, 7*, 1-22.

Chen, M.-H. and Shao, Q.M. (1997). On Monte Carlo Methods for Estimating Ratios of Normalizing Constants. *Annals of Statistics, 25*, 1563-1594.

Chib, S. (1995). Marginal Likelihood from the Gibbs Output. *Journal of the American Statistical Association, 90*, 1313-1321.

Chib, S. and Greenberg, E. (1998). Bayesian Analysis of Multivariate Probit Models. *Biometrika, 85*, 347-361.

Cowles, M.K. (1996). Accelerating Monte Carlo Markov Chain Convergence for Cumulative-Link Generalized Linear Models. *Statistics and Computing, 6*, 101-111.

Cowles, M.K. and Carlin, B.P. (1996). Markov Chain Monte Carlo Convergence Diagnostics: A Comparative Review. *Journal of the American Statistical Association, 91*, 883-904.

Cowles, M.K., Carlin, B.P., Connett, J.E. (1996). Bayesian Tobit Modeling of Longitudinal Ordinal Clinical Trial Compliance Data with Nonignorable Missingness. *Journal of the American Statistical Association, 91*, 86-98.

Choy, S.T.B. (1995). Robust Bayesian Analysis Using Scale Mixture of Normals Distributions. Ph.D. Dissertation, Department of Mathematics, Imperial College, London.

Devroye, L. (1986). *Non-Uniform Random Variate Generation.* Springer-Verlag: New York.

Dey, D.K., Gelfand, A.E., Swartz, T.B., and Vlachos, P.K. (1998). Simulation Based Model Checking for Hierarchical Models. *Test, 7.*

Ganter, L.G., Paulauskas, K.P., and Chen, M.-H. (1995). An Analysis of Master's Degree Programs for Secondary School Mathematics Teachers: A Needs Assessment. *Technical Report, Department of Mathematical Sciences, Worcester Polytechnic Institute.*

Gelfand, A.E. and Smith, A.F.M. (1990). Sampling Based Approaches to Calculating Marginal Densities. *Journal of the American Statistical Association, 85*, 398-409.

Geman, S. and Geman, D. (1984). Stochastic Relaxation, Gibbs Distributions and the Bayesian Restoration of Images. *IEEE Transactions on Pattern Analysis and Machine Intelligence, 6,* 721-741.

Geweke, J. (1991). Efficient Simulation from the Multivariate Normal and Student-t Distributions Subject to Linear Constraints. *Computing Science and Statistics: Proceedings of the Twenty-Third Symposium on the Interface*, 571-578.

Hastings, W.K. (1970). Monte Carlo Sampling Methods Using Markov Chains and Their Applications. *Biometrika, 57*, 97-109.

Kass, R.E. and Raftery, A.E. (1995). Bayes Factors. *Journal of the American Statistical Association, 90*, 773-795.

Liang, K.-Y. and Zeger, S.L. (1986). Longitudinal Data Analysis Using Generalized Linear Models. *Biometrika, 73*, 13-22.

Liu, J.S. (1994). The Collapsed Gibbs Sampler in Bayesian Computations with Applications to a Gene Regulation Problem. *Journal of the American Statistical Association, 89*, 958-966.

Meng, X.L. and Wong, W.H. (1996). Simulating Ratios of Normalizing Constants via a Simple Identity: A Theoretical Exploration. *Statistica Sinica, 6*, 831-860.

Metropolis, N., Rosenbluth, A.W., Rosenbluth, M.N., Teller, A.H. and Teller, E. (1953). Equations of state calculations by fast computing machines. *Journal of Chemical Physics, 21*, 1087-1092.

Nandram, B. and Chen, M.-H. (1996). Accelerating Gibbs Sampler Convergence in the Generalized Linear Models via a Reparameterization. *Journal of Statistical Computation and Simulation, 54*, 129-144.

O'Neill, R. (1971). Algorithm AS47-Function Minimization Using a Simplex Procedure. *Applied Statistics, 20*, 338-345.

Prentice, R.L. (1988). Correlated Binary Regression with Covariate Specific to Each Binary Observation. *Biometrics, 44*, 1033-1048.

Tanner, M.A. (1996). *Tools for Statistical Inference.* Third Edition, New York: Springer-Verlag.

Tierney, L. (1994). Markov Chains for Exploring Posterior Distributions (with discussions). *Annals of Statistics, 22*, 1701-1762.

Vounatsou, P., Smith, A.F.M., and Choy, S.T.B. (1996). Bayesian Robustness for Location and Scale Parameters Using Simulation. *Technical report*, Imperial College London.

Wakefield, J.C., Gelfand, A.E., and Smith, A.F.M. (1991). Efficient Generation of Random Variates via the Ratio-of-Uniforms Method. *Statistics and Computing, 1*, 129-133.

Zeger, S.L. and Liang, K.-Y. (1986). Longitudinal Data Analysis for Discrete and Continuous Outcomes. *Biometrics, 42*, 121-130.

9

Bayesian Methods for Time Series Count Data

Joseph G. Ibrahim
Ming-Hui Chen

ABSTRACT Correlated count data arise often in practice, especially in repeated measures situations or instances in which observations are collected over time. In this chapter, we consider a parametric model for a time series of counts by constructing a likelihood based version of a model similar to that of Zeger (1988). The model has the advantage of incorporating both overdispersion and autocorrelation. We consider a Bayesian approach and discuss a class of informative prior distributions for the model parameters that are useful for variable subset selection. The prior specification is motivated from the notion of the existence of data from similar previous studies, called historical data, which is then quantified into a prior distribution for the current study. We give theoretical and computational properties of the proposed priors, as well as examine properties of the implied posterior distributions. In addition, computational methods for sampling from the posterior distribution of the parameters and computing posterior model probabilities are discussed. The computational methods are based on the idea of hierarchical centering (Gelfand et al., 1996), and are quite efficient for sampling from the posterior distribution for the models considered here. To compute the posterior model probabilities, only posterior samples from the full model are needed to estimate the posterior probabilities for all of the possible subset models. The methodology is motivated from a real data set involving yearly pollen counts, which is also discussed.

1. Introduction

Historical data can be very helpful in interpreting the results of the current study. However, very few methods exist for the formal incorporation of historical data to construct the prior distribution. There is some literature addressing this issue for the linear model and generalized linear models. See for example, Ibrahim and Laud (1994), Laud and Ibrahim (1995), Bedrick, Christensen, and Johnson (1996), Ibrahim, Ryan, and Chen (1998), and Chen, Ibrahim, and Yiannoutsos (1999). In all of these papers, the authors assume a univariate independent response variable. The literature for informative prior elicitation for models with correlated responses is essentially nonexistent.

In this chapter, we present an extension of the priors in Ibrahim, Ryan, and

Chen (1998) and and Chen, Ibrahim, and Yiannoutsos (1999) to time series count data. The prior specification is based on the notion of specifying a prior prediction y_0 for the response vector, y, of the current study, along with a covariate matrix X_0 corresponding to y_0. Then $D_0 = (n_0, y_0, X_0)$ is used to specify an automated parametric informative prior for the regression coefficients. Thus, D_0 is the historical data. The quantity y_0 can be taken as the raw response vector from the historical data, a vector of fitted values based on the historical data, a vector obtained from a theoretical prediction model, or a vector specified from expert opinion or case-specific information. Thus y_0 can be viewed as a prior "prediction" for y, the actual data in the current study. Similarly, X_0 can be taken as the raw covariate matrix based on the historical data or it can be specified in other ways. In any case, taking D_0 to be the raw historical data results in a more natural, interpretable, and automated specification.

The priors discussed here are attractive for variable subset selection, and thus these applications serve as a primary motivation for the priors. Another major advantage of this methodology is that the time series considered here are virtually impossible to fit in the frequentist context, let alone the entire problem of frequentist variable subset selection. If for example, there are k possible covariates, then there are 2^k models to evaluate in the variable subset selection problem. This is a computational nightmare in general, regardless of a frequentist or Bayesian approach. Aside from the computational issues, the methodology in the frequentist paradigm is not well developed for the class of models we consider here. In this chapter, the Monte Carlo methods we discuss facilitate a very fast and efficient way of computing the posterior model probabilities using only a *single* posterior sample from a *single* model, that being the full model. Such a procedure has proved to be quite feasible and powerful in the model selection context (see for example, Ibrahim and Chen (1998) and Chen, Ibrahim, and Yiannoutsos, (1999)).

The rest of the chapter is organized as follows. In Section 2, we present notation for the model, the likelihood, the prior distributions, and the posterior distributions. Computational techniques for sampling from the posterior are discussed in Section 3. In Section 4, we present a real dataset demonstrating the computational feasibility and the strength of our method over GEE and likelihood based methods. We conclude with some general discussion and some possible extensions in Section 5.

2. The Method

2.1 The Likelihood Function

Let $\mathcal{M}$ denote the model space. We enumerate the models in $\mathcal{M}$ by $m = 1, 2, \ldots, \mathcal{K}$, where $\mathcal{K}$ is the dimension of $\mathcal{M}$ and model $\mathcal{K}$ denotes the full model. The full model is defined here as the model containing all of the available covariates in the study. Letting k denote the number of covariates for the full model, our model space, $\mathcal{M}$, then contains 2^k models. Also, let $\beta^{(\mathcal{K})} = (\beta_0, \beta_1, \ldots, \beta_k)'$ denote the regression coefficients for the full model including an intercept, and let $\beta^{(m)}$ denote a $k_m \times 1$ vector of regression coefficients for model m with an intercept, and a specific choice of $k_m - 1$ covariates. We write $\beta^{(\mathcal{K})} = (\beta^{(m)\prime}, \beta^{(-m)\prime})'$, where $\beta^{(-m)}$ is $\beta^{(\mathcal{K})}$ with $\beta^{(m)}$ deleted.

Consider a time series of counts y_t, $t = 1, \ldots, n$, where each y_t has corresponding

$k_m \times 1$ covariate vector $x_t^{(m)}$ under model m. Under model m, conditional on $\beta^{(m)}$ and a stationary unobserved process ϵ_t, the y_t's are assumed to be independent Poisson random variables with mean $\lambda_t = \exp(\epsilon_t + (x_t^{(m)})'\beta^{(m)})$, leading to the conditional density

$$\begin{aligned} & p(y \mid \beta^{(m)}, \epsilon, D^{(m)}) \\ = & \prod_{t=1}^{n} p(y_t \mid \beta^{(m)}, \epsilon_t) \\ = & \prod_{t=1}^{n} \exp\left\{y_t(\epsilon_t + (x_t^{(m)})'\beta^{(m)}) - \exp(\epsilon_t + (x_t^{(m)})'\beta^{(m)}) \; - \log(y_t!)\right\} \\ = & \exp\left\{y'(\epsilon + X^{(m)}\beta^{(m)}) - J_n' Q(\beta^{(m)}, \epsilon) - J_n' C(y)\right\}, \end{aligned} \tag{1}$$

where $y = (y_1, \ldots, y_n)'$, $\epsilon = (\epsilon_1, \ldots, \epsilon_n)$, $X^{(m)}$ is the $n \times k_m$ matrix of covariates with tth row equal to $(x_t^{(m)})'$, J_n is an $n \times 1$ vector of ones, and $Q(\beta^{(m)}, \epsilon)$ is an $n \times 1$ vector with tth element equal to $q_t = \exp\left\{\epsilon_t + (x_t^{(m)})'\beta^{(m)}\right\}$, and $C(y)$ is an $n \times 1$ vector with jth element $\log(y_j!)$. Finally, $D^{(m)} = (n, y, X^{(m)})$ denotes the data for the current study under model m. The latent process ϵ_t is assumed to have normal distribution with mean 0. In particular, we assume that ϵ has a multivariate normal distribution with mean 0 and, covariance matrix $\sigma^2\Sigma$, where the (i, j)th element of Σ has the form $\sigma_{ij} = \rho^{|i-j|}$, where $\rho^{|i-j|}$ is the correlation between (ϵ_i, ϵ_j), and $-1 \leq \rho \leq 1$. The unobserved process ϵ_t is analogous to a "random effect" in a random effects model, with the exception that the latent process is correlated. We note that the mean and variance of ϵ_t do not depend on t. Zeger (1988) constructs a similar model through the mean and covariance of the latent process, which then define the estimating equations. He does not specify a parametric distribution for the latent process as is done here.

The joint density of (y, ϵ) can be written as

$$\begin{aligned} & p(y, \epsilon \mid \beta^{(m)}, \sigma^2, \rho, D^{(m)}) \\ = & (2\pi\sigma^2)^{-n/2}(1-\rho^2)^{-(n-1)/2} \\ \times & \exp\left\{y'(\epsilon + X^{(m)}\beta^{(m)}) - J_n' Q(\beta^{(m)}, \epsilon) - J_n' C(y) - \frac{1}{2\sigma^2}\epsilon'\Sigma^{-1}\epsilon\right\}. \end{aligned} \tag{2}$$

To induce the correlation structure on y, we integrate out ϵ from (2) leading to the "marginal" likelihood of $\beta^{(m)}$, given by

$$p(y \mid \beta^{(m)}, \sigma^2, \rho, D^{(m)}) = \int p(y, \epsilon \mid \beta^{(m)}, \sigma^2, \rho, D^{(m)}) \; d\epsilon \; , \tag{3}$$

where $p(y, \epsilon \mid \beta^{(m)}, \sigma^2, \rho, D^{(m)})$ is given by (2). The marginal likelihood of $\beta^{(m)}$ in (3) does not have a closed form, and thus the integral cannot be evaluated analytically.

The implications of the process ϵ_t on the correlation structure in the y_t's and the regression model is as follows. Note first that $\epsilon_t^* = \exp(\epsilon_t)$ has a log-normal distribution with mean $\alpha = \exp(\frac{1}{2}\sigma^2)$ and variance $\nu^2 = \exp(2\sigma^2) - \exp(\sigma^2)$. It follows upon integration over ϵ_t, that the marginal mean of y_t is given by

$$\begin{aligned} \mu_t^{(m)} &= E(y_t \mid \beta^{(m)}, D^{(m)}) \\ &= \exp((x_t^{(m)})'\beta^{(m)}) \; E(\exp(\epsilon_t)) \\ &= \alpha \; \exp((x_t^{(m)})'\beta^{(m)}) \; , \end{aligned}$$

so that the intercept in the marginal model is $\log(\alpha) + \beta_0$. In addition, we have

$$\text{Var}(y_t \mid \beta^{(m)}, D^{(m)}) = \mu_t^{(m)} + \frac{\nu^2}{\alpha^2} (\mu_t^{(m)})^2 , \tag{4}$$

and

$$\text{Cov}(y_t, y_{t+k}|D^{(m)}) = \mu_t^{(m)} \mu_{t+k}^{(m)} (\exp(\sigma^2 \rho^k) - 1) , \tag{5}$$

so that

$$\text{Corr}(y_t, y_{t+k}|D^{(m)}) = \frac{\exp(\sigma^2\rho^k) - 1}{\left[((\mu_t^{(m)})^{-1} + \nu^2\alpha^{-2}))((\mu_{t+k}^{(m)})^{-1} + \nu^2\alpha^{-2})\right]^{1/2}} .$$

From (4) and (5), we see that the unobserved process ϵ_t allows for overdispersion and autocorrelation into y_t. In addition, the degree of overdispersion depends on μ_t. The autocorrelation in y_t must be less than or equal to that in ϵ_t and the degree of autocorrelation in y_t relative to ϵ_t decreases as μ_t and ν^2 decrease.

2.2 *The Prior Distributions*

Informative prior elicitation is an important part of a Bayesian analysis, especially in the problem of variable subset selection since proper prior distributions are required to compute posterior model probabilities. Here, we present a class of informative priors for the regression coefficients $\beta^{(m)}$. The prior construction for $\beta^{(m)}$ is based on the availability of historical data as motivated in Section 1. Suppose there are N historical data sets and the sample size of the ith historical study is n_{0i}. Let y_{0i} denote the $n_{0i} \times 1$ vector of time series counts for the ith historical study and let $X_{0i}^{(m)}$ denote the $n_{0i} \times k_m$ matrix of covariates corresponding to the ith historical study. In addition, let ϵ_{0i} denote the latent process for the ith historical study, where ϵ_{0i} is an $n_{0i} \times 1$ vector, $i = 1, \ldots, N$, and ϵ_{0i} has an n_{0i} dimensional multivariate normal distribution with mean 0 and covariance matrix $\sigma^2\Sigma_{0i}$, where Σ_{0i} is an $n_{0i} \times n_{0i}$ matrix with $(j, j^*)th$ element equal to $\rho^{|j-j^*|}$. Finally let $D_{0i}^{(m)} = (n_{0i}, X_{0i}^{(m)}, y_{0i})$ denote the data from the ith historical study under model m, and $D_0^{(m)} = (D_{01}^{(m)}, \ldots, D_{0N}^{(m)})$ denotes all of the historical data under model m.

The prior distribution for $\beta^{(m)}$ for the ith historical study takes the form

$$\begin{aligned} &\pi(\beta^{(m)} \mid \sigma^2, \rho, D_{0i}^{(m)}, a_{0i}) \\ &\propto \int p(y_{0i} \mid \beta^{(m)}, \epsilon_{0i}, a_{0i})\, (2\pi\sigma^2)^{-\frac{n_{0i}}{2}}\, (1-\rho^2)^{-\frac{(n_{0i}-1)}{2}} \\ &\quad \times \exp\left\{-\frac{1}{2\sigma^2}(\epsilon_{0i}'\Sigma_{0i}^{-1}\epsilon_{0i})\right\}\, d\epsilon_{0i}, \end{aligned} \tag{6}$$

where

$$\begin{aligned} & p(y_{0i} \mid \beta^{(m)}, \epsilon_{0i}, a_{0i}) \\ = \ & \exp\left\{a_{0i}\left[y_{0i}'(\epsilon_{0i} + X_{0i}^{(m)}\beta^{(m)}) - J_{n_{0i}}' Q(\beta^{(m)}, \epsilon_{0i}) - J_{n_{0i}}' C(y_{0i})\right]\right\}, \end{aligned} \tag{7}$$

and a_{0i} is a scalar prior parameter that controls the weight of the ith historical study relative to the likelihood of the current study. That is, a_{0i} controls the weight of the likelihood function based on the ith historical study. Small values of a_{0i} give less weight whereas large values give more weight. It is most sensible to restrict a_{0i} to $0 \le a_{0i} \le 1$, since we would not want to weight the historical data more than the current data. The parameter a_{0i} can also be interpreted as an overdispersion parameter which takes into account the between and within study variability in the historical data sets.

The prior distribution of $\beta^{(m)}$ based on all of the historical studies is thus given by

$$\begin{aligned} &\pi(\beta^{(m)} \mid \sigma^2, \rho, D_0^{(m)}, a_0) \\ &\propto \prod_{i=1}^{N} \int p(y_{0i} \mid \beta^{(m)}, \epsilon_{0i}, a_{0i})\, (2\pi\sigma^2)^{-\frac{n_{0i}}{2}}\, (1-\rho^2)^{-\frac{(n_{0i}-1)}{2}} \\ &\qquad \times \exp\left\{ -\frac{1}{2\sigma^2} (\epsilon_{0i}' \Sigma_{0i}^{-1} \epsilon_{0i}) \right\} \, d\epsilon_{0i} \,, \end{aligned} \tag{8}$$

where $a_0 = (a_{01}, \ldots, a_{0N})$. From (8), we see that $a_{0i} = 0$ corresponds to no incorporation of historical data, and yields a uniform improper prior for β. The case $a_{0i} = 1$ implies that the historical data and the current data are weighted equally. The prior in (8) does not have a closed form but it has several attractive theoretical and computational properties as shown in Section 3.

The prior specification is completed by specifying priors for (σ^2, ρ, a_0). We take these parameters to be independent a priori. We specify an inverse gamma prior for σ^2, denoted $IG(\delta_0, \gamma_0)$, a scaled beta prior for ρ, denoted scbeta(ν_0, ψ_0), and independent identically distributed beta priors for each a_{0i}, denoted beta(α_0, λ_0). Here, $(\delta_0, \gamma_0, \nu_0, \psi_0, \alpha_0, \lambda_0)$ are specified prior hyperparameters. Thus, the joint prior distribution takes the form

$$\begin{aligned} &\pi(\beta^{(m)}, \sigma^2, \rho, a_0 \mid D_0^{(m)}) \\ &\propto \prod_{i=1}^{N} \Bigg(\int p(y_{0i} \mid \beta^{(m)}, \epsilon_{0i}, a_{0i})\, (2\pi\sigma^2)^{-\frac{n_{0i}}{2}}\, (1-\rho^2)^{-\frac{(n_{0i}-1)}{2}} \\ &\qquad \times \exp\left\{ -\frac{1}{2\sigma^2} (\epsilon_{0i}' \Sigma_{0i}^{-1} \epsilon_{0i}) \right\} \, d\epsilon_{0i} \Bigg) \times \left(\prod_{i=1}^{N} a_{0i}^{\alpha_0 - 1} (1-a_{0i})^{\lambda_0 - 1} \right) \\ &\qquad \times (\sigma^2)^{-(\delta_0+1)} \exp(-\sigma^{-2}\gamma_0)\, (1+\rho)^{\nu_0-1} (1-\rho)^{\psi_0-1}. \end{aligned} \tag{9}$$

We see that our joint prior for $(\beta^{(m)}, \sigma^2, \rho, a_0)$ clearly does not have a closed form in general. However, it has a natural motivation and several appealing interpretations. One motivation for the prior in (9), is that by taking the a_0 random, the tails of the marginal prior distribution for $\beta^{(m)}$ are heavier than those obtained by taking a_0 a fixed hyperparameter. In addition, a prior on a_0 provides great flexibility and allows us to express our uncertainty about it. By allowing different a_{0i}'s for different historical studies, we are able to develop a much more flexible prior that can weight each historical study differently. This would certainly be desirable if one historical study has a much larger sample size than another historical study. Another motivation for (9) is that it mimics the marginal likelihood function of $\beta^{(m)}$ based on the historical data. If, for example $a_0 = 1$, then (9) is precisely the marginal likelihood function of $\beta^{(m)}$ based on the historical data. Thus, our prior can be viewed as a weighted marginal likelihood of $\beta^{(m)}$, which is a natural prior to consider when such historical data is available.

2.3 Specifying the Hyperparameters

In the context of model selection, (ρ, σ^2) are viewed as nuisance parameters, and therefore we take vague choices for their prior hyperparameters. In particular, it is reasonable to take $\nu_0 = \psi_0 = 1$ which yields a uniform prior for ρ on $[-1, 1]$. Also, we take $\delta_0 \to 0$ and $\gamma_0 \to 0$, which yields a noninformative prior for σ^2. For a_{0i}, we recommend that several values of the hyperparameters be chosen and sensitivity analyses conducted. For elicitation purposes, it is easier to work with the prior mean and variance of a_{0i}, given by $\mu_{a_0} = \alpha_0/(\alpha_0 + \lambda_0)$, and $\sigma^2_{a_0} = \mu_{a_0}(1 - \mu_{a_0})(\alpha_0 + \lambda_0 + 1)^{-1}$. It can be shown that a sufficient condition for the propriety of the prior distribution is that $\alpha_0 > k+1$ for the full model. Therefore, a reasonable starting point for the analysis is to choose $\alpha_0 = \lambda_0 = k+2$, which gives $\mu_{a_0} = 1/2$. Then we conduct several sensitivity analyses within a suitable range of the uniform prior, using various values of $(\mu_{a_0}, \sigma^2_{a_0})$. Small and large values of $(\mu_{a_0}, \sigma^2_{a_0})$ should be considered. We do not recommend doing an analysis based on one set of proposed values of $(\mu_{a_0}, \sigma^2_{a_0})$, nor do we propose specifying $(\mu_{a_0}, \sigma^2_{a_0})$ by a one-time automated procedure. The choice of $(\mu_{a_0}, \sigma^2_{a_0})$ depends on the context of the problem and the structure of the historical data. In Section 4, we demonstrate several choices of $(\mu_{a_0}, \sigma^2_{a_0})$ and conduct sensitivity analyses and examine their impact on variable selection.

2.4 Prior Distribution on the Model Space

Let

$$\begin{aligned} & p_0^*(\beta^{(m)}|D_0^{(m)}) \\ = & \int \Bigg\{ \prod_{i=1}^{N} \Bigg(\int p(y_{0i}|\beta^{(m)}, \epsilon_{0i}, a_{0i}) (2\pi\sigma^2)^{-\frac{n_{0i}}{2}} (1-\rho^2)^{-\frac{(n_{0i}-1)}{2}} \\ & \times \exp\left\{ -\frac{1}{2\sigma^2} (\epsilon_{0i}' \Sigma_{0i}^{-1} \epsilon_{0i}) \right\} \, d\epsilon_{0i} \Bigg) \times \left(\prod_{i=1}^{N} a_{0i}^{\alpha_0 - 1} (1-a_{0i})^{\lambda_0 - 1} \right) \\ & \times (\sigma^2)^{-(\delta_0+1)} \exp(-\frac{\gamma_0}{\sigma^2}) \, (1+\rho)^{\nu_0 - 1} (1-\rho)^{\psi_0 - 1} \Bigg\} d\sigma^2 \, d\rho \, da_0. \end{aligned} \tag{10}$$

We see that $p_0^*(\beta^{(m)}|D_0^{(m)})$ is proportional to the marginal prior of $\beta^{(m)}$. We propose to take the prior probability of model m as

$$p(m) = \frac{\int p_0^*(\beta^{(m)}|D_0^{(m)}) \, d\beta^{(m)}}{\sum_{j=1}^{\mathcal{K}} \int p_0^*(\beta^{(j)}|D_0^{(j)}) \, d\beta^{(j)}} . \tag{11}$$

This choice for $p(m)$ in (11) is a natural one since the numerator is just the normalizing constant of the joint prior of $(\beta^{(m)}, a_0, \sigma^2, \rho)$ under model m. The prior model probabilities in (11) are based on coherent Bayesian updating and this results in several attractive interpretations. Firstly, $p(m)$ in (11) corresponds to the posterior probability of model m based on the data $D_0^{(m)}$ using a uniform prior for the previous study, $p_0(m) = 2^{-k}$ for $m \in \mathcal{M}$ as $\alpha_0 \to \infty$. That is, $p(m) \propto p(m \mid D_0^{(m)})$, and thus $p(m)$ corresponds to the usual Bayesian update of $p_0(m)$ using $D_0^{(m)}$ as the data. Secondly, as $\lambda_0 \to \infty$, $p(m)$ reduces to a uniform prior on the model space.

Therefore, as $\lambda_0 \to \infty$, the historical data $D_0^{(m)}$ have a minimal impact in determining $p(m)$. On the other hand, $p(m)$ in (11) has a nice theoretical property, which greatly eases the computational burden for calculating posterior model probabilities using the Markov chain Monte Carlo (MCMC) output.

3. Computation of Model Probabilities

In this section, we state the theoretical properties of the posterior model probabilities based on the choice of the prior model probabilities $p(m)$ given in (11) and then disuss the Monte Carlo implementation procedures to compute posterior model probabilities. A key result that is presented is that we give a formula for the posterior model probability that does not depend directly on $p(m)$. This is due to a cancellation of terms that results from the structure of $p(m)$ given in (11).

The posterior probability of model m is given by

$$p(m|D^{(m)}) = \frac{p(D^{(m)}|m)\ p(m)}{\sum_{j=1}^{\mathcal{K}} p(D^{(j)}|j)\ p(j)}, \tag{12}$$

where $p(D^{(m)}|m)$ denotes the marginal distribution of the data $D^{(m)}$ for the current study under model m, and $p(m)$ denotes the prior probability of model m in (11).

We first obtain an expression for $p(D^{(m)}|m)$. From (9), the joint prior distribution for $(\beta^{(m)}, \sigma^2, \rho)$ is given by

$$\begin{aligned}
&\pi(\beta^{(m)}, \sigma^2, \rho|D_0^{(m)}) \propto p_0^*(\beta^{(m)}, \sigma^2, \rho|D_0^{(m)}) \\
&= \int \Bigg\{ \prod_{i=1}^{N} \Bigg(\int p(y_{0i}|\beta^{(m)}, \epsilon_{0i}, a_{0i})(2\pi\sigma^2)^{-\frac{n_{0i}}{2}}(1-\rho^2)^{-\frac{(n_{0i}-1)}{2}} \\
&\quad \exp\left\{-\frac{1}{2\sigma^2}(\epsilon_{0i}'\Sigma_{0i}^{-1}\epsilon_{0i})\right\}\ d\epsilon_{0i}\Bigg) \times \left(\prod_{i=1}^{N} a_{0i}^{\alpha_0-1}(1-a_{0i})^{\lambda_0-1}\right) \\
&\quad \times(\sigma^2)^{-(\delta_0+1)}\exp(-\sigma^{-2}\gamma_0)\ (1+\rho)^{\nu_0-1}(1-\rho)^{\psi_0-1}\Bigg\} da_0,
\end{aligned} \tag{13}$$

where $p_0^*(\beta^{(m)}, \sigma^2, \rho, \mid D_0^{(m)})$ represents the unnormalized joint prior density function of $(\beta^{(m)}, \sigma^2, \rho, \mid D_0^{(m)})$ under model m. i.e., the right side of equation (9). Then, using (13), the joint posterior distribution of $(\beta^{(m)}, \sigma^2, \rho)$ under model m is given by

$$\begin{aligned}
&p(\beta^{(m)}, \sigma^2, \rho \mid D^{(m)}, D_0^{(m)}) \\
\propto\ & p^*(\beta^{(m)}, \sigma^2, \rho \mid D^{(m)}, D_0^{(m)}) \\
=\ & p(y \mid \beta^{(m)}, \sigma^2, \rho, D^{(m)})\ \pi(\beta^{(m)}, \sigma^2, \rho \mid D_0^{(m)}),
\end{aligned} \tag{14}$$

where $p(y \mid \beta^{(m)}, \sigma^2, \rho, D^{(m)})$ is given by (2), $\pi(\beta^{(m)}, \sigma^2, \rho \mid D_0^{(m)})$ is given by (13) and $p^*(\beta^{(m)}, \sigma^2, \rho \mid D^{(m)}, D_0^{(m)})$ is an unnormalized joint posterior density function. Clearly if the prior is proper, then so is the posterior. Thus

$$p(D^{(m)}|m) = \int p^*(\beta^{(m)}, \sigma^2, \rho \mid D^{(m)}, D_0^{(m)})\ d\sigma^2\ d\rho\ d\beta. \tag{15}$$

Recall that $\beta^{(\mathcal{K})} = (\beta^{(m)\prime}, \beta^{(-m)\prime})'$ where $\beta^{(-m)}$ is $\beta^{(\mathcal{K})}$ with $\beta^{(m)}$ deleted. Then, it can be shown that the posterior probability $p(m|D^{(m)})$ in (12) of model m is given by

$$p(m|D^{(m)}) = \frac{p(\beta^{(-m)} = 0|D^{(\mathcal{K})}, D_0^{(\mathcal{K})})}{\sum_{j=1}^{\mathcal{K}} p(\beta^{(-j)} = 0|D^{(\mathcal{K})}, D_0^{(\mathcal{K})})}\,, \tag{16}$$

$m = 1, \ldots, \mathcal{K}$, where $\beta^{(\mathcal{K})} = (\beta^{(m)\prime}, \beta^{(-m)\prime})'$, and $p(\beta^{(-m)} = 0|D^{(\mathcal{K})}, D_0^{(\mathcal{K})})$ is the marginal posterior density of $\beta^{(-m)}$ evaluated at $\beta^{(-m)} = 0$. In (16), for notational convenience we assume that $p(\beta^{(-\mathcal{K})} = 0|D^{(\mathcal{K})}, D_0^{(\mathcal{K})}) = 1$. The result in (16) is very attractive since it shows that the posterior model probability $p(m|D^{(m)})$ is simply a function of the marginal posterior density functions of $\beta^{(-m)}$ for the full model evaluated at $\beta^{(-m)} = 0$. This formula does not algebraically depend on the prior model probability $p(m)$ since it cancels out in the derivation due to the structure of $p(m)$. This is an important feature since it allows us to compute the posterior model probabilities directly *without* numerically computing the prior model probabilities. This has a clear computational advantage and as a result, allows us to compute posterior model probabilities very efficiently. We note that this computational device works best if all of the covariates are standardized to have mean 0 and variance 1. This is not restrictive since this is a typical transformation taken quite often in practice to numerically stabilize the Gibbs sampler and the adaptive rejection algorithms.

Due to the complexity of our model, the analytical evaluation of $p(\beta^{(-m)} = 0|D^{(\mathcal{K})}, D_0^{(\mathcal{K})})$ does not appear possible. We use the Monte Carlo method developed in Ibrahim and Chen (1998) and Chen, Ibrahim, and Yiannoutsos (1999) to compute posterior model probabilities using a single Markov chain Monte Carlo (MCMC) output from the full model. The hierarchical centering reparameterization technique of Gelfand et al. (1996) is particularly suitable for the implementation of MCMC sampling for this problem. This technique is also very useful in developing an efficient Monte Carlo method for estimating the marginal posterior density $p(\beta^{(-m)} = 0|D^{(\mathcal{K})}, D_0^{(\mathcal{K})})$.

To this end, consider the following reparameterization:

$$\eta = \epsilon + X^{(\mathcal{K})}\beta^{(\mathcal{K})} \tag{17}$$

and

$$\eta_{0i} = \epsilon_{0i} + X_{0i}^{(\mathcal{K})}\beta^{(\mathcal{K})}\,, \tag{18}$$

for $i = 1, 2, \ldots, N$. Let $\eta_0 = (\eta_{01}, \ldots, \eta_{0N})$. Now the reparameterized posterior for the full model is given by

$$\begin{aligned}
&p(\beta^{(\mathcal{K})}, \sigma^2, \rho, a_0, \eta, \eta_0|D^{(\mathcal{K})}, D_0^{(\mathcal{K})}) \propto \sigma^{-n}(1-\rho^2)^{-(n-1)/2}\\
&\times \exp\left\{y'\eta - J_n'Q(\eta) - \frac{1}{2\sigma^2}(\eta - X^{(\mathcal{K})}\beta)'\Sigma^-1(\eta - X^{(\mathcal{K})}\beta)\right\}\\
&\times \left(\prod_{i=1}^{N} \exp\left\{a_{0i}\left[y_{0i}'\eta_{0i} - J_{n_{0i}}'Q_0(\eta_{0i}) - J_{n_{0i}}'C(y_{0i})\right]\right\}\right) \times (2\pi\sigma^2)^{-\frac{n_{0i}}{2}}\\
&\times (1-\rho^2)^{-\frac{(n_{0i}-1)}{2}} \exp\left\{-\frac{1}{2\sigma^2}(\eta_{0i} - X_{0i}^{(\mathcal{K})}\beta)'\Sigma_{0i}^{-1}(\eta_{0i} - X_{0i}^{(\mathcal{K})}\beta)\right\}\\
&\times \left(\prod_{i=1}^{N} a_{0i}^{\alpha_0-1}(1-a_{0i})^{\lambda_0-1}\right)(\sigma^2)^{-(\delta_0+1)}\exp\left(-\frac{\gamma_0}{\sigma^2}\right)\\
&\times (1+\rho)^{\nu_0-1}(1-\rho)^{\psi_0-1}\,,
\end{aligned} \tag{19}$$

where $Q(\eta)$ and $Q_0(\eta_{0i})$ are vectors of length n and n_{0i}, respectively, with jth element equal to $\exp(\eta_j)$, and $\exp(\eta_{0ij})$, $\eta_j = (x_j^{(\mathcal{K})})'\beta^{(\mathcal{K})} + \epsilon_j$, and $\eta_{0ij} = (x_{0ij}^{(\mathcal{K})})'\beta^{(\mathcal{K})} + \epsilon_{0ij}$. Assume that $\{(\beta_{(l)}^{(\mathcal{K})}, \sigma_{(l)}^2, \rho_{(l)}, a_{0(l)}, \eta_{(l)}, \eta_{0(l)}),\ l = 1, 2, \ldots, L\}$ is an MCMC sample from the reparameterized posterior $p(\beta^{(\mathcal{K})}, \sigma^2, \rho, a_0, \eta, \eta_0 | D^{(\mathcal{K})}, D_0^{(\mathcal{K})})$ given in (19). Then, from (19) and following the lines of Chen (1994), $p(\beta^{(-m)} = 0 | D^{(\mathcal{K})}, D_0^{(\mathcal{K})})$ can be estimated by the conditional marginal density estimation (CMDE) method. Gelfand, Smith, and Lee (1992), Chen (1994), and Chen and Shao (1997) have shown that the CMDE is the most efficient Monte Carlo method for estimating marginal posterior densities when a joint posterior density is known up to a normalizing constant. It directly follows from, for example, Chen and Shao (1997) that a simulation consistent estimator of $p(\beta^{(-m)} = 0 | D^{(\mathcal{K})})$ is given by

$$\hat{p}(\beta^{(-m)} = 0 | D^{(\mathcal{K})}) = \frac{1}{L} \sum_{l=1}^{L} N_{k+1-k_m}(\beta^{(-m)} = 0 | \beta_{(l)}^{(-m)}, \hat{\beta}_{(l)}, (B_{(l)})^{-1}), \tag{20}$$

where $N_{k+1-k_m}(\beta^{(-m)} = 0 | \beta_{(l)}^{(-m)}, \hat{\beta}_{(l)}^{(\mathcal{K})}, (B_{(l)})^{-1})$ is the $(k+1-k_m)$-dimensional conditional normal density function of $N_{k+1}(\hat{\beta}_{(l)}^{(\mathcal{K})}, (B_{(l)})^{-1})$ given $\beta_{(l)}^{(m)}$ evaluated at $\beta^{(-m)} = 0$,

$$B_{(l)} = \frac{1}{\sigma_{(l)}^2} \left((X^{(\mathcal{K})})' \Sigma_{(l)}^{-1} X^{(\mathcal{K})} + \sum_{i=1}^{N} (X_{0i}^{(\mathcal{K})})' \Sigma_{0i(l)}^{-1} X_{0i}^{(\mathcal{K})} \right),$$

$$\hat{\beta}^{(\mathcal{K})} = \frac{1}{\sigma_{(l)}^2} \left\{ B_{(l)}^{-1} \left((X^{(\mathcal{K})})' \Sigma_{(l)}^{-1} \eta_{(l)} + \sum_{i=1}^{N} (X_{0i}^{(\mathcal{K})})' \Sigma_{0i(l)}^{-1} \eta_{0i(l)} \right) \right\},$$

$\Sigma_{(l)}$ is an $n \times n$ matrix with $(j, j^*)th$ element equal to $\rho_{(l)}^{|j-j^*|}$, and $\Sigma_{0i(l)}$ is an $n_{0i} \times n_{0i}$ matrix with $(j, j^*)th$ element equal to $\rho_{(l)}^{|j-j^*|}$.

There are several advantages to using the above Monte Carlo procedure. First, as previously mentioned, it is *not* required to compute $p(m)$ for each model. Second, we need only one random draw from $p(\beta^{(\mathcal{K})}, \sigma^2, \rho, a_0, \eta, \eta_0 | D^{(\mathcal{K})}, D_0^{(\mathcal{K})})$. Third, after we obtain an MCMC sample from the posterior distribution of the full model, calculating $\hat{p}(\beta^{(-m)} = 0 | D^{(\mathcal{K})})$ given by (20) is straightforward and almost free of computational time. Fourth, for the purposes of computing posterior model probabilities, it is required only to store a $(k+1)$-dimensional vector $\hat{\beta}_{(l)}$ and a $(k+1) \times (k+1)$ matrix $B_{(l)}$ for each MCMC sampling iteration, which will greatly reduce the computer storage space. This becomes even more advantageous for cases where multiple previous studies are available and each n_{0i} is large. The above features of our Monte Carlo procedure essentially make the computation of Bayesian variable selection feasible in the presence of a large number of covariates (say, $k > 20$).

4. Example: Pollen Data

Pollen allergy is a common disease causing hay fever and respiratory discomfort in approximately 10% of the United States population. Although not a life threatening illness, allergy symptoms seem to be increasingly more troublesome, as well as costing society a great deal of money and resources. Therefore, it is becoming increasingly important to identify the important covariates that help predict pollen levels.

We consider a real data set in which ragweed pollen was collected daily in Kalamazoo, Michigan form 1991 to 1994. Frequentist analyses of these data using standard Poisson regression methods have been conducted by Stark et al. (1997). Our aim here is not to do a detailed data analysis, but rather demonstrate our Bayesian methodology for variable selection. The response variable y, is the pollen count for a particular day in the season for a given year. We take the 1991, 1992, and 1993 data as the historical data and the 1994 data as the current data. The full model contains an intercept and seven covariates, which were extensively discussed and motivated by Stark et al. (1997). These are x_1 = rain, (which is a binary variable taking the value 0 if there were at least three hours of steady rain, and 1 otherwise), x_2 = day in the pollen season , x_3 = log(day). In addition, we consider two covariates that are functions of temperature. These are x_4 which is the `lowess` smoothed function of temperature constructed from a nonparametric estimate of the regression of pollen count on average temperature, and x_5, which denotes the deviation from the daily averages temperature to the `lowess` line. The final two covariates are x_6 = windspeed and x_7 = cold, (which is a binary variable taking the value 0 if the overnight temperature dropped below 50 degrees Fahrenheit, and 1 otherwise). Tables 9.1 and 9.2 summarize the response variable and covariate data for the four years.

The analysis in Stark et al. (1997) is based on a Poisson model assuming independent counts, and thus it does not introduce a latent process nor does it account for the time series structure in the data. It is quite different from the model we consider in (2). We model the pollen counts as a Poisson distribution as in equation (1) with covariates $(x_1, \ldots, x_7)$. The model space $\mathcal{M}$ contains 2^7 models. We specify noninformative priors for ρ and σ^2. Specifically, we take a uniform prior for ρ on $[-1, 1]$ (i.e. $\nu_0 = \psi_0 = 1$) and take $\sigma^2 \sim IG(.005, .005)$.

Table 9.3 give results for the model with the largest posterior probability based on several values of $(\mu_{a_0}, \sigma_{a_0})$. From Table 9.3, we see that the top model in each case is $(x_1, x_2, x_3, x_4, x_5)$. In addition, we see that the posterior model probabilities increase monotonically as more weight is given to the historical data. When we put very small weight on the historical data, such as $(\mu_{a_0}, \sigma_{a_0}) = (.009, .003)$ the $(x_1, x_2, x_3, x_4, x_5)$ model still obtains the largest posterior probability, with value .117. When we put extremely small weight on the historical data such as $(\mu_{a_0}, \sigma_{a_0}) = (.0009, .0003)$, the $(x_2, x_3, x_4, x_5, x_7)$ model obtains the largest posterior probability, with value .122 and the $(x_1, x_2, x_3, x_4, x_5)$ model obtains the fourth largest posterior probability with value .101. Thus, we see that model choice is reasonably robust to the choice of $(\mu_{a_0}, \sigma_{a_0})$, consistently yielding the $(x_1, x_2, x_3, x_4, x_5)$ model as the top model for a suitable range of $(\mu_{a_0}, \sigma_{a_0})$. Based on these analyses, it does not appear that the variables x_6 (windspeed) and x_7 (coldness of temperature) are important predictors of pollen counts.

5. Discussion

We have discussed a class of informative priors for time series count data that are quite natural and useful when historical data is available. The priors have some very attractive properties and are proper under some very general conditions. We have also discussed computational methods for sampling from the posterior distribution

TABLE 9.1. Summaries of Variables for Pollen Data - Part I

year	variable	range	mean	standard deviation
1991	x_2	1.0 - 92.0	46.5	27.7
	x_3	0.0 - 4.5	3.56	0.92
	x_4	53.0 - 73.8	64.6	8.2
	x_5	-13.4 - 16.1	0.18	6.9
	x_6	0.0 - 18.0	11.1	3.9
	y	0.0 - 377	43.1	73.4
1992	x_2	1.0 - 82.0	41.7	23.9
	x_3	0.0 - 4.4	3.4	0.92
	x_4	46.2 - 70.4	61.2	7.7
	x_5	-11.4 - 13.7	0.46	6.1
	x_6	4.0 - 24.0	12.8	3.9
	y	0.0 - 440	53.9	86.3
1993	x_2	1.0 - 87.0	44.0	25.3
	x_3	0.0 - 4.5	3.5	0.92
	x_4	49.9 - 75.2	62.5	9.1
	x_5	-12.6 - 15.5	0.16	6.2
	x_6	0.0 - 15.0	8.6	3.5
	y	0 - 362	47.0	78.8
1994	x_2	1.0 - 79.0	40.8	23.2
	x_3	0.0 - 4.4	3.42	0.94
	x_4	50.5 - 69.3	62.9	6.3
	x_5	-10.0 - 12.8	0.30	6.0
	x_6	4.0 - 18.0	10.47	2.83
	y	0 - 205	32.29	49.1

TABLE 9.2. Summaries of Variables for Pollen Data - Part II

year	variable	value	count	percent
1991	x_1	0	10	10.9
	x_1	1	82	89.1
	x_7	0	27	29.3
	x_7	1	65	70.7
1992	x_1	0	8	9.9
	x_1	1	73	90.1
	x_7	0	30	37.0
	x_7	1	51	63.0
1993	x_1	0	11	12.6
	x_1	1	76	87.3
	x_7	0	29	33.3
	x_7	1	58	66.7
1994	x_1	0	6	8.0
	x_1	1	69	92.0
	x_7	0	25	33.3
	x_7	1	50	66.7

TABLE 9.3. Posterior Model Probabilities For Pollen Data

Model	$(\mu_{a_0}, \sigma_{a_0})$	$p(m \mid D^{(m)})$
$(x_1, x_2, x_3, x_4, x_5)$	(.5, .11)	.142
$(x_1, x_2, x_3, x_4, x_5)$	(.5, .08)	.290
$(x_1, x_2, x_3, x_4, x_5)$	(.5, .06)	.385
$(x_1, x_2, x_3, x_4, x_5)$	(.5, .05)	.420
$(x_1, x_2, x_3, x_4, x_5)$	(.98,.02)	.421

and for computing posterior model probabilities for variable subset selection. The expressions obtained for the posterior model probabilities facilitate a very quick and efficient method of calculation. In addition, the algorithms developed for sampling from the posterior distribution are quite efficient and feasible even for large data sets with a large number of covariates. The examples presented in Section 4 demonstrate the feasibility and the power of our methods. The Bayesian approach proposed here for this class of models appears to have a clear advantage over frequentist based procedures or other Bayesian procedures. Future work includes extending our methodology to multivariate discrete response models and multivariate models for longitudinal data.

References

Bedrick, E. J., Christensen, R., and Johnson, W. (1996). A New Perspective on Priors for Generalized Linear Models. *journal of the American Statistical Association, 91*, 1450-1460.

Chen, M.-H. (1994). Importance-weighted Marginal Bayesian Posterior Density Estimation. *Journal of the American Statistical Association, 89*, 818-824.

Chen, M.-H., Ibrahim, J.G., and Yiannoutsos, C. (1999). Prior Elicitation, Variable Selection, and and Bayesian Computation for Logistic Regression Models. *Journal of the Royal Statistical Society, Series B*, 61, 223-242.

Chen, M.-H. and Shao, Q.-M. (1997). Performance Study of Marginal Posterior Density Estimation via Kullback-Leibler Divergence. *Test, A Journal of the Spanish Society of Statistics and O.R., 6*, in press.

Gelfand, A.E., Sahu, S.K., and Carlin, B.P. (1996). Efficient Parametrisations for Generalized Linear Mixed Models (with discussion). In *Bayesian Statistics 5*, eds. J.M. Bernardo, J.O. Berger, A.P. Dawid and A.F.M. Smith, Oxford: Oxford University Press, 165–180.

Gelfand, A.E., Smith, A.F.M. and Lee, T.M. (1992). Bayesian Analysis of Constrained Parameter and Truncated Data Problems Using Gibbs Sampling. *Journal of the American Statistical Association, 87*, 523-532.

Ibrahim, J.G. and Chen, M-H. (1998). Prior Distributions and Bayesian Computation for Proportional Hazards Models. *Sankhya, Series B*, 60, 48-64.

Ibrahim, J.G., and Laud, P.W. (1994). A Predictive Approach to the Analysis of Designed Experiments. *Journal of the American Statistical Association, 89*, 309-319.

Ibrahim, J. G., Ryan, L. M., and Chen, M.-H. (1998). Use of Historical Controls to Adjust for Covariates in Trend Tests for Binary Data. *Journal of the American Statistical Association*, 93, 1282-1293.

Laud, P.W., and Ibrahim, J.G. (1995). Predictive Model Selection. *Journal of the Royal Statistical Society, Ser.B, 57*, 247-262.

Stark, P. C., Ryan, L. M., McDonald, J. L., Burge, H. A. (1997). Using Meteorologic Data to Predict Daily Ragweed Pollen Levels. *Aerobiologia*, 13, 177-184.

Zeger, S.L., (1988). A Regression Model for Time Series of Counts. *Biometrika*, 75, 621-629.

10

Item Response Modeling

James Albert
Malay Ghosh

ABSTRACT This chapter introduces the Bayesian fitting and checking of a family of item response models. The one and two parameter item response models are described and the models are illustrated using a mathematics placement exam. The choice of prior distributions is discussed. It is shown that some standard noninformative priors will result in improper posterior distributions, and some guidance is provided on the choice of informative priors for the item parameters. Markov chain Monte Carlo algorithms are outlined for simulating from the joint posterior distribution of the item and ability parameters. Bayesian residuals and the posterior predictive distribution are used to assess model fit. The methods are used to examine the difficulty and discrimination characteristics of the items on the mathematics placement test.

1. Introduction

In this chapter, we consider the analysis of test data consisting of a number of multiple-choice questions. In the particular example that will be analyzed, 200 beginning college students were given a mathematics placement test. The test consists of 35 multiple-choice items on topics in intermediate and college algebra. The purpose of this exam is to assess a student's ability in high school algebra towards the goal of placing the student in a suitable mathematics course. The results of the exam are used together with other information such as the student's score on the ACT and their high school grade point average to recommend the best mathematics class for enrollment in the fall semester.

In designing this mathematics placement test, there are several concerns. One is interested in placing items on the exam which have different levels of difficulty. If most of the students do very poorly on the exam, then the items are generally too difficult, and the results of the exam may not be helpful in accurately assessing the students' mathematical ability. A similar problem would be present if the majority of students were able to get all of the questions correct. Generally it is desirable to have a broad range of performances on the exam, ranging from students who get relatively few questions correct to students who get most of the questions correct. This wide variation of student performances will make it possible to more accurately assess the mathematical abilities of the students.

If the test is effective in measuring the students' abilities, then the next concern is whether the individual items on the test are effective in discriminating among

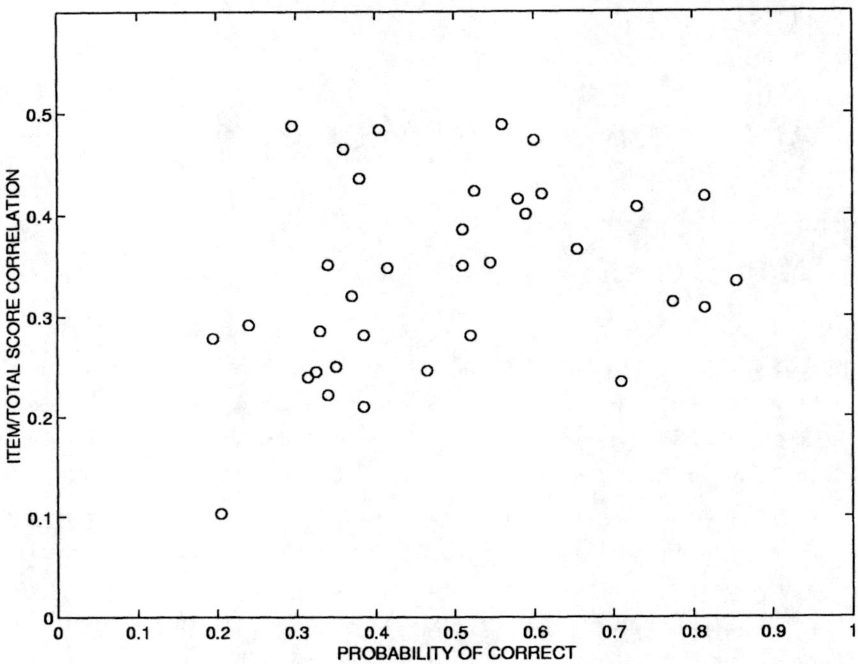

FIGURE 10.1. Scatterplot of observed proportion correct and point-biserial correlation for all items on the mathematics placement test.

students of different abilities. Suppose that one item is incorrectly answered by all of the students taking the exam. Then this particular item is useless in distinguishing between weak and strong students – nothing would be lost if this item was removed from the exam. In contrast, an "ideal" item would be correctly answered by students of above-average mathematics ability and incorrectly answered by students of below-average ability. This ideal item probably will not exist, but the most valuable test items are those that are strongly positively correlated with the students' mathematics proficiency.

In this example, we are interested in learning about the student's performances on the exam. In addition, we wish to learn about the characteristics of each of the 35 items that make up the exam. The items can be characterized by their difficulty levels, and their discriminatory power to distinguish between students of different abilities. One can estimate an item's difficulty by computing the proportion of correct responses for all students in the sample. An item's discrimination can be measured by the point-biserial correlation between the item's response (1 if correct and 0 if incorrect) and the total score on the 35 item test. Figure 10.1 plots the observed proportion correct against the point-biserial correlation for all items. Note that the items appear to vary considerably with respect to their difficulty and also with respect to their discrimination characteristics.

We can learn about the students' abilities and test characteristics by fitting an item response model. This model is a representation of the probability that a student answers a particular item correctly. The probability is a function of three unknown parameters – a single parameter which describes the student's ability, and two parameters which characterize the item's difficulty and discrimination ability.

This chapter will introduce the basic properties of the item response model, and describe the fitting and checking of this model from a Bayesian perspective. Section 2 introduces the basic model in the simple setting where a single student is taking a test with a single item. Section 3 extends this model to the case where n students are taking a multiple choice test with k items. A key issue in the construction of a Bayesian item response model is the choice of priors. Section 4 gives theoretical results which indicate problems with using vague improper priors and give guidance on the choice of subjective prior priors on the sets of item parameters. Algorithms for the Bayesian fitting of one and two-parameter item response models are outlined in Section 5 and methods for criticizing a particular fitted model are described in Section 6. The methods are illustrated in Section 7 for the mathematics placement dataset.

2. An Item Response Curve

Suppose a student is taking a test consisting of a single multiple-choice item. There are two possible results — either the student gets the item correct or incorrect. We let the variable y denote the observed response - $y = 1$ denotes a correct response and $y = 0$ denotes an incorrect correct. The probability of a correct response is represented by the model

$$\Pr(y = 1) = F(a\theta - b),$$

where θ is a parameter describing the ability of the student, a and b are parameters that characterize the particular test item, and F is a known cumulative distribution function. We describe each of these quantities in turn.

One basic assumption of this model is that there exists a single quantity, θ, called a *latent trait* which represents the intrinsic ability of the student to succeed on the one question test. For an algebra item on the math placement test, we are assuming the existence of one continuous quantity θ, which represents the student's basic aptitude in high school algebra. We are unable to directly measure a students aptitude in algebra, but we obtain a *indication* of this trait by the student's performance on the one question exam.

It may be unrealistic to assume that a student's ability can be characteristic by means of a single latent trait. For example, in the mathematics test example, perhaps there exist two distinct latent traits that underlie a student's exam performance, where one trait is an ability to perform arithmetic and basic algebraic operations, and the second trait is an ability to solve mathematical problems using algebra. Although the problem of detecting and measuring multiple latent traits is interesting, we focus here on the assumption of a single latent trait, since this assumption is basic for the majority of applications of item response theory.

The cumulative distribution function F connects the continuous-valued latent trait θ with the probability of a correct response on the item. There are two popular choices for the function F. If F is assumed to be the standard logistic distribution

$$F(x) = \frac{\exp(x)}{1 + \exp(x)},$$

this leads to the *logistic model*

$$\Pr(y = 1) = \frac{\exp(a\theta - b)}{1 + \exp(a\theta - b)}$$

If F is chosen to be the standard normal cdf, denoted $\Phi()$, then we obtain the *probit model*

$$\Pr(y = 1) = \Phi(a\theta - b).$$

The logistic and normal distributions both are symmetric about zero and have similar shapes Thus in typical applications of item response models, the fitted probabilities using the logistic and probit models will be very similar. One advantage of the logistic model is its relative ease in interpretation. The logit, or log odds, of the probability of a correct response is

$$\log \frac{\Pr(y = 1)}{1 - \Pr(y = 1)} = a\theta - b.$$

Here we will focus on the use of the probit model, since there exists an attractive Bayesian algorithm for fitting the model in this case.

The parameters a and b in the model describe the characteristics of the particular test item. To help interpret these parameters, we consider the *item response function*

$$f(\theta) = \Pr(y = 1|\theta) = F(a\theta - b),$$

which plots the probability of a correct response as a function of the latent trait θ. This function is typically plotted for values of θ between -3 and 3 which corresponds to the range of abilities of the population of students.

The parameter b is called the *difficulty* parameter. This parameter controls the general difficulty level of the particular test item. If one fixes the value of the parameter a and increases the value of b, the curve maintains its basic shape but is shifted to the right. A test item with a large negative value of b corresponds to an easy item where even students with below average abilities (corresponding to small values of θ) have a high probability of getting the item correct. In contrast, an item with a large b is difficult — even strong students (large values of θ) have a relatively small probability of a correct response.

The second item parameter a is called the *discrimination* parameter. This parameter controls the slope of the item response function. If one considers an item response curve with fixed b, the curve becomes more steep for increasing values of the parameter a. A steep item response curve corresponds to an item that is highly discriminatory between weak and strong students. The probability of a correct response changes rapidly as the latent trait θ increases in an interval about 0. In contrast an item with a small value of a has a flat shape. This means that the probability of a correct response only changes a small amount as one passes from weaker (small θ) to stronger (large θ) students. An item with a small value of a is a relatively poor discriminator between students of varying abilities.

3. Administering an Exam to a Group of Students

In the previous section, we considered a model for the response of an individual to a single test item. In the mathematics placement test example, there are 200 students taking a test consisting of 35 items. We describe how item response theory is used to model this more general data structure.

In general, suppose that there are n individuals that take an item consisting of k items. Each item is scored as correct or incorrect — we represent these responses by the numbers 1 and 0, respectively. Let y_{ij} denote the response of the ith individual to the jth item on the test. So the variables $y_{11}, ..., y_{ik}$ represent the binary responses of the first individual to items $1, ..., k$ of the test, the variables $y_{21}, ..., y_{2k}$ denote the test responses of the second individual, and so on. We can view the observed data as a matrix of n rows and k columns, where the rows correspond to individuals and the columns to test items.

The first fundamental assumption in this model is that a given individual's performance on the test is dependent on a single unknown latent trait or ability θ_i. We let $\theta_1, ..., \theta_n$ denote the latent traits of the n individuals. In the math placement example, we are assuming that the individuals have distinct latent mathematical abilities that are measured on a continuous scale.

The latent traits of the individuals are not directly observable, but we learn about them by means on the responses to the test items. We assume that the probability that the ith individual obtains the jth item correctly ($y_{ij} = 1$) depends on the individual's ability θ_i and the characteristics of the particular test item. The probability is modeled, conditional on a value of the latent trait, as

$$\mathrm{P}(y_{ij} = 1|\theta_i) = F(a_j\theta_i - b_j), \tag{1}$$

where F is a known cumulative distribution function and a_j and b_j are parameters specific to the particular item. As in the previous section, the parameter b_j measures the difficulty of the jth item and a_j represents the item's ability to discriminate between individuals of different abilities.

Note that $\mathrm{P}(y_{ij} = 1)$ represents the probability that an individual obtains a correct response to a single item. To obtain the probability that an individual gets a particular sequence of responses $y_{i1}, ..., y_{ik}$, we need to make the assumption of *local independence*. This assumption means that the set of k responses of the individual are conditionally independent, given a value of the latent trait θ_i. This assumption does not mean that the individual's responses to different items are independent. It is very possible that the responses to different items are correlated, but this correlation is explained entirely by the latent trait of the individual. Using this assumption, the probability of observing an individual's entire sequence of responses is given by

$$\begin{aligned}\mathrm{P}(y_{i1}, \ldots, y_{ik} \mid \theta_i) &= \mathrm{P}(y_{i1} \mid \theta_i) \times \cdots \times \mathrm{P}(y_{ik} \mid \theta_i) \\ &= \textstyle\prod_{j=1}^{k} F(a_j\theta_i - b_j)^{y_{ij}}[1 - F(a_j\theta_i - b_j)]^{1-y_{ij}}.\end{aligned}$$

To combine the data across all individuals, we assume that the responses of the n individuals are independent. The unknown parameters here are the vector of latent traits $\theta = (\theta_1, ..., \theta_n)$, the vector $\mathbf{a} = (a_1, ..., a_k)$ of item discrimination parameters, and the vector $\mathbf{b} = (b_1, ..., b_k)$ of item difficulty parameters. The likelihood function of $(\theta, \mathbf{a}, \mathbf{b})$ is proportional to the probability of the response patterns for all individuals which is equal to

$$L(\theta, \mathbf{a}, \mathbf{b}) = \prod_{i=1}^{n}\prod_{j=1}^{k} F(a_j\theta_i - b_j)^{y_{ij}}[1 - F(a_j\theta_i - b_j)]^{1-y_{ij}}.$$

The above model is described as a two-parameter item response model, since there are two parameters a_i, b_i associated with each test item. This model implicitly assumes that each item has a unique difficulty level. In addition, since the parameters $a_1, ..., a_n$ are distinct, the model assumes that each item has a unique discrimination ability. One way to considerably simplify the two-parameter model is to assume that the items have equal discriminatory competencies — without loss of generality, we can assume that $a_1 = ... = a_k = 1$. The probability of response of the ith individual to the jth item for this *one-parameter item response model* is then given by the simpler expression

$$P(y_{ij} = 1|\theta_i) = F(\theta_i - b_j).$$

When F is chosen to be the logistic distribution, the one-parameter item response model is called the Rasch model. In the explanation of the fitting and model checking procedures, we will focus on the two-parameter model and indicate how the described methods can be adjusted for the simpler one-parameter model.

4. Prior Distributions

After the likelihood function has been defined, the next task is to define prior distributions for the vector of latent traits θ and the item parameter vectors a and b.

4.1 Noninformative Priors and Propriety of the Posterior Distribution

Much of the existing Bayesian literature on item response models deals with noninformative improper priors for $\{\theta_i\}$, $\{a_j\}$ and $\{b_j\}$. This is especially tempting when the prior information is vague, and the Bayesian paradigm can often be used effectively by the introduction of noninformative priors. However, this leads to the possibility of improper posteriors. Proper posteriors are vital for any Bayesian analysis, since otherwise, there is no meaning of posterior moments, quantiles and credible intervals. The impropriety of posterior is not always easy to recognize unless calculations are done analytically. For instance, it may so happen that all of the conditionals needed for the implementation of MCMC algorithms such as the Gibbs sampler are proper, and yet the posterior of interest is improper.

For the two-parameter item response model, a general class of priors is given by

$$\pi(\theta, \mathbf{a}, \mathbf{b}) \propto \prod_{i=1}^{n} g_{1i}(\theta_i) \prod_{j=1}^{k} g_{2j}(a_j) \prod_{j=1}^{k} g_{3k}(b_j), \tag{2}$$

where the densities $\{g_{1i}\}$, $\{g_{2j}\}$, $\{g_{3j}\}$, are proper or improper pdf's. The joint posterior density of $(\theta, \mathbf{a}, \mathbf{b})$ is given by

$$\pi(\theta, \mathbf{a}, \mathbf{b}|\mathbf{y}) \propto L(\theta, \mathbf{a}, \mathbf{b})\pi(\theta, \mathbf{a}, \mathbf{b}).$$

We now state several theorems which show that some typical improper priors for the parameters will lead to improper posteriors.

Theorem 1. Consider the prior given in (1.2), where at least one of the g_{2j} is improper. Let $\bar{F}(x) = 1 - F(x)$. Then the joint posterior of $(\theta, \mathbf{a}, \mathbf{b})$ is improper.

Proof: Consider the computation of the marginal posterior density of the item parameters $(\mathbf{a}, \mathbf{b})$ if we define

$$I_i(\mathbf{a}, \mathbf{b}) = \int_{-\infty}^{\infty} \prod_{j=1}^{k} [F^{y_{ij}}(a_j\theta_i - b_j)\bar{F}^{1-y_{ij}}(a_j\theta_i - b_j)]g_{1i}(\theta_i)d\theta_i,$$

for $1 \leq i \leq n$, then

$$\pi(\mathbf{a}, \mathbf{b}|\mathbf{y}) \propto \prod_{i=1}^{n} I_i(\mathbf{a}, \mathbf{b}) \prod_{j=1}^{k} g_{2j}(a_j) \prod_{j=1}^{k} g_{3k}(b_j).$$

For definiteness, assume that $g_{21}(a_1)$ is improper. Now for each fixed i, $y_{i1} = 1$ or 0. Hence, each integral $I_i(\mathbf{a}, \mathbf{b})$ contains a term $F(a_1\theta_i - b_1)$ or $1 - F(a_1\theta_i - b_1)$, but not both. In the first case,

$$\begin{aligned} I_i(\mathbf{a}, \mathbf{b}) &\geq \int_0^\infty F(a_1\theta_i - b_1) \prod_{j=2}^k [F^{y_{ij}}(a_j\theta_i - b_j)\bar{F}^{1-y_{ij}}(a_j\theta_i - b_j)] g_{1i}(\theta_i) d\theta_i \\ &\geq F(-b_1) \int_{-\infty}^0 \prod_{j=2}^k [F^{y_{ij}}(a_j\theta_i - b_j)\bar{F}^{1-y_{ij}}(a_j\theta_i - b_j)] g_{1i}(\theta_i) d\theta_i, \end{aligned}$$

while in the second case

$$\begin{aligned} I_i(\mathbf{a}, \mathbf{b}) &\geq \int_0^\infty \bar{F}(a_1\theta_i - b_1) \prod_{j=2}^k [F^{y_{ij}}(a_j\theta_i - b_j)\bar{F}^{1-y_{ij}}(a_j\theta_i - b_j)] g_{1i}(\theta_i) d\theta_i \\ &\geq \bar{F}(-b_1) \int_{-\infty}^0 \prod_{j=2}^k [F^{y_{ij}}(a_j\theta_i - b_j)\bar{F}^{1-y_{ij}}(a_j\theta_i - b_j)] g_{1i}(\theta_i) d\theta_i. \end{aligned}$$

From above, it follows that for each i, there is a nonnegative lower bound for $I_i(\mathbf{a}, \mathbf{b})$ which does not involve a_1. Integrating this lower bound with respect to the marginal (improper) pdf $g_{21}(a_1)$ over $(0, \infty)$, one gets

$$\int_0^\infty \prod_{i=1}^{n} I_i(\mathbf{a}, \mathbf{b}) g_{21}(a_1) da_1 = +\infty.$$

This implies that the posterior density $\pi(\mathbf{a}, \mathbf{b}|\mathbf{y})$ is improper, and hence $\pi(\theta, \mathbf{a}, \mathbf{b}|\mathbf{y})$ is also improper.

It is interesting to note, however, that all of the full conditional posteriors are proper in this case. Thus the impropriety of the posterior may not be detected in the implementation of a Gibbs sampling algorithm. To show that the full conditionals are proper, notice that

$$\pi(\theta_i|\theta_l(l \neq i), \mathbf{a}, \mathbf{b}, \mathbf{y}) \propto \prod_{j=1}^{k} [F^{y_{ij}}(a_j\theta_i - b_j)\bar{F}^{1-y_{ij}}(a_j\theta_i - b_j)] g_{1i}(\theta_i)$$

which is typically integrable with respect to θ_i. Similarly, $\pi(a_j|a_m(m \neq j), \theta, \mathbf{b}, \mathbf{y})$ and $\pi(b_j|b_m(m \neq j), \theta, \mathbf{a}, \mathbf{y})$ may all be proper. A simple example is a normal or a logistic F and g_{1i}, g_{2j} are uniform priors over the real line, while g_{3j} are uniform over $(0, \infty)$.

Swaminathan and Gifford (1985, p. 353) suggest that flat priors for θ, **a** and **b** make the Bayesian analysis comparable to a likelihood-based analysis. But the Bayesian analysis is actually questionable, since the posterior is improper in that case.

The above fact is a consequence of a more general result stated below.

Theorem 2 Consider the prior $\pi(\theta, \mathbf{a}) \propto g(\mathbf{a})$, where g is an arbitrary positive function of a (for example g could be a proper pdf for a). Then the posterior distribution is always improper.

4.2 Choosing an Informative Prior

The theorems in the previous section show that noninformative priors for item response problems often lead to improper posteriors. Thus it is desirable to choose proper prior distributions to ensure that the posterior distribution is also proper. The prior distributions discussed here have two primary purposes. First, the choice of a prior distribution on the latent traits resolves particular identification problems in the two-parameter model. Second, informative prior distributions placed on the item response parameters can be used to reflect the prior belief that the values of the item parameters are not extreme.

The two-parameter item response model, as defined in the previous two sections, is overparameterized. Looking at the expression for the probabilities (1.1), note that we can multiply the latent abilities $\{\theta_i\}$ by a given positive constant, and divide all of the slope parameters $\{a_j\}$ by the same positive constant and preserve the model. One solution to this overparameterization problem is to place a constraint on the latent traits, such as $\sum_{i=1}^{n} \theta_i = 0$. Another solution to this problem is to place a known prior distribution on the vector of latent traits. Following common practice, we assign $\theta_1, .., \theta_n$ a normal distribution with mean 0 and standard deviation 1. This prior is making the implicit assumption that all of the latent abilities of the individuals fall in the interval $(-3, 3)$.

As discussed in the previous section, proper priors need to be chosen for the item parameters to ensure that the posterior distribution is proper. This will be most important in the case where extreme data is observed where students are observed to get correct or incorrect answers to every item. In the common situation where little prior information is available about the difficulty parameters, one can assume that the b_j are independent, where b_j is distributed $\mathrm{N}(0, s_b)$, where s_b is chosen to be a large value. This choice of prior will result in a proper posterior distribution when extreme data is observed, and will have a modest effect on the posterior distribution for nonextreme data.

Proper distributions on the discrimination parameters $\{a_j\}$ can be constructed based on prior knowledge about the shape of the item response curve. One assumption implicit in the graphs of the item response curve is that the probability of a correct response to an item is an increasing function of the latent trait, which means that individuals with higher abilities are more likely to get the item correct. The item response curve for item j will be increasing when the slope parameter a_j is positive. Thus, it is reasonable to state a priori that all of the slope parameters are positive with high probability. One can model this prior belief by the choice of different prior distributions. One could assume that the parameters $a_1, ..., a_k$ are independent, where a_j is assigned a Gamma distribution with known hyperparameter values. Since the support of the Gamma distribution is on positive values, this prior is stating that the parameter is positive with probability one. Alternately, one can assign a_j a normal distribution with mean μ_a and standard deviation s_a. By choosing a positive value for the normal mean μ_a and an appropriate value for s_a, one can model the belief that a_j is positive with high probability. In the following, we will use a normal prior on a_j since it leads to a simple fitting procedure.

5. Bayesian Fitting of Item Response Models

5.1 Fitting of the Two-parameter Model Using Gibbs Sampling

The prior distribution discussed in the previous section is summarized as follows:

- The latent abilities $\theta_1, ..., \theta_n$ are a random sample from a standard normal distribution.

- The item slope parameters $a_1, ..., a_k$ and intercept parameters $b_1, ..., b_k$ are independent random samples from normal(0, s_a) and normal(μ_b, s_b) distributions, respectively.

Combining this prior density with the likelihood function, the posterior density is expressed, up to a proportionality constant, as

$$g(\theta, \mathbf{a}, \mathbf{b}|\text{data})$$

$$= L(\theta, \mathbf{a}, \mathbf{b}) \prod_{i=1}^{n} \phi(\theta_i; 0, 1) \prod_{j=1}^{k} \phi(a_j; 0, s_a)\phi(b_j; 0, s_b),$$

where $\phi(x; \mu, \sigma)$ denotes the normal density with mean μ and standard deviation σ.

One can fit this Bayesian two-parameter item response model by simulating a large sample of values from the joint posterior distribution of the latent abilities and the item parameters $(\theta, \mathbf{a}, \mathbf{b})$. This sample of values is generated by simulation from a Markov Chain Monte Carlo (MCMC) algorithm. This algorithm will produce a correlated sequence of random variables in which the jth value in the sequence, $(\theta^{(j)}, \mathbf{a}^{(j)}, \mathbf{b}^{(j)})$ is dependent on the $(j-1)$st value $(\theta^{(j-1)}, \mathbf{a}^{(j-1)}, \mathbf{b}^{(j-1)})$. The MCMC algorithm is constructed by specifying a starting value for the parameter vector and stating a method for moving from one simulated iterate in the sequence to the next. Under general conditions, the distribution of the jth iterate, as j approaches infinity, will converge to the posterior distribution of interest. If a large number of iterations are performed, the last group of iterates in the sequence, say $\{(\theta^{(j)}, \mathbf{a}^{(j)}, \mathbf{b}^{(j)}), j = m_0, ..., m_0 + m\}$, will approximate a dependent sample from the posterior distribution.

One general strategy for the construction of a MCMC algorithm is based on the notion of Gibbs sampling. This simulation algorithm partitions the random variable into blocks, and iteratively samples each block of parameters from a posterior density which conditions on the most recent simulated values of all parameters not included in the block. In this setting suppose that we partition the parameter set into the vector of latent abilities θ and the vectors of item parameters $(\mathbf{a}, \mathbf{b})$. Then the Gibbs sampling algorithm will iterate from the posterior distribution $[\theta|\mathbf{a}, \mathbf{b}]$ of latent abilities conditional on fixed values of the item parameters, and the posterior distribution $[\mathbf{a}, \mathbf{b}|\theta]$ of item parameters conditional on the latent abilities. We examine each conditional posterior distribution below.

Suppose first that the item parameters are known. Then the posterior density of the latent traits, conditional on the item parameters, is given by

$$g(\theta|\mathbf{a}, \mathbf{b}, \text{data})$$

$$= \prod_{i=1}^{n} \prod_{j=1}^{k} F(a_j\theta_i - b_j)^{y_{ij}}[1 - F(a_j\theta_i - b_j)]^{1-y_{ij}} \prod_{i=1}^{n} \phi(\theta_i; 0, 1).$$

Note that the individual latent traits are conditionally independent, with the ith latent trait distributed according to the density

$$g(\theta_i|\mathbf{a},\mathbf{b},\text{data}) = \prod_{j=1}^{k} F(a_j\theta_i - b_j)^{y_{ij}}[1 - F(a_j\theta_i - b_j)]^{1-y_{ij}}\phi(\theta_i;0,1).$$

Next, let's reverse the roles of the two sets of parameters and suppose that the latent abilities are known. Then the posterior density of the item parameters, conditional on these latent traits, is given by

$$g(\mathbf{a},\mathbf{b}|\theta,\text{data})$$
$$= \prod_{j=1}^{k}\left[\phi(a_j;0,s_a)\phi(b_j;0,s_b)\prod_{i=1}^{n} F(a_j\theta_i - b_j)^{y_{ij}}[1 - F(a_j\theta_i - b_j)]^{1-y_{ij}}\right].$$

We see that this conditional density factors into independent components, where the set of parameters corresponding to the jth item, (a_j, b_j), is distributed according to the density

$$g(a_j, b_j|\theta,\text{data})$$
$$= \prod_{i=1}^{n} F(a_j\theta_i - b_j)^{y_{ij}}[1 - F(a_j\theta_i - b_j)]^{1-y_{ij}}\phi(a_j;0,s_a)\phi(b_j;0,s_b).$$

The convenient factorizations of the two sets of conditional posterior distributions suggests the following Gibbs sampling algorithm for simulating from the joint posterior distribution.

1. (Choose starting values.) Choose reasonable initial values for the item parameters — call these values $\{(a_j^{(0)}, b_j^{(0)}), j = 1, ..., k\}$. These values can be estimated from the data, as illustrated in the discussion of the example below. Set the iteration number $m = 0$.

2. (Simulate the latent traits.) Simulate independent values of the latent traits $\theta_1, ..., \theta_n$, where θ_i is simulated from the conditional posterior density

   $$g(\theta_i|\mathbf{a}^{(m)},\mathbf{b}^{(m)},\text{data})$$
   $$= \prod_{j=1}^{k} F(a_j^{(m)}\theta_i - b_j^{(m)})^{y_{ij}}[1 - F(a_j^{(m)}\theta_i - b_j^{(m)})]^{1-y_{ij}}\phi(\theta_i;0,1).$$

   Denote the vector of simulated latent traits by $\theta = (\theta_1^{(m)}, ..., \theta_n^{(m)})$.

3. (Simulate the item parameters.) Simulate independent k sets of the item parameters, where (a_j, b_j) is simulated from the conditional density

   $$g(a_j, b_j|\theta^{(m)},\text{data})$$
   $$= \prod_{i=1}^{n} F(a_j\theta_i^{(m)} - b_j)^{y_{ij}}[1 - F(a_j\theta_i^{(m)} - b_j)]^{1-y_{ij}}\phi(a_j;0,s_a)\phi(b_j;0,s_b).$$

   Denote the simulated item parameters as $(a_1^{(m)}, b_1^{(m)}), ..., (a_k^{(m)}, b_k^{(m)})$, and the vectors $\mathbf{a}^{(m)} = (a_1^{(m)}, ..., a_k^{(m)})$, $\mathbf{b}^{(m)} = (b_1^{(m)}, ..., b_k^{(m)})$.

4. (Iterate.) Update the counter $m = m + 1$ and return to step 2.

5.2 Implementation of Gibbs Sampling for General F

A general algorithm for simulating from the above conditional posterior distributions, for any choice of the link function F, is based on the Metropolis-random walk method. To describe this simulation algorithm, suppose that one is interested in simulating from the posterior density g which is given by

$$g(\theta) = Kh(\theta),$$

where K is an unknown proportionality constant. Let $\theta^{(m)}$ denote the current simulated value from the density g. We propose a new simulated value θ^p from g, where θ^p is obtained from the current value by adding a normal variate with mean 0 and standard deviation c. Using symbolic notation,

$$\theta^p = \theta^{(m)} + cZ,$$

where Z is a standard normal variate. Next, we compute an acceptance probability $PROB$, which is equal to the minimum of 1 and the quotient of the posterior density evaluated at the proposal value and the posterior density evaluated at the current value.

$$PROB = \min\left(1, \frac{h(\theta^p)}{h(\theta^{(m)})}\right).$$

To complete the algorithm, we simulate a uniform variate U. If U is smaller than $PROB$, we accept the proposal value and the next value of θ in the sequence is $\theta^{(m+1)} = \theta^p$. Otherwise, if $U \geq PROB$, we reject the proposal value and the next simulated value remains at the current value $\theta^{(m+1)} = \theta^{(m)}$.

This basic Metropolis method can be used to simulate from all of the conditional posterior distributions in the fitting algorithm for the two-parameter item response model. In step 2 of the algorithm, values of the latent traits $\theta_1, ..., \theta_n$ are simulated independently using parallel Metropolis steps using normal proposal densities with known scale factors $c_1, ..., c_n$. Values of the k pairs of item parameters $(a_1, b_1), ..., (a_k, b_k)$ are simulated independently from a bivariate version of the above Metropolis algorithm. Given a current value at the j pair, $(a_j^{(m)}, b_j^{(m)})$, a proposal pair can be generated by adding normal random variables to each component:

$$a_j^p = a_j^{(m)} + c_{a_j} Z_1, \; b_j^p = b_j^{(m)} + c_{b_j} Z_2.$$

As in the single variable Metropolis case, the pair (a_j^p, b_j^p) is accepted or rejected as the next simulated value in the sequence. The probability that the pair is accepted depends on the ratio of the posterior density evaluated at the proposal pair and the density evaluated at the current pair.

In the implementation of this Gibbs sampling/Metropolis algorithm for simulating a sample from the posterior density, some care should be taken in the choice of the normal scale parameters $\{c_i\}$ and $\{(c_{a_j}, c_{b_j})\}$. Generally, it is desirable to choose values of these scale parameters so that the acceptance rate in each step of the algorithm is between 25-50 percent. This can be accomplished by choosing the scale to be approximately twice the standard deviation of the conditional posterior distribution being simulated. In practice, one can perform two runs of the algorithm. The first run with default values of the scale parameters can be used to estimate the posterior standard deviations of each parameter. The algorithm is then run again using new values of the scale parameters determined from these estimated standard deviations.

5.3 Gibbs Sampling for a Probit Link Using Data Augmentation

In the case of a probit link function ($F = \Phi$), an alternative MCMC algorithm can be implemented by the use of data augmentation and Gibbs sampling. Corresponding to each binary response y_{ij}, define a continuous variable Z_{ij} such that

$$Z_{ij} \text{ is distributed } \mathcal{N}(a_j\theta_i - b_j, 1),$$

and y_{ij} indicates if the sign of Z_{ij} is positive or negative:

$$y_{ij} = 1 \text{ if } Z_{ij} > 0, \text{ and } y_{ij} = 0 \text{ if } Z_{ij} \leq 0.$$

One can show that this formulation leads to the model $\Pr(y_{ij} = 1) = \Phi(a_j\theta_i - b_j)$. In the mathematics placement test example, one can interpret Z_{ij} as a latent mathematics ability, measured on a continuous scale, that underlies the student's performance on the particular test item. Note that Z_{ij} is not observed, but we get indications of the sign of Z_{ij} by means of the binary response y_{ij}.

Suppose that we augment the current set of parameters $\{\theta_i\}$, $\{a_j\}$, $\{b_j\}$ with the unobserved vector of latent variables $Z = (Z_{11}, ..., Z_{jk})$. The joint posterior density of all model parameters and latent variables is given by

$$\begin{aligned} g(Z, \theta, \boldsymbol{a}, \boldsymbol{b}|\text{data}) \quad \propto \quad & \prod_{i=1}^{n} \prod_{j=1}^{k} \left[\phi(Z_{ij}, m_{ij}, 1) I^*(Z_{ij}, y_{ij})\right] \\ \times \quad & \prod_{i=1}^{n} \phi(\theta_i; 0, 1) \prod_{j=1}^{k} \left[\phi(a_j; \mu_a, s_a)\phi(b_j; 0, s_b)\right], \end{aligned}$$

where the normal mean $m_{ij} = a_j\theta_i - b_j$, and $I^*(c, d)$ is equal to one when $\{c > 0, d = 1\}$ or $\{c < 0, d = 0\}$, and equal to zero otherwise.

On the surface, the inference problem now appears to be more complicated, since the posterior density is a function of $n + 2k$ parameters and nk values of the latent variables. However, this new representation of the posterior distribution leads to a simple Gibbs sampling scheme for simulating from the joint posterior distribution.

Suppose that we partition the unknown quantities into three groups, the latent variables Z, the ability parameters θ, and the item parameters $\mathbf{a}, \mathbf{b}$. The posterior distribution of each group, conditional on values of the remaining parameters, have familiar functional forms and are easy to simulate.

1. (Conditional distribution of latent variables.) Suppose that values of the ability parameters and item parameters are held fixed. Then $Z_{11}, ..., Z_{nk}$ have independent truncated normal distributions. The conditional posterior distribution of Z_{ij} is truncated normal with mean $m_{ij} = a_j\theta_i - b_j$ and standard deviation 1. The truncation of this posterior distribution depends on the value of the corresponding binary observation y_{ij}. If the observation is a success ($y_{ij} = 1$), the truncation of Z_{ij} is from the left at 0; if the observation is a failure ($y_{ij} = 0$), the truncation is from the right at 0.

2. (Conditional distribution of ability parameters.) Suppose that the latent variables and item parameters are fixed. We can write the model for the latent variables as

$$Z_{ij} + b_j = a_j\theta_i + e_{ij},$$

where e_{ij} are independent error terms with a standard normal distribution. Note that for fixed values of Z and $\mathbf{a}, \mathbf{b}$, this is a normal linear model. Combining the normal likelihood with the normal prior, one can show that $\theta_1, ..., \theta_n$ have independent normal posterior distributions.

3. (Conditional distribution of item parameters.) Last, we consider the distribution of the item parameters $\{a_j, b_j\}$ conditional on the values of the latent variables and the latent traits. Rewrite the latent variables model as

$$Z_{ij}^{(t)} = a_j \theta_i^{(t)} - b_j + e_{ij}.$$

Since the values of the latent data $\{Z_{ij}\}$ and the latent traits $\{\theta_i\}$ are fixed, this (for a fixed value of j) can be viewed as a normal linear model with unknown parameter vector (a_j, b_j) and known covariate vector $(\theta_i^{(t)}, -1)$. Again this can be combined with the normal priors on a_j and b_j to obtain independent bivariate normal posterior densities for $\{(a_j, b_j\}), j = 1, ..., k$.

The three tractable conditional posterior distributions lead to an attractive Gibbs sampling algorithm in the probit link case. Each iteration of the algorithm consists of three steps. The first step simulates latent variables Z from independent truncated normal distributions conditioning on the current values of the item and ability parameters. The second step simulates values of the ability parameters and the third step simulates item parameters. In each of the last two steps, simulations are based on the posterior distribution that conditions on the latent variables and the remaining parameters.

This algorithm is somewhat more time-consuming to run than the Gibbs sampling/Metropolis algorithm, due to the extra simulation of the latent variables. However, the algorithm does not depend on the assignment of scale parameter values in the Metropolis algorithm that may determine the rate of convergence of the algorithm. One attractive feature of the probit data augmentation algorithm is that, due to the underlying normal linear model, it is straightforward to generalize this algorithm to a multinomial response regression model where the categories are ordered or unordered.

5.4 Bayesian Fitting of the One-parameter Model

Since the one-parameter item response model is a special case of the two-parameter model with $a_1 = ... = a_k = 1$, the Gibbs sampling algorithm outlined in the previous section can also be used to fit the one-parameter model. The Gibbs algorithm will alternately simulate values of the latent traits conditional on values of the item difficulty parameters, and simulate values of the item parameters conditional on the current values of the latent traits. In the case of a general link function F, the Metropolis within Gibbs algorithm can be used. In the case of a probit link function, the Gibbs/data augmentation scheme provides an attractive scheme for fitting this model.

6. Inferences from the Model

In many applications of item response modeling, the focus of the analysis is on the characteristics of the test items. The jth item on the test is quantified in terms of the slope parameter a_j, which describes the discrimination ability of the item, and the intercept parameter b_j, which describes the item's difficulty. To aid in the interpretation of the item characteristics, the slope and intercept parameters can be transformed to new parameters which are more easily interpreted.

In the case of a probit link function, one useful description of the discrimination ability of the jth item is given by the biserial correlation, defined as

$$r_j = \frac{a_j}{\sqrt{1 + a_j^2}}.$$

This quantity measures the correlation between the binary responses $\{y_{1j}, \ldots, y_{nj}\}$ and the latent traits $\{\theta_1, \ldots, \theta_n\}$. It can be interpreted much like a standard correlation coefficient. If r_j is a large positive number, then this particular item is an effective discriminator between weak and strong students. In contrast, a value of r_j near zero indicates that the test item provides little information regarding the ability of the student.

It can be hard to interpret the difficulty parameter b_j since it is not expressible on the probability scale. An alternative measure of difficulty is the probability p_j that a randomly chosen individual from the population obtains a correct response to the jth question. In the case where the latent traits $\{\theta_i\}$ are assumed distributed from a standard normal distribution

$$p_j = \Phi\left(\frac{-b_j}{\sqrt{1 + a_j^2}}\right).$$

The discrimination and difficulty characteristics of the jth item can be analyzed by inspection of the marginal posterior distributions of the correlation r_j and the probability p_j. In addition, one is interested in learning about the item characteristic curve for the item. For an individual with the latent trait θ, the probability of a correct response on the jth item is given by the probability

$$P_j(\theta) = F(a_j\theta - b_j).$$

One can learn about this probability by means of its marginal posterior distribution. The result of the Bayesian fitting of the item response model is a simulated sample $\{\theta^{(m)}, \mathbf{a}^{(m)}, \mathbf{b}^{(m)}\}$ from the joint posterior distribution of the latent traits and item parameters. From this sample, one can obtain a simulated sample from the marginal posterior distribution of any function $f(\mathbf{a}, \mathbf{b})$ by applying this function to each of the simulated parameter values. In particular, one can obtained simulated samples for the item correlations $\{r_j\}$ and the item difficulty parameters $\{p_j\}$. The posterior distributions for these parameters can be summarized by the use of posterior means and by probability intervals.

7. Model Checking

7.1 Bayesian Residuals

After a particular item response model has been fit, one is interested in assessing the closeness of the observed data with the fitted probabilities from the model. If $p_{ij} = F(a_j\theta_i - b_j)$ denotes the probability that individual i responds correctly to the jth item, one definition of a residual is the difference between the observed binary response y_{ij} and the response probability

$$r_{ij} = y_{ij} - p_{ij}.$$

From a Bayesian perspective, after observing data, the location of the fitted probability p_{ij} is described by its posterior probability distribution and the observed y_{ij} is a constant. Thus the residual r_{ij} has a posterior probability distribution which can be summarized to learn if this particular observation is not well fit by the model. Observations for which the residual distribution is located away from zero may indicate some lack-of-fit.

In practice, it can be difficult to identify unusually large values of r_{ij} due to the binary nature of the response variable. For the purpose of model checking, it can be helpful to group the data by latent ability. Suppose that we group the individuals by their estimated latent abilities. For the latent ability class g with midpoint θ_g, we can compute the proportion of individuals y_g that answer correctly to a particular item. If p_g denotes the probability that an individual with latent ability θ_g obtains a correct answer, then we can inspect the posterior distributions of the group residuals $r_g = y_g - p_g$. Johnson and Albert (1999), Chapter 6 illustrates the use of these group residuals to demonstrate the lack-of-fit of the one-parameter model for a sociological application.

7.2 *Posterior Predictive Checks*

A different strategy for model checking is based on the posterior predictive distribution. In the Bayesian model, data $\mathbf{y}$ is observed from the sampling density $P(\mathbf{y}|\theta, \mathbf{a}, \mathbf{b})$, and the parameters have a prior $g(\theta, \mathbf{a}, \mathbf{b})$. Inferences about the parameters is based on the posterior density $g(\theta, \mathbf{a}, \mathbf{b}|\mathbf{y})$. Suppose that the same mathematics placement test is administered to a new sample of n students. Let $\tilde{\mathbf{y}}$ denote the response matrix of this future sample. The probability function of this future data, called the *posterior predictive density*, is computed by averaging the sampling density of $\tilde{\mathbf{y}}$ over the posterior distribution of the parameters:

$$P(\tilde{\mathbf{y}}|\mathbf{y}) = \int P(\tilde{\mathbf{y}}|\theta, \mathbf{a}, \mathbf{b}) g(\theta, \mathbf{a}, \mathbf{b}|\mathbf{y}) d\mathbf{a}\, d\mathbf{b}\, d\theta.$$

This predictive density represents typical data generated from the fitted model. If the observed data $\mathbf{y}$ is not representative of data $\tilde{\mathbf{y}}$ from the posterior predictive distribution, then one has doubt that $\mathbf{y}$ is generated from the model. Generally, it is hard to detect if $\mathbf{y}$ is a typical value from the distribution $P(\tilde{\mathbf{y}}|\mathbf{y})$ since the data values $\mathbf{y}$ and $\tilde{\mathbf{y}}$ are multidimensional. However, one can often construct a *testing function* $T(\mathbf{y})$ which measures some aspect of the data which may not be consistent with the stated model. One can compute the posterior predictive distribution of $T(\tilde{\mathbf{y}})$ to see what values of T are predicted from the model. If the observed value of T, $T(\mathbf{y})$, is unusual relative to its posterior predictive distribution, then this casts doubt on the suitability of the item response model. One can measure unusualness by the computation of a p-value

$$P(T(\tilde{\mathbf{y}}) \geq T(\mathbf{y})).$$

If this posterior predictive p-value is small, then this indicates that it is unlikely that the assumed model can generate data like the one that was observed.

In practice, the posterior predictive distribution is computed by simulation. We can simulate one set of future data $\tilde{\mathbf{y}}$ by a two-step process: (1) simulate parameters $\theta, \mathbf{a}, \mathbf{b}$ from their posterior distribution and (2) simulate data $\tilde{\mathbf{y}}$ from the sampling density $P(\tilde{\mathbf{y}}|\theta, \mathbf{a}, \mathbf{b})$, where one is conditioning on values of the simulated parameters. If this process is repeated a large number of times, one obtains a sample of simulated values of $\tilde{\mathbf{y}}$, and the posterior predictive distribution of the checking function $T(\tilde{\mathbf{y}})$ can be summarized by means of a histogram. The location of the observed value $T(\mathbf{y})$ on this histogram is informative about the consistency of the observed data with the model. See Gelman, Meng and Stern (119) for a general discussion on the use of the posterior predictive distribution in model checking.

8. The Mathematics Placement Test Example

The two-parameter item response model with a probit link function was fit to the mathematics placement exam described in Section 1. The hyperparameters of the normal distributions were chosen to be $s_a = s_b = 1$ and $\mu_a = 0$. The Gibbs sampling/data augmentation simulation described in Section 5.2 was run for a total of 5000 iterations. We focus on the estimation of the item response curve for each of the 35 items on the test.

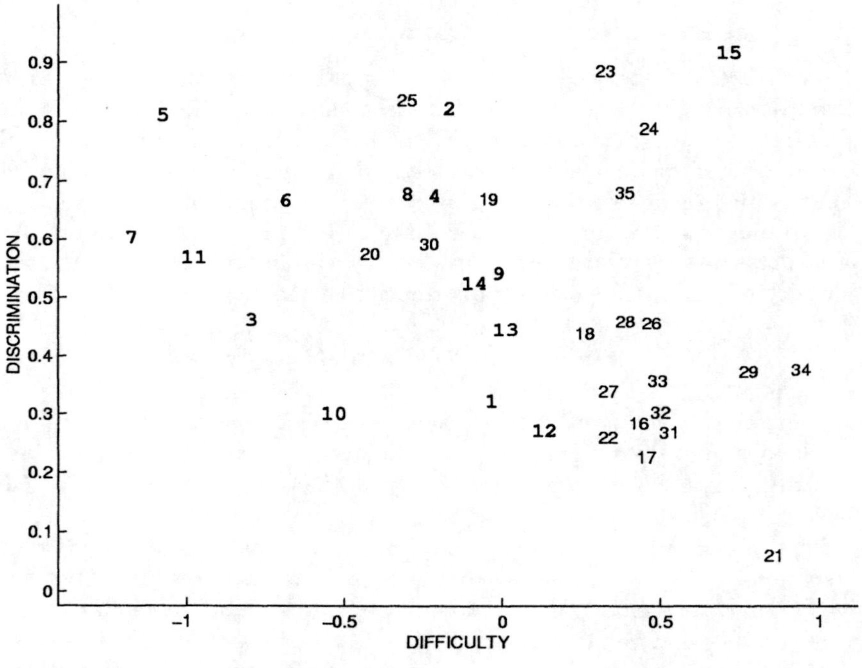

FIGURE 10.2. Scatterplot of posterior means of difficulty parameters $\{b_j\}$ against posterior means of discrimination parameters $\{a_j\}$ for all items.

To summarize the characteristics of all of the test items, Figure 10.2 plots the posterior means of the slope parameters $\{a_j\}$ against the posterior means of the intercept parameters $\{b_j\}$ for all items. The plotting symbol used in the figure is the item number on the test; the bold numbered items are questions on content in intermediate algebra and the remaining points correspond to questions on college algebra. This figure confirms the comments made in the initial exploration of the data in Section 1. The items appear to differ substantially in terms of their difficulty and discrimination levels. The questions on intermediate algebra appear to be

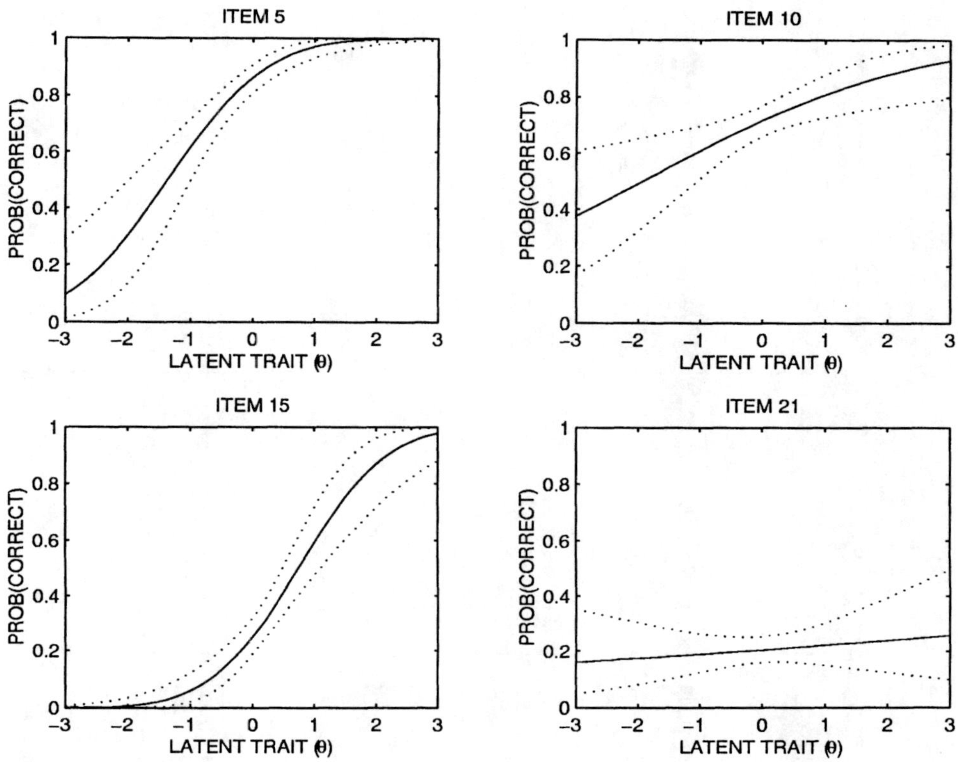

FIGURE 10.3. Posterior medians and 5th and 95th percentiles of the posterior distribution of the probability of a correct response $P_j(\theta)$ for four items using a two-parameter item response model.

generally easier than the questions on college algebra. The lower right point on the graph, corresponding to question 21, has an unusually small estimated discrimination parameter.

Let's focus on four items on the test, questions 5, 10, 15, 21, which have distinctive characteristics in terms of difficulty and discrimination.

Figure 10.3 graphs the posterior median of the probability of a correct response $P_j(\theta)$ for a sequence of values of the latent trait θ between -3 and 3. In addition to the solid line that represents the posterior median, this figure also displays the 5th and 95th percentiles of the posterior distribution of this probability. The fitted response curves of items 5 and 15 correspond to questions that are relatively effective in discriminating between students of different ability. In contrast, items 10 and 21, with relatively flat item response curves, are poor discriminators. It is interesting to note that the confidence bands are wider for extreme values of the latent trait θ — this reflects more uncertainty in estimating the probability of correct response for these extreme students.

The two-parameter model assigns each item a distinctive slope parameter a_j, which assumes that each item possesses a unique ability to discriminate between students of different ability. A far simpler model is the one-parameter item response model which assumes that all items have the same discrimination ability.

This one-parameter model was also fit to the math placement dataset and the fitted item response curves for items 5, 10, 15, 21 are displayed in Figure 10.4.

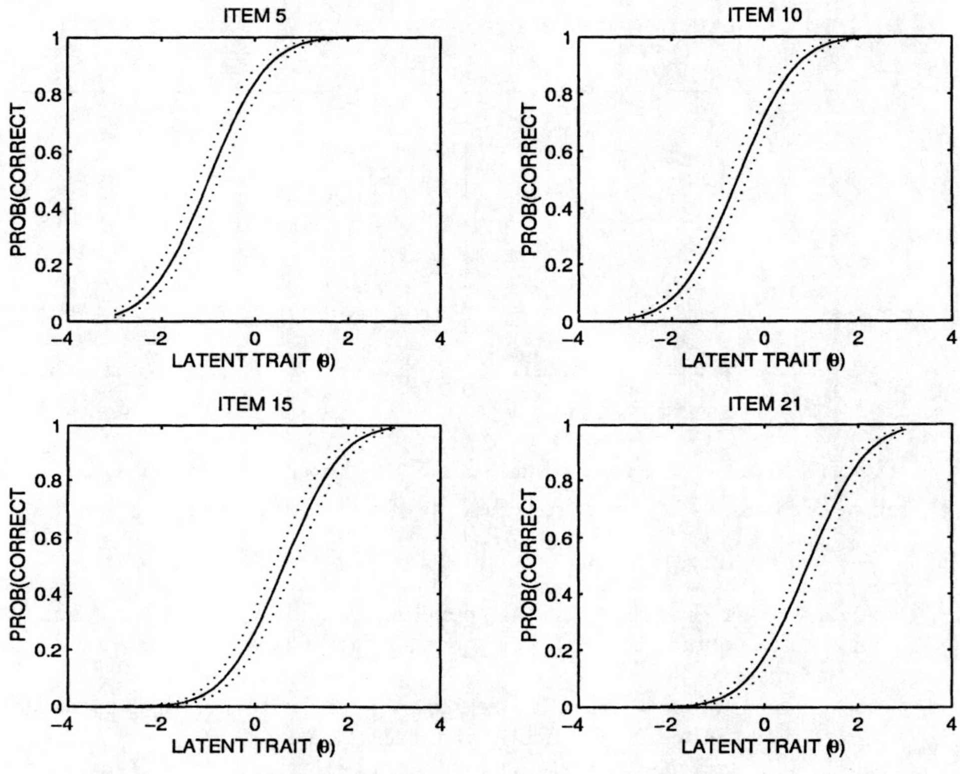

FIGURE 10.4. Posterior medians and 5th and 95th percentiles of the posterior distribution of the probability of a correct response $P_j(\theta)$ for four items using a one-parameter item response model.

These fitted curves are much different in appearance than the corresponding curves in Figure 10.3. The curves all have the same shape and differ only in their location. Also, note that the confidence bands for the probability of a correct response are much narrower than the bands for the fitted two-parameter model, especially for the small and large values of the latent trait θ.

Since the fitted item response curves for the one and two-parameter models are so different for these four items, this raises the issue of model fit. In particular, is the observed dataset consistent with the assumption of a one-parameter model which has only a single discrimination parameter? In Figure 10.1, we noted the high variability of the point-biserial correlations for the 35 items. Is this high variability of these observed discriminations consistent with the assumption of a one-parameter model? We answer this question by use of the posterior predictive distribution. Future samples of test results $\tilde{\mathbf{y}}$ are simulated from the posterior predictive distribution $f(\tilde{\mathbf{y}}|\mathbf{y})$, where the parameters $\{\theta, \mathbf{b}\}$ are simulated from the one-parameter item response model. For each simulated sample of test results, we compute the set of point-biserial correlations, and summarize the variability of these correlations by use of a sample standard deviation. One thousand future datasets were simulated and Figure 10.5 displays a histogram of the standard deviations of the correlations for all of these datasets. Note that a typical value of this standard deviation is .06. The observed standard deviation of the point-biserial correlations, .0935, is in the extreme right-tail of this distribution. The probability of observing a correlation value at least as large as this value is approximately zero. This indicates that the observed variation in the discriminations of the 35 items is inconsistent with the assumption of a one-parameter model. This brief analysis suggests that the two-parameter model may be more suitable than the one-parameter model for describing this mathematics placement dataset.

9. Further Reading

Although this chapter has focused on the Bayesian fitting of item response models, there is a broad literature on the classical fitting and description of this class of models. Hambleton and Swaminathan (1985), Baker (1992) present general reviews of classical methods, and van der Linden and Hambleton (1997) illustrate the generalization of item response theory for categorical responses. Bock and Aitkin (1981) illustrate the use of the EM algorithm for computing maximum likelihood estimates of the item parameters.

Early Bayesian analyses of item response models are found in Swaminathan and Gifford, J. A. (1982, 1985) and Tsutakawa and Lin (1986). The Gibbs sampler (Gelfand and Smith, 1990), and related Metropolis-Hastings type algorithms (Chib and Greenberg, 1995) have revolutionized Bayesian computation. The use of data augmentation and Gibbs sampling to fit probit models is described in Albert and Chib (1993). Albert (1992) and Bradlow et al (1997) illustrate the use of Gibbs sampling for modeling probit item response models and Patz and Junker (1997) demonstrate the use of Metropolis within Gibbs simulation algorithms for fitting item response curves with a logistic link. Ghosh et al (1997) discuss the choice of noninformative prior distributions to ensure propriety of the posterior distribution. There is less literature available on model checking of item response curves from a Bayesian viewpoint. Gelman et al (1995) give a general discussion of the use of the posterior predictive distribution in model criticism, and Johnson and Albert (1998)

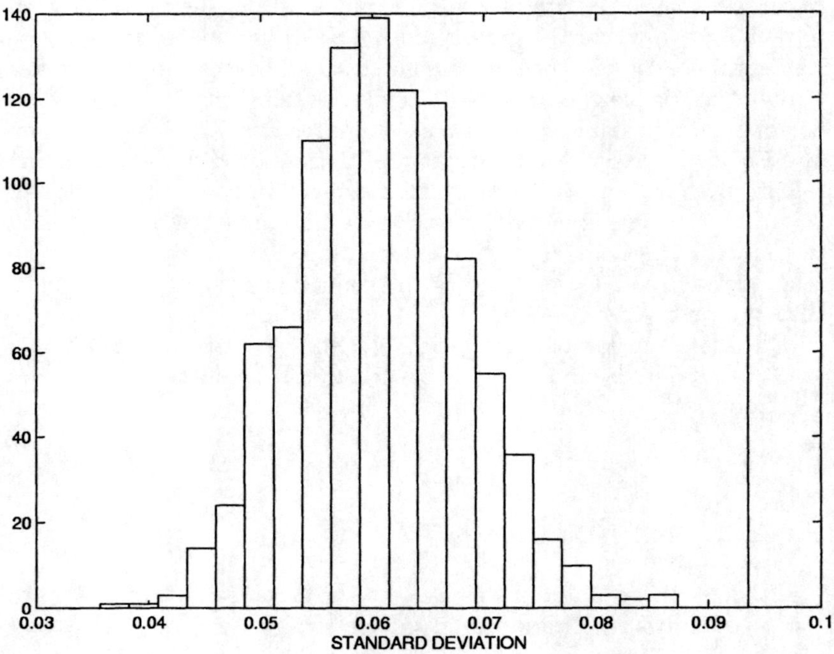

FIGURE 10.5. Histogram of the posterior predictive distribution of the standard deviation of the future point-biserial correlations based on the one-parameter item response model. The observed value of the standard deviation of the correlations is represented by a vertical line.

discuss the use of Bayesian residuals in checking item response models.

References

Albert, J. H. (1992). Bayesian estimation of normal ogive item response curves using Gibbs sampling. *Journal of Educational Statistics*, 17, 261-269.

Albert, J. H. and Chib, S. (1993). Bayesian regression analysis of binary and polychotomous response data. *Journal of the American Statistical Association*, 88, 657-667.

Baker, F. B. (1992). *Item response theory: Parameter estimation techniques.* New York: Marcel Dekker.

Bock, R. D. and Aitkin, M. (1981). Marginal maximum likelihood estimation of item parameters: Application of an EM algorithm. *Psychometrika*, 46, 443-459.

Bradlow, E. T., Wainer, H., and Wang, X. (1997). A Bayesian random effects model for testlets. manuscript.

Chib, S., and Greenberg, E. (1995). Understanding the Metropolis-Hastings algorithm. *The American Statistician*, 49, 327-335.

Gelfand, A. E., and Smith, A. F. M. (1990). Sampling-based approaches to calculating marginal densities. *Journal of the American Statistical Association*, 85, 398-409.

Gelman, A., Carlin, J. B., Stern, H. S., and Rubin, D. B. (1995). *Bayesian Data Analysis.* New York: Chapman and Hall.

Gelman, A., Meng, X. L., and Stern, H. S. (1996). Posterior predictive assessment of model fitness via realized discrepancies. *Statistica Sinica*, 6, 733-807.

Ghosh, M., Ghosh, A., Chen, M., and Agresti, A. (1997). Bayesian estimation for item response models. manuscript.

Hambleton, R. and Swaminathan, H. (1985). *Item response theory: principles and applications.* Boston: Kluwer.

Johnson, V. and Albert, J. (1999). *Ordinal data modeling.* New York: Springer-Verlag.

Patz, R. J. and Junker, B. W. (1997). A straightforward approach to Markov chain Monte Carlo methods for item response models. manuscript.

Swaminathan, H. and Gifford, J. A. (1982). Bayesian estimation in the Rasch model. *Journal of Educational Statistics*, 7, 175-192.

Swaminathan, H. and Gifford, J. A. (1985). Bayesian estimation in the two-parameter logistic model. *Psychometrika*, 50, 349-364.

Tsutakawa, R. K. and Lin, H. Y. (1986). Bayesian estimation of item response curves. *Psychometrika*, 51, 251-267.

van der Linden, W. J. and Hambleton, R. K. (Eds.) (1997). *Handbook of modern item response theory.* New York: Springer-Verlag.

11

Developing and Applying Medical Practice Guidelines Following Acute Myocardial Infarction: A Case Study Using Bayesian Probit and Logit Models

Mary Beth Landrum
Sharon-Lise Normand

ABSTRACT Measuring the quality of care delivered to patient populations comprises one of several foci in health services research. For example, a plethora of studies have indicated that there are large regional variations in the use of many surgical procedures. Observations such as these raise the question as to whether there is overuse (or underuse) of surgical procedures within particular strata. To answer this question, however, it is necessary to define an explicit *standard of care* – who should receive the surgical procedure and who should not? In this Chapter, using ratings elicited from a multidisciplinary panel of medical experts, we estimate a Bayesian ordinal probit model to build an explicit guideline of care for the appropriateness of coronary angiography (an invasive diagnostic procedure) for 890 clinical scenarios following a heart attack. Utilizing the posterior distribution of the appropriateness scores that comprise the guideline, we profile 294 hospitals who treated 5998 Medicare patients who suffered a heart attack. We compare adherence to the expert-derived guidelines across hospitals by estimating the relationship between the use of angiography and the patients' appropriateness for the procedure. We also examine provider and patient characteristics that impact adherence using hierarchical logistic regression models. While we found that most hospitals did treat patients in accordance with expert opinion, we also demonstrated that incorporating uncertainty in guidelines based on expert opinion results in more conservative conclusions regarding quality of care.

1. Background and Significance

Concerns surrounding rising health care costs, an aging US population, and the implementation of many new managed health care organizations have prompted health care consumers to question whether the quality of care they receive has been compromised. Cardiovascular disease, for example, remains one of the leading causes of death in the US and among the costliest. It is estimated in 1998, heart attacks and other cardiovascular diseases will cost the United States $274.2 billion (Boston Globe, 1998). For Medicare beneficiaries, those over the age of 65, the government becomes the primary health care insurer. The nationwide mean Medicare payment to hospitals for acute myocardial infarction (AMI) patients was $9261 in 1990, compared to an overall mean Medicare hospital payment of $463 for non-AMI beneficiaries in the same age range (McClellan, 1995). With so much money at stake, the government wants to stem rising costs while not compromising quality of care.

Profiling, the process of comparing the quality of care delivered by medical care providers to a normative standard, is central to much of the quality assessment effort. Providers of medical care can be assessed according to either *process*-based or *outcome*-based measures. The former relate to the appropriateness of the delivery of treatments and other medical processes whereas the latter relate to patient-specific outcomes of care, such as mortality or patient functioning, that resulted from treatments. Normand, Glickman, and Gatsonis (1997) and Normand, Glickman, and Ryan (1997) describe profiling medical care providers using outcome measures. They compare mortality rates after an acute myocardial infarction (AMI) across hospitals and discuss many of the statistical issues involved in outcome-based profiling, including risk-adjustment, and the development of performance indices. In this Chapter, we present a case-study for profiling medical providers using *process*-based measures. Process measures can be more sensitive than outcome measures, as poor process does not always lead to poor outcomes (Brook et al., 1996). However, process measures are only valid to the extent to which it can be demonstrated that adherence to recommended processes of care leads to better patient outcomes.

A key issue in monitoring quality of care using process-based measures is the need to define a *standard of care*, that is, determining who should receive the treatment and who should not. Subsets of patients for whom a test, procedure, or treatment is likely to benefit are identified, and providers are then judged according to the extent to which these patients receive the therapy. A popular approach to establishing such clinical guidelines for care, pioneered by the RAND corporation (Park et al., 1986), is the consensus panel. Judgments, based on the results of published efficacy and effectiveness research, regarding the appropriateness of a therapy for subsets of patients categorized by their symptoms and prior medical history are elicited from a multispecialty panel of experts. Each expert assigns ratings on an ordinal scale that describes the extent to which the potential benefits of the therapy outweigh the potential risks. Many methodological questions arise in employing practice guidelines developed by an expert panel to monitor quality of care including a) how to combine observed appropriateness ratings into a guideline for care b) how to develop performance measures using the guideline to define a standard of care, and c) how to account for the uncertainty in the guideline and the performance measures.

In this chapter, we address these issues through the use of generalized linear models and illustrate both the development of practice guidelines based on elicited ratings from an expert panel and the subsequent application of the guidelines to assess quality of care across providers. We illustrate our methods using practice guidelines for the use of coronary angiography, an invasive diagnostic procedure used to assess the extent of coronary disease, following a heart attack.

In section 2, we describe the construction of practice guidelines for coronary angiography following an AMI based on elicited appropriateness ratings from an expert panel. We utilize ordinal probit models to combine the elicited ratings into a measure of appropriateness for a set of medical indications and to estimate their associated measures of precision. Albert and Chib (1993), Nandram and Chen (1996), and Johnson (1996) have previously discussed ordinal probit models from a Bayesian perspective. Our methods represent an extension to the class of models proposed by Johnson (1996).

In section 3, we illustrate an application of the practice guidelines to profile the quality of care provided by hospitals based on their adherence to these guidelines. We employ hierarchical logistic regression models (Longford, 1993; Gatsonis et al., 1993, 1995; Normand et al., 1997) to estimate hospital-specific measures of quality and to explain variations in quality according to patient and provider characteristics. We also discuss the importance of accounting for the uncertainty in the standard of care when assessing quality across hospitals.

Finally, in section 4 , we discuss the advantages of a Bayesian approach to profiling the appropriate use of medical technologies, and summarize the policy implications of our analysis.

2. Developing Practice Guidelines

2.1 Elicitation of Appropriateness Ratings

An expert panel was convened in Boston, MA, during October of 1995 to update the 1992 RAND ratings (Bernstein et al., 1992) for the appropriateness of coronary angiography following an AMI. Following a comprehensive review of the literature relevant to the benefits and risks of coronary angiography (Bates et al., 1997), a list of clinically homogeneous strata, referred to as clinical indications, was developed. The clinical indications, which categorize patients in terms of their symptoms, past medical history, and results of previous diagnostic tests, were chosen so that patients within an indication are homogeneous insofar that angiography is equally appropriate or inappropriate. The indications, numbering 888, were separated into two chapters. The first 90 indications, comprising *Chapter 1*, described clinical scenarios for angiography during hospitalization following an AMI; the remaining indications described scenarios for angiography after discharge but within 12 weeks of the AMI.

A panel of nine experts were selected from nominations by specialty societies. Panelist were chosen for diversity in geographic location, practice setting, and specialty. Each expert was asked to rate the appropriateness of the procedure within each indication independently without discussion or contact with other panel members by answering the following question:

> *Do the expected benefits of the procedure outweigh its expected risks by a sufficient margin so that the procedure is worth doing?*

A nine-point scale, where a value of 1 denotes that the risks of the procedure greatly exceed the benefits, a value of 5 indicates that the benefits and risks are equivalent, and a value of 9 denotes that the benefits of the procedure greatly outweigh the risks, was employed.

The results of the baseline elicitation were summarized and discussed at a meeting of the panelists. During this meeting the panelists had the opportunity to revise the list of indications, at which time two indications were added to Chapter 1. At the conclusion of this meeting, the experts confidentially re-rated treatment efficacy in

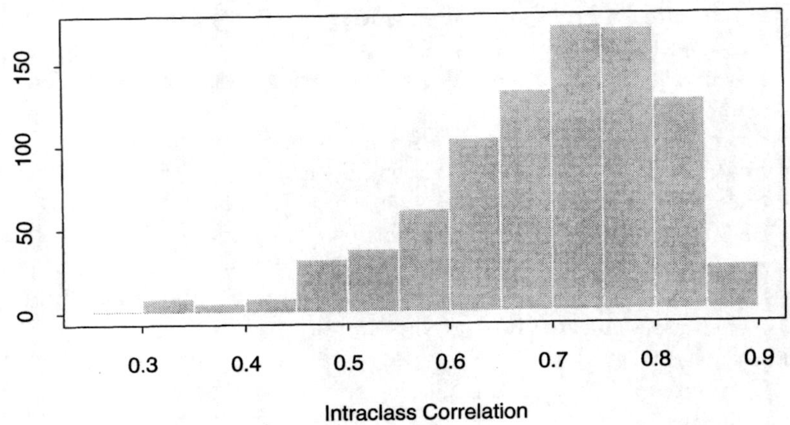

FIGURE 11.1. **Reliability of Expert Ratings.** This histogram displays the distribution of the 890 indication-specific estimates of the intraclass correlation of reliability.

each indication. Indication-specific estimates of the intraclass correlation of reliability (Fleiss, 1986) of the second round ratings are displayed in Figure 11.1. Estimates of the reliability of the expert ratings ranged from 29% to 88%, suggesting clear differences among the experts in their beliefs regarding the appropriateness of coronary angiography for certain indications following AMI even after panel discussion.

2.2 Combining the Angiography Panel Data

The approach commonly employed in the medical literature to combine expert ratings into a guideline is to classify indications into mutually exclusive categories according to the observed median rating and a measure of disagreement in the panel (Bernstein et al., 1992). For example, with 9 experts participating on the panel, angiography would be considered appropriate for indications rated with a median score greater than or equal to 7, as long as fewer than three panelist rated the indication with a score of 3 or below.

There are many shortcomings to the conventional method for producing guidelines for care based on expert appropriateness ratings. First, each indication is classified independently, ignoring the repeated nature of the elicited ratings. In addition, each expert's ratings are weighted equally, ignoring differences in the overall aggressiveness of the experts' beliefs regarding the benefits of interventions. Finally, only a crude measure of disagreement is considered, leading to an understatement of the uncertainty inherent in estimating appropriateness based on a sample of expert opinion.

We developed a Bayesian model-based method for combining the elicited ratings into a guideline (Landrum and Normand, 1999) that exploits the structure of the data by permitting expert-specific threshold parameters as well as indication random effects. Let R_{ir} be the appropriateness rating of the ith indication, $i = 1, \cdots, 890$, made by the rth expert, $r = 1, \cdots, 9$. We assumed there exits a latent appropriateness trait, μ_i, that describes the degree to which the benefits of the procedure outweigh its risk for patients within clinical strata i. Medical Expert r was assumed to have rated indication i by first estimating μ_i using

$$L_{ir} \mid \mu_i, \epsilon_{ir} = \mu_i + \epsilon_{ir}, \tag{1}$$

so that L_{ir} is an unobserved continuous measurement of the appropriateness of indication i, and ϵ_{ir} is the error in the assessment made by Expert r. Expert r then assigned indication i to one of K ordered categories according to his/her assessment of the appropriateness of the procedure and to his/her own set of thresholds via

$$R_{ir} = k \text{ iff } \alpha_{r,k-1} < L_{ir} \leq \alpha_{r,k} \tag{2}$$

where $\boldsymbol{\alpha}_r = \{\alpha_{rk}\}, k = 0, 1, 2, \cdots, K$ $(\alpha_{r0} = -\infty, \alpha_{rK} = \infty)$ are Expert r's threshold parameters. We assumed that the measurement error, ϵ_{ir}, followed a Normal distribution with mean zero and variance σ_r^2, so that

$$P(R_{ir} \leq k \mid \mu_i, \alpha_{r,k}, \sigma_r) = \Phi\left(\frac{\alpha_{r,k} - \mu_i}{\sigma_r}\right).$$

In earlier work (Landrum and Normand, 1999), we estimated three models for the measurement error structure a) a model that assumed constant measurement error across indications and raters b) a model that allowed for heterogeneity across rates in the spread of the measurement error and c) a model that allowed for heterogeneity across indications in the spread of the measurement error. We determined that a model allowing for rater-heterogeneity in the measurement error provided the best fit to the observed appropriateness ratings for coronary angiography and thus focused on the rater-heterogeneity model in this chapter. Identifiability was achieved by fixing the location and the scale of the prior for the appropriateness parameters, $\mu_i \sim \text{N}(0, 1)$. The model was completed by assuming vague but proper priors for the remaining parameters: $\sigma_r^{-2} \sim \text{Gamma}(1, 1)$ and $\alpha_{r,k} \sim \text{Unif}(-10, 10)$, subject to order constraints on the $\alpha_{r,k}$'s.

2.3 Estimation

Gibbs sampling, implemented using specialty functions written in Fortran, was employed to fit the regression models. Model fit was assessed with several posterior predictive checks (Gelman et al., 1995) using 100 replicated data sets, and by examining the posterior distribution of the log-likelihood given the observed data (Dempster, 1974). For details regarding the assessment of convergence and model-fit see Landrum and Normand (1999).

2.4 Defining the Standard of Care

Several indices of appropriateness were estimated for each indication, including a 95% posterior credible interval for μ_i. We also estimated the posterior probability that the underlying appropriateness level exceeds the common standard for defining a procedure to be appropriate, denoted $\hat{P}_{\text{A,i}}|\mathbf{R}$. The probability that angiography is appropriate for indication i was estimated using

$$\hat{P}_{\text{A,i}}|\mathbf{R} = \sum_{r=1}^{9} \frac{\hat{P}_{\text{A,ir}}|\mathbf{R}}{9} \quad \text{where} \quad \hat{P}_{\text{A,ir}}|\mathbf{R} = \sum_{d=1}^{n} \frac{I(\mu_i^{(d)} > \alpha_{r,6}^{(d)})}{n}, \tag{3}$$

where $\mu_i^{(d)}$ is the dth iterate for μ_i, n is the total number of iterates employed for inference, $\alpha_{r,6}$ denotes Expert r's threshold for a cutoff between 6 and 7, and I is the indicator function. $\hat{P}_{\text{A,ir}}|\mathbf{R}$ estimates the probability that the latent appropriateness score of indication i exceeded Expert r's threshold for a rating greater than or equal to 7. We employed the average of this quantity across the nine experts to estimate the probability than an average expert would rate indication i as appropriate.

We compared model-based estimates of appropriateness to the categorization of indications according to the standard methodology employed in the medical literature:

$$\hat{A}_i|\mathbf{R}_i = \begin{cases} 1 & \text{if Median}(\mathbf{R}_i) \geq 7 \text{ and} \sum_r I(R_{ir} \leq 3) < 3 \\ 0 & \text{otherwise.} \end{cases} \tag{4}$$

2.5 *Results*

Figure 11.2 displays appropriateness estimates for 92 clinical indications describing the use of angiography after an AMI before discharge from the hospital. The 92 strata address the potential use of angiography in three time frames: less than 6 hours after symptom onset, 6 to 12 hours after symptom onset, and more than 12 hours after symptom onset before discharge. The remaining indications, not shown in the figure, describe scenarios for angiography after discharge from the hospital, but within 12 weeks of the AMI. Angiography was estimated to be highly appropriate ($\hat{P}_{A,i}|\mathbf{R} > 0.80$) for a large number of indications in all three time frames. In addition, there was good separation of the clinical indications into two classes: those for which the elicited ratings provided evidence of a potential benefit to angiography, and those with little support for the use of angiography. Generally, there was also good agreement between the posterior estimates of appropriateness and the binary classification of indications according to the standard methodology. The agreement between the two estimates was stronger for patients assessed more than 12 hours after symptom onset, suggesting less uncertainty among the experts regarding the benefits of angiography for these patients.

3. Applying the Practice Guidelines

3.1 *Study Population*

In this section, we use the practice guidelines for angiography as a yard stick against which to assess the quality of care delivered to patients following a heart attack. The study population consisted of fee-for-service Medicare beneficiaries between 65 and 89 years of age discharged alive or dead with a heart attack (*International Classification of Diseases, Ninth Revision, Clinical Modification* [ICD-9-CM] principal diagnosis codes of 410.xx) during the period February 1, 1994 to July 31, 1995 from hospitals located in one US state.

Medical record data regarding the index episode of care following the AMI was abstracted by two Clinical Data Abstraction Centers who were under contract with the Health Care Financing Administration. Abstracted data included dates of hospitalization, admission severity, medications, cardiac history, medical therapies, and results of non-invasive and invasive tests. We assigned patients to the first hospital

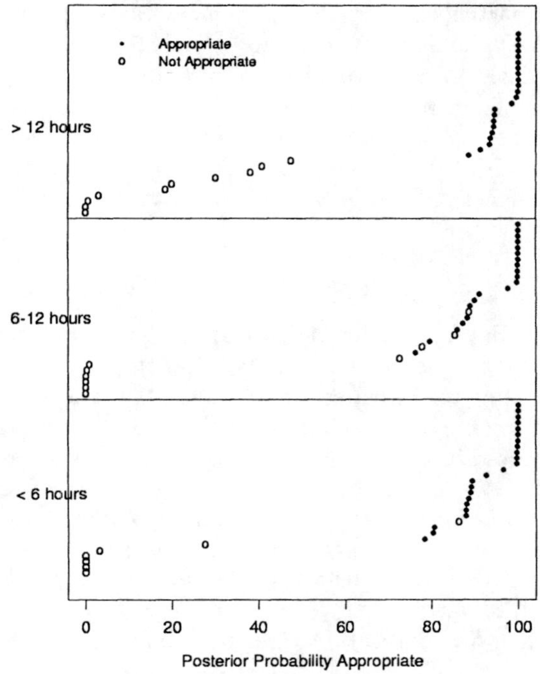

FIGURE 11.2. **Appropriateness Estimates for Chapter 1 Clinical Strata.** The posterior probability that angiography is appropriate, $\hat{P}_{A,i}|\mathbf{R}$, for 92 clinical strata describing scenarios for angiography after an AMI before discharge from the hospital. The binary classification of these indications according to the standard methodology, $\hat{A}_i|\mathbf{R}$, is denoted by filled (appropriate) or open (not appropriate) circles.

to which they were admitted and linked together medical record information from transfer hospitals in order to create a complete inpatient episode of care.

To assess adherence to the expert-derived guidelines for coronary angiography, we focused on a subset of the 890 clinical indications rated by the expert panel that define clinical scenarios for coronary angiography more than 12 hours after symptom onset, before discharge from the hospital. We focused on these indications because 90% of patients in this cohort who underwent angiography in the 90 days following their AMI received the procedure in this time frame. The scenarios classify patients according to their age, the receipt of thrombolytic therapy, and the presence of key cardiac complications, such as persistent chest pain, pulmonary edema, and cardiogenic shock. With the consultation of cardiologists on our research team, we developed an algorithm to assign patients to an indication according to the clinical information abstracted from their medical records. Appropriateness measures were than assigned to each patient according to their clinical indication. Finally, we classified patients who were not candidates for an invasive revascularization procedure because of extreme comorbid conditions (terminal illness, hepatic failure, metastatic cancer, anoxic brain damage, or a do not resuscitate order) as highly inappropriate for coronary angiography.

We obtained hospital characteristics from a detailed survey of the hospitals in which the patients were treated. Hospital information collected in the survey included information regarding the availability of various cardiac services, training programs, and participation in cardiac clinical trials. In this case-study, we focused on the availability of invasive cardiac services at the hospital to which the patient was admitted. We classified hospitals into three categories: those with the capability to perform invasive revascularization procedures - either percutaneous transluminal coronary angioplasty (PTCA) or coronary artery bypass graft surgery (CABG), those with the capability to perform coronary angiography, but not PTCA or CABG, and those with no invasive cardiac capabilities.

3.2 Modeling Adherence to Practice Guidelines

We employed Bayesian hierarchical logistic regression models to study the relationship between the use of angiography in practice and the appropriateness of the procedure according to the expert panel. In the first stage, we modeled the patient's log-odds of undergoing angiography at each hospital as a function of the appropriateness of the procedure and patient demographic characteristics (age and sex). In the second stage, we linked the hospital-specific effects of appropriateness and patient characteristics on the propensity to undergo angiography to the availability of invasive cardiac procedures at the admitting hospital.

Let $Y_{ijh} = 1$ if the jth patient admitted to the hth hospital assigned to clinical stratum i received angiography, $A_{ijh} = 1$ if angiography was appropriate for a patient presenting with indication i (the true, but unknown state of appropriateness according to expert belief). We fitted models of the general form

$$\text{logit}(P(Y_{ijh} = 1|\mathbf{X}, \boldsymbol{\beta})) = \beta_{0_h} + \beta_{1_h} A_{ijh} + \mathbf{X}_{ijh}^T \boldsymbol{\beta}_{2_h}, \tag{5}$$

$$\boldsymbol{\beta}_h \mid \mathbf{Z}, \boldsymbol{\Gamma}, \mathbf{D} = \begin{pmatrix} \beta_{0_h} \\ \beta_{1_h} \\ \boldsymbol{\beta}_{2_h} \end{pmatrix} \sim \text{MVN}(\boldsymbol{\Gamma}\mathbf{Z}_h^T, \mathbf{D}), \tag{6}$$

where $\mathbf{X}_{ijh}$ is a vector of p patient characteristics centered at their mean; $\mathbf{Z}_h$ is a vector of q hospital characteristics; $\boldsymbol{\beta}_{2_h}$ is a vector of p regression coefficients describing the relationship between the patient characteristics contained in $\mathbf{X}$ and the log-odds of receiving angiography at hospital h; and $\boldsymbol{\Gamma}$ is a $(p+2)$xq matrix of population regression coefficients describing the relationship between the hospital-specific regression coefficients and the provider characteristics contained in $\mathbf{Z}$.

Because there is a substantial amount of empirical evidence which suggests that provider characteristics impact utilization of invasive procedures such as coronary angiography (Pilote et al., 1996; Every et al., 1993), we included provider characteristics in the between-hospital model to examine their ability to explain hospital variability, to better specify the model, and to improve the precision of the performance indexes. While there is little empirical evidence to suggest nonexchangeability in the slope coefficients, we also modeled between-hospital variability in the slope parameters as a function of provider characteristics and examined the evidence for nonexchageability in our data.

A_{ijh}, the true state of appropriateness for angiography according to expert opinion, is not known. Rather, we estimated appropriateness for angiography using the elicited ratings from a sample of nine experts. To incorporate our uncertainty regarding the true state of appropriateness we replaced A_{ijh} in Equation (5) with $\hat{P}_{A,ijh}|\mathbf{R}$, the estimated probability, conditional on the observed appropriateness ratings by the expert panel, that the underlying appropriateness level of the procedure for a patient presenting with indication i exceeded the threshold for a rating of 7 or above (as defined by Equation (3)). To compare our analysis with the standard approach, we also fitted models in which we employed $\hat{A}_{ijh}|\mathbf{R}_i$ to estimate the true appropriateness of the procedure for patient j, where $\hat{A}_{ijh}|\mathbf{R}_i = 1$ if indication i was classified as an appropriate indication according to the standard method for combining elicited ratings into a guideline (as defined by Equation (4)).

Vague, but proper priors were specified for remaining parameters: $\gamma_{kl} \sim \mathrm{N}(0, 10^6), \mathrm{k} = 0, 1, \cdots, \mathrm{p}+1; \mathrm{l} = 0, 1, \cdots, \mathrm{q}-1$; $D \sim \mathrm{Wishart}(C, p+2)$, where $p+2$ is the number of regression parameters estimated for each hospital. Values for C, our prior guess for the order of magnitude of the covariance matrix for the hospital-specific regression coefficients, were obtained from a similar sample of patients admitted to 168 hospitals located in a different state.

3.3 Estimation

Gibbs sampling, implemented in BUGS (Gilks et al., 1994), was employed to fit the hierarchical regression models. Convergence of the sampler was assessed using the Potential Scale Reduction (PSR) statistic proposed by Gelman and Rubin (1992). Using this diagnostic, we determined that sampling beyond 10,000 iterations would not improve the precision of the model estimates. Inference for functions of relevant model parameters were obtained by combining iterates from 5 parallel chains after a burn-in period of 5000 iterates.

The fit of the within-hospital model (Equation (5)) was assessed by examining residuals from a patient-level analysis. Goodness-of-fit statistics were also calculated on the fit of the first-stage model to several large hospitals. We examined the adequacy of the second-stage model by plotting posterior estimates of the first-stage parameters against provider characteristics.

3.4 Profiling Hospitals

We evaluated the quality of care provided at each hospital, in terms of adherence to expert opinion regarding the use of coronary angiography, by examining the posterior distribution of selected functions of hospital-specific regression parameters. Specifically, we focused on profiling both the probability that an appropriate patient (a patient for whom $A_{ijh} = 1$) received treatment, and the relative odds of an appropriate patient undergoing therapy compared to a patient not rated appropriate by the expert panel (a patient for whom $A_{ijh} = 0$). The first quantity, the probability that an "average" appropriate patient underwent angiography,

$$\Pi_h = E(Y|\beta_h, A_{ijh} = 1, X_{ijk} = \overline{X}) = \mathrm{logit}^{-1}(\beta_{0_h} + \beta_{1_h}),$$

quantifies the rate at which hospitals provided angiography to patients appropriate for the procedure according to the practice guideline. Although we would not expect all appropriate patients to undergo angiography, we would expect that at least 50% of patients considered appropriate should have in fact received the procedure, and thus considered hospitals with $\hat{\Pi} < 0.50$ to have provided below standard care.

While the likelihood of angiography for appropriate patients identifies hospitals where underuse of angiography may be a problem, it does not measure the ability of the providers to distinguish between patients who are appropriate for angiography from those who are not. To address this issue, we also estimated the odds that an appropriate patient received angiography compared to a similar patient not rated appropriate for the procedure,

$$\theta_h = \exp(\beta_{1_h}).$$

$\theta_h < 1$ indicates non-compliance to the guideline at hospital h, in that the probability of receiving treatment decreased as a function of appropriateness for treatment.

We estimated the posterior distribution of each of these two performance measures by computing each measure conditional on a draw of model parameters from the Gibbs sampler. For example, to estimate θ_h we computed at each iteration d after burn-in, $\theta_h^{(d)} = \exp(\beta_{1_h}^{(d)})$, where $\beta_{1_h}^{(d)}$ is the dth iterate for β_{1_h}. We then estimated the posterior distribution of θ_h using the empirical distribution of the n quantities computed conditional on each draw of model parameters, where n is the total number of iterates employed for inference.

To determine whether the hospital delivered below standard care, we calculated the posterior probability that appropriate patients were less likely to receive angiography compared to patients who were not rated appropriate using

$$\hat{P}(\theta_h < 1 \mid \mathbf{y}) = \frac{1}{n}\sum_{d=1}^{n} I(\exp(\beta_{1_h}^{(d)}) < 1).$$

We then defined hospitals to have provided below standard care, as measured by adherence to the guideline, if this probability was large.

3.5 Explaining Variability in Quality of Care

To explain variability in the quality of care patients received across hospitals, we estimated the posterior distribution of selected functions of the second stage-regression parameters (the elements of $\mathbf{\Gamma}$). For example, to compare adherence across hospital types, we estimated a model of the form described by Equations (5) - (6), with

$$\beta_{k_h} = \gamma_{k0} + \gamma_{k1} z_{1_h} + \gamma_{k2} z_{2_h}; \quad k = 0, \cdots, p+1.$$

z_{1_h} is a binary variable equal to 1 if the admitting hospital had the capability to perform coronary angiography, but not bypass surgery or PTCA, z_{2_h} is a binary variable equal to 1 if the admitting hospital could perform either bypass surgery or PTCA, and p is the number of demographic characteristics contained in the model. We then computed selected odds-ratios that describe the average relationships between angiography, the appropriateness of the patient, and patient demographic characteristics at each hospital type. For example, to estimate the odds that an appropriate patient admitted to a hospital with bypass or angioplasty capabilities received angiography compared to a patient with equivalent demographic characteristics not considered appropriate by the experts, we calculated

$$\theta_{CABG/PTCA} = \exp(\gamma_{10} + \gamma_{12}).$$

Posterior estimates of selected functions such as this were then obtained from the empirical distribution of quantities computed at each iteration.

TABLE 11.1. **Patient and Hospital Characteristics.**

Patient Characteristics (n = 5998)	Mean	Lower Quantile	Upper Quantile
Age	76.3	71	82
% Male	54		
Hospital Characteristics (n = 294)	Mean	Min	Max
Angiography Rate (%)	45.6	0	100.0
Number of AMI Patients	20	1	144
Cardiac Services	% Patients	% Hospitals	
None	26	50	
Cath Only	13	13	
CABG/PTCA	62	37	

3.6 Results

Patient Population

Our patient sample was comprised of virtually all Fee-For-Service Medicare beneficiaries aged between 65 and 89 discharged alive or dead with a principal diagnosis of AMI from hospitals located in a single state during the period February 1, 1994 through July 31, 1995. Patients discharged alive in less than 4 days, patients who underwent emergent angiography (received the procedure 0-12 hours after symptom onset), patients whose residence was outside the US, and patients who died within 1 day of admittance were excluded from the analyses. Patients admitted to a hospital for which the information regarding the availability of invasive cardiac procedures was missing, and patients missing clinical data required to assigned them to a clinical indication were additionally dropped from the analyses, leaving a total sample of 5998 patients admitted to 294 hospitals.

Descriptive statistics regarding the 5998 AMI patients and the 294 hospitals are reported in Table 11.1. Fifty percent of the hospitals, to which 26% of the patients were admitted, did not have the capability to perform coronary angiography. In order for patients admitted to one of these hospitals to have undergone angiography, his/her physicians had to determine that the procedure would benefit the patient enough to warrant their transfer to a facility capable of performing angiography. Thirteen percent of the hospitals could perform coronary angiography, but not angioplasty or bypass surgery, while 37% of the hospital, to which over 60% of the patients were admitted, had the capabilities of providing all invasive cardiac services.

The distribution of two appropriateness measures, the posterior probability angiography was appropriate, $\hat{P}_A|\mathbf{R}$, and the classification of patients as appropriate according to the standard methodology, $\hat{A}|\mathbf{R}$, are displayed in Figure 11.3. Angiography was estimated to be highly appropriate for approximately half of the patients. However, only 50% of these patients received the procedure before discharge. Moreover, a large fraction (42%) of the patients with low estimated probability of being appropriate for angiography received the procedure. As discussed in Section 2.5,

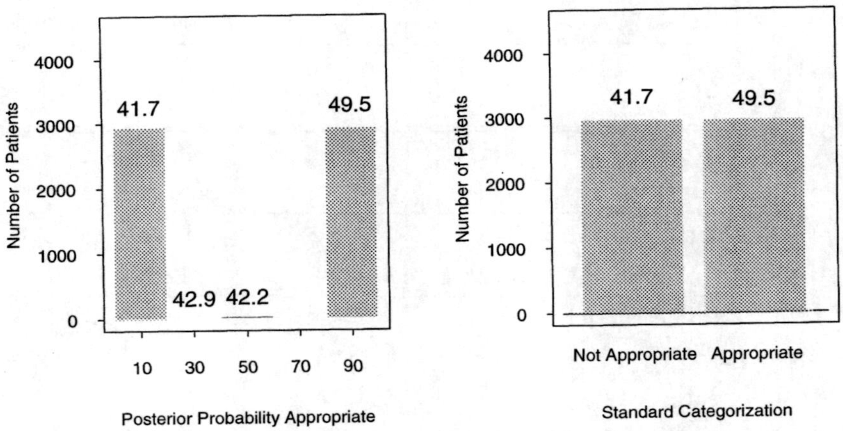

FIGURE 11.3. **Distribution of Appropriateness Estimates Across Patients.** The distribution of two appropriateness measures, the posterior probability angiography was appropriate, $\hat{P}_A|\mathbf{R}$, and the classification of patients as appropriate according to the standard methodology, $\hat{A}|\mathbf{R}$ across the 5998 patients. The proportion of patients receiving angiography in each interval are reported at the top of each bar.

there was a great deal of agreement between the two estimates of the appropriateness of angiography in this time frame.

Modeling Adherence

We estimated two models of the form described by Equations (5) - (6). To examine the relationship between appropriateness for angiography and its use in practice, we employed two different estimates of the true level of angiography appropriateness: $\hat{P}_A|\mathbf{R}$, the estimated probability, conditional on the observed appropriateness ratings by the expert panel, that the underlying appropriateness level exceeded the threshold for a rating of 7 or above, and $\hat{A}|\mathbf{R}$, the categorization according to the standard methodology. To examine the impact of patient characteristics on quality of care, both models adjusted for the patient's age and sex. Finally, both models included hospital variables that reflected the hospital's capability to provide invasive cardiac services in the between-hospital model (Equation (6)). The Appendix reports posterior summaries corresponding to the two models (Table A1).

Profiling Hospitals

Figure 11.4 displays the posterior mean estimates of two hospital-specific measures of quality, the probability that an average appropriate patient received angiography at hospital h, $\hat{\Pi}_h$, and the odds-ratio comparing the likelihood of angiography for an appropriate patient to a patient not rated appropriate by the expert panel, $\hat{\theta}_h$. We estimated these quantities fitting a model which adjusted for patient demographic characteristics and which employed $\hat{P}_{A,i} \mid \mathbf{R}$ to estimate the true level of angiography appropriateness.

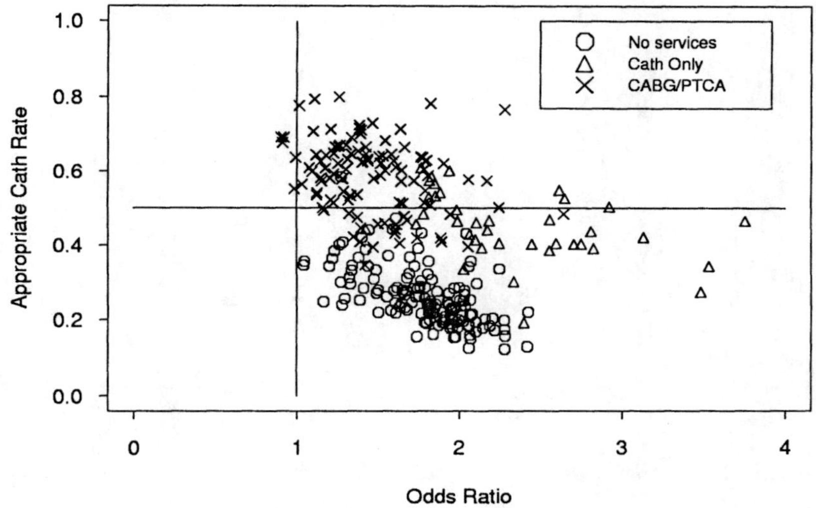

FIGURE 11.4. **Hospital-Specific Estimates of Adherence.** Plots of the posterior mean of the probability that an average appropriate patient received angiography at hospital h, $\hat{\Pi}_h$, versus the posterior mean of the odds-ratio comparing the likelihood of angiography for an appropriate patient to a patient not rated appropriate by the expert panel, $\hat{\theta}_h$. The labels "No services," "Cath Only," and "CABG/PTCA" refer to hospitals without the capacity to perform angiography, hospitals that could perform angiography but not bypass surgery or angioplasty, and hospitals providing either bypass surgery or angioplasty, respectively.

Ideally all hospitals would fall in the upper right quadrant of Figure 11.4, indicating that more than 50% of all appropriate patients received angiography, and that the hospital correctly identified appropriate patients for treatment. Most hospitals without the capacity to perform coronary angiography tended to correctly choose the appropriate patients for transfer to another facility to receive angiography. However, a small fraction of appropriate patients admitted to these hospitals received the procedure, suggesting that hospital of this type should be targeted to reduce underuse of angiography in appropriate patients. Hospitals with bypass or angioplasty capabilities were more likely to provide angiography to more than 50% of the appropriate patients, but also tended to be less discriminating in terms of providing angiography to the most appropriate patients according to the expert panel. Thus high services hospitals should be targeted to reduce overuse of angiography among patients who may not benefit from the procedure. Hospitals with only angiography capabilities tended to best comply with the opinions of the expert panel, but still tended provide angiography to less than 50% of the appropriate patients.

Figure 11.5 displays the distribution of posterior estimates that each hospital's compliance with the guideline was below standard. The histograms display the estimated probability that odds of receiving angiography was lower for appropriate patients compared to those not rated appropriate ($\hat{P}(\theta_h < 1)$). We compared the

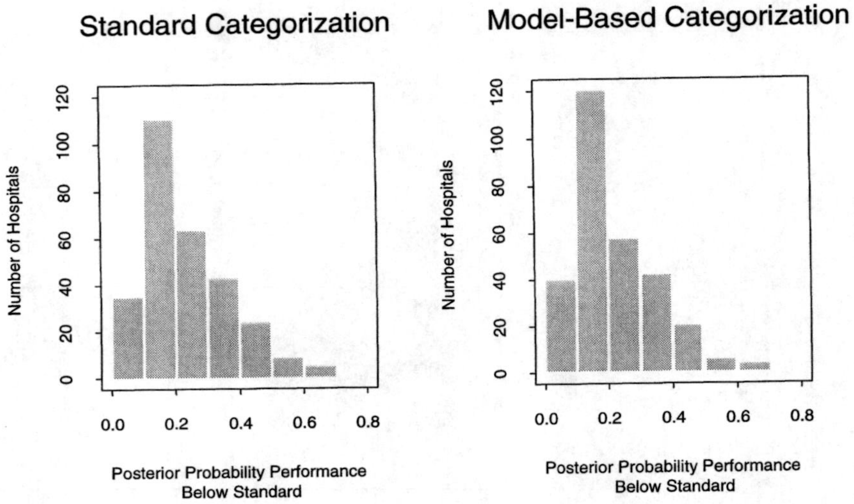

FIGURE 11.5. **Evaluation of Hospitals.** Each histogram displays the distribution of the estimated probability that a hospital's compliance to the guideline was below standard ($\hat{P}(\theta_h < 1)$) across the 294 hospitals.

distribution of the quality measures estimated using the model-based measure of appropriates to estimates obtained using the standard categorization. We conclude that compliance with expert opinion was not significantly below standard at any of the 294 hospitals ($\hat{P}(\theta_h < 1) < 0.80$ for all hospitals). However, as expected, incorporating uncertainty in the assessment of appropriateness according to the expert panel produced more conservative results than in that employing model-based estimates of appropriateness led to smaller estimates of the degree of evidence for poor quality.

Explaining Variability in Quality of Care

Figure 11.6 displays posterior density estimates of the odds ratio comparing the likelihood of an appropriate patient (one for whom $\hat{P}_{A,ijh} = 1$) receiving angiography to a patient of equivalent age and gender who was not rated appropriate by the expert panel ($\hat{P}_{A,ijh} = 0$) at each hospital type. The relationship between a patient's appropriateness for angiography and his/her likelihood to receive the procedure was weak but significant at all three hospital types. The relationship between appropriateness and receipt of the procedure was strongest for patients admitted to hospitals with the capability to perform angiography, but not bypass surgery or angioplasty: patients rated appropriate for the procedure were twice as likely to receive the procedure compared to a similar patient not rated appropriate (95% credible interval for the odds-ratio was equal to (1.3,3.0)). The relationship was weakest among patients admitted to hospitals with the capability to perform angioplasty or bypass surgery: patients rated appropriate for the procedure were only 30% more likely to undergo angiography (95% credible interval for the odds ratio was equal to (1.1,1.6)).

Figure 11.6 also displays the relationship between patients' demographic characteristics and their likelihood of undergoing angiography. The age of a patient was strongly related to their likelihood of receiving the procedure, and patients of older

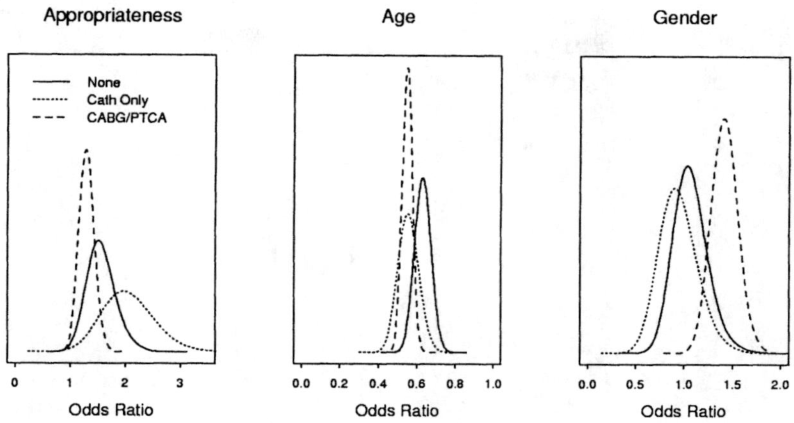

FIGURE 11.6. **Relationship between Hospital and Patient Characteristics and the Likelihood of Angiography.** Smoothed density estimates of the posterior distribution of selected odds-ratios describing the average relationships between angiography, the appropriateness of the patient, patient demographic characteristics, and the availability of cardiac services at the admitting hospital. The three panels plot the estimated odds-ratios comparing appropriate ($\hat{P}_A|\mathbf{R} = 1$) versus not appropriate patients ($\hat{P}_A|\mathbf{R} = 0$), older versus younger patients (an increase in 5 years of age), and males versus females, respectively.

age were less likely to receive angiography. This relationship did not significantly differ among the three types of hospitals. Gender was a less important predictor of angiography in these patients. Males patients admitted to hospitals with bypass surgery or angioplasty capabilities were significantly more likely to undergo angiography compared to females (95% credible interval for the odds-ratio was equal to (1.2,1.7)), however gender was less important at the other hospital types.

Posterior density estimates of the probability of receiving angiography for subsets of patients admitted to each hospital type are displayed in Figure 11.7. The age of the patient and the invasive cardiac services available at the admitting hospital impacted on the likelihood of receiving angiography to a much larger degree than the appropriateness of the procedure according to the panel of experts. The likelihood of receiving angiography varied from 13% for 80 year old males not rated appropriate for the procedure admitted to a hospital without any invasive cardiac services to 87% for 65 year old males rated appropriate for the procedure and admitted to a hospital with the capability to perform bypass surgery or angioplasty. Moreover, an 80 year old had a greater likelihood of receiving angiography if admitted to a high service hospital than a 65 year old admitted to a hospital without the capability to perform any invasive cardiac services.

4. Discussion

In this chapter, we demonstrated an application of Bayesian generalized linear models in the area of health policy. First, using ordinal probit models, we combined ratings from an expert panel in order to *explicitly quantify* a medical guideline. Second, using hierarchical logistic regression modeling, we profiled hospitals with

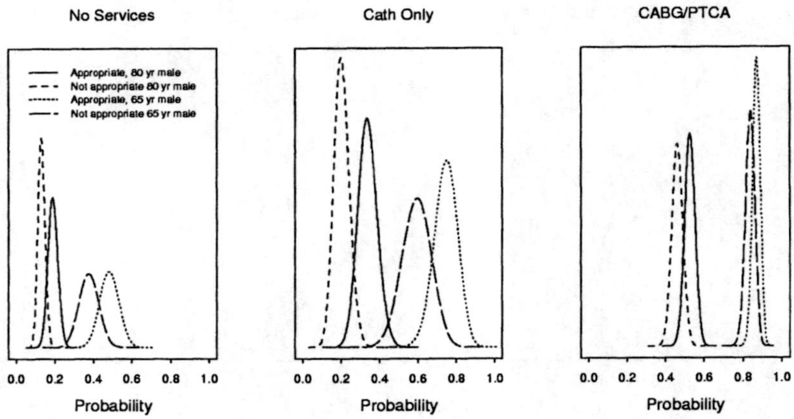

FIGURE 11.7. **Estimate Probability of Angiography for Selected Subsets of Patients.** Smoothed density estimates of the posterior distribution of the likelihood of undergoing angiography for selected subsets of patients.

respect to their adherence to the medical guideline, explained deviations from the guideline, and created several posterior performance indices of policy relevance. We were able to implicitly incorporate the precision of the appropriateness ratings by using the posterior probability that angiography was appropriate for each clinical indication. This represents a major step forward in the quality of care literature which currently employs an all-or-nothing rule. Lastly, by linking our posterior estimates of appropriateness provided to us by the expert panel to patient outcomes, such as mortality and functioning, we will be able to update the panel ratings using the usual prior-to-posterior paradigm in future work.

Substantively, we found compliance rates to the updated coronary angiography practice guidelines among elderly patients with AMI to be acceptable. However, a patient's appropriateness for the procedure was only weakly associated with his/her likelihood of undergoing the procedure, particularly for patients admitted to hospitals with the capability to perform angiography. While the relationship between appropriateness for angiography and its use in this population was weak, age (which may be a proxy for unmeasured disease severity) and the availability of invasive cardiac services at the admitting hospital were strongly related to the use of angiography. Moreover, the availability of the procedure at the admitting hospital dominated the effect of age; an 80 year old had a greater likelihood of receiving angiography if admitted to a high service hospital than a 65 year old admitted to a hospital without any invasive cardiac services.

We found that hospitals tended to do well according to only one of the dimensions of quality we examined. Hospitals with the capability to perform invasive cardiac procedures provided angiography to a large proportion of patients likely to benefit from the procedure. However, they did not discriminate well between patients according to the potential benefits of the procedure. In contrast, hospitals without the capability to perform bypass surgery or angioplasty tended to better "ration" care, in that they better identified patients more likely to benefit from the procedure. However, the rate at which appropriate patients admitted to these hospitals received the procedure was surprisingly low. Our results suggest that both dimensions of quality should be targeted in order to improve the quality of care among elderly AMI patients.

Appendix

In Table A1 the posterior summaries of models of the form described by Equations (5) - (6) in Section 3.6 are reported:

$$\mathrm{logit}(P(Y_{ijh}=1|\mathbf{X},\boldsymbol{\beta})) \;=\; \beta_{0_h}+\beta_{1_h}A^{*}_{ijh}+\mathbf{X}^{T}_{ijh}\boldsymbol{\beta}_{2_h},$$

$$\boldsymbol{\beta}_h \mid \mathbf{Z},\boldsymbol{\Gamma},\mathbf{D} = \begin{pmatrix}\beta_{0_h}\\ \beta_{1_h}\\ \boldsymbol{\beta}_{2_h}\end{pmatrix} \;\sim\; \mathrm{MVN}(\boldsymbol{\Gamma}\mathbf{Z}^{T}_{h},\mathbf{D}).$$

where $A^{*}_{ijh}=\hat{A}|\mathbf{R}$ or $\hat{P}_A|\mathbf{R}$; $\mathbf{X}^{T}_{ijh}=(\mathrm{Age}_{ijh}-\overline{\mathrm{Age}},\mathrm{Male}_{ijh}-\overline{\mathrm{Male}})^{T}$; and $\mathbf{Z}_h=(1,z_{1_h},z_{2_h})$, $z_{1_h}=1$ if hospital h had the capability to perform coronary angiography, but not bypass surgery or PTCA, $z_{2_h}=1$ if hospital h could perform either bypass surgery or PTCA.

Table A1. **Stage 2 Regression Parameter Estimates.** Posterior mean and 95% credible intervals (in parentheses) for the second stage regression parameters (Γ) and posterior mean estimates of the covariance matrix of the first stage regression parameters (D).

	(a) - Standard Categorical Estimates of Appropriateness					
Intercept	-1.58	(-1.84,-1.32)	0.68	(0.24, 1.12)	1.71	(1.39, 2.04)
$\hat{A}\|\mathbf{R}$	0.41	(0.10, 0.70)	0.23	(-0.24, 0.72)	-0.18	(-0.53, 0.18)
Age	-0.09	(-0.12,-0.07)	-0.03	(-0.07, 0.02)	-0.03	(-0.06, 0.004)
Males	0.05	(-0.23, 0.33)	-0.12	(-0.57, 0.36)	0.30	(-0.04, 0.61)
D	$\begin{pmatrix}(0.78)^2 & & & \\ -0.23 & (0.50)^2 & & \\ -0.02 & 0.01 & (0.05)^2 & \\ 0.04 & -0.03 & -0.01 & (0.36)^2\end{pmatrix}$					
	(b) - Model-Based Estimates of Appropriateness					
Intercept	-1.59	(-1.85,-1.34)	0.66	(0.23, 1.11)	1.70	(1.40, 2.02)
$\hat{P}_A\|\mathbf{R}$	0.44	(0.14, 0.75)	0.26	(-0.24, 0.76)	-0.18	(-0.54, 0.17)
Age	-0.09	(-0.12,-0.07)	-0.03	(-0.07, 0.02)	-0.03	(-0.06, 0.002)
Males	0.06	(-0.23, 0.37)	-0.13	(-0.61, 0.33)	0.29	(-0.05, 0.63)
D	$\begin{pmatrix}(0.78)^2 & & & \\ -0.23 & (0.52)^2 & & \\ -0.02 & 0.01 & (0.05)^2 & \\ 0.04 & -0.03 & -0.01 & (0.37)^2\end{pmatrix}$					

Acknowledgments: We are grateful to several colleagues in the Department of Health Care Policy, Harvard Medical School: Barbara J. McNeil for helpful comments; Edward Guadagnoli for use of the hospital survey data; and Margaret Volya and Fung-

Yea Huang, for building and maintaining the Harvard Guideline Study data base. This research was support by grant RO1-HS08071 from the Agency for Health Care Policy and Research, Rockville, MD.

References

Albert, J. H. and Chib, S. (1993). Bayesian analysis of binary and polychotomous response data. *Journal of the American Statistical Association*, **88**, 669-679.

Bates, D. W., Miller, E., Berstein, S. J., Hauptman, P. J. and Leape, L. L. (1997). Coronary angiography and angioplasty after acute myocardial infarction. *Annals of Internal Medicine*, **126**, 539-550.

Berstein, S. J., Laouri, M., Hilborne, L. H. et al. (1992). *Coronary Angiography: A Literature Review and Ratings of Appropriateness and Necessity*. Santa Monica, CA: RAND Corporation.

Boston Globe (January 2, 1998). Costs of cardiovascular disease keep rising, A16.

Brook, R. H., McGlynn, E. A. and Cleary, P. D. (1996). Quality of health care, part 2: measuring quality of care. *New England Journal of Medicine*, **335**, 966-970.

Dempster, A. P. (1974). The direct use of likelihood for significance testing, in *Proceedings of Conference on Foundational Questions in Statistical Inference*. eds. O. Barndorff-Nielsen, P. Blaesild, and G. Schou, Department of Theorectical Statistics, University of Aarhus, 335-352.

Every, N. R., Larson, E. B., Litwin, P. E. et al. (1993). The association between on-site cardiac catheterization facilities and the use of coronary angiography after acute myocardial infarction. *New England Journal of Medicine*, **329**, 546-551.

Fleiss, J. L. (1986). *The Design and Analysis of Clinical Experiments*. Toronto, Canada: Wiley.

Gatsonis, C. A., Epstein, A. M., Newhouse, J. P., Normand, S. T. and McNeil, B. J. (1995). Variations in the utilization of coronary angiography for elderly patients with acute myocardial infarction: an analysis using hierarchical logistic regression. *Medical Care*, **33**, 625-642.

Gatsonis, C. A., Normand, S. T., Liu, C. and Morris, C. (1993). Geographic variation of procedure utilization: a hierarchical model approach. *Medical Care*, **31**, YS54-YS59.

Gelman, A., Carlin, J., Stern, H. and Rubin, D. B. (1995). *Bayesian Data Analysis*. New York, NY: Chapman and Hall.

Gelman, A. and Rubin, D. B. (1992). Inference from iteration simulation using multiple sequences, *Statistical Science*. **7**, 457-472.

Gilks, W. R., Thomas, A. and Spiegelhalter, D. J. (1994). A language and program for complex Bayesian modeling, *The Statistician*. **43**, 169-178.

Johnson, V. E. (1996). On Bayesian analysis of multirater ordinal data: an application to automated essay grading. *Journal of the American Statistical Association*, **91**, 42-51.

Landrum, M. B. and Normand, S. T. (1999). Applying Bayesian ideas to the development of medical guidelines, *Statistics in Medicine.* **18**, 117-137.

Longford, N. (1993). *Random Coefficient Models.* Oxford, U.K.: Oxford University Press.

McClellan, M. (1995). Uncertainty, health care technologies, and health care costs. *American Economic Review Papers and Proceedings*, **85**, 38-44.

Nandram, B. and Chen, M. (1996). Reparameterizing the generalized linear model to accelerate Gibbs sampler convergence. *Journal of Statistical Computations and Simulations*, **54**, 129-144.

Normand, S. T., Glickman, M. E. and Gatsonis, C. A. (1997). Statistical methods for profiling providors of medical care: issues and applications. *Journal of the American Statistical Association*, **92**, 803-814.

Normand, S. T. and Glickman, M. E. and Ryan, T. (1997). Modeling mortality rates for elderly heart attack patients: profiling hospitals in the Cooperative Cardiovascular Project, in *Case Studies in Bayesian Statistics.* eds. C. Gatsonis, J. Hodges and N. Singpurwalla, Springer-Verlag, 155-236.

Park, R. E., Fink, A., Brook R. H. et al. (1986). Physician ratings of appropriate indications for six medical and surgical procedures. *American Journal of Public Health*, **76**, 766-772.

Pilote, L., Miller, D. P., Califf, R. M. et al. (1996). Determinants of the use of coronary angiography and revascularization after thrombolysis for acute myocardial infarction. *New England Journal of Medicine*, **335**, 1198-1205.

Part IV

Semiparametric Approaches

12

Semiparametric Generalized Linear Models: Bayesian Approaches

Bani K. Mallick
David G.T. Denison
Adrian F.M. Smith

ABSTRACT Generalized linear models are one of the most widely used tools of the data analyst. However, the model assumes that the structure of the regression relationship between the response and the covariates is linear on a known transformed scale. We focus here on different methods to perform the same type of analyses. These involve using nonparametric models to determine the relationship between the response and covariates after the usual transformation has been carried out. We demonstrate how such a semiparametric model performs for binary regression.

1. Introduction

Regression techniques are among some of the most widely used methods in applied statistics. Given a response variable Y, and a set of covariates $X = (X_1, X_2, \cdots, X_p)$, one is often interested in estimating an assumed functional relationship between Y and X, and in predicting further responses for new values of the covariates. One way of modeling such a relationship is to present the expected value of Y as

$$E(Y|X) = \mu(X),$$

where, in general, $\mu(\cdot)$ is an unknown function of the covariates. In practice, however, $\mu(\cdot)$ is usually approximated by a simple parametric function $\phi(\cdot;\boldsymbol{\beta})$, where $\boldsymbol{\beta} = (\beta_1, \cdots, \beta_p)$ denotes a vector of unknown parameters. The function $\phi(\cdot;\boldsymbol{\beta})$ is then treated as if it were the true underlying function $\mu(\cdot)$, so the problem is reduced to that of estimating $\boldsymbol{\beta}$. Furthermore, in most applications the probability distribution of the response Y is assumed to belong to an exponential familty. This gives rise to the important class of *generalized linear models* (GLM) (Nelder and Wedderburn, 1972; McCullagh and Nelder, 1989), which we shall find convenient to describe as follows.

Random component: If $Y_1, \cdots, Y_n$ are independent random variables then the probability density function for an individual realization y_i has the form

$$p_{m_i}(y_i|\theta_i, \sigma^2) = b(y_i, \sigma^2/m_i) \exp\left[\frac{m_i}{\sigma^2}\{y_i\theta_i - a(\theta_i)\}\right], \tag{1}$$

for some functions $a(\cdot)$ and $b(\cdot,\cdot)$, where the m_i are known weights. If σ^2 is known then equation (1.1) is a natural exponential family model (see, for example, Morris, 1982) with *canonical* parameter θ_i. In this case $\mu_i = E(Y_i|\theta_i, \sigma^2) = a'(\theta_i)$ and $\text{var}(Y_i|\theta_i, \sigma^2) = \frac{\sigma^2}{m_i} a''(\theta_i)$. The parameter σ^2 is referred to as the *dispersion* parameter of the model. If σ^2 is unknown (1.1) is not a natural exponential family.

Systematic component: For each response y_i, a vector of covariates, $\mathbf{x}_i = (x_{i1}, \cdots, x_{ip})$ of covariates is observed, producing the linear predictor

$$\eta_i = \boldsymbol{\beta}.\mathbf{x}_i, \tag{2}$$

where $\mathbf{a}.\mathbf{b}$ denotes the usual vector product of $\mathbf{a}$ and $\mathbf{b}$.

Link: The random and systematic components are related via a *link function* $g(\cdot)$, such that

$$\eta_i = g(\mu_i). \tag{3}$$

An important particular link is obtained when $g^{-1}(\cdot) = a'(\cdot)$. In this case $\theta_i = \eta_i$, and $g(\cdot)$ is then called the canonical link.

The motivation of this chapter is to discuss the extended version of GLM in semiparametric frameworks. We keep the parametric structure for the random component of the model but use nonparametric methods for the systematic component or the link function, so that the entire enterprise falls within what is referred to as semiparametric regression modeling. Here, we focus on work which takes a Bayesian viewpoint in modeling such semiparametric regression structure. The Bayesian approach to inference is attractive in incorporating prior information into the inference machinery through Bayes theorem, resulting in a unifying, constructive inference methodology.

Classical inference for generalized linear models relies on maximum-likelihood estimation of the parameters and the associated asymptotic distributional properties of the estimates. Fortunately, exponential family models enjoy convexity properties that in many cases guarantee the existence and uniqueness of the maximum-likelihood estimator (Wedderburn, 1976). Typically, however, maximization of the likelihood must be carried out numerically. On the other hand, Markov chain Monte Carlo (MCMC) methods provide a relatively straightforward means of making exact Bayesian inference for a wide class of generalized linear models (Dellaportas and Smith, 1993).

To create nonparametric regression models in a GLM setup we can make the link function g unknown keeping the structure η linear, or we can assume the g is known but create a flexible model for η which could be completely nonlinear or conditionally linear.

2. Modeling the Link Function g

Here we treat the link function g as another unknown in the GLM specification and estimate it jointly with the mean structure. It turns out to be more straightforward to work not directly with the link function, but with a function related to its inverse. This will avoid the need to invert the function to evaluate the likelihood.

2.1 Binary Response Regression

Binary response regression is used when the response takes only one of two values, say 0 and 1. The usual model assumes that the Y_i are independent and

$$P(Y_i = 1|\mathbf{x}_i, \boldsymbol{\beta}) = F(\boldsymbol{\beta}.\mathbf{x}_i),$$

where F is an unknown cumulative distribution function.

Newton, Czado and Chappell (1996) introduce a nonparametric approach to binary regression by assuming that F is a random draw from a Dirichlet process prior independent of $\boldsymbol{\beta}$. They show that with the inclusion of the latent variables $u_i \sim F$ such that $\Pr(Y_i = 1|\boldsymbol{\beta}, \mathbf{x}_i) = \Pr(u_i < \boldsymbol{\beta}.\mathbf{x}_i)$, F can be marginalized out and a straightforward Gibbs sampler arises. For more detailed discussion about this see Basu and Mukhopadhyay (in this volume).

2.2 *General Regression*

The Dirichlet process prior is conveniently employed in binary regression since the inverse link function is a distribution function. However, extensions to generalized linear models with other sampling mechanisms are not obvious.

Mallick and Gelfand (1994) used mixture models for g. They observed that modeling a strictly monotone function g, is equivalent to modeling a cumulative distribution function (c.d.f.). For instance, if the range of g is the real line then $T(g(\cdot))$ where, for example, $T(z) = k_1 e^{k_2 z}(1 + k_1 e^{k_2 z})^{-1}$ or $T(z) = 1 - \exp(-k_1 e^{k_2 z})$, with $k_1, k_2 > 0$, is a c.d.f. Here k_1 and k_2 are not model parameters but constants chosen so that under the transformation the resultant c.d.f. is well-behaved. (see Mallick (1994) for more details).

The mixture model approach models this unknown c.d.f. using a dense class of mixture of known distributions. For instance, Diaconis and Ylvisaker (1985) observed that a discrete mixture of Beta densities provides a dense class of models for densities on [0,1].

In particular, consider

$$T(g(\theta)) = \sum_{l=1}^{r} w_l IB(T(g_0(\theta)); c_l, d_l),$$

where r is the number of mixands, $w_l \geq 0$, $\sum w_l{=}1$ are the mixing weights, $IB(u; c, d)$ denotes the incomplete Beta function associated with the Beta(c, d) density evaluated at u and g_0 is a centering function for g. Mallick and Gelfand (1994) fix r, c_r and d_r so g is specified by $\mathbf{w} = (w_1, \cdots, w_r)$. They suggest experimenting with various values of r while c_l and d_l are chosen to provide a set of Beta densities which blanket [0,1], e.g. $c_l = l$, $d_l = r + 1 - l$, $l = 1, 2, \cdots, r$. If $\mathbf{w}$ is a Dirichlet random variable with distribution $\text{Dir}(\alpha \mathbf{1}_r)$ then Mallick and Gelfand showed that $E\{T(g(\theta))\} = T(g_0(\theta))$, i.e. g is roughly centered about g_0.

Gibbs sampling is trivial as drawing $g|\boldsymbol{\beta}$ and $\boldsymbol{\beta}|g$ is straightforward because drawing g is equivalent to drawing $\mathbf{w}$. Although this is not a fully nonparametric approach however, it is still nonparametric in motivation and does enrich the class of models for g near g_0 as observed by Gelfand (1997).

3. Modeling the Systematic Part η

Modeling the deterministic part η is more popular than modeling g as it enhances the predictive power as well as keeping the interpretation of the regression parameters. If g is unknown then the interpretation of the regression parameters $\boldsymbol{\beta}$ is difficult because they are the unknown slopes of an unknown transformation. In the spirit of Mallick and Gelfand (1994), if we can integrate out all the parameters and are only interested in the prediction of the model then the problem is resolved (also we get rid of the possible identifiability problems). Otherwise there exist several ways of modeling η while keeping the link function g known.

3.1 Model with Random Effects

One way of extending nonparametric GLMs is to include a random intercept term having an unknown distribution into η. This model is advocated by Follman and Lambert (1989) where they used nonparametric mixture estimation to fit the model using the original work of Laird (1978). In this approach the distribution of the random effects is taken to be discrete with sufficiently few atoms to ensure the identifiability of the parameters. Observing these limitations, recently Walker and Mallick (1997) proposed a Bayesian nonparametric model using Polya tree distributions for the random effects. Mukhopadhyay and Gelfand (1997) also proposed a similar type of model using a Dirichlet process as the nonparametric prior distribution for the random effect. These methods certainly create a better fitted model but we shall not expand on them further in this chapter.

3.2 Model with Deterministic Error

Recently Gutierrez-Pena and Smith (1997) introduced an interesting class of models based on the proposal of Blight and Ott (1975). They model the systematic part as

$$g(\mu_i) = \eta_i = \boldsymbol{\beta}.\mathbf{x}_i + \delta(\mathbf{x}_i),$$

where $\delta(\mathbf{x}_i) = g(\mu_i) - \boldsymbol{\beta}.\mathbf{x}_i$ is the nonparametric component of the model. Note that $\delta(\mathbf{x})$ is a deterministic error component of the model rather than the usual random one as $\delta(\mathbf{x}_i) = \delta(\mathbf{x}_j)$ is $\mathbf{x}_i = \mathbf{x}_j$. Usually $\delta(\cdot)$ is modelled by a Gaussian process prior with mean zero and covariance function $\rho^2 K_\lambda(\cdot,\cdot)$ where

$$K_\lambda(\mathbf{x},\mathbf{x}^*) = \Pi_{l=1}^{p} \lambda_l^{|x_l - x_l^*|^\alpha},$$

with $\alpha \in (0,2]$ and $\lambda \in (0,1)$. The rationale of this particular specification is that $\delta(\mathbf{x}_i)$ and $\delta(\mathbf{x}_j)$ are expected to have similar values if the difference between $\mathbf{x}_i$ and $\mathbf{x}_j$ is small, while the degree of similarity must decrease as $\mathbf{x}_i$ and $\mathbf{x}_j$ become further apart. They have used the MCMC technique to find the posterior distributions of $\boldsymbol{\beta}$ and δ.

If we are more interested in exploring the relationships (which may be complicated nonlinear functions) between the covariates and the response as well as to create a model with high predictive power we have to use nonlinear curves and surfaces within the linear-systematic structure. This is the main focus of this chapter.

4. Models Using Curves and Surfaces

Suppose the dependence of the response Y on one of the covariates T is of more interest than its dependence with the other covariates X. Then the usual GLM structure for a single datapoint $(y_i, t_i, \mathbf{x}_i)$ can be modified as

$$g(\mu_i) = \boldsymbol{\beta}.\mathbf{x}_i + h(t_i), \tag{4}$$

where h is an unknown curve to be estimated from the data. For classical analysis of this type of model see Green and Silverman (1994).

Now if we want to express the relationships between the response and all the covariates in such a non-linear manner, the simplest model will be the additive model (Hastie and Tibshirani, 1990) where the model will be extended as

$$g(\mu_i) = h_1(x_{i1}) + h_2(x_{i2}) + \cdots + h_k(x_{ip}), \tag{5}$$

where $h_1, h_2, \cdots h_k$ are unknown curves to be estimated. In this basic additive model formulation we ignore the interactions among covariates which could be an important part of the model.

So the general model should be

$$g(\mu_i) = f(\mathbf{x}_i), \tag{6}$$

where f is a surface with p dimensions. Models in (1.3) and (1.4) are particular cases of this very general model.

Multivariate adaptive regression spline (MARS) methodology introduced by Friedman (1991) is an efficient way to model surfaces. Denison, Mallick and Smith (1998c) developed Bayesian MARS to create random surfaces which will be used here to model f.

5. GLMs using Bayesian MARS

5.1 Classical MARS

The original MARS model of Friedman (1991) was motivated by the recursive partioning methods previously used (Morgan and Sonquist, 1963) such as classification and regression trees (CART) (Brieman *et al.*, 1984; Denison and Mallick, this volume). The MARS model was designed to have the intrepretability associated with CART but to overcome its greatest drawback, the poor predictions associated with fitting an assumed continous regression function with a piecewise constant surface.

The MARS model uses truncated linear splines and their products as basis functions and with this formulation Friedman achieved his desired objectives. However, in the classical context a deterministic search of the candidate model space induced by the basis functions is required to find the final model. These types of searches can be restrictive because, by design, they only cover a subspace of the complete space of models.

In the GLM framework the MARS model is used to estimate the function which is a known transformation of the mean as in (1.6). Thus the nonparametric MARS model becomes embedded in a semiparametric framework. The estimate $\widehat{f}$ to the true regression function f can be written as

$$\widehat{f}(\mathbf{x}) = \sum_{i=1}^{k} \beta_i B_i(\mathbf{x}),$$

where the B_i are the basis functions of the MARS model, of which there is an unknown number k. These basis functions, using the notation of Friedman (1991), can be written as

$$B_i(\mathbf{x}) = \begin{cases} 1, & i = 1 \\ \prod_{j=1}^{J_i} [s_{ji} \cdot (x_{v(j,i)} - t_{ji})]_+, & i = 2, 3, \ldots \end{cases} \tag{7}$$

where $(\cdot)_+ = \max(0, \cdot)$, J_i is the degree of the interaction of basis B_i, the s_{ji}, which we shall call the sign indicators, equal ± 1, the $v(j, i)$ give the indices of the predictor variables corresponding to the *knots* t_{ji}. The $v(j, \cdot)$ $(j = 1, \ldots, J.)$ are constrained to be distinct so each predictor only appears once in each interaction term to maintain the 'linear' nature of the basis functions. See Section 3, Friedman (1991) for a comprehensive illustration of the model. Note that in all the work that follows we shall take the maximum number of interactions in a basis function to be two so that $J_i \leq 2$ for all i. This is sensible in the generalized linear model framework where interpretation of the model is important.

The form of the MARS function is found by starting the algorithm with only B_1 (the constant basis function) in the model and then by stepwise addition of the basis functions which most reduce the chosen lack-of-fit criterion [usually the generalized cross-validation measure (Craven and Wabha, 1979)]. The candidate bases which can be added are found by 'splitting' the bases that are currently in the model; this prevents the candidate search space from becoming unmanageably large even though it is restrictive. After the model has been grown to have many basis functions stepwise deletion takes place. This involves working out the lack-of-fit when each basis function is, in turn, removed and then taking away the one that least degrades the fit. When some minimum of the lack-of-fit is reached, then the process is terminated and the MARS model has been fitted. Note that at each step a suitable algorithm for maximizing the coefficients $\boldsymbol{\beta}$ is undertaken; this is not straightforward for GLM because the errors are not normal as is often the case for simple linear models.

5.2 Bayesian MARS

In this section we describe an extension to the classical MARS model. We attempt to account for model uncertainty by finding a posterior distribution for the unknown parameters in the model given the data, and we also want to perform a wide search of the model space. Both of these things are possible by taking a Bayesian perspective and using standard MCMC simulation methods. A full description of the Bayesian MARS (BMARS) algorithm, which we shall use in the remainder of this chapter, is given in Denison *et al.* (1998c) but here we shall just give a general outline of the model. Note that the Bayesian MARS method is just an extension in many dimensions of the Bayesian curve fitting methodology given in Denison, Mallick and Smith (1998a).

We place prior distributions over the unknown MARS model parameters $(k, \boldsymbol{\theta}^{(k)})$. Here $\boldsymbol{\theta}^{(k)} = \left[C_i, \beta_i, \{s_{ji}, t_{ji}\}_{j=1}^{J_i}\right]_{i=1}^{k}$ where C_i is the *type* of basis function that B_i is classified as. This is determined from the $v(j, i)$ and just classifies all basis functions involving the same covariates as of the same type no matter what ordering of j is used. With this model structure the dimension of $\boldsymbol{\theta}^{(k)}$ is given by $n(k) = \sum_{i=1}^{k} 2(1 + J_i)$.

The priors we use are as follows. We assign a Poisson prior with parameter λ over k with discrete uniform priors over the possible values of s_{ji} and t_{ji}, i.e. over $\{-1, 1\}$ and the marginal predictor values of variable $X_{v(j,i)}$ respectively. The prior over C_i is chosen more carefully. We want the prior to reflect that among all interaction terms and main effects each type of basis is equally likely, but also that main effects are favored over interactions. So, for the C_i which represent main effects ($i = 1, \ldots, p$) the prior probability for C_i is ψ/p and for the other C_i their prior probability of being in the model is $2\psi/\{m(m-1)\}$ where $\psi(> 0.5)$ is the prior proportion of basis functions expected to be main effects. This specifies a prior distribution over the model parameters, Δ say, so $p(\Delta) = p(\theta \backslash \boldsymbol{\beta})$. We create a hierarchical setup to assign the prior to the $\boldsymbol{\beta}$ because $p(\Delta, \boldsymbol{\beta}) = p(\boldsymbol{\beta}|\Delta)p(\Delta)$. We choose the prior for the coefficients given the model to be a normal distribution centered around zero and with dispersion matrix $\Lambda = 0.2\text{diag}(r_2, \ldots, r_k)$ where r_i is the range of the output of the ith basis function. Note that no 'shrinkage' takes place on the constant basis function. This hierachical structure avoids possible problems with model selection, notably 'Lindley's paradox' (Lindley, 1957).

We wish to make inference on the model parameters given the data so our target posterior distribution is $p(k, \theta^{(k)}|\mathcal{D})$, where $\mathcal{D}$ denotes the data matrix (Y, X). Unfortunately, this posterior distribution is analytically intractable so we must use a simulation technique to generate samples from it. The dimension of the posterior is varying so we use a reversible jump MCMC algorithm (Green, 1995) to perform the simulation. This takes the form of a stochastic search over the model space $\Theta = \bigcup_{k=0}^{\infty} \Theta_k$ where Θ_k is the subspace of the Euclidean space, $R^{n(k)}$, corresponding to the vector space spanned by all the elements $\boldsymbol{\theta}^{(k)}$ with k basis functions.

In the context of our problem, with multiple parameter subspaces of different dimensionality, it is necessary to devise different types of moves between the subspaces, Θ_k. These are combined to form what Tierney (1994) calls a hybrid sampler, making random choice between available moves at each transition, in order to traverse freely around the combined parameter space.

We use the following move types: (a) a movement in a knot location; (b) a change in a factor in a basis function; (c) a change in the basis coefficients; (d) the addition of a basis function and (e) the deletion of a basis function. Note that in steps (d) and (e) we are changing the dimension of the model and that we do not add basis functions in pairs as in the standard MARS forward-stepwise procedure; in fact, we depart completely from any sort of recursive partitioning approach. We have found that adding basis functions singly makes our procedure more flexible and the reversibility condition easier to satisfy (Denison, 1997).

Thus the algorithm proceeds as follows.

1. Start with only the constant basis function, B_1, in the model.

2. Set k equal to the number of basis functions in the current model.

3. Generate a random quantity u uniformly on [0,1] and use this to determine which move step to perform. Thus with probability: b_k goto BIRTH step (d), d_k goto DEATH step (e), ρ_k goto MOVE step (a), η_k goto BIGMOVE step (b) and ν_k goto CHANGE step (c).

4. Repeat 2 for a suitable number of iterations and collect every 5th model after the burn-in period is deemed to have ended.

Note that the probabilities with which the different move types are undertaken in step 3 are chosen so that they sum to one. Specifically, we choose $b_k = 0.25 \min\{1, \lambda/(k+1)\}$, $d_k = 0.25 \min\{1, k/\lambda\}$ and the others are chosen so that $\rho_k = \eta_k = 0.5\nu_k$. This type of formulation was suggested in Green (1995).

5.3 Bayesian MARS for GLMs

When we change the MARS structures, as described below, the coefficients of the basis functions $\boldsymbol{\beta}$ in the new model can be found in a variety of ways. For the normal error regression model integrating out the coefficients using a conjugate prior for them is the most elegant method and leads to the acceptance probability just incorporating a Bayes factor (Holmes and Mallick, 1997). Unfortunately for the more difficult GLM examples we must draw the coefficients at each step using a simple Metropolis-Hastings proposal (Metropolis *et al.*, 1953; Hastings, 1970). We propose the new coefficients from a normal distribution centered around their current values with dispersion matrix Λ. Proposing the new coefficients from a normal distribution with the same dispersion matrix as the prior is useful here as it leads to some cancellation in the acceptance probability.

A problem with this specification is that the sampler moves slowly around the probability space because the proposal distribution is not diffuse. Hence, many iterations of the sampler must be made and we take only every 100th one as being in the 'independent' sample. We still use this algorithm, however, because it ensures that the acceptance rates are adequate ($> 10\%$). We have employed other strategies to sample the coefficients and these have produced better results but they are not fully Bayesian leading to the algorithm only performing a stochastic search of the model space.

6. Examples of Bayesian MARS for GLMs

In the examples that follow we ran the sampler in exactly the same way. That is for 500,000 burn-in iterations with every 100th iteration after that taken as being in the generated sample until 10,000 models were collected (another one million iterations). The Poisson mean for the number of basis functions was chosen to be two so that parsimonious models were encouraged. Note that we only demonstrate the model using the logit link for binary regression but any GLM can be handled in exactly the same way with the only difference being the likelihood.

6.1 Motivating Example

We present this example purely to demonstrate how it can be the case that GLMs unrealistically restrict the form of the model and cannot handle some relationships for which nonparametric functions for the transformed mean work well.

We use the `kyphosis` dataset which is available in Splus (Becker *et al.*, 1988). The response is binary and represents the presence or absence of a postoperative deformity known as kyphosis. There are 81 datapoints of which 17 had kyphosis after the operation. The original dataset has three predictors but, for illustrative purposes, we shall do logistic regression using just one of them, that is the `Start` predictor which gives the beginning of the range of vertebrae involved in the operation.

We use the canonical link function for the binary data which is assumed to come from a Bernoulli distribution with an unknown parameter. Thus the g in (1.6) is taken to be the logit function, i.e. $\text{logit}(\mu) = \mu/(1-\mu)$. Other possible link functions for data of this type are the probit and the complementary log-log link functions (McCullagh and Nelder, 1989).

Figure 12.1 compares the standard logistic regression model with the logistic

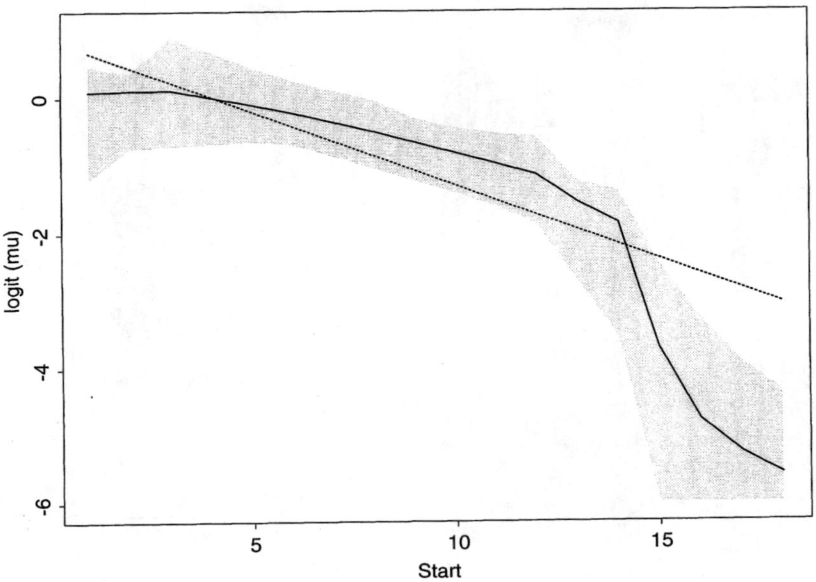

FIGURE 12.1. Comparison of the classical logistic regression model (dotted line) with the logistic regression estimate using BMARS (solid line). The shading corresponds to the 95% Bayesian credible intervals.

BMARS model. It appears that there is a most difference between the estimates when `Start` is over 15. The BMARS model estimates the probability of kyphosis as being significantly lower than the GLM when `Start` is over 15.

The boxplots of `Start` for the two classes, given in Figure 12.2, confirm that when `Start` is high the presence of kyphosis is less likely, as captured by the nonparametric Bayesian model.

6.2 Pima Indian Example

The Pima Indian dataset is another binary regression example but this one concerns the presence or absence of diabetes among Pima Indian woman living near Phoenix, Arizona. There are eight covariates: number of pregnancies; plasma glucose concentration; diastolic blood pressure (mmHg); triceps skin fold thickness (mm); serum insulin (μU/ml); body mass index (kg m^{-2}); diabetes pedigree function and age in years. The data were collected by the US National Institute of Diabetes and Digestive and Kidney Diseases and can be obtained free of charge, from the website `http://markov.stats.ox.ac.uk/pub/PRNN`.

An early study of this data was carried out by Smith *et al.* (1988) and it is extensively studied using a variety of methods, including logistic regression, in Ripley (1996). As in Ripley (1996) we omit the serum insulin predictor and use the 532 complete records and split them into a training set of 200 points with a test set made up of the other 332 points. The best results given in Ripley (1996) have a error rate of about 20% (66 misclassified points) which is much better than the overall rate of diabetes of 33%. The logistic regression model was among those that

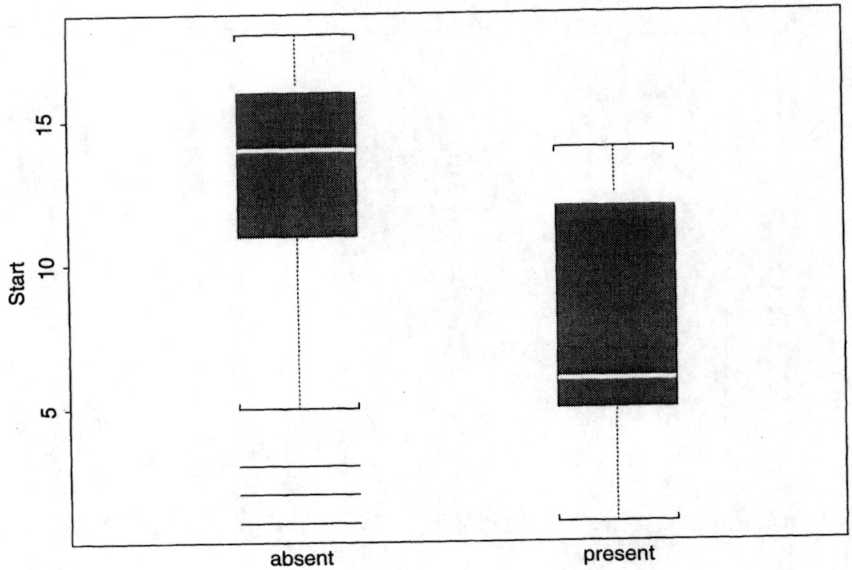

FIGURE 12.2. Boxplot of kyphosis data. Note how kyphosis is not present in any cases for which `Start` is greater than 15.

performed well and had an error rate of 66/332 when fitted with all the predictors or when using stepwise selection to drop two of them (skin thickness and blood pressure). One reason that the standard GLM does so well is that the dataset is thought to be linear.

The Bayesian MARS model has a misclassification rate of 67 points using the average of the whole sample and, as no interaction term is present in more than 4% of the generated sample, it suggests linearity in the predictors. One of the main advantages of the Bayesian method is the measure of uncertainty we obtain over the predicted probabilities. If we need to make a prediction at a new location we find that we can do this with some certainty if the 95% Bayesian credible interval for its predicted probability does not include 0.5 but if it does we may decide that making a prediction is too difficult and therefore not worthwhile: we shall refer to these points as being *unclear*.

In Figure 12.3 we display the results of our Bayesian GLM approach on the test set. It shows how the majority of points are correctly classified and, in particular, demonstrates how the unclear points are not necessarily those which are nearest the 0.5 mark.

Note that we have achieved better predictive results in simulations using Bayesian-motivated stochastic searches of the posterior probability space. These typically involve a crude maximization of the coefficients at each step. Good maximizations without gradient information and in many dimensions take too long to be useful in this iterative procedure. However, just this basic change can reduce the number of misclassified points to a level not yet reported in the literature (i.e. 61) at the expense of losing a truly Bayesian procedure. The improvement is mainly due to the fact that these stochastic searches tend to find many local modes but do not explore the tails of the distributions fully (Chipman, George and McCulloch, 1998;

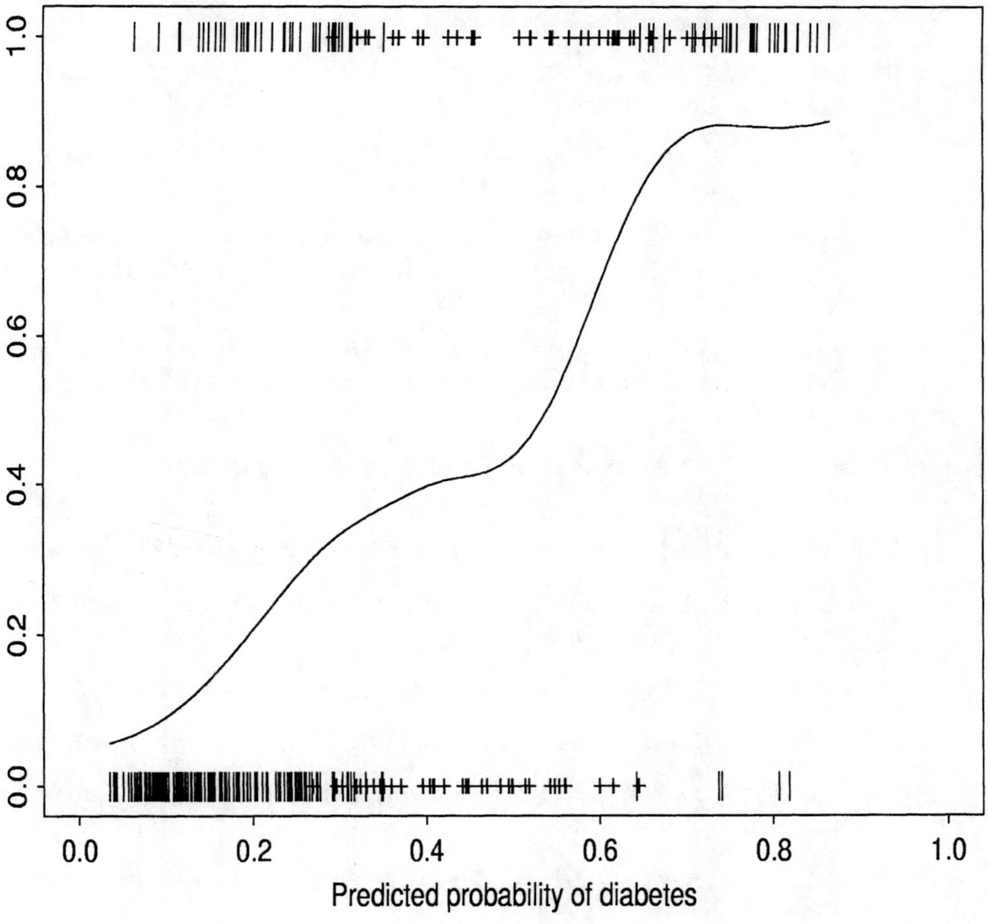

FIGURE 12.3. The 'rug' of ticks give the test datapoints and their predicted probabilities. Those identified with a '+' are the unclear points.

Denison *et al.*, 1998b,d). For simple predictive purposes, where error bounds are not important, this seems to be a good strategy.

References

Becker, R.A., Chambers, J.M., and Wilks, A.R. (1988). *The New S Language.* Pacific Grove, CA: Wadsworth.

Blight, B.J.N. and Ott, L. (1975). A Bayesian approach to model inadequacy for polynomial regression. *Biometrika*, **62**, 79-85.

Breiman, L., Friedman, J.H., Olshen, R. and Stone, C.J. (1984). *Classification and Regression Trees.* Belmont, CA: Wadsworth.

Chipman, H., George, E.I. and McCulloch, R.E. (1998). Bayesian CART model search. *J. Am. Statist. Assoc.* (to appear).

Craven, P. and Wabha, G. (1979). Smoothing noisy data with spline functions. Estimating the correct degree of smoothing by the method of cross-validation. *Numer. Math.*, **31**, 317-403.

Dellaportas, P. and Smith, A.F.M. (1993). Bayesian-inference for generalized linear and proportional hazards models via Gibbs sampling. *Appl. Statist.*, **42**, 443-459.

Denison, D.G.T. (1997). Simulation based Bayesian nonparametric regression methods. *Unpublished PhD Thesis.* Imperial College, London.

Denison, D.G.T. and Mallick, B.K. (1998). Classification trees. This volume.

Denison, D.G.T., Mallick, B.K. and Smith, A.F.M. (1998a). Automatic Bayesian curve fitting. *J. Roy. Statist. Soc.* B, **60**, 333-350.

——— (1998b). A Bayesian CART algorithm. *Biometrika*, **85**, 363-377.

——— (1998c). Bayesian MARS. *Statistics and Computing* (to appear).

——— (1998d). Discussion of Chipman, George, McCulloch. *J. Am. Statist. Assoc.* (to appear).

Diaconis, P. and Ylvisaker, D. (1985) Conjugate priors for Exponential families. *Annals of Statistics*, **7**, 269-281.

Follman, D.A. and Lambert, D. (1989). Generalising logistic regression by nonparametric mixing. *J. Amer. Statist. Assoc.*, **84**, 295-300.

Friedman, J.H. (1991). Multivariate adaptive regression splines. *The Annals of Statistics*, **19**, 1-141.

Gelfand, A.E. (1997). Approaches for semiparametric Bayesian regression. *Technical report.* University of Connecticut, CT.

Green, P.J. and Silverman, B.W. (1994). Nonparametric regression and generalized linear models: a roughness penalty approach. London: Chapman and Hall.

Guitierrez-Pena, E. and Smith, A.F.M. (1997). Aspects of smoothing and model inadequacy in generalized regression. *Journal of Statistical Planning and Inference* (to appear).

Hastie, T.J. and Tibshirani, R.J. (1990). *Generalized additive models.* London: Chapman and Hall.

Hastings, W.K. (1970). Monte Carlo sampling methods using Markov chains and their applications. *Biometrika*, **57**, 97-109.

Holmes, C.C. and Mallick, B.K. (1997). Bayesian wavelet networks for nonparametric regression. *Technical report.* Imperial College, London.

Laird, N. (1978). Nonparametric maximum likelihood estimation of a mixing distribution. *J. Am. Statist. Assoc.*, **73**, 805-811.

Lindley, D.V. (1957). A statistical paradox. *Biometrika*, **44**, 187-192.

McCullagh, P. and Nelder, J.A. (1989). *Generalized Linear Models.* London: Chapman and Hall.

Mallick, B.K. (1994). Bayesian semiparametric modeling using mixtures. *Unpublished PhD Thesis.* University of Connecticut, CT.

Mallick, B.K. and Gelfand, A.E. (1994). Generalised linear models with unknown link functions. *Biometrika*, **81**, 237-245.

Metropolis, N., Rosenbluth, A.W., Rosenbluth, M.N., Teller, A.H. and Teller, E. (1953). Equations of state calculations by fast computing machines. *J. Chem. Phys.*, **21**, 1087-91.

Morgan, J.N. and Sonquist, J.A. (1963). Problems in the analysis of survey data and a proposal. *J. Am. Statist. Assoc.*, **58**, 415-434.

Morris, C. (1982). Natural Exponential families with quadratic variance functions. *Ann. Statist.*, **10**, 65-80.

Nelder, J.A. and Wedderburn, R.M. (1972). Generalized linear models. *J. Roy. Statist. Soc.* A , **135**, 370-384.

Newton, M.A., Czado, C. and Chappell, R. (1996). Bayesian inference for semiparametric binary regression. *J. Amer. Statist. Assoc.*, **91**, 1996.

Ripley, B.D. (1996). *Pattern Recognition and Neural Networks.* Cambridge: Cambridge University Press.

Smith, J.W., Everhart, J.E., Dickson, W.C., Knowler, W.C. and Johannes, R.S. (1988). Using the ADAP learning algorithm to forecast the onset of diabetes mellitus. In *Proceedings of the Symposium on Computer Applications in Medical Care (Washington, 1988).* (Ed. R.A. Greenes), pp. 261-265. Los Alamitos, CA: IEEE Computer Society Press.

Tierney, L. (1994). Markov chains for exploring posterior distributions (with discussion). *Ann. Statist.*, **22**, 1701-1762.

Walker, S.G. and Mallick, B.K. (1997). Hierarchical generalized linear models and frailty models with Bayesian nonparametric mixing. *J. Roy. Statist. Soc.* B, **59**, 845-869.

Wedderburn, R.M. (1976). On the existance and uniqueness of the maximum likelihood estimates for certain generalized linear models. *Biometrika*, **63**, 27-32.

13

Binary Response Regression with Normal Scale Mixture Links

Sanjib Basu
Saurabh Mukhopadhyay

ABSTRACT Binary response regression is a useful technique for analyzing categorical data. Popular binary models use special link functions such as the logit or the probit link. We propose Bayesian binary regression models where the inverse link functions H are scale mixtures of normal cumulative distribution functions. Two such models are described: (i) H is a finite scale mixture with a Dirichlet distribution prior on the mixing distribution, and (ii) H is a general scale mixture with the mixing distribution having a Dirichlet process prior. Bayesian analyses of these models use data augmentation and Gibbs sampling. Model diagnostics by cross validation of the conditional predictive distributions are proposed. The application of these models and the model diagnostic techniques are illustrated in student retention data.

1. Introduction

In many areas of applications of statistical principles and procedures one encounters observations that take one of two possible forms. Such binary data are often measured with covariates or explanatory variables that are either continuous or discrete or categorical. The relation between the response and the covariates is usually modeled by assuming that the probability of a "positive response", after a suitable transformation, is linear in the covariates. Let $(y_i, \underset{\sim}{x}_i)$, $i = 1, \ldots, N$ be the set of observations where $\underset{\sim}{x}_i^T = (x_{1i}, \ldots, x_{ki})$ is the set of covariates and the binary response y_i is either 0 or 1. Binary regression models assume that the random variables $Y_1, \ldots, Y_N$ are independent and

$$P(Y_i = 1) = H(\underset{\sim}{x}_i^T \underset{\sim}{\beta}), \quad i = 1, \ldots, N. \tag{1}$$

Here $\underset{\sim}{\beta} = (\beta_1, \ldots, \beta_k)^T$ is a vector of unknown parameters and the function H is usually assumed to be known. In the terminology of Generalized Linear Models (McCullagh and Nelder (1989)), H is the *inverse link* function. For ease of exposition, we refer to H as the link function in this article. Popular probit and logit models are obtained if H is chosen as the standard normal cumulative distribution function (cdf) Φ or the cdf of the standard logistic distribution respectively. Such choices of H are often done for convenience and on an ad hoc basis.

Prentice (1976), Aranda-Ordaz (1981), Guerrero and Johnson (1982), Stukel (1988) and many others studied binary regression with a parametric family of link functions (instead of a single fixed link) from a non-Bayesian viewpoint. Their work shows that these extended models can significantly improve fits. Recent Bayesian works on binary and polychotomous response regression with an extended class of link functions include Doss (1994), Albert and Chib (1993), Newton, Czado, and Chappell (1996), and Erkanli, Stangl, and Müller (1993).

Notice that (1) requires the range of H to be $[0, 1]$. Usually it is also preferable to have a nondecreasing smooth function H. These requirements match exactly with a smooth continuous cumulative distribution function (cdf). Let $\mathcal{F}$ be the class of all cdfs on $\Re$. The family $\mathcal{F}$ includes cdfs which are often undesirable as choices for H, for example, cdfs of discrete distributions. Instead, Basu and Mukhopadhyay (1998) (Basu and Mukhopadhyay (1998) from now on) consider the subclass of normal scale mixture cdfs $\mathcal{F}_N = \{ F_N(\cdot) = \int_{[0,\infty)} \Phi(\frac{\cdot}{\sigma})\, dG(\sigma),\ G \text{ is a cdf on } [0, \infty)\}$ as possible choices for the function H. The class of normal scale mixtures allows a variety of functional forms and varying tail structures (including normal, all t distributions, Logistic, Double Exponential, and Cauchy), thus presenting us with a wide array of choices for the link H.

For $H(\cdot) = F_N(\cdot) = \int \Phi(\frac{\cdot}{\sigma})\, dG(\sigma)$, our binary regression model (1) becomes

$$P(Y_i = 1) = \int \Phi(\{\underset{\sim}{x}_i^T \underset{\sim}{\beta}\}/\sigma)\, dG(\sigma), \quad i = 1, \ldots, N, \tag{2}$$

which includes two unknowns, the regression coefficient $\underset{\sim}{\beta}$ and the mixing distribution G. To complete the Bayesian model specification, a prior $\pi(\underset{\sim}{\beta}, G)$ is assumed on the unknowns. Typically, $\underset{\sim}{\beta}$ and G are assumed independent a priori, i.e., $\pi(\underset{\sim}{\beta}, G) = \pi_1(\underset{\sim}{\beta})\, \pi_2(G)$.

The posterior distribution $\pi(\underset{\sim}{\beta}, G|\underset{\sim}{y})$ which combines the prior $\pi(\underset{\sim}{\beta}, G)$ and the sampling model of (2) is analytically intractable. In the spirit of Albert and Chib (1993), Basu and Mukhopadhyay (1998) describe a data-augmented Gibbs sampling method (Gelfand and Smith (1990),Tanner and Wong (1987)) which obtains Monte Carlo estimates of the posterior and other quantities of interest.

Basu and Mukhopadhyay (1998) assume a finite mixing distribution with prespecified mixing location. In this article, we extend that finite mixture model to a general mixture model. We assume that the mixing distribution G is an arbitrary member of the class $\mathcal{F}^+ = \{ F \text{ is a cdf on } [0, \infty)\}$. To complete our Bayesian model specification, we assume a Dirichlet process prior $DP(\alpha \cdot \nu)$ on G. Here $\alpha > 0$ is the concentration parameter and ν is a cdf on $[0, \infty)$. The implementation of this model is discussed in section 3.

Model checking is an integral part of any model building. We consider several cross validation model checking criteria that are suggested by Gelfand, Dey, and Chang (1992). Computations of these criteria in our models are described in section section 4. In section section 5, we illustrate an application of these models to student retention data from the University of Arkansas. Conclusions are given in section 6.

A word about notations. Throughout this article, we use upper case letters, Y, Z, to denote random variables and lower case letters, y, z, to denote observed values or running variables. Also, we use F, G to denote distributions and cdfs.

2. The Finite Mixture Model

In this section, we briefly describe the finite normal scale mixture link considered by Basu and Mukhopadhyay (1998). In a typical binary regression setup, the observed data consist of $(y_i, \underset{\sim}{x}_i)$, $i = 1, \ldots, N$, where y_i is the binary response and $\underset{\sim}{x}_i$ is a set of covariates which are either continuous or categorical. The binary response Y_i's are assumed to be independent Bernoulli (θ_i). The normal scale mixture link model assumes $\theta_i = \int \Phi(\{\underset{\sim}{x}_i^T \underset{\sim}{\beta}\}/\sigma)\, dG(\sigma)$ as in (2).

In their finite mixture model, Basu and Mukhopadhyay (1998) assume that the mixing distribution G is discrete with finite support, i.e., $G = \sum_{j=1}^{s} p_j\, \delta_{\{\tau_j\}}$ where $0 \leq p_j \leq 1$, $\sum_{j=1}^{s} p_j = 1$, $\delta_{\{\tau_j\}}$ is the degenerate distribution at τ_j, and the support points $0 < \tau_1 < \ldots < \tau_s < \infty$ are prespecified. Basu and Mukhopadhyay (1998) assume a Dirichlet distribution prior $\pi_2(G) = DD(\underset{\sim}{\nu})$ on the mixing proportions $\underset{\sim}{p} = (p_1, \ldots, p_s)^T$. They further assume an independent prior $\pi_1(\underset{\sim}{\beta})$ on $\underset{\sim}{\beta}$.

The posterior distribution of $\underset{\sim}{\beta}$ and G under this model specification is $\pi(\underset{\sim}{\beta}, G|\underset{\sim}{y}) =$ constant $\cdot\, \pi_1(\underset{\sim}{\beta})\, \pi_2(G) \prod_{i=1}^{N} \{\int \Phi(\{\underset{\sim}{x}_i^T \underset{\sim}{\beta}\}/\sigma)\, dG(\sigma)\}$. This posterior distribution is analytically intractable. The Gibbs sampler provides a simulation based computational approach which enables one to obtain Monte Carlo estimates of posterior quantities of interest. General discussions about the Gibbs sampler, its implementation, and its convergence can be found in Gelfand and Smith (1990), Gelman and Rubin (1992), Casella and George (1992), Tierney (1994) and in the recent book by Gilks et al (1998). In the finite normal mixture setup Basu and Mukhopadhyay (1998) show that the implementation of the Gibbs sampler is further facilitated by introduction of latent random variable $\underset{\sim}{Z} = (Z_1, \ldots, Z_N)^T$ and $\underset{\sim}{\sigma} = (\sigma_1, \ldots, \sigma_N)^T$ (as in Albert and Chib (1993)). The complete structure of the Basu and Mukhopadhyay (1998) model along with the distribution of the latent variables is described below:

(*a*) Given $\underset{\sim}{\beta}$ and the latent $\underset{\sim}{\sigma}$, the latent variables $Z_1, \ldots, Z_N$ are independent with $Z_i \sim N(\underset{\sim}{x}_i^T \underset{\sim}{\beta}, \sigma_i^2)$.

(*b*) Given $\underset{\sim}{Z}$; $Y_1, \ldots, Y_N$ are completely determined with $Y_i = 1$ if $Z_i > 0$, and $Y_i = 0$ otherwise. Note that marginalized over Z_i, we have $P(Y_i = 1) = \Phi(\{\underset{\sim}{x}_i^T \underset{\sim}{\beta}\}/\sigma_i)$. The latent Z_i's thus provide a flexible way to introduce the probit structure in the binary link. This flexible conditional representation of the probit model was first introduced by Albert and Chib (1993).

(*c*) Given G, the latent variables $\sigma_1, \ldots, \sigma_N$ are i.i.d. $\sim G$. These latent σ_i's incorporate the normal scale mixture structure.

(*d*) The mixing distribution G is discrete with finite support, i.e., $G = \sum_{j=1}^{s} p_j\, \delta_{\{\tau_j\}}$. Here the support points $0 < \tau_1 < \ldots < \tau_s < \infty$ are user specified and $\underset{\sim}{p} = (p_1, \ldots, p_s)^T$ has a Dirichlet distribution prior $\pi_2(\underset{\sim}{p}) = DD(\underset{\sim}{\nu})$.

(*e*) The regression parameter $\underset{\sim}{\beta}$ is independent of G and has a prior $\pi_1(\underset{\sim}{\beta})$.

To recapitulate, $P(Y_i = 1|\underset{\sim}{\sigma}, \underset{\sim}{\beta}, G) = P(Z_i > 0|\underset{\sim}{\sigma}, \underset{\sim}{\beta}, G) = \Phi(\{\underset{\sim}{x}_i^T \underset{\sim}{\beta}\}/\sigma_i)$. Integrated over σ_i, $P(Y_i = 1|\underset{\sim}{\beta}, G) = \int \Phi(\{\underset{\sim}{x}_i^T \underset{\sim}{\beta}\}/\sigma_i)\, dG(\sigma_i)$, which is our model of (2).

For this finite mixture model, Basu and Mukhopadhyay (1998) obtain the full conditional distributions of each unobserved variable given the observed $\underset{\sim}{y}$ and the remaining variables. These distributions are needed for the Gibbs generation. Introduction of the latent variables $\underset{\sim}{Z}$ and $\underset{\sim}{\sigma}$ substantially simplifies the calculation of the conditional distributions. Each conditional density is obtained in a closed form and is, in fact, in the form of a common distribution and hence is easy to simulate from.

3. General Mixtures and a Dirichlet Process Prior

In the finite mixture model of Basu and Mukhopadhyay (1998), the mixing distribution $G = \sum_{j=1}^{s} p_j \delta_{\{\tau_j\}}$ is discrete with finite support. The user needs to choose the support points of the mixing distributions, $\tau_1 < \tau_2 < \ldots < \tau_s$. In this section, we provide further flexibility in the specification of G and make the specification process more automated than the finite mixture model. The mixing G is allowed to be a general distribution $\in \mathcal{F}^+$ where $\mathcal{F}^+$ is the class of all distributions on $[0, \infty)$. As a natural extension of the Dirichlet distribution prior of section 2, we assume a Dirichlet process DP$(\alpha \cdot G_0)$ prior (see Ferguson (1974)) on G over the domain $\mathcal{F}^+$. DP priors have received considerable attention in recent Bayesian literature. They provide a flexible way to incorporate the user's prior knowledge about G into the analysis. The hyperparameter G_0 is the prior mean of G and represent the user's guess about G. The "concentration" parameter $\alpha > 0$ reflects the degree of "closeness" of the random G to the prior mean G_0 and represent the user's degree of belief about G_0.

DP priors were first introduced by Ferguson (1974). These priors introduces flexibility in the model; however the computations get more intensive and closed form expressions become hard to obtain. The recent attention in Bayesian Modeling with DP priors stemmed from the MCMC based computational breakthroughs obtained by Escobar (1994), MacEachern (1994) and others. For different applications of DP priors, see Escobar and West (1995), Basu (1996), Mukhopadhyay and Gelfand (1997) and the references therein.

The structure of the mixture model that we develop in this section is a generalization of the finite finite mixture model described in section 2. The generality is introduced in the generic structure of the mixing distribution G and in the specification of its prior distribution. Thus, parts (a), (b) and (e) of the same as in section 2. Parts (c) and (d) become more general as described below. For convenience in latter calculations, we reparametrize the models in terms of $\lambda_i = 1/\sigma_i^2$.

(c^*) Given G; the latent variables $\lambda_1 = 1/\sigma_1^2, \ldots, \lambda_N = 1/\sigma_N^2$ are i.i.d. $\sim G$.

(d^*) The random distribution G follows a Dirichlet process prior, i.e., $G \sim DP(\alpha \cdot G_0)$.

We use a methodology based on Gibbs sampling to compute posterior quantities from this model. Note that to implement the Gibbs sampler, one needs to generate Monte Carlo samples from the conditional distributions of each unobserved variable given the observed $\underset{\sim}{y}$ and the other unobserved variables iteratively. However, generating from $\pi(G \mid \underset{\sim}{y}, \underset{\sim}{\beta}, \underset{\sim}{Z}, \underset{\sim}{\sigma})$ is almost impossible, since this will require generating a whole distribution G on $[0, \infty)$. Instead, we consider the marginalized (integrated) model over G. Notice that, conditional on G, $\lambda_1, \ldots, \lambda_N$ are i.i.d $\sim G$. This implies that marginally each $\lambda_i \sim G_0$. However, when marginalized over G,

λ_i's are no longer independent, in fact, λ_i has a positive probability of being equal to the previous λ_j's, $1 \le j < i$. This fact was first pointed out by Antoniak (1974)). The joint distribution of the λ_i's can be obtained from their successive conditional distributions

$$\lambda_i \mid \lambda_1, \ldots, \lambda_{i-1} \sim \frac{\alpha}{\alpha + i - 1} G_0(\cdot) + \frac{1}{\alpha + i - 1} \sum_{j=1}^{i-1} \delta_{\{\lambda_j\}}(\cdot). \tag{3}$$

The successive conditional posterior distributions of the λ_i's, i.e., $\lambda_i \mid \underset{\sim}{y}, \lambda_1, \ldots, \lambda_{i-1}$, however, do not have such a simple structure. Escobar (1994) showed (in the context of a Normal means problem) that if one considers instead the full conditional posterior distribution of λ_i given all the other λ_j's, i.e., $\lambda_i \mid \underset{\sim}{y}, \lambda_{(-i)} = (\lambda_1, \ldots, \lambda_{i-1}, \lambda_{i+1}, \ldots, \lambda_N)$, then this distribution has a simple form and is often easy to simulate from. Note that these full conditional distributions are the ones required for Gibbs sampling. MacEachern (1994) later obtained a modification of Escobar's Gibbs sampling algorithm over a collapsed state space. Our implementation of Escobar's algorithm is similar to the one used by Erkanli, Stangl, and Müller (1993).

The full conditional distributions of the other parameters and latent variables are relatively easy to obtain.

(*i*) Given $\underset{\sim}{y}, \underset{\sim}{\beta}, \underset{\sim}{\lambda}$ and G, the latent variables $Z_1, \ldots, Z_N$ are conditionally independent. Moreover, each Z_i conditionally follows a truncated $N(\underset{\sim}{x}_i^T \underset{\sim}{\beta}, \sigma_i^2)$ distribution. The distribution is truncated at left by 0 if $y_i = 1$, and is truncated at right by 0 if $y_i = 0$.

For the following conditional distributions in (*ii*) – (*iv*) below, we assume that the given $\underset{\sim}{y}$ and $\underset{\sim}{z}$ satisfy $(y_i - \frac{1}{2})\, z_i > 0,\ i = 1, \ldots, N$.

(*ii*) Given $\underset{\sim}{y}, \underset{\sim}{\beta}, \underset{\sim}{Z}$ and the other $\lambda_{(-i)} = (\lambda_1, \ldots, \lambda_{i-1}, \lambda_{i+1}, \ldots, \lambda_N)$, the full conditional distribution of λ_i is $c \cdot \sqrt{\lambda_i}\, \phi(\{Z_i - \underset{\sim}{x}_i^T \underset{\sim}{\beta}\}\sqrt{\lambda_i})\, \{\alpha\, G_0(\cdot) + \sum_{j \ne i} \delta_{\{\lambda_j\}}(\cdot)\}$. Here c is a normalizing constant.

(*iii*) If we assume a customary diffuse prior $\pi_1(\underset{\sim}{\beta}) \equiv 1$, then the full conditional distribution of the regression parameter $\underset{\sim}{\beta}$ given data $\underset{\sim}{y}$ and the latent variables $\underset{\sim}{z}, \underset{\sim}{\lambda}$ is $N_k(\widehat{\underset{\sim}{\beta}}, (\boldsymbol{X}^T \boldsymbol{W} \boldsymbol{X})^{-1})$ where $\widehat{\underset{\sim}{\beta}} = (\boldsymbol{X}^T \boldsymbol{W} \boldsymbol{X})^{-1} \boldsymbol{X}^T \boldsymbol{W} \underset{\sim}{z}$, $\boldsymbol{W} = \text{diagonal}(\lambda_i)$, and $\boldsymbol{X} = [\underset{\sim}{x}_1, \ldots, \underset{\sim}{x}_N]^T$ is the design matrix (we assume rank$(\boldsymbol{X}) = k$). See, for example, Pilz (1991). For other prior choices of $\pi_1(\underset{\sim}{\beta})$, the full conditional distribution of $\underset{\sim}{\beta}$ can be similarly calculated (from Bayesian Linear model theory) and is sometimes obtained in a common form.

The Gibbs sampler starts from an initial guess $(\underset{\sim}{Z}^\circ, \underset{\sim}{\lambda}^\circ, \underset{\sim}{\beta}^\circ)$ and successively simulates Monte Carlo samples from each of these conditional distributions by conditioning on the latest sampled value of the other variables at each stage. The simplest step in the Gibbs iterations is the generation of the regression vector $\underset{\sim}{\beta}$ from the multivariate normal conditional distribution in (*iii*). The generation of Z_i's from the conditional distribution in (*i*) is also relatively straightforward. Numerical problems may arise if the truncation moves to the tail of the normal distribution. See Daganpur (1988) for simulation techniques from truncated distributions.

The simulation of λ_i in (*ii*) may not be so immediate. However, in typical applications, one chooses G_0 to be conjugate to the normal likelihood, $\sqrt{\lambda_i}\phi(\{Z_i -$

$\underset{\sim}{x}_i^T \underset{\sim}{\beta}\}\sqrt{\lambda_i})$. For example, one may choose G_0 to be a Gamma$(\nu/2, \nu/2)$ distribution (with density proportional to $\lambda^{\nu/2-1} \exp(-\nu\ \lambda/2)$). In such a case, the "marginal" $A_i = \int \sqrt{\lambda_i}\phi(\{Z_i - \underset{\sim}{x}_i^T \underset{\sim}{\beta}\}\sqrt{\lambda_i})dG_0(\lambda_i)$ can be obtained in closed from. The simulation of λ_i is then done in two stages. For each $j \neq i$, the random variable λ_i equals a previously generated λ_j with probability $\sqrt{\lambda_i}\phi(\{Z_i - \underset{\sim}{x}_i^T \underset{\sim}{\beta}\}\sqrt{\lambda_j})/\{A_i + \sum_{j \neq i} \sqrt{\lambda_i}\phi(\{Z_i - \underset{\sim}{x}_i^T \underset{\sim}{\beta}\}\sqrt{\lambda_j})\}$, the random variable λ_i. With the remaining probability of $A_i/\{A_i + \sum_{j \neq i} \sqrt{\lambda_i}\phi(\{Z_i - \underset{\sim}{x}_i^T \underset{\sim}{\beta}\}\sqrt{\lambda_j})\}$, the random variable λ_i follows the distribution which would result as the posterior distribution of λ_i if $\sqrt{\lambda_i}\phi(\{Z_i - \underset{\sim}{x}_i^T \underset{\sim}{\beta}\}\sqrt{\lambda_i})$ is the likelihood and $G_0(\lambda_i)$ is the prior.

Notice that at any particular Gibbs iteration, the number of distinct λ's is often much less than N, and therefore the computational effort can be reduced by a significant amount if we keep track of distinct values of λ's and their frequencies.

4. Model Diagnostic

4.1 Basic Goal

Model checking is an integral part of any data analysis. Common logit and probit models provide G^2 and χ^2 statistics as summary measures of the overall quality of fit. Residuals comparing observed and fitted values are also used. As Agresti (1990) writes, "such diagnostic analyses help show whether lack of fit is due to an inappropriate link function or perhaps due to nonlinearity in effects of explanatory variables."

In this section, we describe diagnostic tools for our Bayesian binary response model. Many eminent Bayesians including Geisser and Eddy (1979), Box (1980), Berger (1985), and Gelfand, Dey, and Chang [GDC] (1992) argue for using the predictive distribution in such diagnostics. Basu and Mukhopadhyay (1998) followed GDC (1992) in combining cross validation with this predictive approach. The diagnostic tools developed by Basu and Mukhopadhyay (1998) compare predictive distributions conditioned on the observed data with a single data point deleted against observed responses. We briefly describe these diagnostics tools below.

In binary regression, typically many independent binary responses, Y_i's are observed under the same covariate vector $\underset{\sim}{x}_i$. Let L be the number of distinct $\underset{\sim}{x}_i$'s, and we denote them by $\underset{\sim}{x}_1^\star, \ldots, \underset{\sim}{x}_L^\star$. Let n_k be the total number of binary Y's observed under $\underset{\sim}{x}_k^\star$ ($\sum_{k=1}^{L} n_k = N$) out of which T_k ($= \sum_{i:\underset{\sim}{x}_i = \underset{\sim}{x}_k^\star} Y_i$) are 1's. According to our sampling model, $T_1, \ldots, T_L$ are independent and $T_k \sim$ Binomial(n_k, θ_k) where $\theta_k = H(\underset{\sim}{x}^{\star T} \underset{\sim}{\beta})$. For model checking, Basu and Mukhopadhyay (1998) cross validate the sufficient statistics $T_1, \ldots, T_L$ (instead of the Y's). Let t_k be the observed value of T_k, $\underset{\sim}{t}$ be the $L \times 1$ observed data vector, and let $\underset{\sim}{t}_{(-k)}$ denote the $(L-1) \times 1$ vector with kth observation t_k deleted. Also, let $\omega = (\underset{\sim}{\beta}, G, \underset{\sim}{Z}, \underset{\sim}{\sigma})$ denote the set of unobserved variables. We use customary notations; f denote predictive distributions (e.g. $f(T_k \mid \underset{\sim}{t}_{(-k)})$) as well as sampling distributions ($f(T_k \mid \omega)$), and π denote priors ($\pi(\omega)$) as well as posteriors ($\pi(\omega \mid \underset{\sim}{t})$). We assume all relevant integrals in the following exist.

4.2 Diagnostic Tools

Basu and Mukhopadhyay (1998) develop diagnostic tools from a cross validated predictive approach and examine $f(T_k \mid \underset{\sim}{t}_{(-k)})$, i.e., the predictive distribution of the random variable T_k conditioned on the remaining observations $\underset{\sim}{t}_{(-k)}$. Following GDC, a random T_k from the predictive distribution $f(T_k \mid \underset{\sim}{t}_{(-k)})$ is compared against the observed value t_k. This comparison is done by the following two criteria. See GDC (1992) for other checking criteria and more details.

(a) d_{1k} = expected difference between the observed t_k and the random T_k, i.e., $t_k - \mu_k$ where μ_k is the mean of the distribution $f(T_k \mid \underset{\sim}{t}_{(-k)})$. d_{1k} are thus the familiar residuals. The studentized residuals as $d'_{1k} = \frac{d_{1k}}{s_k}$ where $s_k^2 = \text{Var}[T_k \mid \underset{\sim}{t}_{(-k)}]$ can also be used for diagnostic. Basu and Mukhopadhyay (1998) use the quantity $Q_1 = \sum (d'_{1k})^2$ as a summary index of model fit.

(b) $d_{2k} = f(t_k \mid \underset{\sim}{t}_{(-k)})$, i.e., the likelihood of observing $T_k = t_k$ given the remaining observations $\underset{\sim}{t}_{(-k)}$. Small values of d_{2k} critize the model. Following the suggestion of Geisser and Eddy (1979) and GDC (1992), Basu and Mukhopadhyay (1998) use $Q_2 = \prod_{k=1}^{L} d_{2k}$ as the second summary index of model fit. Notice that Q_2 can be interpreted as a joint pseudo–marginal likelihood of the observed $\underset{\sim}{t}$.

4.3 Computational Methods

To compute d'_{1k} and d_{2k}, we need the mean μ_k, the variance s_k^2, and the value $f(t_k \mid \underset{\sim}{t}_{(-k)})$ of the predictive distribution $f(T_k \mid \underset{\sim}{t}_{(-k)})$. Notice $f(T_k \mid \underset{\sim}{t}_{(-k)}) = \int f(T_k \mid \omega)\, \pi(\omega \mid \underset{\sim}{t}_{(-k)})$. One possible strategy to approximate $f(T_k \mid \underset{\sim}{t}_{(-k)})$ is as follows : (*i*) delete t_k from the observed data vector $\underset{\sim}{t}$ to obtain $\underset{\sim}{t}_{(-k)}$; (*ii*) use Gibbs sampling (sections 2. and 3.) to generate R Monte Carlo samples of ω_r from $\pi(\omega \mid \underset{\sim}{t}_{(-k)})$; and (*iii*) approximate $f(T_k \mid \underset{\sim}{t}_{(-k)})$ by the Monte Carlo sum $\frac{1}{R} \sum_{r=1}^{R} f(T_k \mid \omega_r)$. Since we need $f(T_k \mid \underset{\sim}{t}_{(-k)})$ for every $k = 1, \ldots, L$, this strategy would require L separate Gibbs sampling runs. However, this can be substantially simplified using some neat tricks and manipulations involving the relevant posterior distributions and marginals. Basu and Mukhopadhyay (1998) show that all the relevant quantities needed for the computation of the diagnostic tools d'_{1k} and d_{2k} can be efficiently calculated from a single Gibbs run. We refer the reader to Basu and Mukhopadhyay (1998) for further details.

5. Application: Student Retention at the University of Arkansas

In this section, we illustrate our general mixture model using student retention data from the University of Arkansas. We further compare the performance of our general mixture model with the finite mixture model proposed by Basu and Mukhopadhyay (1998). Two other interesting applications of the finite mixture model can be found in Basu and Mukhopadhyay (1998), one on Beetle mortality data and the other in Challenger o-ring distress data.

The student retention data of this example is obtained from the *Retention and Graduation Report* of Donnelly and Line (1992) on the undergraduate students of the University of Arkansas. Freshmen classes entering the University of Arkansas in 1981, 1984-86, 1988, and 1989, consisting of $N = 13890$ students, are tracked in this report. One variable of interest is third year retention, i.e., $Y = 1$ if the student is still enrolled in his/her 3rd year fall semester, $Y = 0$ if s/he has dropped out. Retention is a concern to the University of Arkansas because of the relatively high dropout rate. The explanatory variables we use are ACT (American College Test given to graduating high school students) score as a continuous variable and (starting) year as a categorical variable. Preliminary plots suggest a second degree model in terms of ACT. We consider the model $P(Y_i = 1) = H(\underset{\sim}{x}_i^T \underset{\sim}{\beta})$ for the ith student where $\underset{\sim}{x}_i$ is an 8×1 vector with $x_{i1} = 1$, $x_{i2} = \text{ACT}$, $x_{i3} = (\text{ACT})^2$, $x_{i4} = 1$ if 1981, $= 0$ otherwise and so on.

The 13890 students are divided into 48 groups according to their ACT scores and starting years. We use n_k to denote the number of students in kth group and t_k to denote the number retained in the 3rd year.

A preliminary maximum likelihood logistic regression analysis shows that both ACT and $(\text{ACT})^2$ have "significant" effect (in Frequentist terminology) on $P(Y = 1)$ whereas the categorical covariate Year is "insignificant". However, since our models use different link functions and since the frequentist findings may not immediately translate to the Bayesian models, we keep all 8 x-covariates in our models. We explore two Bayesian models. The two models differ in their link structures, $P(Y_i = 1) = H(\underset{\sim}{x}_i^T \underset{\sim}{\beta})$. The first model is the finite mixture model (MF) of Basu and Mukhopadhyay (1998) where the link function H is assumed to be a finite scale mixture of normal cdfs, i.e., $H(\cdot) = \sum_{j=1}^{s_1} p_j \, \Phi(\cdot/\tau_j)$ and the mixing probabilities has a Dirichlet distribution, i.e., $\underset{\sim}{p} \sim DD(\underset{\sim}{\nu})$. The second model is our proposed general mixture model (MG) where $H(\cdot) = \int \Phi(\cdot \sqrt{\lambda}) \, dG(\lambda)$ and the mixing distribution G has a Dirichlet process prior, i.e., $G \sim DP(\alpha \cdot G_0)$.

In MF, we choose the set of τ_j's as $\mathcal{T} = \{0.5, 1, 2, 3, 4, 6, 8, 10, 15, 20, 30\}$. This choice reflects our subjective feeling that there should be some small τ values, a good representation of moderate τ values and some large τ values. We choose equal values for the Dirichlet distribution parameter, $\nu_1 = \ldots = \nu_{11} = \alpha/11$ where $\alpha > 0$ is a constant. For model MG, G_0 is taken to be the Gamma$(1,1)$ distribution. Several different values of α are tried in both models MF and MG. The summary criteria that we use for model fitting and checking (described below) differ for different choices of α, but not to any significant extent. The "best" results are obtained from $\alpha = 1$ in MF and $\alpha = 8$ in MG

For the finite mixture model MF, we use the sampling based method proposed by Basu and Mukhopadhyay (1998). For the general mixture model MG proposed in this article, we use the Gibbs sampling method described in section 3. The results are shown in Table ?? and Figure 13.1.

We use the sum of squared differences between the observed and the expected counts, i.e., $\text{SSE} = \sum_{j=1}^{L} (t_k - n_k \, E[H(\underset{\sim}{x}_i^T \underset{\sim}{\beta}) \, | \underset{\sim}{t}])^2$ as a summary index of model fit. The values of SSE for the two models MF and MG are listed in Table ??. In this table, we also list the summarized diagnostic indices $Q_1 = \sum {d'_{1k}}^2$ and $Q_2 = \prod d_{2k}$. Note that small values of SSE, Q_1 and large values of Q_2 are preferable. The

TABLE 13.1. Student retention data : SSE and summary model diagnostic measures

	Finite mixture model MF	General mixture model MG
SSE	2890.63	2769.44
$Q_1 = \sum_{k=1}^{48} d'^2_{1k}$	53.91	51.23
$Q_2 = \prod_{k=1}^{48} d_{2k}$	1.49×10^{-70}	8.84×10^{-70}

general mixture model MG does better in all the summary criteria. For each of the two models MF MG, we also plot the diagnostic criteria d'_{1k} and d_{2k} for each observation. These plots are shown in Figure 13.1. For each model, these diagnostic plots identify several observations which have large d'_{1k} and large negative $\log(d_{2k})$ (hence small d_{2k}) values, for example, the 2nd, 4th, 19th, 21st and 41st observations. These are the possible *influential* observations of the models and may require further analysis.

6. Discussion

In this article, we have extended the finite normal scale mixture link model of Basu and Mukhopadhyay (1998) to a Bayesian model of binary response regression which can use general normal scale mixtures as the link function H. This extension provides further flexibility to the user and at the same time, make the analysis more automated and adaptive to data. We describe a Gibbs sampling based method which can be used to efficiently calculate various posterior quantities of interest. Further, we provide several model diagnostic measures and show how they can be easily computed from the same Gibbs sampler run.

Note that the normal scale mixture links which we use (either in the finite mixture or the general mixture model) always produce symmetric links due to the symmetry of the normal distribution. Several authors have described data where asymmetric links are called for. One possible way to generate asymmetric links is by considering both location and scale mixtures of normals, as in Erkanli, Stangl, and Müller (1993). Basu and Mukhopadhyay (1998) provide another alternative asymmetric link structure by considering scale mixtures of truncated normal distributions. This asymmetric link structure has many attractive features but also brings in some computational hurdles. We refer the reader to Basu and Mukhopadhyay (1998) for further details.

References

Agresti, A. (1990). *Categorical Data Analysis.* John Wiley.

Albert, J.H. and Chib, S. (1993). Bayesian Analysis of Binary and Polychotomous Response Data. *J. Amer. Statist. Assoc.*, 88, 669-679.

Antoniak, C.E. (1974). Mixture of Dirichlet Processes with Applications to Bayesian Nonparametric problems. *Ann. Statist..* 2,1152-1174.

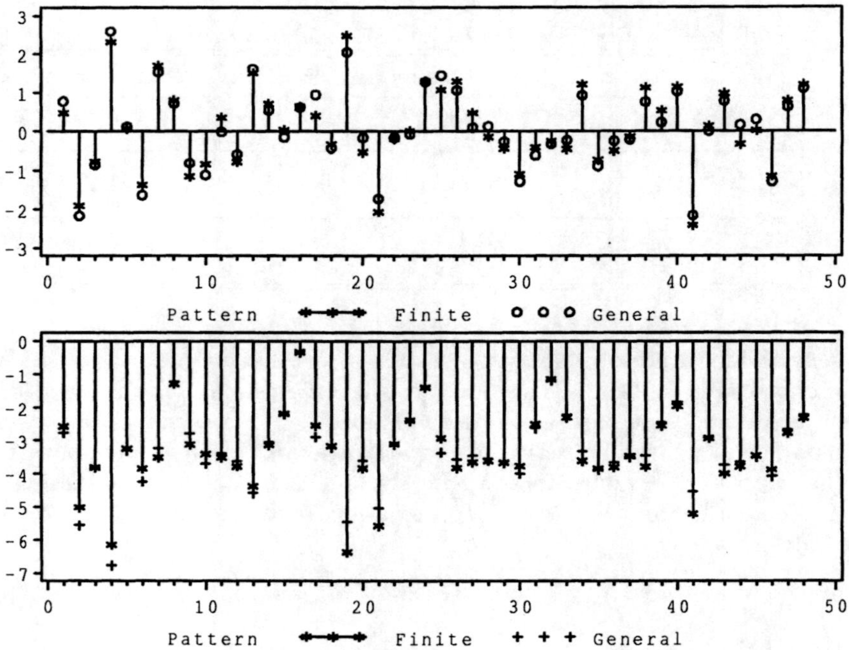

FIGURE 13.1. Model diagnostics of Student retention data for the MF and MG models. The top figure plots d'_{1k} against observation number (1–48). The bottom figure plots $\log(d_{2k})$ against observation number

Aranda-Ordaz F.J. (1981). On two families of transformations to additivity for Binary Response data. *Biometrika*, 68, 357-363.

Basu, S. (1996). Bayesian tests for unimodality. Tech. Rept., Univ. of Arkansas.

Basu, S. and Mukhopadhyay S. (1998). Bayesian analysis of binary regression using symmetric and asymmetric links. Manuscript.

Berger, J. (1985). *Statistical Decision Theory and Bayesian Analysis.* Springer-Verlag, New York.

Box, G. (1980). Sampling and Bayes' inference in scientific modeling and robustness (with discussion). *J. R. Statist. Soc., Ser. A*, 143, 382-430.

Casella, G. and George, E.I. (1992). Explaining the Gibbs Sampler. *Amer. Statistician.* 46, 167-174.

Daganpur, J. (1988). *Principles of Random Variate Generation.* Oxford University Press, Oxford.

Devroye, L. (1986). *Nonuniform Random variate Generation.* Springer-Verlag, New York.

Donnelly, S. and Line, K. (1992). *Retention and Graduation Report.* Division of Student Services, Univ. of Arkansas.

Doss, H. (1994). Bayesian Nonparametric Estimation for Incomplete Data via Successive Substitution Sampling. *Ann. Statist.*, 22, 1763-1786.

Escobar, M.D. (1994). Estimating Normal Means with a Dirichlet Process Prior. *J. Amer. Statist. Assoc.*, 89, 268-277.

Escobar, M.D. and West, M. (1995). Bayesian Density Estimation and Inference Using Mixtures. *J. Amer. Statist. Assoc.*, 90, 577-588.

Erkanli, A., Stangl, D., and Müller, P. (1993). A Bayesian analysis of ordinal data using mixtures. (*to appear in Canad. J. Statist.*).

Ferguson, T. (1974). Prior Distributions on Spaces of Probability Measures. *Ann. Statist.*, 2, 615-629.

Geisser, S. and Eddy, W. (1979). A predictive approach to model selection, *J. Amer. Statist. Assoc.*. 74, 153-160.

Gelfand, A.E., Dey, D.K., and Chang, H. (1992). Model determination using predictive distributions with implementations via sampling-based methods. In *Bayesian Statistics 4*, J.M. Bernardo, et. al. (Eds.), Oxford University Press, Oxford.

Gelfand, A.E. and Smith, A.F.M. (1990). Sampling-Based Approaches to Calculating Marginal Distributions. *J. Amer. Statist. Assoc.*, 85, 398-409.

Gelman, A. and Rubin, D. (1992). Inference from iterative simulation using multiple sequences. *Statist. Sci.*, 7, 457-476.

Gilks, W.R. , Richardson, S., and Spiegelhalter, D.J. (1996). *Markov Chain Monte Carlo in Practice.* Chapman and Hall, Glasgow.

Guerrero, V.M. and Johnson, R. (1982). Use of the Box-Cox transformation with Binary Response models. *Biometrika*, 69, 309-314.

MacEachern, S.N. (1994). Estimating normal means with a conjugate style Dirichlet process prior. *Comm. Statist.-Simula.*, 23, 727-741.

McCullagh, P. and Nelder, J. (1989). *Generalized Linear Models.* 2nd ed., Chapman and Hall.

Mukhopadhyay, S. and Gelfand, A. E. (1997). Dirichlet Process Mixed Generalized Linear Models. *J. Amer. Statist. Assoc.*, 92, 633-639.

Newton, M.A., Czado, C., and Chappell, R. (1996). Semiparametric Bayesian inference for binary regression. *J. Amer. Statist. Assoc.*, 91, 142-153.

Pilz, J. (1991). *Bayesian Estimation and Experimental Design in Linear Regression Models.* John Wiley.

Prentice, R.L. (1976). A Generalization of the Probit and Logit models for Dose Response Curves. *Biometrics*, 32, 761-768.

Stukel, T.A. (1988) Generalized Logistic Models. *J. Amer. Statist. Assoc.*. 83, 426-431.

Tanner, T.A. and Wong, D.H. (1987). The Calculation of Posterior Distributions by Data Augmentations. *J. Amer. Statist. Assoc.*, 82, 528-549.

Tierney, L. (1994). Markov chains for exploring posterior distributions (wtih discussions). *Ann. Statist.*, 22, 1701-1762.

14

Binary Regression Using Data Adaptive Robust Link Functions

Rubén A. Haro-López
Bani K. Mallick
Adrian F. M. Smith

ABSTRACT We present binary regression models that choose an arbitrary link function from a set of different inverse cumulative functions produced by a particular family of parametric distributions. This 'data adaptive' Bayesian analysis is implemented using cumulative functions of scale mixtures of normal distributions. The scale mixtures of normal distributions are symmetrical families of distributions that provide more parameters than the usual location and scale parameters that characterize the normal distribution. 'Data adaptive' Bayesian analysis of the binary regression model is performed using the shape parameter of the exponential power distribution. This methodology is implemented applying Markov chain Monte Carlo (MCMC) simulation methods.

1. Introduction

Binary response data, measured with covariates, are often modeled by assuming that the probability of a positive response, after a suitable transformation, is linear in the covariates. This transformation, usually known as link function, connects the probability to a linear predictor and is normally assumed to be a known function, for instance the inverse cumulative function of a certain distribution. McCullagh and Nelder (1989) provides a comprehensive discussion of the use of link functions to model the relation between a binary response and linear covariates.

The most commonly used binary regression models involve logit or probit link functions. The probit is obtained when the normal cumulative function is utilized and the logit when the logistic cumulative function is used. Bayesian analysis of the logit regression model using the Gibbs sampler is described in Dellaportas *et al.* (1993) and Albert and Chib (1993) extended the Bayesian analysis of the probit regression model to models with link functions defined through the scale mixtures of normal distributions. The latter, introduced in Andrews and Mallows (1974), are families of symmetrical distributions that provide more parameters than the usual location and scale parameters that characterize the normal distribution. The exponential power distribution is an example of a family of scale mixtures of normal distributions, where the extra parameter controls the distributional shape.

In this article, we present binary regression models that choose an arbitrary link function from a set of different inverse cumulative functions, the latter produced by a particular family of scale mixtures of normal distributions that possess a shape parameter. This 'data adaptive' Bayesian analysis is a method of obtaining automatic robustness with respect to the link function choice. This methodology does not constrain the statistical analysis to a fixed link function just as in Dellaportas *et al.* (1993) and Albert and Chib (1993).

'Data adaptive' Bayesian analysis of the binary regression model is implemented using the Gibbs sampler. As in Albert and Chib (1995), we define Bayesian residuals which possess a continuous-valued posterior distribution and use these to study potential outlying observations. Analysis of fitted posterior probabilities is used to compare different adjusted models. Finally, this methodology is illustrated with the analysis of the low birth weight in infants dataset given in Hosmer and Lemeshow (1989).

2. The Binary Regression Model

The 'data adaptive' Bayesian analysis for the binary regression model after observing the set of n independent binary (0 or 1) observations $\boldsymbol{w} = (w_1, \ldots, w_n)^T$ associated with k-dimensional row vectors of regressors $x_i = (x_{i1}, \ldots, x_{ik})$, $i = 1, \ldots, n$, is obtained as in Albert and Chib (1993) by constructing random variables y_i such that

$$w_i = \begin{cases} 1 & y_i > 0 \\ 0 & y_i \leq 0 \end{cases},$$

for $i = 1, \ldots, n$, with

$$y_i|\lambda_i \sim N_1\left(y_i\,|x_i\boldsymbol{\beta}, h(\lambda_i)\right) \qquad \text{and} \qquad \lambda_i \sim \Pi(\lambda_i|\alpha); \tag{1}$$

i.e. y_i is distributed as a scale mixture of normal distributions, where $N_k(x|\mu, \Sigma)$ denotes the k-dimensional normal distribution with probability density function

$$f_{N_k}(x|\mu, \Sigma) = \frac{|\Sigma|^{-1/2}}{(2\pi)^{k/2}} \exp\left(-\frac{1}{2}(x-\mu)^T\Sigma^{-1}(x-\mu)\right),$$

$\Pi(\lambda_i|\alpha)$ is the scale mixture parameter distribution for $\lambda_i \in \mathbb{R}^+$, $\alpha \in (a, b)$ is the shape parameter, $\boldsymbol{\beta} = (\beta_1, \ldots, \beta_k)^T \in \mathbb{R}^k$ is the column vector of unknown regression coefficients and $h(\cdot)$ is a positive function.

The random quantities y_i will produce the link function with the linear predictors and will imply the following posterior distribution

$$\pi(\boldsymbol{\beta}, \alpha|\boldsymbol{w}) \propto \prod_{i=1}^{n} \int_{A_i} \int_0^{\infty} f_{N_1}\left(y_i\,|x_i\boldsymbol{\beta}, h(\lambda_i)\right)\, \pi(\lambda_i|\alpha)\, dy_i\, d\lambda_i\, \pi(\boldsymbol{\beta}, \alpha), \tag{2}$$

where there is a linear relationship between y_i and x_i, and

$$A_i = \begin{cases} (-\infty, 0] & \text{if } w_i = 0 \\ (0, \infty) & \text{if } w_i = 1 \end{cases},$$

for $i = 1, \ldots, n$. The prior distribution is specified as

$$\pi(\boldsymbol{\beta}, \alpha) = \pi_{N_k}(\boldsymbol{\beta}|\boldsymbol{\nu}, \boldsymbol{C})\, \pi_U(\alpha|a, b),$$

with known hyperparameters $\boldsymbol{\nu} \in \mathbb{R}^k$ and $\boldsymbol{C} \in \mathcal{M}(k)$, where $\mathcal{M}(k)$ denotes the set of all positive definite $k \times k$ matrices and $U(\alpha|a, b)$ denotes the uniform distribution in (a, b).

Remark 1 To eliminate sources of confounding, the scale mixture of normal distributions utilized in the model (1) should have known location and scale parameters. Otherwise, the location can be confounded with the intercept term β_1 and the scale can be confounded with the overall scale of the regression coefficients. It is therefore natural to restrict this family of link functions to a fixed interquartile range or a smooth (bounded) variance. See Newton *et al.* for a related discussion.

Defining $U^{-1} = \sum_{i=1}^{n} h(\lambda_i)^{-1} x_i^T x_i + C^{-1}$, $\lambda = (\lambda_1, \ldots, \lambda_n)^T$ and $\boldsymbol{y} = (y_1, \ldots, y_n)^T$, analysis of equation (2) can be performed applying the Gibbs sampler simulation technique to obtain a sample from the marginal distribution $\pi(\boldsymbol{\beta}, \alpha|\boldsymbol{w})$ of $\pi(\boldsymbol{\beta}, \alpha, \boldsymbol{\lambda}, \boldsymbol{y}|\boldsymbol{w})$; see, for example, Smith and Roberts (1993). The Gibbs sampler structure is given by the following full conditional densities,

$$\begin{aligned} f(\boldsymbol{\beta}|\boldsymbol{w}, \boldsymbol{\lambda}, \boldsymbol{y}, \alpha) &= f_{N_k}\left(\boldsymbol{\beta} \middle| \boldsymbol{U}\left(\sum_{i=1}^{n} \frac{y_i}{h(\lambda_i)} x_i^T + \boldsymbol{C}^{-1}\boldsymbol{\nu}\right), \boldsymbol{U}\right), \\ f(y_i|\boldsymbol{w}, \boldsymbol{\beta}, \boldsymbol{\lambda}, \alpha, y_{j,i\neq j}) &\propto f_{N_1}(y_i|x_i\boldsymbol{\beta}, h(\lambda_i))\left(1\left(\begin{smallmatrix} y_i > 0 \\ \boldsymbol{w_i = 1} \end{smallmatrix}\right) + 1\left(\begin{smallmatrix} y_i \leq 0 \\ \boldsymbol{w_i = 0} \end{smallmatrix}\right)\right), \\ f(\alpha|\boldsymbol{w}, \boldsymbol{\beta}, \boldsymbol{\lambda}, \boldsymbol{y}) &\propto \pi_U(\alpha|a, b) \prod_{i=1}^{n} \pi(\lambda_i|\alpha)\, 1_{\alpha \in (a,b)} \end{aligned}$$

and

$$f(\lambda_i|\boldsymbol{w}, \boldsymbol{\beta}, \boldsymbol{y}, \alpha, \lambda_{j,j\neq i}) \propto \frac{1}{h(\lambda_i)^{1/2}} \exp\left(-\frac{\theta_{y_i}}{2\, h(\lambda_i)}\right) \pi(\lambda_i|\alpha)\, 1_{\lambda_i \in \mathbb{R}^+}, \tag{3}$$

for $i = 1, \ldots, n$, where $\boldsymbol{\theta_{y_i} = (y_i - x_i\beta)^2}$ and

$$1_{x\in A} = \begin{cases} 1 & x \in A \\ 0 & x \notin A \end{cases}.$$

To sample from a truncated normal distribution, we use the inversion method discussed in Devroye (1986) with the normal distribution quantiles approximation in Derenzo (1977). The last two full conditional distributions will be determined in the next subsections for two different binary regression models.

Remark 2 When working with symmetric and parametric families of link functions, it is convenient to construct link functions with a family of distributions that has a smooth (bounded) variance function. The exponential power distribution defined below has this property.

Normal Distribution

The probit regression model is obtained using the normal distribution, $N_1(y_i| \boldsymbol{x_i\beta, 1})$, which can be expressed as a scale mixture of normal distributions by setting $h(\lambda_i)$=1 and $\pi(\{1\}|\alpha) = 1$ as the scale mixture parameter distribution. We know that there is no shape parameter in this family of distributions. Therefore, it will not provide a "data adaptive" Bayesian analysis. The Gibbs sampler structure outlined in (3) uses only the two first conditional densities.

Exponential Power Distribution

The exponential power binary regression model is obtained using the exponential power distribution, $E_1(y_i|x_i\beta, 1, \alpha)$, with probability density function

$$f_{EP_1}(x|\mu,\sigma^2,\alpha) = \frac{\alpha\sqrt{c_\alpha}}{\sqrt{2}\Gamma\left(\frac{1}{2\alpha}\right)\sigma}\exp\left(-\left(\frac{c_\alpha(x-\mu)^2}{2\sigma^2}\right)^\alpha\right),$$

and variance function

$$\text{var}(x) = \exp(2(1-\alpha)\log 4)\sigma^2, \tag{4}$$

where

$$c_\alpha = \frac{2\,\Gamma\left(\frac{3}{2\alpha}\right)}{\Gamma\left(\frac{1}{2\alpha}\right)}\exp\left(2(\alpha-1)\log 4\right).$$

Scale mixture of normal distributions can be obtained by setting $h(\lambda_i) = 1/\lambda_i$ and

$$\pi(\lambda_i|\alpha) = \frac{\alpha\sqrt{\pi}}{\Gamma\left(\frac{1}{2\alpha}\right)c_\alpha}\left(\frac{c_\alpha}{\lambda_i}\right)^{1/2} f_S\left(\frac{\lambda_i}{c_\alpha}\middle|\,\alpha,1\right), \tag{5}$$

where $S(x|\alpha,1)$ denotes the standardized positive stable distribution with characteristic exponent $\alpha \in (0,1)$; see West (1987) for the proof. We note that the shape parameter α is the device that enables us to obtain a "data adaptive" Bayesian analysis.

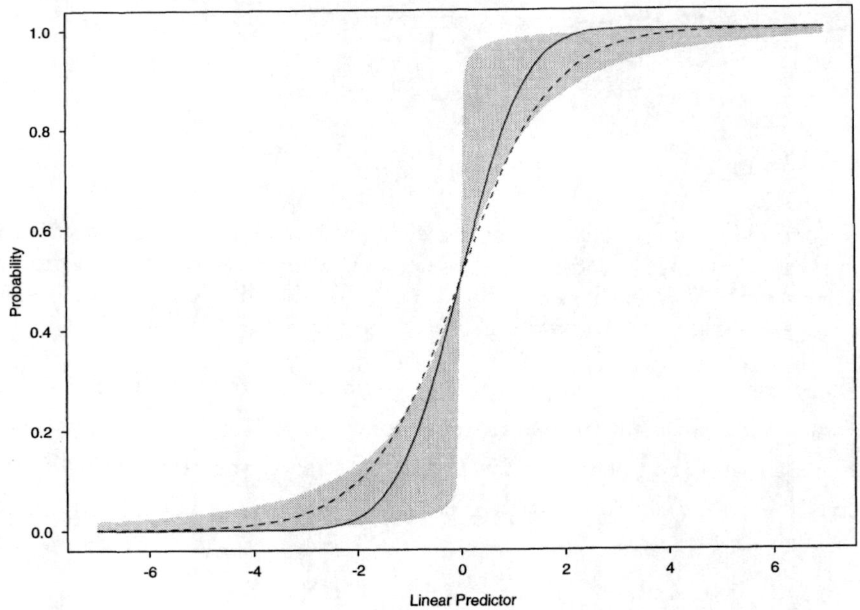

FIGURE 14.1. Set that contains all the prior link functions generated from the exponential power distribution with variance given in equation (4). This set is compared with the probit link function and the logistic link function standardized to match the interquartile range of the Laplace (exponential power with $\alpha = 1/2$) link function.

Figure 14.1 displays the set that contains all prior link functions produced by the family of exponential power distributions with variance given in equation (4).

Applying to equation (5) the standardized positive stable distribution representation given by Ibragimov and Chernin (1959),

$$f_S(\lambda|\alpha,1) = \frac{a_\alpha}{\lambda^{a_\alpha+1}} \int_0^1 t_\alpha(s) \exp\left(-\frac{t_\alpha(s)}{\lambda^{a_\alpha}}\right) ds,$$

for $\alpha, s \in (0,1)$, where

$$t_\alpha(s) = \left(\frac{\sin(\alpha\,\pi\,s)}{\sin(\pi\,s)}\right)^{a_\alpha} \frac{\sin((1-\alpha)\,\pi\,s)}{\sin(\pi\,s)} \quad \text{and} \quad a_\alpha = \frac{\alpha}{1-\alpha};$$

we can see that λ_i and s_i in our problem follow a bivariate distribution with probability density function

$$\pi(\lambda_i, s_i|\alpha) =$$
$$\frac{\sqrt{\pi}\kappa_\alpha}{c_\alpha^{3/2}}\, t_\alpha(s_i) \left(\frac{c_\alpha}{\lambda_i}\right)^{a_\alpha+3/2} \exp\left(-t_\alpha(s_i)\left(\frac{c_\alpha}{\lambda_i}\right)^{a_\alpha}\right) 1_{s_i\in(0,1)}\, 1_{\lambda_i\in\mathbb{R}^+},$$

for $i = 1, \ldots, n$, where

$$\kappa_\alpha = \frac{\alpha^2\sqrt{c_\alpha}}{\Gamma\left(\frac{1}{2\alpha}\right)(1-\alpha)}.$$

To complete the full conditional distributions for the exponential power binary regression model outlined in (3), we transform λ_i so that $\psi_i = \log\lambda_i$ and use instead the following full conditional density

$$f(\psi_i|\boldsymbol{w}, \boldsymbol{\beta}, \boldsymbol{y}, \alpha, \boldsymbol{s}, \lambda_{j,j\neq i}) \propto \exp\left(-a_\alpha\psi_i - \frac{\theta_{y_i}e^{\psi_i}}{2} - \frac{t_\alpha(s_i)\,c_\alpha^{a_\alpha}}{e^{a_\alpha\psi_i}}\right),$$

for $i = 1, \ldots, n$. This full conditional distribution is log-concave, so we can sample from it using the adaptive rejection sampling method for log-concave distributions given in Gilks and Wild (1992). The value for λ_i^* is obtained via the inverse transformation $\lambda_i^* = \exp(\psi_i^*)$.

The full conditional density of s_i,

$$p(s_i|\boldsymbol{w}, \boldsymbol{\beta}, \boldsymbol{\lambda}, \boldsymbol{y}, \alpha, s_{j,j\neq i}) \propto t_\alpha(s_i)\left(\frac{c_\alpha}{\lambda_i}\right)^{a_\alpha} \exp\left(-t_\alpha(s_i)\left(\frac{c_\alpha}{\lambda_i}\right)^{a_\alpha} + 1\right),$$

for $i = 1, \ldots, n$, has a unique maximum at $t_\alpha(s_i) = \max((\lambda_i/c_\alpha)^{a_\alpha}, \alpha^{a_\alpha}(1-\alpha))$. This can be confirmed by observing that $t_\alpha(s)$ is a monotonic function for $s \in (0,1)$ with $\lim_{s\to 0} t_\alpha(s) = \alpha^{a_\alpha}(1-\alpha)$ and $\lim_{s\to 1} t_\alpha(s) = \infty$; see Haro-López *et al.* for details. The knowledge of this maximum makes the adaptive histogram rejection sampling algorithm a suitable candidate to sample from this full conditional distribution (see appendix).

The full conditional distribution of α has the following density

$$p(\alpha|\boldsymbol{w}, \boldsymbol{\beta}, \boldsymbol{\lambda}, \boldsymbol{y}, \boldsymbol{s}) \propto \prod_{i=1}^{n} \kappa_\alpha\, t_\alpha(s_i)\left(\frac{c_\alpha}{\lambda_i}\right)^{a_\alpha} \exp\left(-\sum_{i=1}^{n} t_\alpha(s_i)\left(\frac{c_\alpha}{\lambda_i}\right)^{a_\alpha} + n\right).$$

To sample from this full conditional distribution, we implement the adaptive rejection Metropolis sampling algorithm in Gilks *et al.* (1995), Gilks *et al.* (1997.

Computation of the maximum of this density is crucial for obtaining good candidate points to construct the initial adaptive rejection envelope and this is pursued using the bisection root finder method. In practice for $\alpha < 0.9$, the full conditional distribution of α is nearly log-concave. Therefore, it is convenient to use the algorithm in Gilks *et al.* (1995) because it reduces to the algorithm in Gilks and Wild (1992) when we are sampling from a log-concave distribution.

3. Detection of Outliers and Model Comparison

Once "within-model" inference is completed we can use statistical analysis for the detection of extreme observations. We can find outliers in the binary regression model as in Albert and Chib (1995) by using the following 'latent' Bayesian residual for the unobserved y_i,

$$r_i = \frac{(y_i - x_i\boldsymbol{\beta})^2}{h(\lambda_i)},$$

which possesses a continuous-valued posterior distribution.

In Chaloner and Brant (1988) an outlier is defined as an observation with a large random error, generated by the linear regression model under consideration. This paper establishes that outliers can be detected by examining the posterior distribution of the random errors.

In the context of scale mixtures of normal distributions, the conditional distribution of the residual r_i given $\boldsymbol{\beta}$ and λ_i is χ_1^2, a chi-square distribution with one degree of freedom. For detecting outlying observations, this conditional distribution can be used as the basis with which to compare the posterior distribution of r_i.

For scale mixtures of normal distributions the knowledge of the conditional distribution of w_i given $\boldsymbol{\beta}$ and λ_i can be used to calculate the posterior probabilities fitted by the model. Then, the Rao-Blackwellized estimate, introduced in Gelfand and Smith (1990), of the posterior probability of obtaining the w_i response on the ith observation is given by

$$f(w_i|\boldsymbol{w}_{(-i)}) \approx \frac{1}{R}\sum_{r=1}^{R} \Phi\left(\frac{x_i\boldsymbol{\beta}^{(r)}}{h\left(\lambda_i^{(r)}\right)^{1/2}}\right)^{A_i} \Phi\left(-\frac{x_i\boldsymbol{\beta}^{(r)}}{h\left(\lambda_i^{(r)}\right)^{1/2}}\right)^{1-A_i}, \qquad (6)$$

$i = 1, \ldots, n$, where $\Phi(\cdot)$ is the standard normal cumulative function, $\boldsymbol{w}_{(-i)} = (w_1 \ldots, w_{i-1}, w_{i+1}, \ldots, w_n)$ and $\{\lambda_i^{(r)}\}$ and $\{\boldsymbol{\beta}^{(r)}\}$, $r = 1, \ldots, R$, are posterior distribution samples. Analysis of these fitted posterior probabilities can be used to compare different adjusted models.

4. Numerical Illustration

To illustrate our methodology, we use the low birth weight in infants dataset given in Hosmer and Lemeshow (1989). In this dataset, the binary outcome (w_i) is whether the infant birth weight is less than 2.5 kg. in births from $n = 189$ women. The covariates in x_i are a constant, the age of the mother, the weight of the mother (lbs.) at last menstrual period, race (white, black or other), smoking status during pregnancy (0 or 1), previous premature labors (0 or 1), history of hypertension (0 or 1), presence of uterine irritability (0 or 1) and number of physician visits in the first trimester (0, 1 or 2+).

Running four different Gibbs sampler chains and using the following values for the hyperparameters, $\boldsymbol{\nu} = (0, \ldots, 0)^T$ and with the choice

$$\boldsymbol{C}^{-1} = \begin{pmatrix} 0.001 & 0 & \ldots & 0 \\ 0 & 0.001 & \ldots & 0 \\ \vdots & & \ddots & \\ 0 & 0 & \ldots & 0.001 \end{pmatrix},$$

we fitted normal and exponential power binary regression models with variance as in equation (4). The starting value of α in each chain was randomly chosen.

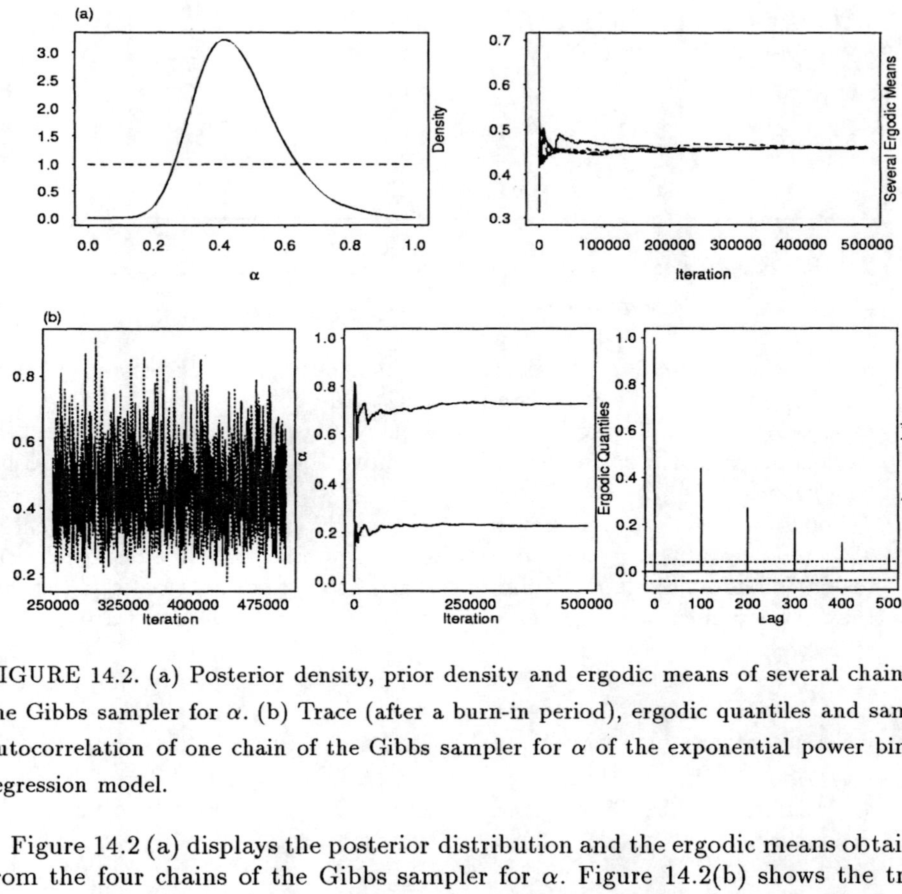

FIGURE 14.2. (a) Posterior density, prior density and ergodic means of several chains of the Gibbs sampler for α. (b) Trace (after a burn-in period), ergodic quantiles and sample autocorrelation of one chain of the Gibbs sampler for α of the exponential power binary regression model.

Figure 14.2 (a) displays the posterior distribution and the ergodic means obtained from the four chains of the Gibbs sampler for α. Figure 14.2(b) shows the trace (after a burn-in period), ergodic quantiles and posterior sample autocorrelation of one Gibbs sampler chain for α. Observing that convergence is attained around the $50,000^{\text{th}}$ and $250,000^{\text{th}}$ iterations, with low autocorrelation reached at every 10th and 100th iterations for the normal and the exponential power models respectively, we used this information to evaluate the posterior means and posterior standard errors of α and β which are summarized in Table 14.1.

Detection of extreme observations is carried out by using 'latent' Bayesian residuals r_i. For normal and exponential power binary regression models, and for each

Variable	Exponential Power Model Post. Mean	Post. Stand. Error	Normal Model Post. Mean	Post. Stand. Error
α	0.4590	0.1235	–	–
Constant	1.1265	1.4489	0.5096	0.7237
Age	-0.0317	0.0443	-0.0244	0.0228
Mother Weight	-0.0200	0.0089	-0.0094	0.0040
Race Black	1.2990	0.5924	0.7309	0.3209
Race Other	0.8201	0.5255	0.4583	0.2699
Smoke	0.7515	0.4807	0.4786	0.2481
Premature	1.5074	0.5698	0.8420	0.2920
Hypertension	2.2382	0.8580	1.1737	0.4321
Irritability	0.6366	0.5310	0.4246	0.2804
One Visit	-0.4978	0.5547	-0.2928	0.2806
More Visits	0.2084	0.5173	0.0889	0.2698

TABLE 14.1. Posterior mean and standard errors of α and β for exponential power and normal binary regression models.

of the 189 women, Fig 14.3displays the box plots of the Bayesian residual posterior distributions. These are compared with the 50% and 95% quantiles of the χ_1^2 conditional distribution of r_i. An extreme observation will have a Bayesian residual posterior distribution substantially above the 95% quantile of its Chi-square conditional distribution.

Looking at the normal Bayesian residuals r_i, it is clear that woman 155 and 183 are likely to be extreme observations. Comparing both models by the Bayesian residuals and applying the definition of an outlying observation in [?], we conclude that the exponential power fits better the data than the normal binary (probit) regression model.

Figure 14.3 also shows for exponential power and normal binary regression models, the Rao-Blackwellized estimates of the posterior probability of obtaining the w_i response on the ith individual. They are calculated using equation (6) with posterior distribution samples $\{\lambda_i^{(r)}\}$ and $\{\beta^{(r)}\}$, $r = 1, \ldots, R = 8000$. Comparing the posterior probabilities given by each model, we definitely conclude that the exponential power model adjusts more satisfactorily to the data than the normal model.

5. Discussion

In Haro-López and Smith (1996) it is noted that the Laplace distribution, included in the exponential power family of distributions, has a similar tail behavior to that of the logistic distribution. Therefore, it may be attractive to think of using the

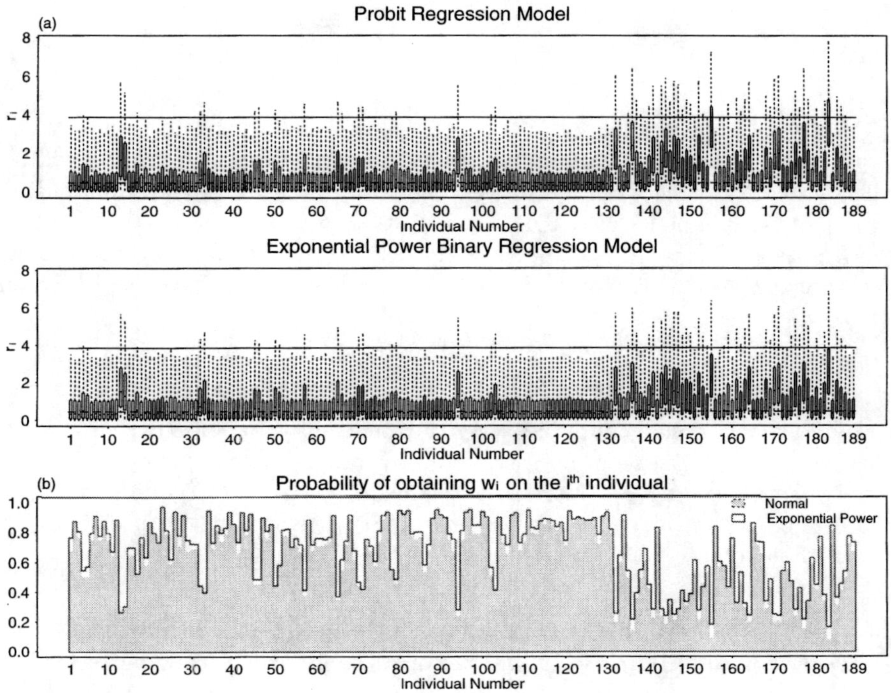

FIGURE 14.3. (a) Box plots of the Bayesian residual posterior distributions for normal and exponential power binary regression models. They are compared with the 95% quantiles of the χ_1^2. (b) Posterior estimate of the probability of obtaining the w_i response under normal (shaded area) and exponential power binary regression models for the ith woman.

exponential power family of link functions' routinely in any Bayesian procedure where logit binary regression models might conventionally be used.

Our methodology permitted the data studied previously to find the necessary degree of robustness by choosing a link function in the neighbourhood of the Laplace distribution from the exponential power link functions family. Without coercing the analysis to a specific model our model was consistent with the logit regression model applied in Venables and Ripley (1994). The numerical example pointed out that the study of the Bayesian residuals and the analysis of fitted posterior probabilities are useful and informative for choosing among different models.

Acknowledgments

The first author gratefully acknowledge the financial support from the National Council of Science and Technology of Mexico (CONACYT) and the Overseas Research Student Awards Scheme of the Committee of Vice-Chancellors and Principals (CVCP), United Kingdom.

Appendix

Adaptive Histogram Rejection Sampling

This rejection algorithm, motivated by the work done in [?], is designed to sample from the following type of densities. Let $f(\theta)$ be a unimodal probability density function with kernel $h(\theta)$ and $\theta \in (a, b)$. Let θ^M be the position where the supremum of $h(\theta)$ is attained and $S_n = \{\theta_i; i = 0, \ldots, n+1\}$ denote the current set of abscissae in ascending order, where $\theta_0 = a$ and $\theta_{n+1} = b$. Define a piecewise linear function,

$$g_n(\theta) = \begin{cases} h(\theta_i) & \theta_{i-1} < \theta < \theta_i \text{ for } \theta_i \leq \theta^M \\ h(\theta_i) & \theta_i < \theta < \theta_{i+1} \text{ for } \theta_i \geq \theta^M \end{cases}, \qquad i = 1, \ldots, n,$$

where we notationally suppress the dependence of $g_n(\theta)$ on S_n. If $\theta_j = \theta^M$ is in S_n for some $j \in (1, \ldots, n)$, then $g_n(\theta)$ is an adaptive envelope for $h(\theta)$.

To sample from $f(\theta)$, initialize abscissae S_n with $\theta_1 = \theta^M$ and carry on with the following steps using the sampling probability density function $g(\theta) \propto g_n(\theta)$ until a point θ is accepted.

- Sample θ from $g(\theta)$ and $u \sim U(u|0, 1)$.
- Evaluate $h(\theta)$ and $g(\theta)$, and perform the following rejection test: if

  $$u \leq \frac{h(\theta)}{g(\theta)},$$

  then accept θ. Otherwise reject θ and set $S_{n+1} = S_n \cup \{\theta\}$ with relabeled points in ascending order. Increment n and update $g(\theta)$.

By continually adding rejected points to S_n, the adaptive rejection envelope $g(\theta)$ will gradually converge to the true probability density $f(\theta)$ and efficiency of sampling will increase.

References

Albert, J.H. and Chib, S. (1993). Bayesian analysis for binary and polychotomous response data. *Journal of the American Statistical Association*, **88**, 669-679.

Albert, J.H. and Chib, S. (1995). Bayesian residual analysis for binary response regression models. *Biometrika*, **82**:746–759.

Andrews, D.F. and Mallows, C.L. (1974). Scale mixtures of normal distributions. *Journal of the Royal Statistical Society* Ser. B, **36**, 99-102.

Buckle, D.J. (1995). Bayesian inference for stable distributions. *Journal of the American Statistical Association*, **90**, 605-613, 1995.

Chaloner, K. and Brant, R. (1988). A Bayesian approach to outlier detection and residual analysis. *Biometrika*, **75**, 651-659.

Derenzo, S.E. (1977) Approximations for hand calculators using small integer coefficients. *Mathematics of Computation*, **31**, 214-222.

Devroye, L. (1986). *Non-Uniform Random Variate Generation.* Springer-Verlag, New York, 1986.

Dellaportas, P. and Smith, A.F.M (1993). Bayesian inference for generalized linear and proportional hazards models via Gibbs sampling. *Applied Statistics*, **42**, 443–459.

Gelfand, A.E. and Smith, A.F.M. (1990). Sampling-based approaches to calculating marginal densities. *Journal of the American Statistical Association*, **85**, 398-409.

Gilks, W.R., Best, N.G. and Tan, K.K.C. (1995). Adaptive rejection Metropolis sampling within Gibbs sampling. *Applied Statistics*, **44**, 455-472.

Gilks, W.R., Neal, R.M., Best, N.G. and Tan, K.K.C. (1997). Corrigendum: Adaptive rejection Metropolis sampling. *Applied Statistics*, **46**, 541-542.

Gilks, W.R. and Wild, P. (1992). Adaptive rejection sampling for Gibbs sampling. *Applied Statistics*, **41**, 337-348.

Haro-López, R.A., Mallick, B.K. and Smith, A.F.M (1998). Data adaptive Bayesian analysis using scale mixtures of normal distributions. *Technical Report.* Imperial College, University of London.

Haro-López, R.A. and Smith, A.F.M (1996). On robust Bayesian analysis for the location and scale parameters. Under revision for the *Journal of Multivariate Analysis.*

Hosmer, D.W. and Lemeshow, S. (1989). *Applied Logistic Regression.* Wiley, New York.

Ibragimov, I.A. and Chernin, K.E. (1959). On the unimodality of stable laws. *Theory of Probability and its Applications*, **4**, 417-419.

McCullagh, P. and Nelder, J.A. (1989). *Generalized Linear Models.* Chapman and Hall, New York.

Newton, M.A., Czado, C. and Chappell, R. (1996). Bayesian inference for semiparametric binary regression. *Journal of the American Statistical Association*, **91**, 142-153.

Smith, A.F.M. and Roberts, G.O. (1993). Bayesian computation via the Gibbs sampler and related Markov chain Monte Carlo methods. *Journal of the Royal Statistical Society* Ser. B, **55**, 3-23.

Venables, W.N. and Ripley, B.D. (1994). *Modern Applied Statistics with S-Plus.* Springer-Verlag, New York.

West, M. (1987). On scale mixtures of normal distributions. *Biometrika*, **74**, 646-648.

15

A Mixture-Model Approach to the Analysis of Survival Data

Lynn Kuo
Fengchun Peng

ABSTRACT We study a mixture model for survival data where covariates may influence both the incidence probabilities and their conditional latency distributions. The data may include the exactly observed, the right-censored, and the interval-censored failure times. We apply the EM algorithm to find the maximum likelihood estimate. We also carry out a Markov chain Monte Carlo algorithm for Bayesian inference. Model selection methods based on the predictive density for cross-validated data are developed. These methods allow us to assess whether simpler models would suffice as opposed to the mixture models. The potential of the methods is illustrated with the flour beetle data given by Hewlett (1974).

1. Introduction

In some analyses of failure-time data, it has been observed that a group of subjects may not react to treatments. A mixture model that incorporates different latency distributions for different groups seems to be appropriate. In this chapter we consider a mixture model where covariates may influence both the incidence probabilities and the conditional latency distributions. Let $\mathbf{x}$ denote the covariate vector $(1 \times q)$ associated with a subject of lifetime T. Then the mixture model assumes that the density of T is

$$f(t|\mathbf{x},\boldsymbol{\theta}) = \sum_{j=1}^{J} p_j(\mathbf{x},\boldsymbol{\rho}) f_j(t|\mathbf{x},\boldsymbol{\phi}_j), \tag{1}$$

where $\sum_{j=1}^{J} p_j(\mathbf{x},\boldsymbol{\rho}) = 1$. Let Y be an index variable for the subpopulations. We use $p_j(\mathbf{x},\boldsymbol{\rho})$ to denote the mixing probability, $P(Y = j|\mathbf{x},\boldsymbol{\rho})$, also called the incidence probability for the jth subpopulation. We use $f_j(t|\mathbf{x},\boldsymbol{\phi}_j)$ to denote $f_j(t|Y = j,\mathbf{x},\boldsymbol{\phi}_j)$, the conditional probability density (conditional latency density) function of the failure time for the jth subpopulation. We assume f_j is a continuous density and indexed by an unknown parameter $\boldsymbol{\phi}_j$ that can be a vector. Moreover, we use $\boldsymbol{\theta} = (\boldsymbol{\phi}_1,\ldots,\boldsymbol{\phi}_J,\boldsymbol{\rho})$ to denote the collection of all unknown parameters. The covariate $\mathbf{x}$ may include 1, the dosage (or log-dose) level, and other explanatory variables. In this study we only deal with time-independent explanatory variables. Logistic regression links can be chosen for the incidence probabilities. For example, we can choose

$$p_j(\mathbf{x}, \boldsymbol{\rho}) = \frac{e^{\xi_j}}{\sum_{l=1}^{J} e^{\xi_l}}, \tag{2}$$

where $\xi_j = \mathbf{x}\boldsymbol{\rho}_j^T$, with $\boldsymbol{\rho}_j^T$ to be the transpose of $\boldsymbol{\rho}_j$ $(1 \times q)$ and $\boldsymbol{\rho} = (\boldsymbol{\rho}_1, \ldots, \boldsymbol{\rho}_J)$. In addition to the logistic link, other models such as the normalized probit link or the normalized complementary log-log link (McCullagh and Nelder, 1989) can be considered for the incidence probabilities.

The cumulative distribution function for T is

$$F(t|\mathbf{x}, \boldsymbol{\theta}) = \sum_{j=1}^{J} p_j(\mathbf{x}, \boldsymbol{\rho}) F_j(t|\mathbf{x}, \boldsymbol{\phi}_j), \tag{3}$$

where F_j is the cumulative distribution function for f_j.

This mixture model has been studied by Farewell (1982), Larson and Dinse (1985), Pack and Morgan (1990), Boos and Brownie (1991), Kuk and Chen (1992), and Taylor (1995). The paper of Boos and Brownie considers the failure times to be independent and identically distributed (i.i.d.) as in (1). The rest of the papers include censored failure times. Pack and Morgan consider interval-censored and right-censored time-to-response data for quantal assay. Larson and Dinse and Kuk and Chen consider proportional hazards models for the conditional latency distributions. Taylor considers Kaplan-Meier type formulations for the conditional latency distributions.

In this chapter, we consider data that include both right-censored and interval-censored observations. Our methodologies can be applied to left-censored observations as well. We provide two methodologies for inference. One is an EM algorithm or a Monte Carlo EM algorithm (MCEM) to obtain the maximum likelihood estimate of $\boldsymbol{\theta}$. The other is a Markov chain Monte Carlo (MCMC) algorithm for Bayesian inference.

To apply the EM algorithm, we need to consider a complete likelihood of a product of components for i.i.d. observations as opposed to the mixture likelihood with censored observations. This is done by augmenting the original data with two classes of latent variables. One is the truncated random variable denoted by $\mathbf{w}$ that allows us to consider a likelihood for i.i.d. observations without censoring. The other is the index variable $\mathbf{z}$ that converts the mixture model to a model of independent components. Although the E-step with respect to the latent variables $\mathbf{w}$ is often difficult to evaluate, this is not the case for $\mathbf{z}$ when it is conditioned on $\mathbf{w}$. We employ a MCEM algorithm (Wei and Tanner, 1990) to approximate the E-step, where integration with respect to $\mathbf{w}$ is done by Monte Carlo integration and integration with respect to $\mathbf{z}$ given $\mathbf{w}$ is evaluated exactly. Then we apply a maximization step to update the parameters until convergence.

The MCMC algorithm consists of generating the two latent variables $\mathbf{w}$, $\mathbf{z}$ and the unknown parameter $\boldsymbol{\theta}$ iteratively. Let $\mathbf{t}$ denote the data set that contains censored observations. Starting at the initial choices of the variates, $\mathbf{w}^{(0)}$, $\mathbf{z}^{(0)}$, and $\boldsymbol{\theta}^{(0)}$, sample $\mathbf{w}^{(1)}$ from $f(\mathbf{w}|\mathbf{z}^{(0)}, \boldsymbol{\theta}^{(0)}, \mathbf{t})$; sample $\mathbf{z}^{(1)}$ from $f(\mathbf{z}|\mathbf{w}^{(1)}, \boldsymbol{\theta}^{(0)}, \mathbf{t})$; and sample $\boldsymbol{\theta}^{(1)}$ from $f(\boldsymbol{\theta}|\mathbf{w}^{(1)}, \mathbf{z}^{(1)}, \mathbf{t})$. Continue iteration until convergence. The f function here is a generic name for the conditional densities.

Both the MCEM and the MCMC algorithms use data augmentation with latent variables to simplify the computation. The techniques used in this paper are related to that in many other papers. We only mention a few here. Dempster, Laird, and Rubin (1977) introduced the EM algorithm for analysis with aggregated data or with missing data. McLachlan and Jones (1988) developed the EM algorithm for grouped and truncated data. Wei and Tanner (1990) proposed MCEM when the E-step is difficult to derive. Jordan and Jacobs (1994) showed how the parameters of the mixtures-of-experts (as well as the hierarchical mixtures-of-experts) architecture can be estimated using the EM algorithm. The MCMC algorithm was introduced by Geman and Geman (1984) for image restoration and was generalized by Tanner and Wong (1987) for missing value models. Gelfand and Smith (1990) applied it to Bayesian analysis. Diebolt and Robert (1994) developed a Gibbs sampler for i.i.d. observations for the usual mixture model, that is a simpler version of (1) without the covariates. Dey, Kuo, and Sahu (1995) studied a predictive approach to selecting the number of components in the mixture model. We are extending the two papers to censored data. The basic principle in the extension can be seen in Gelfand, Smith, and Lee (1992) and Kuo and Smith (1992). Peng, Jacobs, and Tanner (1996) present a hierarchical mixtures-of-experts architecture using both EM and a Gibbs sampler in a classification problem.

The rest of the paper is organized as follows. Section 2 describes the likelihood. Section 3 describes the EM and MCEM algorithms. Section 4 describes the MCMC algorithm. Model selection based on the predictive approach is discussed in Section 5. An illustration using the flour beetle data given by Hewlett (1974) is given in Section 6.

2. Likelihood

We assume that failure times are independently distributed, also independent from the censoring process. For each individual we observe the time to failure, a right-censored, or an interval-censored failure time. That is, for the right-censored individual we only know that the time to failure is greater than the censored time, while for the interval-censored individual, the time to failure is only known to be between two time points.

The likelihood is then (Cox and Oakes, 1984)

$$\begin{aligned} L(\boldsymbol{\theta}|\mathbf{t},\mathbf{X}) = &\prod_{i\in\mathcal{U}} f(t_i|\mathbf{x}_i,\boldsymbol{\theta}) \prod_{i\in\mathcal{C}} \{1 - F(t_i^+|\mathbf{x}_i,\boldsymbol{\theta})\} \\ &\times \prod_{i\in\mathcal{I}} \{F(t_{iU}|\mathbf{x}_i,\boldsymbol{\theta}) - F(t_{iL}|\mathbf{x}_i,\boldsymbol{\theta})\}, \end{aligned} \tag{1}$$

where $\mathbf{t}$ denotes the data set with the ith entry denoted by t_i if the ith subject is exactly observed, by t_i^+ if it is right-censored at t_i, and by (t_{iL}, t_{iU}) if the ith subject is interval-censored between t_{iL} and t_{iU}; the sets $\mathcal{U}$, $\mathcal{C}$ and $\mathcal{I}$ are the sets of indices for the exactly observed, the right-censored, and the interval-censored subjects, respectively. We assume there are N subjects in the study. Therefore, we use $\mathbf{X}$ to denote $(\mathbf{x}_1^T, \ldots, \mathbf{x}_N^T)$, the matrix of all covariates.

3. EM and Monte Carlo EM

In this section, we develop a Monte Carlo EM algorithm to maximize $\boldsymbol{\theta}$ in the likelihood. First observe that part of the likelihood in (1) for the censored data is a product of *integrals* of the mixture density in (1). This incomplete expression makes maximization difficult even for simple latency distributions such as normal, gamma, etc. For each subject i, $i = 1, \dots, N$, we consider the latent variable w_i that is truncated between (t_{iL}, t_{iU}) if it is interval-censored, truncated at t_i^+ if it is right-censored, and is t_i if it is exactly observed. Then the joint density of the data and the latent variables w's has i.i.d. components, each distributed as (1). This will alleviate the difficulties in maximization with incomplete likelihood as discussed earlier. More specifically, if the ith subject is interval-censored between (t_{iL}, t_{iU}), then we can generate a latent variable w_i from the truncated density $\{f(w_i)/[F(t_{iU}) - F(t_{iL})]\}\mathrm{I}(t_{iL} < w_i < t_{iU})$, where f is given in (1). This can be done by setting

$$w_i = F^{-1}\{F(t_{iL}|\mathbf{x}_i, \boldsymbol{\theta}) + U[F(t_{iU}|\mathbf{x}_i, \boldsymbol{\theta}) - F(t_{iL}|\mathbf{x}_i, \boldsymbol{\theta})]\}, \tag{1}$$

where $U \sim \text{uniform}(0, 1)$ and F^{-1} is the inverse function of F. Similarly, if the ith subject is right-censored at t_i^+, then we can generate a latent variable w_i from the truncated density $\{f(w_i)/[1 - F(t_i^+)]\}\mathrm{I}(t_i^+ < w_i < \infty)$ by setting

$$w_i = F^{-1}\{F(t_i^+|\mathbf{x}_i, \boldsymbol{\theta}) + U[1 - F(t_i^+|\mathbf{x}_i, \boldsymbol{\theta})]\}. \tag{2}$$

More details about Equations (1) and (2) are given by Devroye (1986, pp. 38-39). Note from (1) that w_i can be generated as the unique root of $F(w_i) = F(t_{iL}|\mathbf{x}_i, \boldsymbol{\theta}) + U[F(t_{iU}|\mathbf{x}_i, \boldsymbol{\theta}) - F(t_{iL}|\mathbf{x}_i, \boldsymbol{\theta})]$ and similarly for (2). They can be easily implemented using the IMSL subroutine "ZREAL" in FORTRAN. If the ith subject is exactly observed at t_i, then we set $w_i = t_i$.

For each i, we generate w_i independently as discussed above. Let $\mathbf{w}$ denote $(w_1, \dots, w_N)$. Then the likelihood of $\boldsymbol{\theta}$ given $\mathbf{w}, \mathbf{t}, \mathbf{X}$ is

$$L(\boldsymbol{\theta}|\mathbf{w}, \mathbf{t}, \mathbf{X}) = \prod_{i=1}^{N} f(w_i \mid \mathbf{x}_i, \boldsymbol{\theta}) = \prod_{i=1}^{N} \left(\sum_{j=1}^{J} p_j(\mathbf{x}_i, \boldsymbol{\rho}) f_j(w_i \mid \mathbf{x}_i, \boldsymbol{\phi}_j) \right). \tag{3}$$

Next, we make use of the fact that a mixture model can always be expressed by a product of its components. It is done by considering another latent variable $\mathbf{z}_i = (z_{i1}, \dots, z_{iJ})$, where $z_{ij} = 1$ if w_i is considered as coming from the subpopulation having pdf f_j and $\sum_{j=1}^{J} z_{ij} = 1$. The vector $\mathbf{z}_i$ follows a multinomial distribution, i.e.,

$$\mathbf{z}_i|\boldsymbol{\theta}, \mathbf{w}, \mathbf{t}, \mathbf{X} \sim Mult(1, h_{i1}, \dots, h_{iJ}), \tag{4}$$

with h_{ij}, the abbreviation for $h_{ij}(\boldsymbol{\theta}|w_i, \mathbf{t}, \mathbf{X})$, given by

$$h_{ij}(\boldsymbol{\theta}|w_i, \mathbf{t}, \mathbf{X}) = \frac{p_j(\mathbf{x}_i, \boldsymbol{\rho}) f_j(w_i|\mathbf{x}_i, \boldsymbol{\phi}_j)}{\sum_j p_j(\mathbf{x}_i, \boldsymbol{\rho}) f_j(w_i|\mathbf{x}_i, \boldsymbol{\phi}_j)}. \tag{5}$$

Let $\mathbf{z} = (\mathbf{z}_1, \dots, \mathbf{z}_N)$. Then the augmented likelihood for the completely imputed data $\mathbf{z}$ and $\mathbf{w}$ is

$$L(\boldsymbol{\theta}|\mathbf{z}, \mathbf{w}, \mathbf{t}, \mathbf{X}) = \prod_{i=1}^{N} \prod_{j=1}^{J} \{p_j(\mathbf{x}_i, \boldsymbol{\rho}) f_j(w_i|\mathbf{x}_i, \boldsymbol{\phi}_j)\}^{z_{ij}}. \tag{6}$$

The complexity of fitting the mixture model is greatly simplified by working with the augmented likelihood because we only need to evaluate $p_j(\mathbf{x}_i, \boldsymbol{\rho}) \times f_j(w_i|\mathbf{x}_i, \boldsymbol{\phi}_j)$ when $z_{ij} = 1$.

Suppose the current EM iteration is at the kth stage. The E-step in the E-M algorithm consists of evaluating

$$\begin{aligned} Q(\boldsymbol{\theta}, \boldsymbol{\theta}^{(k)}) &= \int \log[L(\boldsymbol{\theta}|\mathbf{z}, \mathbf{w}, \mathbf{t}, \mathbf{X})] f(\mathbf{z}|\mathbf{w}, \boldsymbol{\theta}^{(k)}, \mathbf{t}, \mathbf{X}) f(\mathbf{w}|\boldsymbol{\theta}^{(k)}, \mathbf{t}, \mathbf{X}) d\mathbf{z} d\mathbf{w} \\ &= \sum_{i=1}^{N} \sum_{j=1}^{J} E\{h_{ij}(\boldsymbol{\theta}^{(k)}|w_i, \mathbf{t}, \mathbf{X}) \log[p_j(\mathbf{x}_i, \boldsymbol{\rho}) f_j(w_i|\mathbf{x}_i, \boldsymbol{\phi}_j)]\} \end{aligned} \quad (7)$$

where the expectation on the right-hand side of the second equality is with respect to the truncated random variable $\mathbf{w}$. The M-step requires maximizing the above expression as a function of $\boldsymbol{\theta}$. The maximization for $\boldsymbol{\phi}_j$ and $\boldsymbol{\rho}$ can be done separately. For many standard probability distributions, statistical packages are available to handle the maximization. We continue this iteration using (7) until the two consecutive $\boldsymbol{\theta}$'s are close enough. Mixture models are usually multimodal; hence the EM solutions are not unique and depend on the starting values. Therefore exploration with different starting points is needed to obtain the global maximum.

Now we extend the EM to the MCEM algorithm. When the expectation is hard to evaluate with respect to the truncated random variables $\mathbf{w}$, we circumvent it by multiple imputation by generating M copies of $\mathbf{w}$, where the mth copy of $\mathbf{w}$ denoted by $\mathbf{w}^m$ equals $(w_1^m, \ldots, w_N^m)$. Then for each $\mathbf{w}^m$, we generate $\mathbf{z}^m = (\mathbf{z}_1^m, \ldots, \mathbf{z}_N^m)$, where $\mathbf{z}_i^m$ is generated by (4) with $h_{ij}^m(\boldsymbol{\theta}^{(k)}|w_i^m, \mathbf{t}, \mathbf{X})$ defined in (5), where w_i is replaced by w_i^m and $\boldsymbol{\theta}$ is replaced by $\boldsymbol{\theta}^{(k)}$. Therefore the E-step in (7) can be approximated by

$$\hat{Q}(\boldsymbol{\theta}, \boldsymbol{\theta}^{(k)}) = \frac{1}{M} \sum_{m=1}^{M} \sum_{i=1}^{N} \sum_{j=1}^{J} h_{ij}^m(\boldsymbol{\theta}^{(k)}|w_i^m, \mathbf{t}, \mathbf{X}) \{\log[p_j(\mathbf{x}_i, \boldsymbol{\rho}) f_j(w_i^m|\mathbf{x}_i, \boldsymbol{\phi}_j)]\}.$$

4. Gibbs Sampler

Our next objective is a general Bayesian approach to mixture models for the survival data with censoring. Our sampling-based approach enables us to construct credible regions for parameters or functionals of interest for any sample size. Moreover, by plotting the entire posterior distribution of $\boldsymbol{\theta}$, we can detect interesting features of the posterior distribution, such as skewness, heavy tailedness, etc. Let $\pi(\boldsymbol{\theta})$ denote the prior density for $\boldsymbol{\theta}$; the posterior density is proportional to $L(\boldsymbol{\theta}|\mathbf{t}, \mathbf{X}) \times \pi(\boldsymbol{\theta})$. We first consider the data augmentation idea of the latent variables $\mathbf{z}$ discussed earlier, so updating on the $\boldsymbol{\theta}$ can be made simple in the MCMC by means of considering a product of the components. Now we extend these techniques to handle censored data. In addition to $\mathbf{z}$, we augment the data with another class of latent variables $\mathbf{w}$ that treats the unobserved failure times due to censoring as missing. This enables us to consider the augmented likelihood as given in (6). Then simulating $\boldsymbol{\theta}$ given $\mathbf{w}, \mathbf{z}, \mathbf{t}$, and $\mathbf{X}$ can be performed either from standard distributions or by using the Metropolis algorithm (Metropolis *et al.*, 1953).

Now let us briefly discuss the Gibbs sampler. Start with some initial values of $\boldsymbol{\theta}^{(0)}, \mathbf{z}^{(0)}, \mathbf{w}^{(0)}$. One of the values could be the EM or the MCEM solution of the mixture model. We can also perturb it with small errors for the starting points for replications of the Markov chain. We iteratively generate $\mathbf{w}$, $\mathbf{z}$, and $\boldsymbol{\theta}$ from the following full conditional distributions:

Step 1: Given $\boldsymbol{\theta}^{(0)}, \mathbf{t}, \mathbf{X}$, we generate $w_i^{(1)}$ by (1) if the ith subject is interval censored; generate $w_i^{(1)}$ by (2), if it is right censored; set it to t_i, if it is exactly observed.

We do this independently for $i = 1, \ldots, N$.
Step 2: Given $\mathbf{w}^{(1)}$, $\boldsymbol{\theta}^{(0)}, \mathbf{t}, \mathbf{X}$, we generate $\mathbf{z}_i^{(1)} \sim Mult(1, h_{i1}, \ldots, h_{iJ})$, where $h_{ij} = h_{ij}(\boldsymbol{\theta}^{(0)}|w_i^{(1)}, \mathbf{t}, \mathbf{X})$ as given in (5). This is done for all $i = 1, \ldots, N$.
Step 3: Given $\mathbf{z}^{(1)}, \mathbf{w}^{(1)}, \mathbf{t}, \mathbf{X}$, we generate $\boldsymbol{\theta}^{(1)}$ from $p(\boldsymbol{\theta}|\mathbf{w}^{(1)}, \mathbf{z}^{(1)}, \mathbf{t}, \mathbf{X}) \propto \pi(\boldsymbol{\theta})L(\boldsymbol{\theta}| \mathbf{z}^{(1)}, \mathbf{w}^{(1)}, \mathbf{t}, \mathbf{X})$, where $L(\boldsymbol{\theta}|\mathbf{z}, \mathbf{w}, \mathbf{t}, \mathbf{X})$ is given in (6).
Then we continue to iterate by repeating steps 1-3 where the superscripts in the stage indicators are incremented. Convergence can be monitored by the Gelman and Rubin (1992) method.

5. Model Selection

Given the mixture model in (1), we need to know what is the best number of mixtures to fit the data. We also have many choices for the mixing probabilities that include the logistic link and the probit link, and many choices for the latency distributions that include normal, log-normal, exponential, and Weibull distributions. We explore a predictive approach for model selection.

The basic idea is simple. We use part of the data to fit the model and the remaining part to test it. We divide our data into G groups of roughly equal size. For the data in the gth group, we evaluate the predictability of a model by evaluating the predictive likelihood of the gth group relative to the posterior distribution of $\boldsymbol{\theta}$ given the rest of the data. Let $\mathbf{t}_g$ denote the data set for the gth group that may include censored data. Similarly, we use $\mathbf{X}_g$ to denote the covariates for the gth group. Then the remaining data are denoted by $\mathbf{t}_{-g}$ and $\mathbf{X}_{-g}$. The conditional predictive ordinate for the gth group (GCPO) is defined by

$$\begin{aligned} p(\mathbf{t}_g|\mathbf{t}_{-g}, \mathbf{X}_{-g}) &= \int p(\mathbf{t}_g|\boldsymbol{\theta}, \mathbf{X}_g)\pi(\boldsymbol{\theta}|\mathbf{t}_{-g}, \mathbf{X}_{-g})d\boldsymbol{\theta} \qquad (1) \\ &= \int L(\boldsymbol{\theta}|\mathbf{t}_g, \mathbf{X}_g)\pi(\boldsymbol{\theta}|\mathbf{t}_{-g}, \mathbf{X}_{-g})d\boldsymbol{\theta}. \end{aligned}$$

This GCPO assesses the predictability of our model for the new data $\mathbf{t}_g$. We do this for all g, $g = 1, \ldots, G$. The pseudo-marginal predictive likelihood for the whole data is defined to be the product of these GCPO's,

$$PGCPO_{(\mathbf{t},\mathbf{X})} = \prod_{g=1}^{G} p(\mathbf{t}_g|\mathbf{t}_{-g}, \mathbf{X}_{-g}). \qquad (2)$$

The predictive pseudo-likelihood depends on the model. We use $PGCPO_{(\mathbf{t},\mathbf{X})}(M)$ to denote the predictive pseudo-likelihood for the model M. Therefore, we select the best model that has the largest $\widehat{PGCPO}_{(\mathbf{t},\mathbf{X})}(M)$ among the class of models $\{M \in \mathcal{M}\}$.

Usually, if the data size is large, a G of two may be sufficient. (cf. p. 240 of Efron and Tibshirani, 1993, for a related problem on estimating the prediction error). If the data size is small, we may need more groups. If G is N, then the method is just leave-one-out cross-validation. Then (2) reduces to the usual pseudo-marginal likelihood as defined in Gelfand, Dey, and Chang (1992).

On the computation for (2), if G is small, we might use a brute-force Monte Carlo integration. For the gth group, we apply the Gibbs sampler as described in Section 4 using the new data set with the gth group deleted. Then we apply

Monte Carlo integration to (1) by averaging $L(\boldsymbol{\theta}|\mathbf{t}_g, \mathbf{X}_g)$ over the $\boldsymbol{\theta}$ sampled from the Gibbs sampler applied to the new data set $(\mathbf{t}_{-g}, \mathbf{X}_{-g})$. If $G = N$, it would be too computing intensive to repeat the Gibbs sampler N times, every time using a different data set. Then a harmonic mean estimator of the CPO (GCPO in (1 with $G = N$)) can be obtained from the Gibbs sample based on the full data. Let $\mathcal{K}$ denote an index set that counts the usable variates $\boldsymbol{\theta}^{(k)}$ generated by the Gibbs sampler applied to the full data as in Section 4. These variates can be a second half of a single chain, or a second half of all multiple chains. Let K denote the size of $\mathcal{K}$. Then

$$\hat{p}(t_i|\mathbf{t}_{-i}, \mathbf{X}_{-i}) = \begin{cases} K\{\sum_{k\in\mathcal{K}} f(t_i|\mathbf{x}_i, \boldsymbol{\theta}^{(k)})]^{-1}\}^{-1} & \text{for } i \in \mathcal{U}; \\ K\{\sum_{k\in\mathcal{K}}[1 - F(t_i^+|\mathbf{x}_i, \boldsymbol{\theta}^{(k)})]^{-1}\}^{-1} & \text{for } i \in \mathcal{C}; \\ K\{\sum_{k\in\mathcal{K}}[F(t_{iU}|\mathbf{x}_i, \boldsymbol{\theta}^{(k)}) - F(t_{iL}|\mathbf{x}_i, \boldsymbol{\theta}^{(k)})]^{-1}\}^{-1} & \text{for } i \in \mathcal{I}. \end{cases}$$

6. Example

As an illustration, consider the data set in Table 15.1 given in Hewlett (1974). It was also used by Diggle and Gratton (1984), Pack and Morgan (1990), and several other authors. The data set consists of 317 male adult flour beetles that were exposed separately to pyrethrum, a plant-based insecticide. Among the 317 males, 144, 69, 54, and 50 males were sprayed of pyrethrum at concentrations of 0.20, 0.32, 0.50, 0.80 mg/cm^2, respectively. The equivalent log concentrations (denoted by log-doses) are -1.61, -1.14, -0.69, and -0.22. We will also analyze a similar data set for the females. That data set is omitted here, but it can be found in Hewlett (1974) and Pack and Morgan (1990, p.750). The group sizes for the female data set corresponding to the four dosages are 152, 81, 44, and 47.

Let $x_c = \log(d_c)$, where d_c is the dosage given to the beetles in the cth column. The survival time of a beetle given dosage x_c is modeled by a mixture of normal and exponential densities:

$$f(t|x_c, \boldsymbol{\theta}) = p_1(x_c, \gamma, \tau) f_1(t|x_c, \alpha, \beta, \sigma) + p_2(x_c, \gamma, \tau) f_2(t|\lambda); \tag{3}$$

where

$$\begin{aligned} p_1(x_c, \gamma, \tau) &= \frac{e^{\gamma+\tau x_c}}{1 + e^{\gamma+\tau x_c}}; \\ f_1(t|x_c, \alpha, \beta, \sigma) &= \frac{1}{\sqrt{2\pi}\sigma} \exp\left\{-\frac{(t - \alpha - \beta x_c)^2}{2\sigma^2}\right\}; \\ f_2(t|\lambda) &= \lambda \exp\{-\lambda t\} \mathrm{I}\{t > 0\}; \text{ and} \\ p_2(x_c, \gamma, \tau) &= 1 - p_1(x_c, \gamma, \tau). \end{aligned}$$

This model suggests a normal distribution for the failure times in the susceptible group with the mean to be a linear function of the log dose as in Boos and Brownie (1991). Because the other group is assumed to be more resistant to treatment, we choose an exponential distribution independent of the dosages. Other selections of standard distributions that might fit the data as well will be explored later. Because the mixing probabilities are directly related to the dosages, we consider logistic regression links for the incidence probabilities. Pack and Morgan (1990) ignore the contribution from the subpopulation of the long-term survivors for $t \leq 13$ in their numerical examples. Our model includes contributions from both the susceptible and the long-term survivor groups over the entire range of the survival times. These contributions are weighted by the mixing probabilities that correspond to the proportions of the subgroups. These two aspects are the advantages of using a mixture model.

Table 15.1: Male Flour Beetle (*Tribolium castaneum*) Data. The numbers indicate the number dead per day, the row 14 gives the number survived after day 13.

r/c	Log-dose(mg/cm^2)			
	-1.61	-1.14	-0.69	-0.22
1	3	7	5	4
2	11	10	8	10
3	10	11	11	8
4	7	16	15	14
5	4	3	4	8
6	3	2	2	2
7	2	1	1	1
8	1	0	1	0
9	0	0	0	0
10	0	0	0	1
11	0	0	0	0
12	1	0	0	0
13	1	0	0	0
14	101	19	7	2
Total	144	69	54	50

6.1 EM Algorithm for the Specific Example

Now we derive the EM algorithm for this example. Let n_{rc} denote the cell count in the rth row and the cth column, where $c = 1, \ldots, 4$ is the index for the dose levels and $r = 1, \ldots, 14$ is the index on the days. For simplicity, we write $\sum_{rc}$ for $\sum_{c=1}^{4} \sum_{r=1}^{14}$. We treat the data in the rth row as interval censored for $r \leq 13$, that is, beetles in the rth row die between the $r-1^{th}$ and r^{th} days; and data in the 14th row as right-censored with survival until at least the 13th day.

For this example, we are able to evaluate the E-step in closed form. Therefore, no MCEM is needed. A related EM algorithm can be found in McLachlan and Jones (1988). Suppose the current EM iteration is at the kth stage. Let $\boldsymbol{\theta}^{(k)} = \left(\alpha^{(k)}, \beta^{(k)}, (\sigma^{(k)})^2, \lambda^{(k)}, \gamma^{(k)}, \tau^{(k)}\right) = \left(\boldsymbol{\phi}_1^{(k)}, \boldsymbol{\phi}_2^{(k)}, \boldsymbol{\rho}_1^{(k)}\right)$, where $\boldsymbol{\phi}_1$ denotes $(\alpha, \beta, \sigma^2)$, $\boldsymbol{\phi}_2$ denotes λ, and $\boldsymbol{\rho}_1$ denotes (γ, τ). Note we set $\boldsymbol{\rho}_2 = (0, 0)$. For the E-step, we need to evaluate the expectation in (7) with respect to the truncated random variable $\mathbf{w}$. Because of the grouped data, the summation over i in (7) can be written as $\sum_{rc}$ with w_i replaced by W_{rc}, where W_{rc} is distributed as a truncated version of (3) with the current value of $\boldsymbol{\theta}^{(k)}$. We use (a_{rc}, b_{rc}) to denote the range of W_{rc}. We write $p_j^{(k)}$ for $p_j^{(k)}(x_c, \boldsymbol{\rho})$ for short. For $j = 1$ or 2, let

$$h_j(W_{rc}; \boldsymbol{\theta}^{(k)}) = \frac{p_j^{(k)} f_j^{(k)}(W_{rc}|x_c, \boldsymbol{\phi}_j^{(k)})}{\sum_j p_j^{(k)} f_j^{(k)}(W_{rc}|x_c, \boldsymbol{\phi}_j^{(k)})}.$$

Therefore, for the E-step, we need to evaluate the following expectations:

$$A_{s,jrc}^{(k)} = E^{(k)}[h_j(W_{rc}; \boldsymbol{\theta}^{(k)}) W_{rc}^s],$$

for $s = 0, 1$ and

$$A_{2,1rc}^{(k)} = E^{(k)}[h_1(W_{rc};\boldsymbol{\theta}^{(k)})(W_{rc} - \mu_1^{(k+1)})^2].$$

For simplicity, we write μ_1, instead of $\mu_1(x_c)$, for $\alpha + \beta x_c$. It is straightforward to verify the following equalities for $c = 1, \ldots, 4$ and $r = 1, \ldots, 14$, where $b_{14c} = \infty$, for all c:

$$A_{s,jrc}^{(k)} = \frac{p_j^{(k)} G_{s,jrc}^{(k)}}{F^{(k)}(b_{rc}) - F^{(k)}(a_{rc})}, \quad (s = 0, 1, 2),$$

where

$$\begin{aligned}
G_{0,jrc}^{(k)} &= H_{0,jrc}^{(k)}, \\
G_{1,1rc}^{(k)} &= \mu_1^{(k)} H_{0,1rc}^{(k)} - (\sigma^{(k)})^2 H_{1,1rc}^{(k)}, \\
G_{1,2rc}^{(k)} &= \exp\{-\lambda^{(k)} a_{rc}\}(a_{rc} + 1/\lambda^{(k)}) - \exp\{-\lambda^{(k)} b_{rc}\}(b_{rc} + 1/\lambda^{(k)}), \\
G_{2,1rc}^{(k)} &= (\sigma^{(k)})^2 [H_{0,1rc}^{(k)} + (2\mu_1^{(k+1)} - \mu_1^{(k)}) H_{1,1rc}^{(k)} - H_{2,1rc}^{(k)}] + (\mu_1^{(k+1)} - \mu_1^{(k)})^2 H_{0,1rc}^{(k)},
\end{aligned}$$

and where

$$\begin{aligned}
H_{0,jrc}^{(k)} &= F_j^{(k)}(b_{rc}) - F_j^{(k)}(a_{rc}), \\
H_{1,jrc}^{(k)} &= f_j^{(k)}(b_{rc}) - f_j^{(k)}(a_{rc}), \\
H_{2,1rc}^{(k)} &= b_{rc} f_1^{(k)}(b_{rc}) - a_{rc} f_1^{(k)}(a_{rc}).
\end{aligned}$$

Therefore, we obtain the following equations for the M-step:

$$\alpha^{(k+1)} = \frac{\sum_{rc} n_{rc} A_{1,1rc}^{(k)} - \beta^{(k+1)} \sum_{rc} n_{rc} A_{0,1rc}^{(k)} x_c}{\sum_{rc} n_{rc} A_{0,1rc}^{(k)}},$$

$$\beta^{(k+1)} = \frac{\sum_{rc} n_{rc} A_{1,1rc}^{(k)} x_c \sum_{rc} n_{rc} A_{0,1rc}^{(k)} - \sum_{rc} n_{rc} A_{1,1rc}^{(k)} \sum_{rc} n_{rc} A_{0,1rc}^{(k)} x_c}{\sum_{rc} n_{rc} A_{0,1rc}^{(k)} x_c{}^2 \sum_{rc} n_{rc} A_{0,1rc}^{(k)} - (\sum_{rc} n_{rc} A_{0,1rc}^{(k)} x_c)^2},$$

$$(\sigma^{(k+1)})^2 = \frac{\sum_{rc} n_{rc} A_{2,1rc}^{(k)}}{\sum_{rc} n_{rc} A_{0,1rc}^{(k)}},$$

$$\lambda^{(k+1)} = \frac{\sum_{rc} n_{rc} A_{0,2rc}^{(k)}}{\sum_{rc} n_{rc} A_{1,2rc}^{(k)}},$$

and the M-step for γ and τ is computed using the Newton-Raphson method (Boos and Brownie, 1991, and Jordan and Xu, 1993). Let

$$\boldsymbol{\rho}_1^{T(k+1)} = \boldsymbol{\rho}_1^{T(k)} + g(R^{(k)})^{-1} \mathbf{e}^{(k)},$$

where the gradient vector at the kth iteration is

$$\mathbf{e}^{(k)} = \sum_{rc} n_{rc} \left(E^{(k)}[h_1(W_{rc};\boldsymbol{\theta}^{(k)})] - p_1^{(k)} \right) \frac{\partial \xi_1}{\partial \boldsymbol{\rho}_1^{T(k)}},$$

where $\xi_1 = \mathbf{x}\boldsymbol{\rho}_1^T$ and $\frac{\partial \xi_1}{\partial \boldsymbol{\rho}_1^{T(k)}} = \{1, x_c\}^T$. The Hessian matrix at the kth iteration is

$$R^{(k)} = \sum_{rc} n_{rc} p_1^{(k)} (1 - p_1^{(k)}) \frac{\partial \xi_1}{\partial \boldsymbol{\rho}_1^{T(k)}} \frac{\partial \xi_1}{\partial \boldsymbol{\rho}_1^{(k)}},$$

where g is a learning rate.

6.2 Gibbs Samplers for the Specific Example

In the prior specification, we assume the components of $\boldsymbol{\theta}$ are independent. Flat priors are chosen for α, β, λ, and σ^2 because we lack precise knowledge about these parameters. Although the priors are improper, they result in proper posteriors in most situations. We use normal priors for γ and τ, both with mean zero and a large known variance σ_0^2, that is, $\gamma \sim N(0, \sigma_0^2)$ and $\tau \sim N(0, \sigma_0^2)$. Proper priors are chosen here due to the concern about the convergence of the posterior sample if flat priors were used for the parameters in the mixing probability (Diebold and Robert, 1994, p. 367).

We choose the covariate vector $\mathbf{x} = (1, x)$ where $x = \log(\text{dose level})$. As in the EM method, we treat each entry of Table 15.1 as interval censored except that the row labeled 14 is treated as right-censored. Therefore, reading from the first column, the data set t_1 to t_{144} consists of $\{\underbrace{(0,1),\ldots,(0,1)}_{3\ copies}, \underbrace{(1,2),\ldots,(1,2)}_{11\ copies}, \ldots, (12,13), \underbrace{13^+,\ldots,13^+}_{101\ copies}\}$ with the same covariate $\mathbf{x}_i = (1, x_i) = (1, \log .2) = (1, -1.61)$ for $i = 1, \ldots, 144$. Similarly, we can list the rest of the data for the remaining dosages.

On implementing the Gibbs sampler, we follow steps 1–3 in Section 4. We first follow steps 1-2 to generate latent variable for w_i and $\mathbf{z}_i = (z_{i1}, z_{i2})$ for each i. Then we follow step 3 to update the parameters. Now we give more details for step 3, where the joint conditional density of $\boldsymbol{\theta}$ can be implemented sequentially by updating one variable at a time. Because given $\mathbf{w}$ and $\mathbf{z}$ (we omit "given $\mathbf{t}, \mathbf{x}$" for simpler notation), the parameters $\boldsymbol{\rho}_1 = (\gamma, \tau)$, $\boldsymbol{\phi}_1 = (\alpha, \beta, \sigma^2)$, and $\boldsymbol{\phi}_2 = \lambda$ are independent, we have some simplifications in generating the inner loop. That is, we generate $\alpha^{(1)}$ given $\beta^{(0)}, \sigma^{(0)}, \mathbf{w}^{(0)}, \mathbf{z}^{(0)}$; generate $\beta^{(1)}$ given $\alpha^{(1)}, \sigma^{(0)}, \mathbf{w}^{(0)}, \mathbf{z}^{(0)}$; generate $\sigma^{2^{(1)}}$ given $\alpha^{(1)}, \beta^{(1)}, \mathbf{w}^{(0)}, \mathbf{z}^{(0)}$; generate $\lambda^{(1)}$ given $\mathbf{w}^{(0)}$ and $\mathbf{z}^{(0)}$; and independently generate $\gamma^{(1)}$ and $\tau^{(1)}$ given $\mathbf{w}^{(0)}$ and $\mathbf{z}^{(0)}$. Then continue back to step 1 in Section 4 of the iteration. Let $z_{+1} = \sum_{i=1}^N z_{i1}$. Now we just list the conditional densities used in the inner loop of step 3.

$$\alpha|\ldots \sim N\left(\frac{\sum_{i=1}^N z_{i1}(w_i - \beta x_i)}{z_{+1}}, \frac{\sigma^2}{z_{+1}}\right); \tag{4}$$

$$\beta|\ldots \sim N\left(\frac{\sum_{i=1}^N z_{i1}x_i(w_i - \alpha)}{\sum_{i=1}^N z_{i1}{x_i}^2}, \frac{\sigma^2}{\sum_{i=1}^N z_{i1}{x_i}^2}\right); \tag{5}$$

$$\sigma^2|\ldots \sim IG\left(\frac{z_{+1}}{2}, \frac{2}{C}\right), \tag{6}$$

where IG denotes the inverse gamma distribution with mean $C/(z_{+1} - 2)$, where $C = \sum_{i=1}^N z_{i1}(w_i - \alpha - \beta x_i)^2$;

$$\lambda|\ldots \sim \Gamma\left(N - z_{+1} + 1, \frac{1}{\sum_{i=1}^N w_i(1 - z_{i1})}\right), \tag{7}$$

where Γ denotes the gamma distribution with mean $(N - z_{+1} + 1)/(\sum_{i=1}^N w_i(1 - z_{i1}))$; and

$$g(\gamma, \tau|\ldots) \propto \prod_{i=1}^N \left(\frac{\exp\{\gamma z_{i1} + \tau x_i z_{i1}\}}{1 + \exp\{\gamma + \tau x_i\}}\right) \exp\left\{-\frac{\gamma^2 + \tau^2}{2{\sigma_0}^2}\right\}. \tag{8}$$

The variates γ and τ are generated by the Metropolis algorithm with (8) as the target distribution.

In addition to the normal distribution for the latency distribution of the subgroup 1, we also consider the log-normal distribution. Moreover, we consider probit links as opposed to logistic links for the mixing probabilities. We would like to answer the question: "Do we really need the mixture model? Would a simpler model with only one population suffice?" Therefore, we also fit the data with just the normal or the log-normal model. These models are the usual linear or log-linear models with Gaussian errors. We summarize the six models in Table 15.3.

On implementing for Model 2, we only need to change the original data to the log scale, and change the step in (7) to

$$\lambda | \ldots \sim \Gamma\left(N - z_{+1} + 1,\ \frac{1}{\sum_{i=1}^{N} e^{w_i}(1 - z_{i1})}\right).$$

For Models 3 and 4, we assume $p_1(\mathbf{x}, \boldsymbol{\rho}) = \Phi(\gamma + \tau x)$, and $p_2(\mathbf{x}, \boldsymbol{\rho}) = 1 - p_1(\mathbf{x}, \boldsymbol{\rho})$, where Φ is the standard cumulative normal distribution function. We follow the same procedures as before, except (8) is replaced by

$$g(\gamma, \tau | \ldots) \propto \prod_{i=1}^{N} \Phi(\gamma + \tau x_i)^{z_{i1}} (1 - \Phi(\gamma + \tau x_i))^{z_{i2}} \exp\left\{-\frac{\gamma^2 + \tau^2}{2\sigma_0{}^2}\right\}.$$

Then γ and τ can either be generated by the Metropolis algorithm or by using another data augmentation with two classes of latent normal variates; one is right-truncated at $\gamma + \tau x_i$, the other is left-truncated at the same point. For Models 5 and 6, we follow steps 1 and 3 of the Gibbs sampler in Section 4, where step 1 depends on a single model. Step 2 is not needed because we have a single linear or log-linear model. Then step 3 follows the usual Bayesian linear model updating.

6.3 Numerical Results

Most of our results are based on the male data given in Table 15.1. Table 15.2 lists the point estimates of the parameters computed from the EM algorithm and the MCMC algorithm for the model 1 given by (3). The 90% and 95% highest posterior density (HPD) intervals are computed from the Gibbs sampler. The Gibbs sampler is iterated 10,000 times and the results of the last 5,000 iterations are kept as a final sample from the joint posterior distribution. The incidence probabilities $p_1(x_c, \gamma, \tau) = \exp\{\gamma + \tau x_c\}/(1 + \exp\{\gamma + \tau x_c\})$ for the susceptible group are estimated by the Gibbs sampler at .266, .623, .872, .967 for the four dosages. The results are quite comparable to (.299, .724, .870, .960)=(43/144,50/69,47/54,48/50), the crude estimates of the proportion of beetles that react to the treatments (killed before day 13). Observe the 90% HPD interval of β includes zero; that is, there are no appreciable dose effects on survival times for the susceptible group. This finding is consistent with previous studies by Hewlett (1974) and Pack and Morgan (1990). However, dosage does affect mixing probabilities upon observing the point and the interval estimates of τ. Moreover, the mean survival times for the male beetles for each of the four dosages $\mu(x_c) = p_1(x_c, \gamma, \tau)(\alpha + \beta x_c) + p_2(x_c, \gamma, \tau)/\lambda$ are estimated at 125.8, 66.0, 24.2, and 8.5 from the Gibbs sampler.

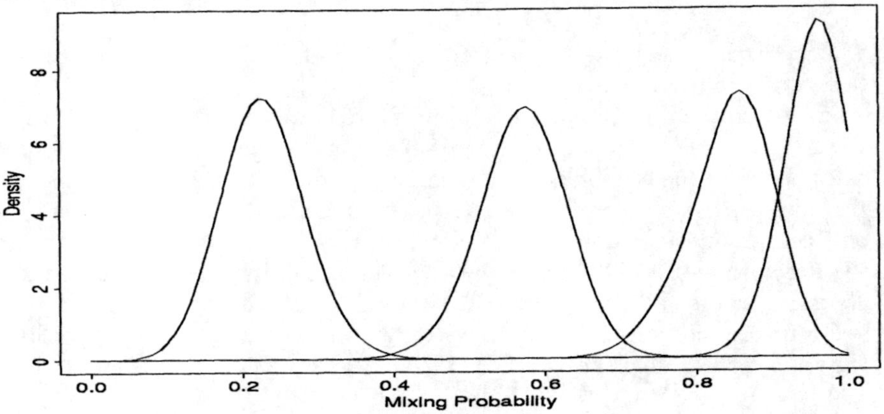

FIGURE 15.1. Plot of the posterior densities of the incidence probability $p_1(x_c, \gamma, \tau)$ for the subpopulation 1 evaluated at the four dosages. The plot is the overlay of the four histograms for the incidence probabilities for the susceptible group each evaluated at the log-dose level -1.61, -1.14, -0.69 and -0.22 mg/cm^2, respectively.

Table 15.2: Parameter Estimates for the Male Data.

Parameter	MLE (EM)	Post. Mean (MCMC)	90% HPD Int.	95% HPD Int.
α	2.937	2.982	(2.350, 3.425)	(2.002, 3.729)
β	0.254	0.198	(-0.462, 0.652)	(-0.673, 0.721)
σ^2	2.217	2.253	(1.559, 2.924)	(1.148, 3.256)
λ	0.006	0.006	(0.004, 0.008)	(0.003, 0.009)
γ	4.213	4.197	(3.146, 5.318)	(2.986, 5.729)
τ	3.241	3.238	(2.466, 4.051)	(2.237, 4.124)

Figure 1 overlays the four histograms of the posterior incidence probabilities $p_1(x_c, \gamma, \tau)$ evaluated at the four log-dosages. Clearly it is monotonically increasing in the doses. The mixture model gives more weight to the normal distribution than to the exponential distribution for higher dose levels. This is consistent with fewer long-term survivors being observed at higher dosages. Figure 2 overlays the four posterior density plots for $\mu(x_c)$ for the four dosages respectively. The four densities reading from left to right correspond to the mean survival times for the beetles in the fourth column to the first column. The figure shows that all densities are unimodal. Their modes are quite comparable to the means given above. The mean decreases as the dosage increases.

We also analyzed the data set for the female beetles. The mean survival times for the females were estimated to be 163.3, 124.7, 74.4, and 34.0 for the four dosages. Apparently, there are big differences in the means between the sexes. The strong sex effect is also observed by Hewlett (1974), Diggle and Gratton (1984), and Pack and Morgan (1990). It is also interesting to note $\hat{\beta}$ is -.318, with (-.734, .099) as the 90% HPD interval. This shows that the dose has a stronger effect for the female than the male on the latency distribution for the susceptible group. As expected, the mean latent time of survival for the susceptible group decreases as dose increases.

For each of the six models, we evaluate the two-fold and four-fold cross-validated GCPOs. For the two-fold cross-validated GCPO, we randomly divide the data into two halves by randomly selecting the data points associated with half of the ran-

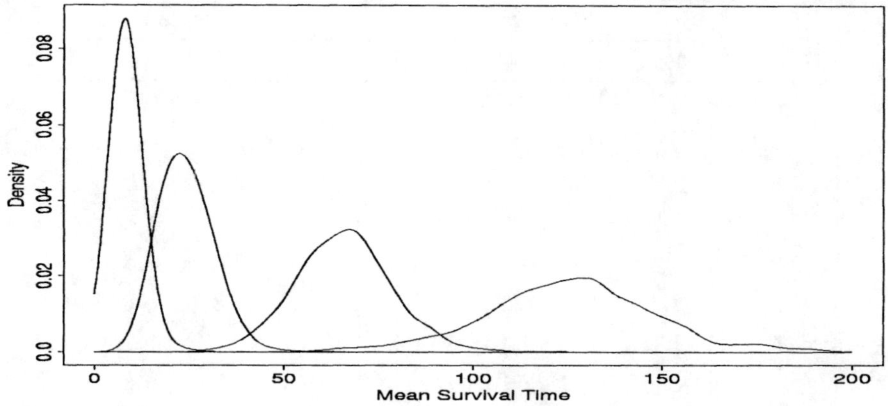

FIGURE 15.2. Plot of the posterior densities of the mean survival time $\mu(x_c)$ evaluated at the four dosages. The plot is the overlay of the four histograms (from left to right) for the mean survival times evaluated at the log-dose level -.22, -.69, -1.14, and -1.61 mg/cm^2, respectively. Note the mean survival time increases as the dosage decreases.

TABLE 15.1. Model Comparisons

Model	p_1	f_1	f_2	log PGCPO two-fold	log PGCPO four-fold
1	logistic	normal	exponential	-523.09	-522.73
2	logistic	log-normal	exponential	-524.09	-523.88
3	probit	normal	exponential	-526.66	-529.17
4	probit	log-normal	exponential	-527.33	-529.49
5	1	normal		-726.93	-773.79
6	1	log-normal		-587.18	-590.06

domly chosen labels in 1–317. Similarly, we divide the data randomly into four groups of approximately equal sizes.

Table 15.3 lists the log product of the GCPO for each of the six models for the male data. The results show that the mixture models outperform the single-component model by a huge margin. Model 1 (3) is the best among the six models. Model 2 for a log-normal distribution is slightly inferior. The logistic link fits the data much better than the normal link. Finally, both two-fold and four-fold cross-validation yield similar results.

Table 15.4 gives the fitted values and the residuals in absolute values of the cell count for Model 1. To check the overall model fitting, we use the summation of absolute deviances, that is, $D_{abs} = \sum_{r=1}^{14}\sum_{c=1}^{4} |\text{observed} - \text{fitted}|$ for all the cells in Table 15.1.

For model 1, this sum based on the first 13 days is 51.4, contrasting to 58.3 fitted by Pack and Morgan. If we break the sum into the absolute deviance for each dosage level, we see the absolute deviances are 14.2, 14.5, 10 and 12.7 for our model 1, versus 10.7, 16.3, 14.9 and 16.4 for the Pack and Morgan model for the four dosage levels respectively. That is our model 1 improves upon the Pack and

TABLE 15.2. Fitted Values and Absolute Residuals

r/c	Fitted value				|Residual|			
	-1.61	-1.14	-0.69	-0.22	-1.61	-1.14	-0.69	-0.22
1	5.8	5.4	5.3	5.0	2.8	1.6	0.3	1.0
2	8.0	8.0	8.2	7.8	3.0	2.0	0.2	2.2
3	10.5	11.2	12.0	12.1	0.5	0.2	1.0	4.1
4	9.0	10.0	11.2	11.8	2.0	6.0	3.8	2.2
5	5.3	5.8	6.6	7.3	1.3	2.8	2.6	0.7
6	2.4	2.3	2.6	2.9	0.6	0.3	0.6	0.9
7	1.1	0.8	0.8	0.8	0.9	0.2	0.2	0.2
8	0.8	0.3	0.2	0.2	0.2	0.3	0.8	0.2
9	0.7	0.3	0.1	0.1	0.7	0.3	0.1	0.1
10	0.7	0.2	0.1	0.1	0.7	0.2	0.1	0.9
11	0.7	0.2	0.1	0.1	0.7	0.2	0.1	0.1
12	0.6	0.2	0.1	0.1	0.4	0.2	0.1	0.1
13	0.6	0.2	0.1	0.0	0.4	0.2	0.1	0.0
14	97.8	24.2	6.5	1.7	3.2	5.2	0.5	0.3

Morgan model in prediction for all levels except at the log dose of -1.61 level. We can also compare the residuals using graphical methods. The graphs omitted here show that Model 1 improves upon the Pack and Morgan model for essentially most of the cells.

References

D. D. Boos and C. Brownie. Mixture models for continuous data in dose-response studies when some animals are unaffected by treatment. *Biometrics* vol. 47, pp. 1489-1504, 1991.

D.R. Cox and D. Oakes. *Analysis of Survival Data.* Chapman and Hall: London, 1984.

A. Dempster, N. Laird, N and D. Rubin. Maximum likelihood from incomplete data via the EM algorithm. *Journal of the Royal Statistical Society* Series B vol. 39, pp.1-38, 1977.

L. Devroye. *Non-Uniform Random Variate Generation.* Springer-Verlag: New York, 1986.

D. Dey, L. Kuo, and S. Sahu. A Bayesian predictive approach to determining the number of components in a mixture distribution. *Statistics and Computing* vol. 5, pp. 297-305, 1995.

J. Diebolt and C. Robert. Estimation of finite mixture distributions through Bayesian sampling. *Journal of the Royal Statistical Society* Series B vol. 56, pp. 363-375, 1994.

P.J. Diggle and R.J. Gratton. Monte Carlo methods of inference for implicit statistical models. *Journal of the Royal Statistical Society* Series B vol. 46, pp. 193-227, 1984.

B. Efron and R. Tibshirani. *An Introduction to the Bootstrap.* Chapman & Hall: New York, 1993.

V.T. Farewell. The use of mixture models for the analysis of survival data with long-term survivors. *Biometrics* vol. 38, pp. 1041-1046, 1982.

A. E. Gelfand, D. K. Dey, and H. Chang. Model determination using predictive distributions with implementation via sampling-based methods (with Discussion). In *Bayesian Statistics 4,* J. M. Bernardo, J. O. Berger, A.P. Dawid, and A. F. M. Smith (eds), pp. 147-169, Oxford University Press: Oxford, 1992.

A.E. Gelfand and A. F. M. Smith. Sampling based approaches to calculating marginal densities. *Journal of the American Statistical Association* vol. 85, pp. 398-409, 1990.

A.E. Gelfand, A.F.M. Smith, and T.M. Lee. Bayesian analysis of constrained parameter and truncated data problems using Gibbs sampling. *Journal of the American Statistical Association* vol. 87, pp. 523-532, 1992.

A. Gelman, and D. Rubin. Inference from iterative simulation using multiple sequences. *Statistical Science* vol. 7, pp. 457-472, 1992.

S. Geman and D. Geman. Stochastic relaxation, Gibbs distributions, and the Bayesian restoration of images. *IEEE Transactions on Pattern Analysis and Machine Intelligence* vol. 6, pp. 721-741, 1984.

P. S. Hewlett. Time from dosage to death in beetles Tribolium castaneum, treated with pyrethrins or DDT, and its bearing on dose-mortality relations. *Journal of Stored Product Research* vol. 10, pp. 27-41, 1974.

M. I. Jordan and R. A. Jacobs. Hierarchical mixtures of experts and the EM algorithm. *Neural Computation* vol. 6, pp. 181-214, 1994.

M. I. Jordan, and L. Xu. Convergence results for the EM approach to mixtures of experts architectures. *Neural Networks,* vol. 8, pp. 1409-1431, 1995.

A.Y.C. Kuk and C.H. Chen. A mixture model combining logistic regression with proportional hazards regression. *Biometrika* vol. 79, pp. 531-541, 1992.

L. Kuo and A. F. M. Smith. Bayesian computations in survival models via the Gibbs samplers (with Discussion). In *Survival Analysis: State of the Art,* J. P. Klein and P. K. Goel (eds), pp. 11-14. Kluwer Academic: Dordrecht, 1992.

M. G. Larson and G. Dinse. A mixture model for the regression analysis of competing risks data. *Applied Statistics* vol. 34, pp. 201-211, 1985.

P. McCullagh and J. A. Nelder. *Generalized Linear Models.* Chapman and Hall: London, 1989.

G. J. McLachlan and P. N. Jones. Fitting mixtures models to grouped and truncated data via the EM algorithm. *Biometrics* vol. 44, pp. 571-578, 1988.

N. Metropolis, A. W. Rosenbluth, M. N. Rosenbluth, A. H. Teller, and E. Teller. Equation of state calculations by fast computing machines. *Journal of Chemical Physics* vol. 21, pp. 1087-1092, 1953.

S. E. Pack and B. J. T. Morgan, B.J.T..A mixture model for interval-censored time-to-response quantal assay data. *Biometrics* vol. 46, pp. 749-757, 1990.

F. Peng, R.A. Jacobs, and M. A. Tanner. Bayesian inference in mixtures-of-experts and hierarchical mixtures-of-experts models with an application to speech recognition. *Journal of the American Statistical Association* vol. 91, pp. 953-960, 1996.

M. A. Tanner and W. H. Wong. The calculation of posterior distributions by data augmentation. *Journal of the American Statistical Association* vol 82, pp. 528-550, 1987.

J. Taylor. Semi-parametric estimation in failure time mixture models. *Biometrics* vol. 51, pp. 899-907, 1995.

C. G. Wei and M. Tanner. A Monte Carlo implementation of the EM algorithm and the poor man's data augmentation algorithm. *Journal of the American Statistical Association*, 85, pp. 699-714, 1990.

Part V

Model Diagnostics and Variable Selection in GLMs

16

Bayesian Variable Selection Using the Gibbs Sampler

Petros Dellaportas
Jonathan J. Forster
Ioannis Ntzoufras

ABSTRACT Specification of the linear predictor for a generalized linear model requires determining which variables to include. We consider Bayesian strategies for performing this variable selection. In particular we focus on approaches based on the Gibbs sampler. Such approaches may be implemented using the publically available software BUGS. We illustrate the methods using a simple example. BUGS code is provided in an appendix.

1. Introduction

In a Bayesian analysis of a generalized linear model, model uncertainty may be incorporated coherently by specifying prior probabilities for plausible models and calculating posterior probabilities using

$$f(m|\boldsymbol{y}) = \frac{f(m)f(\boldsymbol{y}|m)}{\sum\limits_{m\in\mathcal{M}} f(m)f(\boldsymbol{y}|m)}, \qquad m \in \mathcal{M} \tag{1}$$

where m denotes the model, $\mathcal{M}$ is the set of all models under consideration, $f(m)$ is the prior probability of model m. The observed data $\boldsymbol{y}$ contribute to the posterior model probabilities through $f(\boldsymbol{y}|m)$, the marginal likelihood calculated using $f(\boldsymbol{y}|m) = \int f(\boldsymbol{y}|m,\boldsymbol{\beta}_m)f(\boldsymbol{\beta}_m|m)d\boldsymbol{\beta}_m$ where $f(\boldsymbol{\beta}_m|m)$ is the conditional prior distribution of $\boldsymbol{\beta}_m$, the model parameters for model m and $f(\boldsymbol{y}|m,\boldsymbol{\beta}_m)$ is the likelihood of the data $\boldsymbol{y}$ under model m.

In particular, the relative probability of two competing models m_1 and m_2 reduces to

$$\frac{f(m_1|\boldsymbol{y})}{f(m_2|\boldsymbol{y})} = \frac{f(m_1)}{f(m_2)} \frac{\int f(\boldsymbol{y}|m_1,\boldsymbol{\beta}_{m_1})f(\boldsymbol{\beta}_{m_1}|m_1)\, d\boldsymbol{\beta}_{m_1}}{\int f(\boldsymbol{y}|m_2,\boldsymbol{\beta}_{m_2})f(\boldsymbol{\beta}_{m_2}|m_2)\, d\boldsymbol{\beta}_{m_2}} \tag{2}$$

which is the familiar expression relating the posterior and prior odds of two models in terms of the Bayes factor, the second ratio on the right hand side of (2).

The principal attractions of this approach are that (1) allows the calculation of posterior probabilities of all competing models, regardless of their relative size or structure, and this model uncertainty can be incorporated into any decisions or predictions required (Draper, 1995, gives examples of this).

Generalized linear models are specified by three components, distribution, link and linear predictor. Model uncertainty may concern any of these, and the approach outlined above is flexible enough to deal with this. In this chapter, we shall restrict attention to variable selection problems, where the models concerned differ only in the form of the linear predictor. Suppose that there are p possible covariates which are candidates for inclusion in the linear predictor. Then each $m \in \mathcal{M}$ can be naturally represented by a p-vector $\boldsymbol{\gamma}$ of binary indicator variables determining whether or not a covariate is included in the model, and $\mathcal{M} \subset \{0,1\}^p$. The linear predictor for the generalized linear model determined by $\boldsymbol{\gamma}$ may be written as

$$\boldsymbol{\eta} = \sum_{i=1}^{p} \gamma_i \boldsymbol{X}_i \boldsymbol{\beta}_i \tag{3}$$

where $\boldsymbol{\beta}$ is the "full" parameter vector with dimension p, and $\boldsymbol{X}_i$ and $\boldsymbol{\beta}_i$ are the design sub-matrix and parameter vector, corresponding to the ith covariate. This specification allows for covariates of dimension greater than 1, for example terms in factorial models.

There has been a great deal of recent interest in Bayesian approaches for identifying promising sets of predictor variables. See for example Brown *et al.*(1998) and Chipman (1996, 1997), Clyde *et al.*(1996), Clyde and DeSimone-Sasinowska (1997), George *et al.*(1996), George and McCulloch (1993, 1996, 1997), Geweke (1996), Hoeting *et al.*(1996), Kuo and Mallick (1998), Mitchell and Beauchamp (1988), Ntzoufras *et al.*(1997), Smith and Kohn (1996) and Wakefield and Bennet (1996).

Most approaches require some kind of analytic, numerical or Monte Carlo approximation because the integrals involved in (2) are only analytically tractable in certain restricted examples. A further problem is that the size of the set of possible models $\mathcal{M}$ may be extremely large, so that calculation or approximation of $f(\boldsymbol{y}|m)$ for all $m \in \mathcal{M}$ is very time consuming. One of the most promising approaches has been Markov chain Monte Carlo (MCMC). MCMC methods enable one, in principle, to obtain observations from the joint posterior distribution of $(m, \boldsymbol{\beta}_m)$ and consequently estimate $f(m|\boldsymbol{y})$ and $f(\boldsymbol{\beta}_m|m, \boldsymbol{y})$.

In this chapter we restrict attention to model determination approaches which can be implemented by using one particular MCMC method, the Gibbs sampler. The Gibbs samper is particularly convenient for Bayesian computation in generalized linear models, due to the fact that posterior distributions are generally log-concave (Dellaportas and Smith, 1992). Furthermore, the Gibbs sampler can be implemented in a straightforward manner using the BUGS software (Spiegelhalter *et al.*, 1996a). To facilitate this, we provide BUGS code for various approaches in Appendix A.

The rest of the chapter is organized as follows. Section 2 describes several variable selection strategies that can be implemented using the Gibbs sampler Section 3 contains an illustrative example analysed using BUGS code. We conclude this chapter with a brief discussion in Section 4.

2. Gibbs Sampler Based Variable Selection Strategies

As we are assuming that model uncertainty is restricted to variable selection, m is determined by $\boldsymbol{\gamma}$. We require a MCMC approach for obtaining observations from the joint posterior distribution of $f(m, \boldsymbol{\beta}_m)$. The Gibbs sampler achieves this by generating successively from univariate conditional distributions, so, in principle, the Gibbs sampler is determined by $f(m, \beta_m)$. However, flexibility in the choice of parameter space, likelihood and prior has led to a number of different Gibbs sampler variable selection approaches being proposed.

The first method we shall discuss is a general Gibbs sampler based model determination strategy. The others have been developed more specifically for variable selection problems.

2.1 Carlin and Chib's Method

This method, introduced by Carlin and Chib (1995) is a flexible Gibbs sampling strategy for any situation involving model uncertainty. It proceeds by considering the extended parameter vector $(m, \beta_k; k \in \mathcal{M})$. If a sample can be generated from the joint posterior density for this extended parameter, a sample from the required posterior distribution $f(m, \beta_m)$ can be extracted easily.

A joint prior distribution for m and $(\beta_k; k \in \mathcal{M})$ is required. Here, $(\beta_k; k \in \mathcal{M})$ contains the model parameters for every model in $\mathcal{M}$. Carlin and Chib (1995) specify the joint prior distribution through the marginal prior model probability $f(m)$ and prior density $f(\beta_m|m)$ for each model, as above, together with independent "pseudoprior" or linking densities $f(\beta_{m'}|m \neq m')$ for each model.

The conditional posterior distributions required for the Gibbs sampler are

$$f(\beta_{m'}|m,\ \{\beta_k : k \in \mathcal{M} \setminus \{m'\}\}, y,) \propto \begin{cases} f(y|m, \beta_m) f(\beta_m|m) & m' = m \\ f(\beta_{m'}|m) & m' \neq m \end{cases} \tag{4}$$

$$f(m|\{\beta_k : k \in \mathcal{M}\}, y) = \frac{A_m}{\sum\limits_{k \in \mathcal{M}} A_k}. \tag{5}$$

where

$$A_m = f(y|m, \beta_m) \prod_{s \in \mathcal{M}} [f(\beta_s|m)] f(m), \quad \forall \ m \in \mathcal{M}.$$

Therefore, when $m' = m$, we generate from the usual conditional posterior for model m, and when $m' \neq m$ we generate from the corresponding pseudoprior, $f(\beta_{m'}|m)$. The model indicator m is generated as a discrete random variable using (5).

The pseudopriors have no influence on $f(\beta_m|m)$, the marginal posterior distribution of interest. They act as a linking density, and careful choice of pseudoprior is essential, if the Gibbs sampler is to be sufficiently mobile. Ideally, $f(\beta_{m'}|m \neq m')$ should resemble the marginal posterior distribution $f(\beta_{m'}|m', y)$, and Carlin and Chib suggest strategies to achieve this.

The flexibility of this method lies in the facility to specify pseudopriors which help the sampler run efficiently. This may also be perceived as a drawback in problems where there are a large number of models under consideration, such as variable selection involving a moderate number of potential variables. Then, specification of efficient pseudopriors may become too time-consuming. A further drawback of the method is the requirement to generate every $\beta_{m'}$ at each stage of the sampler. (This may be avoided by using a 'Metropolis-Hastings' step to generate m, but is outside the scope of the current chapter; see Dellaportas *et al.*, 1997, for details).

Examples which show how BUGS can be used to perform this method can be found in Spiegelhalter *et al.*(1996b).

2.2 Stochastic Search Variable Selection

Stochastic Search Variable Selection (SSVS) was introduced by George and McCulloch (1993) for linear regression models and has been adapted for more complex models such as pharmacokinetic models (Wakefield and Bennett, 1996), construction of stock portfolios in finance (George and McCulloch, 1996), generalized linear models (George *et al.*, 1996, George and McCulloch, 1997), log-linear models (Ntzoufras *et al.*, 1997) and multivariate regression models (Brown *et al.*, 1998).

The difference between SSVS and other variable selection approaches is that the parameter vector $\boldsymbol{\beta}$ is specified to be of full dimension p under all models, so the linear predictor is

$$\boldsymbol{\eta} = \sum_{i=1}^{p} \boldsymbol{X}_i \boldsymbol{\beta}_i. \tag{6}$$

Therefore $\boldsymbol{\eta} = \boldsymbol{X}\boldsymbol{\beta}$ for all models, where $\boldsymbol{X}$ contains all the potential explanatory variables. The indicator variables γ_i are involved in the modelling process through the prior

$$\boldsymbol{\beta}_i|\gamma_i \sim \gamma_i N(0, c_i^2 \Sigma_i) + (1-\gamma_i) N(0, \Sigma_i) \tag{7}$$

for specified c_i and Σ_i. The prior parameters c_i and Σ_i in (7) are chosen so that when $\gamma_i = 0$ (covariate is "absent" from the linear predictor) the prior distribution for $\boldsymbol{\beta}_i$ ensures that $\boldsymbol{\beta}_i$ is constrained to be "close to $\mathbf{0}$". When $\gamma_i = 1$ the prior is diffuse, assuming that little prior information is available about $\boldsymbol{\beta}_i$.

The full conditional posterior distributions of $\boldsymbol{\beta}_i$ and γ_i are given by

$$f(\boldsymbol{\beta}_i|\boldsymbol{y}, \boldsymbol{\gamma}, \boldsymbol{\beta}_{\setminus i}) \propto f(\boldsymbol{y}|\boldsymbol{\gamma}, \boldsymbol{\beta}) f(\boldsymbol{\beta}_i|\gamma_i)$$

and

$$\frac{f(\gamma_i = 1|\boldsymbol{y}, \boldsymbol{\gamma}_{\setminus i}, \boldsymbol{\beta})}{f(\gamma_i = 0|\boldsymbol{y}, \boldsymbol{\gamma}_{\setminus i}, \boldsymbol{\beta})} = \frac{f(\boldsymbol{\beta}|\gamma_i = 1, \boldsymbol{\gamma}_{\setminus i})\, f(\gamma_i = 1, \boldsymbol{\gamma}_{\setminus i})}{f(\boldsymbol{\beta}|\gamma_i = 0, \boldsymbol{\gamma}_{\setminus i})\, f(\gamma_i = 0, \boldsymbol{\gamma}_{\setminus i})} \tag{8}$$

where $\boldsymbol{\gamma}_{\setminus i}$ denotes all terms of $\boldsymbol{\gamma}$ except γ_i.

If we use the prior distributions for $\boldsymbol{\beta}$ and $\boldsymbol{\gamma}$ defined by (7) and assume that $f(\gamma_i = 0, \boldsymbol{\gamma}_{\setminus i}) = f(\gamma_i = 1, \boldsymbol{\gamma}_{\setminus i})$ for all i, then

$$\frac{f(\gamma_i = 1|\boldsymbol{y}, \boldsymbol{\gamma}_{\setminus i}, \boldsymbol{\beta})}{f(\gamma_i = 0|\boldsymbol{y}, \boldsymbol{\gamma}_{\setminus i}, \boldsymbol{\beta})} = c_i^{-d_i} exp\left(0.5 \frac{c_i^2 - 1}{c_i^2} \boldsymbol{\beta}_i^T \Sigma_i^{-1} \boldsymbol{\beta}_i\right) \tag{9}$$

where d_i is the dimension of $\boldsymbol{\beta}_i$.

The prior for $\boldsymbol{\gamma}$ with each term present or absent independently with probability 1/2 may be considered non-informative in the sense that it gives the same weight to all possible models. George and Foster (1997) argue that this prior can be considered as informative because it puts more weight on models of size close to $p/2$. However, posterior model probabilities are most heavily dependent on the choice of the prior parameters c_i^2 and Σ_i. One way of specifying these is by setting $c_i^2 \Sigma_i$ as a diffuse prior (for $\gamma_i = 1$) and then choosing c_i^2 by considering the the value of $|\boldsymbol{\beta}_i|$ at which the densities of the two components of the prior distribution are equal. This can be considered to be the smallest value of $|\boldsymbol{\beta}_i|$ at which the term is considered of practical significance. George and McCulloch (1993) applied this approach. Ntzoufras *et al.*(1997) considered log-linear interaction models where $\boldsymbol{\beta}_i$ terms are multidimensional.

2.3 Unconditional Priors for Variable Selection

Kuo and Mallick (1998) advocated the use of the linear predictor $\boldsymbol{\eta} = \sum_{i=1}^{p} \gamma_i \boldsymbol{X}_i \boldsymbol{\beta}_i$ introduced in (3) for variable selection. They considered a prior distribution $f(\boldsymbol{\beta})$ which is independent of $\boldsymbol{\gamma}$ (and therefore $\mathcal{M}$) so that $f(\boldsymbol{\beta}_i|\boldsymbol{\beta}_{\setminus i}, \boldsymbol{\gamma}) = f(\boldsymbol{\beta}_i|\boldsymbol{\beta}_{\setminus i})$

Therefore, the full conditional posterior distributions are given by

$$f(\boldsymbol{\beta}_i|\boldsymbol{y}, \boldsymbol{\gamma}, \boldsymbol{\beta}_{\setminus i}) \propto \begin{cases} f(\boldsymbol{y}|\boldsymbol{\gamma}, \boldsymbol{\beta}) f(\boldsymbol{\beta}_i|, \boldsymbol{\beta}_{\setminus i}) & \gamma_i = 1 \\ f(\boldsymbol{\beta}_i|\boldsymbol{\beta}_{\setminus i}) & \gamma_i = 0 \end{cases} \tag{10}$$

and

$$\frac{f(\gamma_i = 1|\boldsymbol{y}, \boldsymbol{\gamma}_{\setminus i}, \boldsymbol{\beta})}{f(\gamma_i = 0|\boldsymbol{y}, \boldsymbol{\gamma}_{\setminus i}, \boldsymbol{\beta})} = \frac{f(\boldsymbol{y}|\gamma_i = 1, \boldsymbol{\gamma}_{\setminus i}, \boldsymbol{\beta})}{f(\boldsymbol{y}|\gamma_i = 0, \boldsymbol{\gamma}_{\setminus i}, \boldsymbol{\beta})} \frac{f(\gamma_i = 1, \boldsymbol{\gamma}_{\setminus i})}{f(\gamma_i = 0, \boldsymbol{\gamma}_{\setminus i})}. \tag{11}$$

The advantage of the above approach is that it is extremely straightforward. It is only required to specify the usual prior on $\boldsymbol{\beta}$ (for the full model) and the conditional prior distributions $f(\boldsymbol{\beta}_i|\boldsymbol{\beta}_{\setminus i})$ replace the pseudopriors required by Carlin and Chib's method. However, this simplicity may also be a drawback, as there is no flexibility here to alter the method to improve efficiency. In practice, if, for any $\boldsymbol{\beta}_i$, the prior is diffuse compared with the posterior, the method may be inefficient.

2.4 Gibbs Variable Selection

Dellaportas *et al.*(1997) considered a natural hybrid of SSVS and the "Unconditional Priors" approach of Kuo and Mallick (1998). The linear predictor is assumed to be of the form of (3) where unlike SSVS, variables corresponding to $\gamma_i = 0$ are genuinely excluded from the model. The prior for $(\boldsymbol{\gamma}, \boldsymbol{\beta})$ is specified as $f(\boldsymbol{\gamma}, \boldsymbol{\beta}) = f(\boldsymbol{\gamma}) f(\boldsymbol{\beta}|\boldsymbol{\gamma})$. Consider the partition of $\boldsymbol{\beta}$ into $(\boldsymbol{\beta}_{\boldsymbol{\gamma}}, \boldsymbol{\beta}_{\setminus\boldsymbol{\gamma}})$ corresponding to those components of $\boldsymbol{\beta}$ which are included ($\gamma_i = 1$) or not included ($\gamma_i = 0$) in the model, then the prior $f(\boldsymbol{\beta}|\boldsymbol{\gamma})$ may be partitioned into model prior $f(\boldsymbol{\beta}_{\boldsymbol{\gamma}}|\boldsymbol{\gamma})$ and pseudoprior $f(\boldsymbol{\beta}_{\setminus\boldsymbol{\gamma}}|\boldsymbol{\beta}_{\boldsymbol{\gamma}}, \boldsymbol{\gamma})$.

The full conditional posterior distributions are given by

$$f(\boldsymbol{\beta}_{\boldsymbol{\gamma}}|\boldsymbol{\beta}_{\setminus\boldsymbol{\gamma}}, \boldsymbol{\gamma}, \boldsymbol{y}) \propto f(\boldsymbol{y}|\boldsymbol{\beta}, \boldsymbol{\gamma}) f(\boldsymbol{\beta}_{\boldsymbol{\gamma}}|\boldsymbol{\gamma}) f(\boldsymbol{\beta}_{\setminus\boldsymbol{\gamma}}|\boldsymbol{\beta}_{\boldsymbol{\gamma}}, \boldsymbol{\gamma}) \tag{12}$$

$$f(\boldsymbol{\beta}_{\setminus\boldsymbol{\gamma}}|\boldsymbol{\beta}_{\boldsymbol{\gamma}}, \boldsymbol{\gamma}, \boldsymbol{y}) \propto f(\boldsymbol{\beta}_{\setminus\boldsymbol{\gamma}}|\boldsymbol{\beta}_{\boldsymbol{\gamma}}, \boldsymbol{\gamma}) \tag{13}$$

and

$$\frac{f(\gamma_i = 1|\boldsymbol{\gamma}_{\setminus i}, \boldsymbol{\beta}, \boldsymbol{y})}{f(\gamma_i = 0|\boldsymbol{\gamma}_{\setminus i}, \boldsymbol{\beta}, \boldsymbol{y})} = \frac{f(\boldsymbol{y}|\boldsymbol{\beta}, \gamma_i = 1, \boldsymbol{\gamma}_{\setminus i})}{f(\boldsymbol{y}|\boldsymbol{\beta}, \gamma_i = 0, \boldsymbol{\gamma}_{\setminus i})} \frac{f(\boldsymbol{\beta}|\gamma_i = 1, \boldsymbol{\gamma}_{\setminus i})}{f(\boldsymbol{\beta}|\gamma_i = 0, \boldsymbol{\gamma}_{\setminus i})} \frac{f(\gamma_i = 1, \boldsymbol{\gamma}_{\setminus i})}{f(\gamma_i = 0, \boldsymbol{\gamma}_{\setminus i})}. \tag{14}$$

This approach is simplified if it is assumed that the prior for $\boldsymbol{\beta}_i$ depends only on γ_i and is given by

$$f(\boldsymbol{\beta}_i|\gamma_i) = \gamma_i N(0, \Sigma_i) + (1 - \gamma_i) N(\tilde{\mu}_i, S_i). \tag{15}$$

This prior, where $f(\boldsymbol{\beta}_i|\boldsymbol{\gamma}) = f(\boldsymbol{\beta}_i|\gamma_i)$ potentially makes the method less efficient and is most appropriate in examples where $\boldsymbol{X}$ is orthogonal. In prediction, rather than inference about the variables themselves is of primary interest, then $\boldsymbol{X}$ may always be chosen to be orthogonal (see Clyde *et al.*, 1996).

There is a similarity between this prior and the prior used in SSVS. However, here the full conditional posterior distribution is given by

$$f(\boldsymbol{\beta}_i|\boldsymbol{\gamma}, \boldsymbol{\beta}_{\setminus i}, \boldsymbol{y}) \propto \begin{cases} f(\boldsymbol{y}|\boldsymbol{\gamma}, \boldsymbol{\beta}) N(0, \Sigma_i) & \gamma_i = 1 \\ N(\tilde{\mu}_i, S_i) & \gamma_i = 0 \end{cases}$$

and a clear difference between this and SSVS is that the pseudoprior $f(\beta_i|\gamma_i = 0)$ does not affect the posterior distribution and may be chosen as a "linking density" to increase the efficiency of the sampler, in the same way as the pseudopriors of Carlin and Chib's method. Possible choices of $\tilde{\mu}_i$ and S_i may be obtained from a pilot run of the full model; see, for example Dellaportas and Forster (1999).

2.5 Summary of Variable Selection Strategies

The similarities and differences between the three Gibbs sampling variable selection methods presented in Sections 2.2, 2.3 and 2.4 may easily be summarized by inspecting the conditional probabilities (8), (11) and, in particular, (14).

In SSVS, $f(\boldsymbol{y}|\boldsymbol{\beta},\boldsymbol{\gamma})$ is independent of $\boldsymbol{\gamma}$ and so the first ratio on the right hand side of (14) is absent in (8). For the 'Unconditional Priors' approach of Kuo and Mallick (1998), the second term on the right hand side of (14) is absent in (11) as $\boldsymbol{\beta}$ and $\boldsymbol{\gamma}$ are *a priori* independent. For Gibbs Variable Selection, both likelihood and prior appear in the variable selection step. These differences are also evident by looking at the graphical representations of the three methods in Figure 16.1.

The key differences between the methods (including Carlin and Chib's method) are in their requirements in terms of prior and/or linking densities. Carlin and Chib's method and GVS both require linking densities whose sole function is to aid the efficiency of the sampler. GVS is less expensive in requirement of pseudopriors, but correspondingly less flexible. The prior parameters in SSVS all have an impact on the posterior, and therefore the densities cannot really be thought of linking densities. The simplest method that described by Kuo and Mallick (1988) does not require one to specify anything other than the usual priors for the model parameters.

3. Illustrative Example: 2 × 2 × 2 Contingency Table

We present an analysis of the data in Table 16.1, taken from Healy (1988). This is a three-way table with factors A,B and C. Factor A denotes the condition of the patient (more or less severe), factor B denotes if the patient was accepting antitoxin medication and the (response) factor C denotes whether the patient survived or not.

		Survival(C)	
Condition (A)	Antitoxin (B)	No	Yes
More Severe	Yes	15	6
	No	22	4
Less Severe	Yes	5	15
	No	7	5

TABLE 16.1. Example Dataset.

Purely for illustration purposes, and to present the BUGS code in Appendix A, we model the above data using both log-linear and logistic regression models.

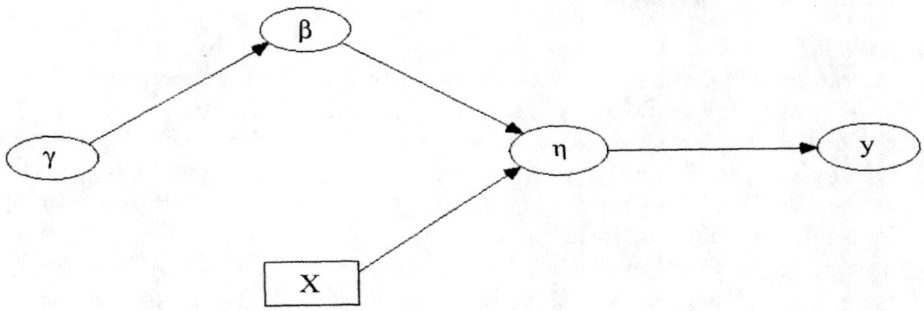

SSVS Graphical Model

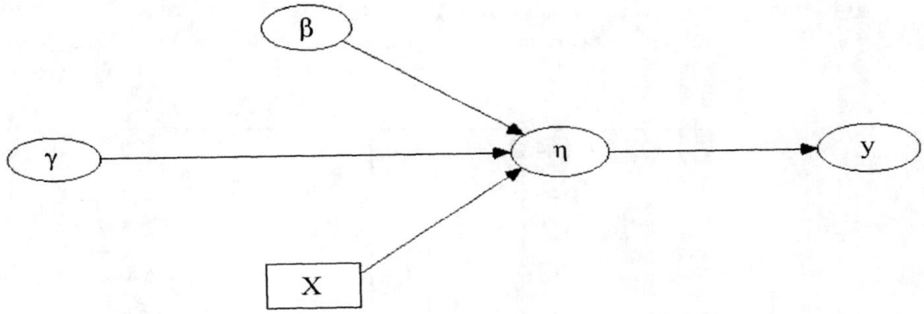

Kuo and Mallick Graphical Model

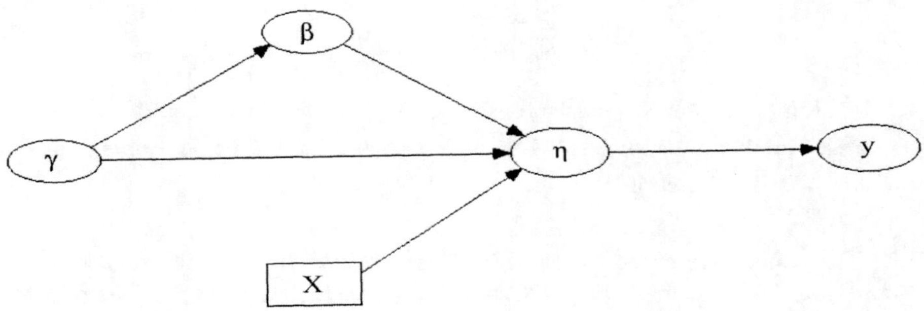

Gibbs Variable Selection Graphical Model

FIGURE 16.1. Graphical Model Representation for Stochastic Search Variable Selection, Kuo and Mallick Sampler and Gibbs Variable Selection [Squares denote Constants; Circles denote Stochastic Nodes].

3.1 Log-Linear models

We focus attention on hierarchical models including the main effects focusing our interest on associations between model factors and the corresponding interaction terms in the models. Here, $i \in \{1, A, B, C, AB, AC, BC, ABC\}$ so $p = 8$. The prior specification for model vector γ is $\gamma_i \sim Bernoulli(\pi)$ with $\pi = 1/9$ if $i = ABC$, $\pi = 1$ if $i \in \{1, A, B, C\}$ and $\gamma_i|\gamma_{ABC} \sim Bernoulli(\pi)$ with $\pi = 0.5(1 - \gamma_{ABC}) + \gamma_{ABC}$ for the two factor interactions ($i \in \{AB, AC, BC\}$). This specification implies that the prior probability of including a two factor interaction in the model is 0.5 if the three factor interaction is excluded from the model and 1 if it is included in the model. Hence the prior probabilities for all 9 possible hierarchical models are 1/9 and and non-hierarchical models are not considered.

For the model coefficients we used the prior specification suggested by Dellaportas and Forster (1999) for log linear models which results in $\Sigma_i = 2$ in (15) when the β_i are considered to be the usual 'sum-to-zero' constrained model parameters For SSVS we used $c_i^2\Sigma_i = 2$ and $c_i = 10^3$ in (7), as suggested by Ntzoufras *et al.*(1997).

		SSVS	KM	GVS
Models	$A+B+C$	0.1	0.2	0.2
	$AB+C$	0.0	0.1	0.1
	$AC+B$	25.1	25.7	25.6
	$BC+A$	0.3	0.6	0.6
	$AB+AC$	7.9	7.5	7.3
	$AB+BC$	0.1	0.2	0.2
	$AC+BC$	58.9	58.4	58.9
	$AB+BC+CA$	6.4	6.6	6.4
	ABC	1.0	0.8	0.6

TABLE 16.2. Posterior model probabilities (%) for log-linear models. SSVS: Stochastic Search Variable Selection; KM: Kuo and Mallick's Unconditional Priors approach; GVS: Gibbs Variable Selection.

The results are based on 100,000 iterations for Gibbs variable selection and Kuo and Mallick's method, and 400,000 iterations for SSVS which seemed to be less efficient. For all methods we discarded 10,000 iterations as a burn-in period. The pseudoprior densities for Gibbs variable selection were constructed from the sample moments of a pilot run of the full model of size 1,000 iterations. All three methods give similar results supporting the same models with very similar posterior probabilities.

3.2 Logistic Regression Models

When we consider binomial logistic regression models for response variable C and explanatory factors A and B, there are 5 possible nested models, 1, A, B, $A+B$ and AB. Priors are specified by setting $c_i^2\Sigma_i = 4 \times 2$ in (7) and $\Sigma_i = 4 \times 2$ in (15) which is equivalent to the prior used above for log-linear model selection. The pseudoprior parameters were specified as before, through a pilot chain, and finally we set $\gamma_{ABC} \sim Bernoulli(1/5)$ and $\gamma_i|\gamma_{AB} \sim Bernoulli(\pi)$, with $\pi = 0.5(1 - \gamma_{AB}) + \gamma_{AB}$ for

$i \in \{A, B\}$. The resulting prior probabilities for all models are 1/5. The results in table (16.3) are based on 500,000 iterations for SSVS and Kuo and Mallick's method and 100,000 iterations for Gibbs variable selection, with burn-in period of 10,000 iterations. Again, the results are very similar, although Gibbs variable selection seemed to be most efficient.

The equivalent log-linear models in Table 16.2 are those which include the AB term, so the results can be seen to be in good agreement.

		SSVS	KM	GVS
Models	1	0.2	0.5	0.5
	A	48.0	49.2	49.3
	B	1.0	1.2	1.2
	$A+B$	45.3	44.0	43.9
	AB	5.5	5.2	5.1

TABLE 16.3. Posterior model probabilities (%) for logistic regression. SSVS: Stochastic Search Variable Selection; KM: Kuo and Mallick's Unconditional Priors approach; GVS: Gibbs Variable Selection.

4. Discussion

We have reviewed a number of Bayesian variable selection strategies based on the Gibbs sampler. Their major practical advantage is that they can be easily applied with a Gibbs sampling software such as BUGS.

It is impossible to provide a general recommendation for a method of computation for a class of problems as large as variable selection in generalized linear models. The methods we have discussed range from the "Unconditional Priors approach" which is extremely easy to implement, but may be insufficiently flexible for many practical problems, to the approach of Carlin and Chib, which is very flexible, but requires a lot of careful specification.

We have only discussed methods based on the Gibbs sampler. Of course other extremely flexible MCMC methods exist, such the reversible jump approach introduced by Green (1996). All MCMC methods require careful implementation and monitoring, and other approaches should also be considered. For many model selection problems involving generalized linear models, an alternative approach is through asymptotic approximation. Raftery (1996) has provided a series of Splus routines for this kind of calculation. Such methods can be used in conjunction with the Gibbs sampler approaches discussed here.

Any Bayesian model selection requires careful attention to prior specification. For discussion of elicitation of prior distributions for variable selection, see Garthwaite and Dickey (1992) and Ibrahim and Chen (1998).

5. Appendix: BUGS CODES

Code and data files are freely available in the web adress *http://www.stat-athens.aueb.gr/~jbn/* or by electronic mail request.

5.1 Code for Log-linear Models for 2^3 Contingency Table

```
model loglinear;
#
#       2x2x2 LOG-LINEAR VARIABLE SELECTION WITH BUGS
#       (c) OCTOBER 1996 FIRST VERSION
#       (c) OCTOBER 1997 FINAL VERSION
#           WRITTEN BY IOANNIS NTZOUFRAS
#       ATHENS UNIVERSITY OF ECONOMICS AND BUSINESS
#
#           SSVS: Stochastic Search Variable Selection
#           KM  : Kuo and Mallick Gibbs sampler
#           GVS : Gibbs Variable Selection
#
const
    N = 8;  # number of Poisson cells
var
        include,     # conditional prior probabability for gi
        pmdl[9],     # model indicator vector
        mdl,         # code of model
        b[N],        # model coefficients
        mean[N],     # mean used in pseudoprior   (GVS only)
        se[N],       # st.dev. used in pseudoprior(GVS only)
        bpriorm[N],  # prior mean for b depanding on g
        tau[N],      # model coefficients precision
#       c,           # precision multiplicator    (SSVS only)
        x[N,N],      # design matrix
        z[N,N],      # matrix used in likelhood
        n[N],        # Poisson cells
        lambda[N],   # Poisson mean for each cell
        g[N];        # term indicator vector
data n,x in "ex1log.dat", mean, se in 'prop1ll.dat';
inits in "ex1ll.in";
{
#       c<-1000.0 # SSVS only
#
#       calculation of the z matrix used in likelihood
        for (i in 1:N) { for (j in 1:N) {
               z[i,j]<-x[i,j]*b[j]*g[j]    # For GVS/KM
#               z[i,j]<-x[i,j]*b[j];       # For SSVS
               }}
#
#       model configuration
        for (i in 1:N) {
                log(lambda[i])<-sum(z[i,]);
                n[i]~dpois(lambda[i])    }
#       defining model code
#       0 for [A][B][C], 1 for [AB][C],      2 for [AC][B],
#       3 for [AB][AC],  4 for [BC][A],      5 for [AB][BC],
#       6 for [AC][BC],  7 for [AB][BC][CA],15 for [ABC].
#
        mdl<-g[5]+2*g[6]+4*g[7]+8*g[8];
        for (i in 0:7) { pmdl[i+1]<-equals(mdl,i) }
        pmdl[9]<-equals(mdl,15)
#
#       Prior for b model coefficient
        tau[1]<-0.1;
        bpriorm[1]<-0.0;
        b[1]~dnorm(bpriorm[1],tau[1]);
        for (i in 2:N) {
#
```

```
#                    GVS using se,mean from pilot run
#                ------------------------------------------
                 tau[i]<-g[i]/2+(1-g[i])/(se[i]*se[i]);
                 bpriorm[i]<-mean[i]*(1-g[i]);
#
#                Kuo and Mallick (prior indepedent of g[i])
#                ------------------------------------------
#                 tau[i]<-1/2;
#                 bpriorm[i]<-0.0;
#
#
#                           SSVS PRIOR SET-UP
#                ------------------------------------------
#                 tau[i]<-pow(c,2-2*g[i])/2;
#                 bpriorm[i]<-0.0;
#
                 b[i]~dnorm(bpriorm[i],tau[i]);
         }
#
#        defining prior information for gi in such way that
#        allow only hierarhical models with equal probability.
#
         include<-(1-g[8])*0.5+g[8]*1.0;
         g[8]~dbern(0.1111111);
         g[7]~dbern(include);
         g[6]~dbern(include);
         g[5]~dbern(include);
         for (i in 1:4) { g[i]~dbern(1.0)}}
```

5.2 Code for Logistic Models with 2 Binary Explanatory Factors

```
model Binomial;
#
#        LOGISTIC REGRESSION VARIABLE SELECTION WITH BUGS
#        (c) OCTOBER 1996 FIRST VERSION
#        (c) OCTOBER 1997 FINAL VERSION
#            WRITTEN BY IOANNIS NTZOUFRAS
#        ATHENS UNIVERSITY OF ECONOMICS AND BUSINESS
#
#            SSVS: Stochastic Search Variable Selection
#            KM  : Kuo and Mallick Gibbs sampler
#            GVS : Gibbs Variable Selection
#
const
   N = 4;  # number of binomial experiments
var
      include,    # conditional prior probabability for gi
      pmdl[5],    # model indicator vector
      mdl,        # code of model
      b[N],       # model coefficients
      mean[N],    # mean used in pseudoprior    (GVS only)
      se[N],      # st.dev, used in pseudoprior (GVS only)
      bpriorm[N],# prior mean for b depanding on g
      tau[N],     # model coefficients precision
#     c,          # precision multiplicator    (SSVS only)
      x[N,N],     # design matrix
      z[N,N],     # matrix used in likelhood
      r[N],       # number of successes in binomial
      n[N],       # total number of observations for binomial
      p[N],       # probability of success for binomial model
      g[N];       # term indicator vector
data r,n,x in "ex1logit.dat", mean, se in 'prop1.dat';
inits in "ex1.in";
{
#         c<-1000 # SSVS only
#
#        calculation of the z matrix used in likelihood
         for (i in 1:N) { for (j in 1:N) {
```

```
                z[i,j]<-x[i,j]*b[j]*g[j]  # for GVS
#               z[i,j]<-x[i,j]*b[j];      # for SSVS
                }}
#
#       model configuration
        for (i in 1:N) {
                r[i]~dbin(p[i],n[i]);
                logit(p[i])<-sum(z[i,]) }
#       defining model code
#       0 constant, 1 for [A], 2 for [B],
#       3 for [A][B], and 6 for [AB]
#
        mdl<-g[2]+2*g[3]+3*g[4];
        pmdl[1]<-equals(mdl,0)
        pmdl[2]<-equals(mdl,1)
        pmdl[3]<-equals(mdl,2)
        pmdl[4]<-equals(mdl,3)
        pmdl[5]<-equals(mdl,6)
#
#       Prior for b model coefficient
        tau[1]<-0.1;
        bpriorm[1]<-0.0;
        b[1]~dnorm(bpriorm[1],tau[1]);
        for (i in 2:N) {
#
#               GVS using se,mean from pilot run
#               --------------------------------
#
                tau[i]<-g[i]/8+(1-g[i])/(se[i]*se[i]);
                bpriorm[i]<-mean[i]*(1-g[i]);
#
#               Kuo and Mallick proposal is indedent of g[i]
#               --------------------------------------------
#
#                tau[i]<-1/8;
#                bpriorm[i]<-0.0;
#
#                           SSVS PRIOR SET-UP
#                ------------------------------------------
#                tau[i]<-pow(c,2-2*g[i])/8;
#                bpriorm[i]<-0.0;
#
                b[i]~dnorm(bpriorm[i],tau[i]);
        }
#
#       defining prior information for gi in such way that
#       allow only hierarhical models with 0.2 probability.
#
        g[4]~dbern(0.2);
        include<-(1-g[4])*0.5+g[4]*1.0
        g[2]~dbern(include);
        g[3]~dbern(include);
        g[1]~dbern(1.0)              }
```

References

Agresti, A. (1990). *Categorical Data Analysis.* John Wiley and Sons, New York.

Brown, P.J., Vannucci, M. and Fearn, T. (1998). Multivariate Bayesian Variable Selection and Prediction. *Journal of Royal Statistical Society, B, 60,* 627–641.

Carlin, B.P. and Chib, S. (1995). Bayesian Model Choice via Markov Chain Monte Carlo Methods. *Journal of Royal Statistical Society, B, 57,* 473–484.

Chipman, H. (1996). Bayesian Variable Selection with Related Predictors. *Canadian Journal of Statistics, 24*, 17–36.

Chipman, H., Hamada, M., Wu, C.F.J. (1997). A Bayesian Variable-Selection Approach for Analysing Designed Experiments with Complex Aliasing. *Technometrics, 39*, 372–381.

Clyde, M. and DeSimone-Sasinowska, H. (1997). Accounting for Model Uncertainty in Poisson Regression Models: Does Particulate Matter? *Technical Report*, Institute of Statistics and Desicion Sciences, Duke University, USA.

Clyde, M., DeSimone, H. and Parmigiani, G. (1996). Prediction via Orthognalized Model Mixing. Journal of the American Statistical Association, 91, 1197–1208.

Dellaportas, P. and Forster, J.J. (1999). Markov Chain Monte Carlo Model Determination for Hierarchical and Graphical Models. *Biometrika*, to appear.

Dellaportas, P., Forster, J.J. and Ntzoufras, I.(1997). On Bayesian Model and Variable Selection Using MCMC. *Technical Report*, Department of Statistics, Athens University of Economics and Business, Greece.

Draper, D. (1995). Assesment and Propogation of Model Uncertainty (with discussion). *Journal of the Royal Statistical Society, B, 57*, 45–97.

Garthwaite, P.H. and Dickey, J.M. (1992). Elicitation of Prior Distributions for Variable-Selection Problems in Regression. *The Annals of Statistics, 20*, 1697–1719.

George, E.I. and Foster, D.P. (1997). Calibration and Empirical Bayes Variable Selection. *Technical Report*, University of Texas at Austin and University of Pennsylvania, USA.

George, E.I. and McCulloch, R.E. (1993). Variable Selection via Gibbs Sampling. *Journal of the American Statistical Association, 88*, 881–889.

George, E.I. and McCulloch, R.E. (1996). Stochastic Search Variable Selection. *Markov Chain Monte Carlo in Practice*, eds. W.R.Gilks , S.Richardson and D.J.Spiegelhalter, Chapman and Hall, London, UK, 203–214.

George, E.I., McCulloch, R.E. and Tsay R.S. (1996). Two Approaches for Bayesian Model Selection with Applications. *Bayesian Analysis in Statistics and Econometrics*, eds. D.A.Berry, M.Chaloner and J.K.Geweke, John Wiley and Sons, New York, USA, 339–348.

George, E.I. and McCulloch, R.E. (1997). Approaches for Bayesian Variable Selection. *Statistica Sinica, 7*, 339–373.

Geweke, J. (1996). Variable Selection and Model Comparison in Regression. *Bayesian Statistics 5*, eds. J.M.Bernardo, J.O.Berger, A.P.Dawid and A.F.M.Smith, Claredon Press, Oxford, UK, 609–620.

Green, P.J. (1996). Reversible Jump Markov Chain Monte Carlo Computation and Bayesian Model Determination. *Biometrika, 82*, 711–732.

Healy, M.J.R. (1988). *Glim: An Introduction.* Claredon Press, Oxford, UK.

Hoeting, J.A., Madigan, D., and Raftery, A.E. (1996). A Method for Simultaneous Variable Selection and Outlier Identification in Linear Regression. *Journal of Computational Statistics and Data Analysis, 22*, 251-270.

Ibrahim, J.G. and Chen, M.H. (1998). Prior Elicitation and Variable Selection for Generalized Mixed Models. *Generalized Linear Models: A Bayesian Perspective*, eds. D.K. Dey, S. Ghosh and B. Mallick, Marcel Dekker Publications.

Kuo, L. and Mallick, B. (1998). Variable Selection for Regression models. *Sankhya, B, 60*, Part 1, 65–81.

Madigan, D. and Raftery, A.E. (1994). Model Selection and Accounting for Model Uncertainty in Graphical Models Using Occam's Window. *Journal of the American Statistical Association, 89*, 1535–1546.

Mitchell, T.J. and Beauchamp, J.J. (1988). Bayesian Variable Selection in Linear Regression. *Journal of the American Statistical Association, 83*, 1023–1036.

Ntzoufras, I., Forster, J.J. and Dellaportas, P. (1997). Stochastic Search Variable Selection for Log-linear Models. *Technical Report*, Faculty of Mathematics, Southampton University, UK.

Raftery, A.E. (1996). Approximate Bayes Factors and Accounting for Model Uncertainty in Generalized Linear Models. *Biometrika, 83*, 251–266.

Smith M. and Kohn R. (1996). Nonparametric Regression Using Bayesian Variable Selection. *Journal of Econometrics, 75*, 317–343.

Spiegelhalter, D., Thomas, A., Best, N. and Gilks, W.(1996a), *BUGS 0.5: Bayesian Inference Using Gibbs Sampling Manual.* MRC Biostatistics Unit, Institute of Public health, Cambridge, UK.

Spiegelhalter, D., Thomas, A., Best, N. and Gilks, W.(1996b). *BUGS 0.5: Examples Volume 2*, MRC Biostatistics Unit, Institute of Public health, Cambridge, UK.

Wakefield, J. and Bennett, J. (1996). The Bayesian modelling of Covariates for Population Pharmacokinetic Models. *Journal of the American Statistical Association, 91*, 917–927.

17

Bayesian Methods for Variable Selection in the Cox Model

Joseph G. Ibrahim
Ming-Hui Chen

ABSTRACT We consider the problem of Bayesian variable selection for proportional hazards regression models with right censored data. We discuss a semi-parametric approach in which a nonparametric prior is specified for the baseline hazard rate and a fully parametric prior is specified for the regression coefficients. For the baseline hazard, we investigate the Extended Gamma (EG) process priors and discuss choices of prior parameters suitable for variable selection. For the regression coefficients and the model space, we consider a semi-automatic parametric informative prior specification that focuses on the observables rather than the parameters. We demonstrate that our prior specification is quite useful and flexible for the variable selection problem under a wide variety of situations. To implement the methodology, we use a Markov chain Monte Carlo method to compute the posterior model probabilities. In particular, the computational method presented in this chapter only requires Gibbs sampling from the full model, and thus is highly efficient for the variable selection problem. A simulated example is given to demonstrate the methodology.

1. Introduction

In the analysis of regression models for censored survival data, one often wishes to assess the importance of certain prognostic factors such as age, gender, or race in predicting survival outcome. This is a general problem which is encountered in most clinical trials research in cancer and AIDS. Typically, proportional hazards regression models using Cox's partial likelihood (Cox, 1975) are used to address this problem. Current techniques used for variable selection include asymptotic procedures based on score tests, Wald tests, and other approximate chi-square procedures. These procedures rely on Cox's partial likelihood and therefore do not use the full likelihood function to do variable selection. As is well known, Cox's partial likelihood is an approximation to the full likelihood. Frequentist variable selection based on the full likelihood requires joint estimation of the baseline hazard and the regression coefficients. In this case, a nonparametric estimate of the baseline hazard rate or a fully parametric specification of the survival model would be required. Again, one needs to rely on asymptotics to obtain variable selection criteria. Joint estimation of the baseline hazard and the regression coefficients can be a very difficult task. We have not seen any methods in the statistical literature that address this in the variable selection context.

Bayesian analyses of proportional hazards models using the full likelihood function are becoming computationally feasible due to modern technology and recent advances in computing techniques such as the Gibbs sampler (Gelfand and Smith, 1990) and other Markov chain Monte Carlo (MCMC) methods. The literature on Bayesian variable selection for survival models, however, is still quite sparse at best. A recent article includes Raftery, Madigan and Volinsky (1995). However, this article does not directly address the joint modelling or computations for the regression coefficients and the hazard rate. There are some articles addressing Bayesian analysis of survival models using MCMC methods. These include Clayton (1991), Skene and Wakefield (1990), Kuo and Smith (1990), and Gray (1994). A somewhat related article focusing on variable selection in generalized linear models is by George, McCulloch and Tsay (1995). The potential advantage of using Bayesian methods to jointly model the baseline hazard and the regression coefficients is that one can accurately compute posterior model probabilities and their standard errors using MCMC simulation techniques. However, there still remains the chore of specifying meaningful prior distributions and doing intensive computations. We present a methodology for this here and give the technical details in Sections 2 and 3.

Ibrahim and Laud (1994), and Laud and Ibrahim (1995, 1996) advocate a predictive approach to variable selection for the linear model by adopting the philosophy in Geisser (1993). The predictive approach they recommend is based on the notion of specifying a prior prediction y_0 for the response vector, and a scalar precision parameter c_0 which quantifies one's prior belief in y_0. Then, (y_0, c_0), along with the design matrix for model m, are used to specify an automated parametric informative prior for the regression coefficients $\beta^{(m)}$. The motivation behind this approach is that the investigator often has prior information on the observables from either past studies assuming the form of replicate experiments or from case-specific information on the subjects in the current study. This information is often quantifiable in the form of a vector of prior predictions for the response vector of the current study. The predictive approach is appealing for variable selection problems, since there are so many parameters arising from the different models, and all have different physical meaning, therefore making direct informative prior elicitation for the model parameters generally quite difficult.

This predictive methodology seems to be well suited for cancer and AIDS clinical trials research. In cancer clinical trials, for example, current studies often use treatments that are very similar or slight modifications of treatments used in past studies for a particular disease. The survival times (possibly censored) from a previous study, denoted y_0, can then be viewed as a prior prediction for the survival times of the current study. These prior predictions can then be used along with some design matrix $X_{0,m}$ for model m and a scalar c_0, to elicit a prior distribution for the regression coefficients in the current study. If y_0 is data from a previous study, one can use a formal Bayesian updating approach to the problem by using the posterior of the previous study as the prior distribution for the current study. We elaborate further on advantages and disadvantages of this approach for this problem in Section 2.4. In the framework we develop here, the prior prediction y_0 is quite general in the sense that it does not have to rely on the same design matrix as the current study. Also, y_0 need not be of the same dimension as the vector of survival times in the current experiment and $X_{0,m}$ need not involve covariate values from the current experiment. Specifically, when there is a previous study of sample size n_0 that measures the same covariates as the current study, y_0 consists of the $n_0 \times 1$ raw data vector of survival times in the previous study, ν_0 is the vector of censoring indicators corresponding to y_0, and $X_{0,m}$ is the design matrix from the previous study.

The remainder of the chapter is organized as follows. In Section 2.2 and Section 2.3, we introduce the priors for the baseline hazard and the approximate likelihood that arises from the induced prior on the hazard rate. The prior distributions for the regression coefficients and a prior distribution for the model space are discussed in Sections 2.4 and 2.5. In Sections 3.1 and 3.2, we present a novel method for estimating the marginal distribution of the data and a novel Gibbs sampling scheme for the parameters that facilitates the computation of posterior model probabilities.

In Section 4, we conduct a detailed simulation study. We conclude this chapter with a discussion section.

2. The Method

2.1 Model and Notation

A proportional hazards model is defined by a hazard function of the form

$$h(t,x) = h_b(t)\exp(x'\beta) \ , \tag{1}$$

where $h_b(t)$ denotes the baseline hazard function at time t, x denotes the covariate vector for an arbitrary individual in the population, and β denotes a vector of regression coefficients. Let m denote a specific model in the model space $\mathcal{M}$. A model throughout refers to a subset of covariates from the full model. Under model m, the likelihood function for a set of right censored data on n individuals for the current study in a proportional hazards model based on (1) is given by

$$L\left(\beta^{(m)}, h_b(t)\right) = \prod_{i=1}^{n} \left[h_b(t_i)\exp(\eta_i^{(m)})\right]^{\nu_i} \left(S_b(t_i)^{exp(\eta_i^{(m)})}\right) , \tag{2}$$

where $\eta_i^{(m)} = x_i^{(m)'}\beta^{(m)}, t_i$ is an observed failure time or censoring time for the ith individual and ν_i is the indicator variable taking on the value 1 if t_i is a failure time, and 0 if it is a censoring time. Moreover, $x_i^{(m)}$ is a $k_m \times 1$ vector of covariates for the ith individual under model m, and X_m denotes the $n \times k_m$ covariate matrix of rank k_m. Further, $F_b(\cdot)$ denotes the baseline cumulative distribution function, and $S_b(\cdot) = 1 - F_b(\cdot)$ is the baseline survivor function, which, since we consider continuous survival distributions, is related to $h_b(\cdot)$ by $S_b(t) = \exp\left(-\int_0^t h_b(u)\, du\right)$.

2.2 Prior Distribution for $h_b(\cdot)$

Since the baseline hazard rate plays such a key role in survival analysis, it is appropriate to specify a model for it. We consider a nonparametric prior over the collection of absolutely continuous hazard rates. Specifying a nonparametric prior for the baseline hazard rate seems natural for this problem in order to allow for a large and flexible class of hazards. Moreover, it will also facilitate comparison with Cox's partial likelihood. Specifically, we use the extended gamma (EG) process priors of Dykstra and Laud (1981) for the baseline hazard rate. This prior is suitable for both right censored and exact observations and has several attractive properties.

The motivation for specifying the EG process prior over the hazard rates is that it has the advantage of placing the prior probability on absolutely continuous rather than on discrete distributions, as is the case with the Dirichlet process prior. Though in many situations the discreteness of the Dirichlet process is not a drawback, it can produce serious difficulties in the proportional hazards setting. Additionally, the EG process priors are not piecewise constant, as are some other priors, therefore giving added flexibility in modeling the hazard rate.

There has been some previous work on modeling the cumulative baseline hazard rate. Kalbfleisch (1978) considers gamma process priors for the cumulative baseline hazard and Clayton (1991) illustrates Gibbs sampling techniques for frailty models using gamma process priors on the cumulative baseline hazard rate. The prior specifications of Kalbfleisch (1978) and Clayton (1991) are entirely different from the ones considered here. As Kalbfleisch (1978) points out, the assumption of independent increments for the prior on the baseline cumulative hazard may not be a very satisfactory representation of the prior distribution for the baseline hazard. Alternatively, we directly specify a prior distribution on the baseline hazard rate.

The EG process prior of Dykstra and Laud (1981) can be described as follows. Let $G(\alpha, \lambda)$ denote the gamma distribution with shape parameter $\alpha \geq 0$ and scale parameter λ. For $\alpha = 0$, we define this distribution to be degenerate at 0. For $\alpha > 0$, its density with respect to Lebesgue measure is

$$f(x|\alpha, \lambda) = \begin{cases} \frac{\lambda^\alpha}{\Gamma(\alpha)} x^{\alpha-1} e^{-\lambda x}, & x > 0 \\ 0 & \text{otherwise} \end{cases} .$$

Let $\alpha(t), t \geq 0$, be a non-decreasing left continuous function such that $\alpha(0) = 0$, and let $Z(t), t \geq 0$, be a gamma process with parameter $\alpha(\cdot)$. That is, $Z(0) = 0, Z(t)$ has independent increments, and for $t > s, Z(t) - Z(s)$ is $G(\alpha(t) - \alpha(s), 1)$. Now let $\lambda(t), t \geq 0$ be a positive right continuous function with left hand limits existing bounded away from 0 and ∞, and define

$$h_b(t) = \int_0^t [\lambda(s)]^{-1} \, dZ(s) ,$$

where the integration is with respect to the sample paths of the $Z(t)$ process. The process $\{h_b(t), t \geq 0\}$ is called the Extended Gamma (EG) process and is denoted by

$$h_b(t) \sim \Gamma\left(\alpha(\cdot), \lambda(\cdot)\right) .$$

Provided that $\alpha(s)$ is not 0, the sample paths of an EG process are well defined increasing hazard rates corresponding to absolutely continuous distributions. Since the sample paths of the EG process are almost surely increasing functions, we are placing our prior probability entirely within the class of distributions with increasing hazard rates. We note here that the EG process prior is thus not piecewise constant and much different from the prior considered by Gray (1994).

We have chosen to place a prior distribution over the class of continuous hazard rates for several reasons. A model for a discrete survival distribution implies an extremely bumpy hazard, something which we seek to avoid. The assumption of an underlying continuous process provides a sound mathematical underpinning for the approximation to which we turn and allows an important flexibility in practice. There are many situations in which an increasing hazard rate is appropriate, as in the myeloma study for example; see Ibrahim and Chen (1998). However, there are also situations in which the increasing hazard rate assumption is inappropriate, for example, when a significant proportion of the patients are "cure". We note that such studies require sufficient follow-up. We demonstrate in Section 4, however, that posterior model probabilities and model choice in general are not sensitive to the increasing hazard rate assumption, and this assumption appears to have little impact on the analysis. For the variable selection problem, the baseline hazard rate is viewed as a nuisance parameter since our main goal is to make inference on the regression coefficients. Thus, the prior distribution and the modeling of the baseline hazard rate play a secondary role to our primary aim of variable selection. In any case, we recommend that the techniques presented in this section be used only when the assumption is deemed appropriate.

To define an approximation to the EG process prior on $h_b(t)$ that is computationally tractable, we first construct a finite partition of the time axis. Let $0 \leq s_0 <$

$s_1 < \ldots < s_M$ denote this finite partition, with $s_M > t_j$, for all $j = 1, \ldots, n$. Let

$$\delta_i = h_b(s_i) - h_b(s_{i-1})$$

denote the increment in the baseline hazard in the interval $(s_{i-1}, s_i], i = 1, \ldots, M$. The δ_i's are random variables since the baseline hazard is assumed random. The δ_i's are independent a priori, and have a distribution induced by the underlying EG process. The density of the δ_i's is most easily expressed in terms of its Laplace transform, which is given by

$$L_h(u) = \exp\left\{-\int_0^t \log\left(1 + \frac{u}{\lambda(s)}\right) d\alpha(s)\right\} . \tag{3}$$

An expression for the density of the δ_i with respect to Lebesgue measure is not available for a general $\lambda(s)$. However, the prior reduces to a gamma process prior when $\lambda(s) = \lambda_i$ on $(s_{i-1}, s_i]$. That is, the δ_i's have independent gamma distributions with shape parameters $\alpha(s_i) - \alpha(s_{i-1})$ and scale parameters λ_i. We give more details on how to do computations with EG process priors in Section 3.2. The variance of the EG process is controlled by choosing $\lambda(s_i)$ large (small) to reflect sharp (vague) prior beliefs at various time intervals. Letting $\Delta = (\delta_1, \ldots, \delta_M)$, the prior density of Δ is given by

$$\pi(\Delta) = \prod_{i=1}^{M} f(\delta_i) , \tag{4}$$

where $f(\delta_i)$ is the prior density of δ_i whose Laplace transform is given by (3). The prior parameters of the δ_i's may, in principle, be chosen to depend on m. This dependence may come from the covariates in the current study or other model specific quantities. We do not attempt such a specification here. We essentially view Δ as a nuisance parameter in the variable selection problem, and thus its dependence on m is not as crucial as the dependence of the regression coefficients on m.

Choices of prior parameters for Δ can be made in several ways. Viewing the baseline hazard rate as a nuisance parameter in the variable selection problem, one may take naive choices of prior parameters for the δ_i's such as $\alpha(s) = s_i - s_{i-1}$ for $s_{i-1} \leq s \leq s_i$, and $\lambda(s) = 1$ for all s. This choice of prior parameters results in independent $G(s_i - s_{i-1}, 1)$ priors for the δ_i's with the shape parameter consisting of the interval width and scale equal to 1. An obvious advantage of this choice is that it simplifies calculations. It may be suitable if there is little prior information available on the baseline hazard rate.

A more elaborate choice of prior parameters for the δ_i's can be made by taking into account the mean and the variance of the EG process. The mean and variance of the process are increasing functions denoted by $(\mu(s), \sigma^2(s))$, where

$$\mu(s) = \int_0^s \lambda(u)\alpha'(u)\, du \tag{5}$$

$$\sigma^2(s) = \int_0^s \lambda^2(u)\alpha'(u)\, du , \tag{6}$$

and $\alpha'(u) = d\alpha(u)/du$. As discussed in Dykstra and Laud (1981), it seems reasonable to assign as $\mu(s)$ the "best" guess of the hazard rate and use $\sigma^2(s)$ to model the amount of uncertainty or variation in the hazard rate at point s.

For the variable selection problem, we consider a method of specifying more informative choices for $(\alpha(s), \lambda(s))$ by incorporating the prior prediction y_0 into the elicitation process. Notice that when $\lambda(u) = 1$, $\mu(s) = \sigma^2(s) = \alpha(s)$ from (5) and (6), and therefore $\alpha(s)$ is both the mean and the variance of the process. Moreover, each δ_i has a gamma prior with shape parameter $\alpha_i = \alpha(s_i) - \alpha(s_{i-1})$. In this case, a suitable choice of $\alpha(s)$ would be an increasing estimate the baseline hazard rate. To construct such an estimate, we can fit a Weibull model via maximum likelihood using $D_0 = (n_0, y_0, X_0, \nu_0)$ as the data. Often, the fit will result in a strictly increasing hazard. We denote such a hazard by $h_b(s \mid y_0)$ and set $\mu(s) = h_b(s \mid y_0)$. With the convention that $\lambda(s) = \lambda$ for all s, we solve to obtain $\alpha(s) = \lambda h_b(s \mid y_0)$. A slight generalization of this parameter choice is to take $\lambda(s)$ piecewise constant over the M subintervals and take $\alpha(s) = \lambda(s)\ h_b(s \mid \mu_0)$, where $\lambda(s) = s_i - s_{i-1}$, for $s_{i-1} < s \le s_i$. In the event that the fitted Weibull model results in a constant or decreasing hazard, doubt is cast on the appropriateness of the EG process as a model for the hazard, and we do not recommend the method discussed here.

There are numerous other approaches to selecting this baseline hazard. Alternative classes of parametric models may be fit to y_0 or a nonparametric method such as that of Padgett and Wei (1980) may be used to construct an increasing hazard. Once the hazard is selected, we have specified $\mu(s)$. With a constant $\lambda(s)$, this determines $\alpha(s)$ and the EG process is specified. With non-constant, but relatively (relative to the interval widths) slowly varying $\lambda(s)$, a solid strategy is to take $\lambda(s)$ piecewise constant and proceed to solve for $\alpha(s)$. Whatever details are used in these choices, the advantage of this predictive approach is that we can use the prior information to construct estimates to be used as prior parameters.

2.3 *The Likelihood Function*

We now construct the approximate likelihood function for $(\beta^{(m)}, \Delta)$ for any model $m \in \mathcal{M}$. Let $x^{(m)}$ denote the $k_m \times 1$ vector of covariates for an arbitrary individual under model m. Then, the cumulative distribution function for the proportional hazards model at time s is given by

$$\begin{aligned} F(s) &= 1 - \exp\left\{-\exp\{\eta^{(m)}\} \int_0^s h_b(t)\ dt\right\} \\ &\simeq 1 - \exp\left\{-\exp\{\eta^{(m)}\} \left((s - s_0)^+ h_b(s_0) + \sum_{i=1}^{M} \delta_i (s - s_{i-1})^+\right)\right\}, \end{aligned} \tag{7}$$

where $(t)^+ = t$ if $t > 0$, 0 otherwise, and $\eta^{(m)} = x^{(m)\prime}\beta^{(m)}$. We assume here that $h_b(s_0) = 0$, and $F(s) = 1$ for $s > s_M$, so that (7) is slightly simplified. This first approximation arises since the specification of Δ does not specify the entire hazard rate, but only the δ_i. For purposes of approximation, we take the increment in the hazard rate, δ_i, to occur immediately after s_{i-1}. Let p_i denote the probability of a failure in the interval $(s_{i-1}, s_i], i = 1, \ldots, M$. Using the fact that $h_b(s_0) = 0$, we have

$$\begin{aligned} p_i &= F(s_i) - F(s_{i-1}) \\ &\simeq \exp\left\{-\exp\{\eta^{(m)}\} \sum_{j=1}^{i-1} \delta_j (s_{i-1} - s_{j-1})\right\} \end{aligned}$$

$$\times \left[1 - \exp \left\{ - \exp\{\eta^{(m)}\}(s_i - s_{i-1}) \sum_{j=1}^{i} \delta_j \right\} \right] .$$

Thus, in the *ith* interval $(s_{i-1}, s_i]$, the contribution to the likelihood function for an exact observation (i.e., a failure) is p_i and $1 - F(s_i)$ for a right censored observation. Let d_i be the number of failures, and c_i be the number of right censored observations in the *ith* interval, respectively, $i = 1, \ldots, M$. For ease of exposition, we order the observations so that in the *ith* interval the first d_i are failures and the remaining c_i are right censored, $i = 1, \ldots, M$. Let $x_{ik}^{(m)}$ denote the vector of covariates for the kth individual in the ith interval under model m, and define

$$\begin{aligned}
u_{ik}(\beta^{(m)}) &= \exp\{x_{ik}^{(m)\prime} \beta^{(m)}\} , \\
a_i &= \sum_{j=i+1}^{M} \sum_{k=1}^{d_j} u_{jk}(\beta^{(m)})(s_{j-1} - s_{i-1}) , \\
b_i &= \sum_{j=i}^{M} \sum_{k=d_j+1}^{d_j+c_j} u_{jk}(\beta^{(m)})(s_j - s_{i-1}) , \\
T_i(\Delta) &= (s_i - s_{i-1}) \sum_{j=1}^{i} \delta_j .
\end{aligned}$$

The approximate likelihood function, given the data D for the current study, over all M intervals is given by

$$\begin{aligned}
L(\beta^{(m)}, \Delta \mid D) &= \left\{ \prod_{i=1}^{M} \exp\left\{ -\delta_i(a_i + b_i) \right\} \right\} \\
&\quad \times \left\{ \prod_{i=1}^{M} \prod_{k=1}^{d_i} \left(1 - \exp\{-u_{ik}(\beta^{(m)}) T_i(\Delta)\} \right) \right\} . \qquad (8)
\end{aligned}$$

We note that this likelihood involves a second approximation. Instead of conditioning on exact event times, we condition on the intervals in which events occur, and thus we approximate continuous right censored data by interval censored data.

2.4 Prior Distribution for the Regression Coefficients

In general, for most problems, there are no firm guidelines on the method of prior elicitation. Typically, one tries to balance sound theoretical ideas with practical and computationally feasible ones. A common situation that arises in statistical practice is that data D_0 from a previous study is available to use as prior information for the current study. The issue of how to incorporate D_0 into the current study has no obvious solution since it depends in large part of how similar the two studies are. In most clinical trials, for example, no two studies will ever be identical. In many cancer clinical trials, the patient populations typically differ from study to study even when the same regimen is used to treat the same cancer. In addition, other factors may make the two studies heterogeneous. These include conducting the studies at different institutions or geographical locations, using different physicians, using different measurement instruments and so forth. Due to these differences, an analysis which combines the data from both studies may not be desirable. In this case, it may be more appropriate to "weight" the data from the previous study so as to control its impact on the current study. Thus, it is desirable for the investigators to have a prior distribution that summarizes the prior data D_0 in an efficient and useful manner and allows them to tune or weight D_0 as they see fit in order to control its impact on the current study. In addition to this, it is always desirable to have a prior distribution that is easy to interpret and has a convenient workable closed form. The prior distribution we consider here satisfies these conditions. It is also a very practical and useful prior which approximates a more elaborate prior based on formal Bayesian updating as discussed in more detail below.

We assume a priori independence between the the baseline hazard rate and the regression coefficients, and thus the joint prior density of $(\beta^{(m)}, \Delta)$ under model m is given by

$$\pi(\beta^{(m)}, \Delta) = \pi(\beta^{(m)})\ \pi(\Delta)\ . \tag{9}$$

The assumption of prior independence between $\beta^{(m)}$ and Δ is a sensible specification, since we are viewing the hazard rate as a nuisance parameter for our problem. We consider a fully parametric multivariate normal prior for $\beta^{(m)}$, since the normal prior has proved to be a flexible and useful class of priors for many regression problems (see Geisser, 1993). Thus let $No_p(\mu, T)$ denote the p dimensional multivariate normal distribution with mean μ and *precision* matrix T. Thus, under model m, we take

$$\beta^{(m)} \sim No_{k_m}(\mu^{(m)}, c_0\ T_m)\ , \tag{10}$$

where c_0 is a scalar quantifying the degree of prior belief one wishes to attach to $\mu^{(m)}$. Under model m, we have the prior information $D_0 = (n_0, y_0, X_{0,m}, \nu_0)$ and c_0, where $X_{0,m}$ is an $n_0 \times k_m$ design matrix, y_0 is an $n_0 \times 1$ vector of prior predictions, and ν_0 is the corresponding $n_0 \times 1$ vector of censoring indicators. We take the prior mean of $\beta^{(m)}$ to be the solution to Cox's partial likelihood equations for $\beta^{(m)}$ using D_0 as data. Suppose there are r failures and $n_0 - r$ right censored values in y_0. Cox's partial likelihood for $\beta^{(m)}$ based on D_0 is given by

$$L^*(\beta^{(m)}) = \prod_{i=1}^{r} \left\{ \frac{\exp\{x_{0i}^{(m)'} \beta^{(m)}\}}{\sum_{\ell \in \mathcal{R}(y_{0i})} \exp\{x_{0\ell}^{(m)'} \beta^{(m)}\}} \right\}\ , \tag{11}$$

where $x_{0i}^{(m)'}$ is the *ith* row of $X_{0,m}$, $(y_{01}, \ldots, y_{0r})$ are the ordered failures and $\mathcal{R}(y_{0i})$ is the set of labels attached to the individuals at risk just prior to y_{0i}. Now we take $\mu^{(m)}$ to be the solution to

$$\frac{\partial \log\left(L^*(\beta^{(m)})\right)}{\partial \beta_j^{(m)}} = 0\ , \tag{12}$$

$j = 1, \ldots, k_m$. The matrix T_m is taken to be the Fisher information matrix of $\beta^{(m)}$ based on the partial likelihood in (11). Thus

$$T_m = \left[\frac{-\partial^2}{\partial \beta_i^{(m)} \partial \beta_j^{(m)}} \log(L^*(\beta^{(m)})) \right] \Bigg|_{\beta^{(m)} = \mu^{(m)}}\ . \tag{13}$$

An attractive feature of the priors for $\beta^{(m)}$ is that they are semi-automatic in the sense that one only needs a one time input of (D_0, c_0) to generate the prior distributions for all $m \in \mathcal{M}$. The priors defined by (9) and (10) represent a summary of the prior data D_0 through $(\mu^{(m)}, T_m)$ which are obtained via Cox's partial likelihood. This is a practical and useful summary of the data D_0 as indicated by many authors including Cox (1972, 1975), and Tsiatis (1981).

The prior given by (9) and (10) has several advantages. First, it has a closed form and is easy to interpret. Second, the prior elicitation is straightforward in the sense that $(\mu^{(m)}, T_m)$ and c_0 completely determine the prior for $\beta^{(m)}$ for all $m \in \mathcal{M}$. Third, (9) and (10) are computationally feasible and relatively simple. Fourth, our prior assumes a priori independence between $(\beta^{(m)}, \Delta)$, which further simplifies interpretations as well as the elicitation scheme.

In addition, (9) and (10) provide a reasonable asymptotic approximation to the prior for $(\beta^{(m)}, \Delta)$ that is obtained from a formal Bayesian update of the data D_0. To see this, let $\pi_0(\beta^{(m)}, \Delta)$ denote the "original" joint prior for $(\beta^{(m)}, \Delta)$ for the previous study, and let $L(\beta^{(m)}, \Delta \mid D_0)$ denote the likelihood function of $(\beta^{(m)}, \Delta)$ for the previous study. Then, for the current study, a formal Bayesian update leads to a prior distribution that is equal to the posterior distribution based on the previous study. That is, the prior distribution of $(\beta^{(m)}, \Delta)$ for the current study based on a formal Bayesian update of the original prior would take the form

$$\pi_1(\beta^{(m)}, \Delta) \propto L(\beta^{(m)}, \Delta \mid D_0)\, \pi_0(\beta^{(m)}, \Delta) . \qquad (14)$$

To simplify the specification, we can take $(\beta^{(m)}, \Delta)$ to be independent in the original prior so that $\pi_0(\beta^{(m)}, \Delta) = \pi_0(\beta^{(m)})\, \pi_0(\Delta)$. Moreover, it is reasonable to take $\pi_0(\beta^{(m)}, \Delta)$ to be a vague proper prior. The original prior $\pi_0(\beta^{(m)}, \Delta)$ does not depend on D_0, and thus $\pi(\beta^{(m)}, \Delta)$ depends on D_0 only through $L(\beta^{(m)}, \Delta \mid D_0)$. One of the major drawbacks with using the prior in (14) is that it does not allow the investigator to directly assign a weight to the prior data D_0. That is, (14) does not allow a weight for $L(\beta^{(m)}, \Delta \mid D_0)$. For example, if $\pi_0(\beta^{(m)}, \Delta)$ is chosen to be non-informative, then (14) effectively weights D_0 and D equally, which may be undesirable, especially in situations where the two studies are not very similar. To control the impact of the prior data D_0 in (14), we can weight $L(\beta^{(m)}, \Delta \mid D_0)$ in an appropriate fashion. We thus modify (14) to take the form

$$\pi_2(\beta^{(m)}, \Delta) \propto \left[L(\beta^{(m)}, \Delta \mid D_0)\right]^{c_0} \pi_0(\beta^{(m)})\, \pi_0(\Delta) , \qquad (15)$$

where $c_0 \geq 0$ is a scalar prior parameter. The case $c_0 = 1$ essentially corresponds to combining the prior data D_0 and the current data D. The case $c_0 = 0$ can be used if no previous study exists and the investigator wishes to specify only an original prior $\pi_0(\beta^{(m)}, \Delta)$. Although (15) may seem like an attractive prior to use, it is relatively difficult to interpret and does not have a closed form. For example, it is not clear what the prior moments of (15) are. Moreover, (15) is quite computationally difficult to work with. Our priors for $(\beta^{(m)}, \Delta)$ given by (9) and (10) provide a reasonable asymptotic approximation to (15). In particular, our prior is based on a normal approximation to $\left[L(\beta^{(m)}, \Delta \mid D_0)\right]^{c_0}$ with $\pi_0(\beta^{(m)})$ taken as a vague proper prior. This is a sensible and practical approximation, since for most inference problems, likelihoods are well approximated by normal distributions for large samples under suitable regularity conditions.

Since the prior parameters $(\mu^{(m)}, T_m)$ in (10) do not depend on Δ, $\beta^{(m)}$ and Δ are independent a priori. This seems to be a reasonable and computationally practical assumption. Moreover, it facilitates a more straightforward semi-automatic prior elicitation which is easy to interpret. Also, as shown in Section 3, our suggested priors have several computational advantages over (14) or (15). Although our methods do not formally allow an exact combining of the raw data from the previous and current study, (10) is a reasonable approximation to combining the data from the previous study if one chooses $c_0 = 1$. One advantage of our priors is that they provide a useful summary of D_0 even if D_0 is elicited from expert opinion. That is, one does not need an actual previous study to use these priors. The priors considered in this section give us a clear specification of our beliefs immediately before incorporating data from the current study. We view our simplification of the prior as an almost necessary step to facilitate the description of the analysis.

If we have a previous study with the same set of covariates as the current study then the choice of D_0 is straightforward. In this case, n_0 would be taken as the sample size from the previous study, $X_{0,m}$ is taken as the design matrix from the previous study under model m, y_0 is taken as the raw data vector of survival times from the previous study along with its censoring indicators ν_0. The parameter c_0 resembles the one from Zellner's g-priors (Zellner, 1986). Possible choices of c_0 include $c_0 = n_0/n$, or $c_0 = \log(n_0)$ where n is the sample size of the current study. Other potential choices of c_0 are discussed in Zellner (1986). If the set of covariates in the current study is a subset of the covariates in the previous study, then we can construct a submatrix $X_{0,m}$ by omitting those columns corresponding to covariates not in the current study.

If the set of covariates in the previous study is a subset of the set of covariates in the current study, then one possibility is to use a combination of prior data and expert opinion to construct the prior. Alternatively, the investigator may specify vague prior information for the regression coefficients of the new covariates. In any case, a prior for $\beta^{(m)}$ can be obtained as follows. Let X_{new} denote the $n \times s$ matrix of covariates not measured in the previous study with corresponding regression coefficients β_{new}. Let $X_{new,m}$ be the $n \times s_m$ matrix of new covariates under model m. Thus, the columns of $X_{new,m}$ are a subset of those of X_{new}. Also, let $\beta_{new}^{(m)}$ denote the $s_m \times 1$ vector of regression coefficients corresponding to $X_{new,m}$. We partition $\beta^{(m)}$ into

$$\beta^{(m)} = \begin{pmatrix} \beta_{old}^{(m)} \\ \\ \beta_{new}^{(m)} \end{pmatrix} ,$$

where $\beta_{old}^{(m)}$ is an $r_m \times 1$ vector which represents the regression coefficients corresponding to the set of covariates X_{old} common to the two studies. Thus, $\beta^{(m)}$ is a $k_m \times 1$ vector of regression coefficients in the current study, where $k_m = r_m + s_m$. In our prior specification, we assume that the new covariates have small or negligible correlation to the old covariates, that is, $\text{Corr}(X_{old}, X_{new}) \approx 0$. This seems to be a sensible assumption if in fact the new set of covariates in the current study are being scientifically investigated for the first time. The prior distribution of $\beta^{(m)} = \begin{pmatrix} \beta_{old}^{(m)} \\ \\ \beta_{new}^{(m)} \end{pmatrix}$ is given by

$$\beta^{(m)} \sim No_{k_m}(\mu^{(m)}, \tilde{T}_m) , \tag{16}$$

where

$$\mu^{(m)} = \begin{pmatrix} \mu_{old}^{(m)} \\ \mu_{new}^{(m)} \end{pmatrix}$$

and

$$\tilde{T}_m = \begin{pmatrix} c_0 T_{m,old} & A_0^{(m)} \\ A_0^{(m)\prime} & b_0 R_{0,m} \end{pmatrix} .$$

Here, $A_0^{(m)}$ is an $r_m \times s_m$ matrix, $R_{0,m}$ is the prior precision matrix for the new regression coefficients, b_0 is a scalar, and $\mu_{new}^{(m)}$ is the prior mean for $\beta_{new}^{(m)}$. The investigator can specify $A_0^{(m)}$, $R_{0,m}$, b_0, and $\mu_{new}^{(m)}$ by expert opinion or by using case specific information from the current study. For example, $A_0^{(m)}$ and $R_{0,m}$ can be constructed from the covariates of the current study. Vague choices of these parameters are also possible by taking $A_0^{(m)} = 0$, $b_0 = .01$ or $b_0 = .001$, $\mu_{new}^{(m)} = 0$, and $R_{0,m}$ to be a diagonal matrix with diagonal elements equal to the inverses of the sample variances of the new covariates. In any case, we always recommend that sensitivity analyses be conducted for several values of these parameters. Note that taking $A_0^{(m)} = 0$ implies that $\beta_{new}^{(m)}$ and $\beta_{old}^{(m)}$ are independent a priori, and taking $R_{0,m}$ to be diagonal implies that the components of $\beta_{new}^{(m)}$ are independent a priori.

If a relevant previous study does not exist, D_0 may be specified as follows. The prior prediction y_0 may be elicited from expert opinion or by using case-specific information about each individual in the current study, and perhaps by using a design matrix X_{m^*} based on a given model m^* to obtain a point prediction of the form

$$y_0 = g(X_{m^*}),$$

where $g(.)$ is a specified function. The vector y_0 may also be obtained by a theoretical model giving forecasts for the survival times. Also, in this case one could take $X_{0,m} = X_m$ and $n_0 = n$. Thus, we use the design matrix of the current study to obtain the prior distribution for $\beta^{(m)}$ in (10).

The elicitation scheme is less automated than the one in which a previous study exists. There are a number of other ways one could elicit y_0, and in general, the elicitation depends on the context of the problem. We do not attempt to give a general elicitation scheme for y_0 in this setting, but rather mention some general possibilities. In any case, the cleanest specification of D_0 is to use data from a previous study for which y_0 would be taken to be the vector of survival times from the previous study and $X_{0,m}$ is taken as the design matrix from the previous study under model m.

2.5 *Prior Distribution on the Model Space*

The prior prediction y_0, whether the actual result of a previous study or a prediction based on expert opinion, can always be viewed as prior data. Viewing y_0 this way, we specify an original prior $\pi_0(\beta^{(m)}, \Delta)$ for the model parameters as described in Section 2.4. We also specify an original prior for the model space, denoted $p_0(m)$. Our strategy is to specify vague proper original priors, and then update $p_0(m)$ by Bayes theorem to obtain prior model probabilities, $p(m)$, for the current study.

To this end, we specify a uniform original prior on $\mathcal{M}$, that is, $p_0(m) = 2^{-k}$ for all $m \in \mathcal{M}$. Moreover, we take $\pi_0(\beta^{(m)}, \Delta) = \pi_0(\beta^{(m)})\pi_0(\Delta)$ to be a vague proper prior as discussed in Section 2.4. Given the prior prediction y_0, the prior probability of model m for the current study based on an update of y_0 via Bayes theorem is given by

$$p(m) \equiv p(m|D_0) = \frac{p(D_0|m)\ p_0(m)}{\sum_{m\in\mathcal{M}} p(D_0|m)\ p_0(m)} , \tag{17}$$

where

$$p(D_0|m) = \int\int L(\Delta, \beta^{(m)}|D_0)\ \pi_0(\beta^{(m)})\ \pi_0(\Delta)\ d\beta^{(m)}\ d\Delta\ , \tag{18}$$

and $L(\Delta, \beta^{(m)}|D_0)$ is the likelihood function of the parameters based on D_0. Thus, the prior probability of model m for the current study is precisely $p(m) \equiv p(m|D_0)$.

This framework for specifying $p(m)$ is most natural if we actually had a previous study that yielded survival times y_0. In this case, $L(\Delta, \beta^{(m)}|D_0)$ is just the likelihood function of the parameters based on the data D_0 from the previous study, $\pi_0(\beta^{(m)})$ is the original prior distribution of the regression coefficients from the previous study and $\pi_0(\Delta)$ is the original prior for the baseline hazard rate for the previous study. Then, Bayes theorem is used in the usual way to update.

The original prior $\pi_0(\beta^{(m)}, \Delta)$ is mainly viewed here as a necessary device to calculate $p(D_0 \mid m)$, and hence the prior model probabilities $p(m)$. The main intent is to pick a vague proper prior $\pi_0(\beta^{(m)}, \Delta)$ that yields a reasonable set of prior model probabilities for $\mathcal{M}$. The choice of $\pi_0(\beta^{(m)})$ we consider here is a multivariate normal prior with mean 0 (i.e., no regression) and a diagonal precision matrix having small diagonal elements. Thus, with a slight abuse of notation, we write

$$\pi_0(\beta^{(m)}) = No_{k_m}(0, d_0^{(m)} W_{0,m})\ ,$$

where $W_{0,m}$ is a diagonal matrix and $d_0^{(m)}$ is a positive scalar changing with each m. The ith diagonal element of $W_{0,m}$ can be chosen to be equal to the ith diagonal element of T_m defined in (13). If the covariates are all standardized or are measured on the same scale, then we can take $W_{0,m} = I$.

The parameter $d_0^{(m)}$ plays a major role in assigning the prior probability to model m. Large values of $d_0^{(m)}$ will tend to increase the prior probability for model m. It seems reasonable to let $d_0^{(m)}$ depend on k_m in some automated and consistent fashion. Following Laud and Ibrahim (1996), we let

$$v_0^{(m)} = \frac{d_0^{(m)}}{1 + d_0^{(m)}}\ ,$$

and take $v_0^{(m)}$ to be of the form

$$v_0^{(m)} = b\ a^{\frac{1}{k_m}}\ , \tag{19}$$

where b and a are specified prior parameters in $[0, 1]$. As Laud and Ibrahim (1996) point out, the decomposition of $v_0^{(m)}$ into the form in (19) provides flexibility in assigning a prior for $\mathcal{M}$. The parameters a and b play an important role in this formulation, especially in determining prior probabilities for models of different size. One method of specifying a and b is as follows. Let k denote the number of covariates for the full model, and suppose, for example, the investigator wishes to assign prior weights in the range $r_1 = .10$ for the single covariate model (*i.e.*, $k_m = 1$) to $r_2 = .15$ for the full model (*i.e.*, $k_m = k$). This leads to the equations

$$r_1 = ba = .10 \tag{20}$$

and

$$r_2 = ba^{\frac{1}{k}} = .15 \tag{21}$$

for which a closed form solution for a and b can be obtained. The choice of a and b, and hence $d_0^{(m)}$, is made with the goal of letting, within each model, the likelihood govern the posterior. This suggests taking $d_0^{(m)}$ small, i.e., r_1 close to r_2. On the other hand, the models in $\mathcal{M}$ have different dimensions, and so taking $d_0^{(m)}$ too small will lead to a posterior that favors models of small dimension very strongly. In practice, the values of a and b obtained from the procedure defined in (20) and (21) provides only a reasonable starting point. In actual practice, we recommend an iterative process to select the final values of a and b. Thus to select the final values, we compute the prior model probabilities given in (17) for various sets of (a, b). Final selection of (a, b) is made to ensure that a reasonable set of models receive non-negligible probability. Special attention must be paid to models of very large and very small dimension. If the covariate sets from the previous study and the current study are not identical, then we make the modifications to $\pi_0(\beta^{(m)})$ in a similar way as was done for $\pi(\beta^{(m)})$ in Section 2.4.

We take the original prior density for Δ, $\pi_0(\Delta)$, to be a product of M_0 independent gamma densities each with parameters (f_i, g_i), $i = 1, \ldots, M_0$, which leads to

$$\pi_0(\Delta) \propto \prod_{i=1}^{M_0} \delta_i^{f_i - 1} \exp\{-\delta_i g_i\} \ .$$

Again, we take a vague proper prior for $\pi_0(\Delta)$, and thus choose g_i very small. Again, in practice we recommend an iterative process for selecting a final set of (f_i, g_i). We start with a specific choice, such as f_i equal to the ith subinterval width and g_i small. Then, we compute the prior model probabilities under several choices using the procedure discussed above to decide on a final set. We note here that we have allowed the construction of $\pi_0(\Delta)$ to depend on a partition, M_0, of the time axis which may be different from the partition used to construct $\pi(\Delta)$. This flexibility is especially useful and more practical when D_0 is based on a previous study, since the data from the previous study may behave quite differently from the data in the current study. Whenever possible, we try to select $M = M_0$.

Finally, once $\pi_0(\beta^{(m)})$ and $\pi_0(\Delta)$ are specified, we use Bayes theorem to update and obtain (17). We emphasize here that $\pi_0(\beta^{(m)}, \Delta)$ only serves as a device for generating a sensible set of prior model probabilities, and thus plays a minimal role otherwise.

3. Computational Implementation

In this section, we discuss the method for computing the posterior model probabilities. The method involves computing the marginal distribution of the data via ratios of normalizing constants. The method requires posterior samples only from the full model for computing posterior probabilities for all possible models. The method is thus very efficient for variable selection. In addition, we devise a novel method for Gibbs sampling from the joint posterior distribution of the parameters by introducing latent variables and by showing that the required posterior densities are log-concave.

3.1 Computing the Marginal Distribution of the Data

The posterior probability of model m (for the current study) is given by

$$p(m|D) = \frac{p(D|m)\, p(m)}{\sum_{m \in \mathcal{M}} p(D|m)\, p(m)} \ , \tag{22}$$

where $p(D|m)$ denotes the marginal distribution of the data given model m, and $p(m)$ denotes the prior probability of model m given by (17). The marginal density $p(D|m)$ corresponds precisely to the normalizing constant of the joint posterior density of $(\Delta, \beta^{(m)})$. That is,

$$p(D|m) = \int\int L(\beta^{(m)}, \Delta \mid D)\ \pi(\beta^{(m)})\ \pi(\Delta)\ d\beta^{(m)}\ d\Delta. \tag{23}$$

Recently, many methods have been developed for estimating normalizing constants of posterior distributions. These include Chen and Shao (1997a), Geyer (1994), Chib (1995), Meng and Wong (1996), and Gelman and Meng (1994, 1996). The basic idea in all of these methods is that samples from the posterior distribution are used to estimate the normalizing constant. Since it is required to simultaneously estimate all posterior model probabilities given in (22), the aforementioned methods are either computationally expensive or inapplicable. However, two recent methods developed by Chen and Shao (1997b, 1998) are very attractive and can be adapted to estimate all of the posterior model probabilities $p(m|D)$.

Let $\beta = (\beta_1, \beta_2, \ldots, \beta_k)'$ denote the regression coefficients for the *full* model and enumerate the models in $\mathcal{M}$ by $m = 1, 2, \ldots, \mathcal{K}$ where $\mathcal{K}$ is the dimension of $\mathcal{M}$ and model $\mathcal{K}$ denotes the *full* model. We also write $\beta = (\beta^{(m)\prime}, \beta^{(-m)\prime})'$ where $\beta^{(-m)}$ is β with $\beta^{(m)}$ deleted. Finally, we let $p(\beta, \Delta|D)$ denote the posterior distribution of the *full* model, that is,

$$p(\beta, \Delta|D) \ \propto \ L(\beta, \Delta \mid D)\ \pi(\beta)\ \pi(\Delta)\ .$$

Suppose that under the full model, we have a posterior sample $\{(\beta_{(j)}, \Delta_{(j)}),\ j = 1, \ldots, N\}$. We explain in detail in Section 3.2 how to obtain these posterior samples. Using the result given in Chen and Shao (1997b), we have the key identity

$$\begin{aligned}\frac{p(D|m)}{p(D|\mathcal{K})} &= E\left(\frac{L(\beta^{(m)}, \Delta \mid D)\pi(\beta^{(m)})\pi(\Delta)w(\beta^{(-m)}|\beta^{(m)}, \Delta)}{L(\beta, \Delta \mid D)\pi(\beta)\pi(\Delta)}\right) \\ &= E\left(\frac{L(\beta^{(m)}, \Delta \mid D)\pi(\beta^{(m)})w(\beta^{(-m)}|\beta^{(m)}, \Delta)}{L(\beta, \Delta \mid D)\pi(\beta)}\right),\end{aligned} \tag{24}$$

where the expectation is taken with respect to the posterior density $p(\beta,\ \Delta\ |D)$ under the full model. Since Δ does not depend on m, we get the reduction above. In (24), the weight function $w(\beta^{(-m)}|\beta^{(m)}, \Delta)$ is a *completely* known conditional density of $\beta^{(-m)}$ given $(\beta^{(m)}, \Delta)$. Chen and Shao (1997a) show that the best choice of $w(\beta^{(-m)}|\beta^{(m)}, \Delta)$ is the conditional density of $\beta^{(-m)}$ given $(\beta^{(m)}, \Delta)$ with respect to the full model posterior distribution $p(\beta, \Delta|D)$. Since the closed form expression of this conditional density is not available, we follow an empirical procedure provided by Chen (1994) to select a $w(\beta^{(-m)}|\beta^{(m)}, \Delta)$. Specifically, using the posterior sample $\{(\beta_{(j)}, \Delta_{(j)}), j = 1, \ldots, N\}$, we construct the posterior mean and covariance matrix, denoted by $(\tilde{\beta}, \tilde{\Sigma})$, and then we choose $w(\beta^{(-m)}|\beta^{(m)}, \Delta)$ to be the conditional density of the normal distribution $No_k(\tilde{\beta}, \tilde{\Sigma}^{-1})$ for $\beta^{(-m)}$ given $\beta^{(m)}$. Since the dimension of β is often smaller than the dimension of Δ for our problem, we may also simply choose $w(\beta^{(-m)}|\beta^{(m)}, \Delta)$ as the conditional density of $\beta^{(-m)}$ given $\beta^{(m)}$ with respect to the prior $\pi(\beta)$. Note that here we choose w independent of Δ since the correlation between β and Δ is small or negligible.

We have the following Monte Carlo scheme to simultaneously estimate $p(m|D)$ for $m = 1, 2, \ldots, \mathcal{K}$. It can be easily observed that using (24), (22) can be rewritten as

$$p(m|D) = \frac{E\left(\frac{L(\beta^{(m)},\Delta|D)\pi(\beta^{(m)})w(\beta^{(-m)}|\beta^{(m)},\Delta)}{L(\beta,\Delta|D)\pi(\beta)}\right)p(m)}{\sum_{r=1}^{\mathcal{K}} E\left(\frac{L(\beta^{(r)},\Delta|D)\pi(\beta^{(r)})w(\beta^{(-r)}|\beta^{(r)},\Delta)}{L(\beta,\Delta|D)\pi(\beta)}\right)p(r)}. \tag{25}$$

Then, following the Monte Carlo method of Chen and Shao (1997b) and using (25) along with the posterior sample $\{(\beta_{(j)}, \Delta_{(j)}), j = 1, \ldots, N\}$, the posterior probability of model m can be estimated by

$$\hat{p}(m|D) = \frac{\frac{1}{N}\sum_{j=1}^{N}\left(\frac{L(\beta_{(j)}^{(m)},\Delta_{(j)}|D)\pi(\beta_{(j)}^{(m)})w(\beta_{(j)}^{(-m)}|\beta_{(j)}^{(m)},\Delta_{(j)})}{L(\beta_{(j)},\Delta_{(j)}|D)\pi(\beta_{(j)})}\right)p(m)}{\sum_{r=1}^{\mathcal{K}}\frac{1}{N}\sum_{j=1}^{N}\left(\frac{L(\beta_{(j)}^{(r)},\Delta_{(j)}|D)\pi(\beta_{(j)}^{(r)})w(\beta_{(j)}^{(-r)}|\beta_{(j)}^{(r)},\Delta_{(j)})}{L(\beta_{(j)},\Delta_{(j)}|D)\pi(\beta_{(j)})}\right)p(r)}, \tag{26}$$

for $m = 1, 2, \ldots, \mathcal{K}$, where $\beta_{(j)} = (\beta_{(j)}^{(m)\prime}, \beta_{(j)}^{(-m)\prime})'$. According to the discussion above, we use a normal distribution

$$No_k(\tilde{\beta}, \tilde{\Sigma}^{-1}) \tag{27}$$

to construct the weight density $w(\beta^{(-m)}|\beta^{(m)}, \Delta)$ where $No_k(\tilde{\beta}, \tilde{\Sigma}^{-1})$ may be either taken to be the prior distribution $\pi(\beta)$ or it can be obtained by using method of moments estimates based on the posterior sample $\{(\beta_{(j)}, \Delta_{(j)}), j = 1, \ldots, N\}$. Therefore, in (26), $w(\beta^{(-m)}|\beta^{(m)}, \Delta)$ can be calculated by

$$\begin{aligned} w(\beta^{(-m)}|\beta^{(m)}, \Delta) &= (2\pi)^{-\frac{(k-k_m)}{2}}|\tilde{\Sigma}_{11.2m}|^{-\frac{1}{2}} \\ &\times \exp\left\{-\frac{1}{2}(\beta^{(-m)} - \tilde{\mu}_{11.2m})'\tilde{\Sigma}_{11.2m}^{-1}(\beta^{(-m)} - \tilde{\mu}_{11.2m})\right\}, \end{aligned} \tag{28}$$

where

$$\tilde{\Sigma}_{11.2m} = \tilde{\Sigma}_{11m} - \tilde{\Sigma}_{12m}\tilde{\Sigma}_{22m}^{-1}\tilde{\Sigma}_{12m}',$$

$\tilde{\Sigma}_{11m}$ is the covariance matrix from the marginal distribution of $\beta^{(-m)}$, $\tilde{\Sigma}_{12m}$ consists of the covariances between $\beta^{(-m)}$ and $\beta^{(m)}$, and $\tilde{\Sigma}_{22m}$ is the covariance matrix of the marginal distribution of $\beta^{(m)}$ with respect to the joint normal distribution $No_k(\tilde{\beta}, \tilde{\Sigma}^{-1})$ for the whole vector β. Also in (28),

$$\tilde{\mu}_{11.2m} = \tilde{\mu}^{(-m)} + \tilde{\Sigma}_{12m}\tilde{\Sigma}_{22m}^{-1}(\beta^{(m)} - \tilde{\mu}^{(m)}),$$

where $\tilde{\mu}^{(-m)}$ is the mean of the marginal distribution of $\beta^{(-m)}$ implied by (27) and $\tilde{\mu}^{(m)}$ is the mean of the marginal distribution of $\beta^{(m)}$ implied by (27). Once w is evaluated, $\hat{p}(m|D)$ in (26) can be computed.

There are several advantages of the above Monte Carlo procedure. First, we need only one random draw from the posterior distribution for the *full* model, which will greatly ease the computational burden. Second, it will be more computationally efficient since there are a lot of common terms that cancel out in the ratios of the two densities. In particular, we see that $\pi(\Delta)$ completely cancels out in the calculation of $\hat{p}(m|D)$. The cancellation of $\pi(\Delta)$ in (26) is especially attractive since the normalizing constant for $\pi(\Delta)$ under the EG process prior is typically difficult to obtain for an arbitrary value of $\lambda(s)$. Even for the simple case where $\lambda(s) = s$, one needs to use the inversion formula for the Laplace transform for evaluating the closed form of the prior density of Δ. Third, this procedure is almost automatic. Finally, note that in (26), $p(\beta, \Delta|D)$ plays the role of the ratio importance sampling density (see Chen and Shao, 1997a) which needs to be known only up to a normalizing constant, since this constant is cancelled out in the calculation.

3.2 Sampling from the Posterior Distribution of $(\beta^{(m)}, \Delta)$

Here we describe in more detail the computational aspects for the Bayesian model discussed in Section 2. Our main objective is to sample from the joint posterior distribution of $(\beta^{(m)}, \Delta)$, and once these samples are obtained, we use the Monte Carlo procedure presented in Section 3.1 to estimate $p(m|D)$. Using D_0 as the data, we use an almost identical procedure to estimate $p(m)$.

To obtain posterior samples with the EG process priors, we follow an algorithm similar to Laud, Smith, and Damien (1996). In their paper, they provide an algorithm for sampling from the EG process, but they do not introduce covariates nor discuss proportional hazards models. We adapt their algorithm here for the proportional hazards model given in Section 2.1 by using a Gibbs sampling procedure. We denote the distribution of a random vector X by $[X]$. For convenience, we order the observations so that the first d_i are exact (i..e, failures) and the remaining c_i are right censored for a total of $n_i = d_i + c_i$ observations in the ith interval.

To obtain samples from $[\beta, \Delta|D]$, we describe a Gibbs sampling strategy for sampling from $[\Delta|\beta, D]$ and $[\beta|\Delta, D]$. The posterior density of $\Delta|\beta, D$ is given by

$$p(\Delta|\beta, D) \propto \left\{\prod_{i=1}^{M} \exp\left(-\delta_i(a_i + b_i)\right)\right\} \times \left\{\prod_{i=1}^{M}\prod_{k=1}^{d_i} (1 - \exp\left\{-u_{ik}(\beta)T_i(\Delta)\right\})\right\} \left\{\prod_{i=1}^{M} \pi(\delta_i)\right\} , \qquad (29)$$

where $u_{ik}(\beta) = \exp\{x_{ik}'\beta\}$, $a_i = \sum_{j=i+1}^{M} \sum_{k=1}^{d_j} \exp\{x_{jk}'\beta\}(s_{j-1} - s_{i-1})$, and $b_i = \sum_{j=i}^{M} \sum_{k=d_j+1}^{d_j+c_j} \exp\{x_{jk}'\beta\}(s_j - s_{i-1})$. Following, Laud et al. (1996), an efficient method of sampling from (29) is to define latent variables in order to make the components of Δ independent a posteriori. We do this by first defining

$$e_i = (e_{i1}, \ldots, e_{id_i}) , \qquad i = 1, \ldots, M$$

to be independent exponential random variables truncated at 1 with mean equal to $(T_i(\Delta)u_{ik}(\beta))^{-1}$. Thus each e_{ik} has density

$$f(e_{ik}) = \begin{cases} \left(\frac{T_i(\Delta)u_{ik}(\beta)}{1-\exp\{-T_i(\Delta)u_{ik}(\beta)\}}\right) \exp\left\{-e_{ik}T_i(\Delta)u_{ik}(\beta)\right\} , & e_{ik} \le 1 \\ 0 & \text{otherwise} \end{cases} .$$

Letting $e = (e_1, \ldots, e_M)$, we can write the posterior distribution of $[\Delta \mid \beta,\ e,\ D]$ as

$$p(\Delta|\beta, e, D) \propto \left\{\prod_{i=1}^{M} (T_i(\Delta))^{d_i}\right\} \left\{\exp\left\{-\sum_{i=1}^{M}\sum_{k=1}^{d_i} e_{ik}T_i(\Delta)u_{ik}(\beta)\right\}\right\} \times \left\{\prod_{i=1}^{M} \exp\left\{-\delta_i(a_i + b_i)\right\}\right\} \left\{\prod_{i=1}^{M} \pi(\delta_i)\right\} .$$

Next we consider additional latent variables

$$q_i = (q_{i1}, \ldots, q_{ii}) , \qquad i = 1, \ldots, M ,$$

where q_i are independent multinomials. Each q_i is an i-cell multinomial of d_i independent trials with probability of the kth cell defined to be $p_k = \frac{\delta_k}{\sum_{j=1}^{i} \delta_j}$. Letting

$q = (q_1, \ldots q_M)$, we are led to

$$\begin{aligned} p(\Delta|\beta, e, q, D) &\propto \prod_{i=1}^{M}\prod_{k=1}^{i}\left\{\frac{\delta_k^{q_{ik}}}{\sum_{j=1}^{i}\delta_j^{q_{ik}}}\right\}\left\{\prod_{i=1}^{M}\left[(s_i - s_{i-1})\sum_{j=1}^{i}\delta_j\right]^{di}\right\} \\ &\times \left\{\exp\left\{-\sum_{i=1}^{M}\sum_{k=1}^{di} e_{ik}T_i(\Delta)u_{ik}(\beta)\right\}\right\} \\ &\times \left\{\prod_{i=1}^{M}\exp\{-\delta_i(a_i + b_i)\}\right\}\left\{\prod_{i=1}^{M}\pi(\delta_i)\right\}. \end{aligned} \quad (30)$$

Equation (30) can be simplified further. Define $w_{ik}(\beta) = u_{ik}(\beta)e_{ik}$ and let $w_{i+}(\beta) = \sum_{k=1}^{d_i} w_{ik}(\beta)$. Since

$$\begin{aligned} &\exp\left\{-\sum_{i=1}^{M}\sum_{j=1}^{i}\delta_j w_{i+}(\beta)(s_i - s_{i-1})\right\} \\ &= \prod_{i=1}^{M}\exp\left\{-\delta_i\sum_{k=i}^{M} w_{k+}(\beta)(s_k - s_{k-1})\right\}, \end{aligned}$$

we can write (30) as

$$\begin{aligned} p(\Delta|\beta, e, q, D) &\propto \left\{\prod_{i=1}^{M}\delta_i^{\sum_{k=i}^{M} q_{ki}}\right\} \\ &\times \left\{\prod_{i=1}^{M}\exp\left\{-\delta_i\left(a_i + b_i + \sum_{k=i}^{M} w_{k+}(\beta)(s_k - s_{k-1})\right)\right\}\right\}\left\{\prod_{i=1}^{M}\pi(\delta_i)\right\}. \end{aligned} \quad (31)$$

We see that from (31) that given the latent variables (e, q), the posterior density of Δ consists of a product of the marginal posterior densities of the δ_i's, thus implying independence. We use (31) in the Gibbs sampling scheme to sample Δ.

The posterior density of $\beta|\Delta, D$ is given by

$$\begin{aligned} p(\beta|\Delta, D) &\propto \left\{\prod_{i=1}^{M}\exp\{-\delta_i(a_i + b_i)\}\right\} \\ &\times \left\{\prod_{i=1}^{M}\prod_{k=1}^{d_i}(1 - \exp\{-u_{ik}(\beta)T_i(\Delta)\})\right\}\pi(\beta). \end{aligned} \quad (32)$$

Therefore, to obtain samples from $[\beta, \Delta \mid D]$, we use a Gibbs sampling scheme to sample from the following four distributions:

a) $[\Delta|\beta, e, q, D]$
b) $[e|\beta, \Delta, q, D]$
c) $[q|\beta, \Delta, e, D]$
d) $[\beta|\Delta, D]$.

As pointed out by Laud et al. (1996), the prior density of Δ is infinitely divisible, and thus the distribution in a) has the form of a gamma density times an infinitely divisible density. Bondesson (1982) and Damien, Laud and Smith (1995) propose an algorithm for sampling from infinitely divisible distributions. Here, we use Bondesson's algorithm to sample from the prior density of Δ, and follow the same basic steps as Laud et al. (1996) to obtain samples from a). Thus to sample from a), we

1. Simulate N independent exponential variables with parameter $\lambda^* = \alpha(s_i) - \alpha(s_{i-1})$.

2. Define an N-vector $T = (x_1, x_1 + x_2, x_1 + x_2 + x_3, \ldots, x_1 + \ldots + x_N)$, where x_i are *i.i.d* exponential random variables with parameter λ^*, $i = 1, \ldots, N$.

3. Sample N independent random variables $V_1, \ldots, V_N$ with distribution proportional to $\alpha(s)$ restricted to $(s_{i-1}, s_i]$. Thus the V_i's have density of the form

$$f_{v_i}(s) = \begin{cases} \frac{\alpha(s)}{\int_{s_{i-1}}^{s_i} \alpha(u)du}, & s_{i-1} < s \leq s_i \\ 0 & \text{otherwise} \end{cases} .$$

4. Sample N independent exponential variables $Z_i, i = 1, \ldots, N$ where each Z_i has an exponential distribution with parameter $\lambda(V_i)\exp(T_i)$, $i = 1, \ldots N$. Note that $\lambda(V_i)$ is the scale parameter of the EG process prior.

5. Define $X = \sum_{i=1}^{N} Z_i$. Then $X \sim \Gamma(\alpha(s_i), \lambda(s_i))$.

6. Given X, we use a rejection algorithm with the gamma density as the envelope to obtain a sample from the distribution in a). The gamma density used in the rejection algorithm proportional to

$$\prod_{i=1}^{M} \delta_i^{\sum_{k=i}^{M} q_{ki}} \times \prod_{i=1}^{M} \exp\left\{ -\delta_i \left(a_i + b_i + \sum_{k=i}^{M} w_{k+}(\beta)(s_k - s_{k-1}) \right) \right\} , \tag{33}$$

which has mode equal to

$$\tilde{m} = \frac{\sum_{k=i}^{M} q_{ki}}{a_i + b_i + \sum_{k=i}^{M} w_{k+}(\beta)(s_k - s_{k-1})} .$$

Now, we use the gamma envelope in (33) along with the mode $\tilde{m}$ in a standard rejection algorithm to decide upon acceptance or rejection of X from $[\Delta|\beta, e, q, D]$.

Distributions b) and c) are quite straightforward to sample from since they correspond to truncated exponential and multinomial distributions, respectively. Specifically,

$$p(e|\beta, \Delta, q, D) \propto \exp\left\{ -\sum_{i=1}^{M} \sum_{k=1}^{d_i} e_{ik} u_{ik}(\beta) T_i(\Delta) \right\} \left\{ \prod_{i=1}^{M} \prod_{k=1}^{d_i} I(e_{ik} \leq 1) \right\} ,$$

where $I(.)$ is the indicator function and

$$p(q|\beta, \Delta, e, D) \propto \prod_{i=1}^{M} \delta_i^{\sum_{k=1}^{M} q_{ki}} .$$

Cycling through a), b), and c) via Gibbs will yield samples from $[\Delta|\beta, D]$. Once a sample of Δ is obtained from $[\Delta|\beta, D]$, we complete the Gibbs cycle by sampling from $[\beta|\Delta, D]$. To obtain a sample β from this distribution, we first observe that $[\beta|\Delta, D]$ is log-concave in each component of β. Therefore, we may directly use the algorithm of Gilks and Wild (1992) to sample from this posterior distribution. To show that $[\beta|\Delta, D]$ is log-concave in each component of β, it suffices to show that

$$\frac{\partial^2 \log p(\beta|\Delta, D)}{\partial \beta_r^2} \leq 0$$

for all $r = 1, \ldots, k$. Letting $A_{ik}(\beta, \Delta) = u_{ik}(\beta)T_i(\Delta)$, $B_{ik}(\beta, \Delta) = 1 - \exp\{-A_{ik}(\beta, \Delta)\}$, and $C_{ik}(\beta, \Delta) = 1 - A_{ik}(\beta, \Delta) - \exp\{-A_{ik}(\beta, \Delta)\}$ we get

$$\frac{\partial^2 \log(p(\beta|\Delta, D))}{\partial \beta_r^2}$$

$$= \sum_{i=1}^{M} \sum_{k=1}^{d_i} \{x_{ikr}^2 \; A_{ik}(\beta, \Delta) \; B_{ik}^{-2}(\beta, \Delta) \; \exp\{-A_{ik}(\beta, \Delta)\} \, C_{ik}(\beta, \Delta)\} \tag{34}$$

$$- \sum_{i=1}^{M} \sum_{j=i+1}^{M} \sum_{k=1}^{d_j} \{\delta_i \; x_{jkr}^2 \exp\{x'_{jk}\beta\}(s_{j-1} - s_{i-1})\} \tag{35}$$

$$- \sum_{i=1}^{M} \sum_{j=i}^{M} \sum_{k=d_j+1}^{d_j+c_j} \{\delta_i \; x_{jkr}^2 \exp\{x'_{jk}\beta\}(s_j - s_{i-1})\} + \frac{\partial^2 \log \pi(\beta)}{\partial \beta_r^2}. \tag{36}$$

We first note that since we are using a normal prior for β, it is well known that $\frac{\partial^2 \log \pi(\beta)}{\partial \beta_r^2} < 0$. Second, we clearly see that (35) and (36) are negative. Thus, to show that $\frac{\partial^2 \log(p(\beta|\Delta, D))}{\partial \beta_r^2} \leq 0$, it is enough to show that (34) is negative. It suffices to show that $C_{ik}(\beta, \Delta)$ in (34) is negative, since all of the other terms in the summand of (34) are positive. We see that $C_{ik}(\beta, \Delta)$ is of the form $f(x) = 1 - e^x - e^{-e^x}$. Clearly, when $x > 0$, $f(x) < 0$. For $x \in (-\infty, 0)$, we see that $f(x)$ is a monotonic decreasing function and $lim_{x \to -\infty} f(x) = 0$. Thus $f(x) < 0$ for all $x \in R^1$, and thus $C_{ik}(\beta, \Delta) \leq 0$ for all (β, Δ). Thus $[\beta|\Delta, D]$ is log-concave in each component of β.

4. Example: Simulation Study

We consider a simulation to illustrate the methodology presented in Sections 2 and 3. Our main goal in this example is to investigate the behavior of the prior and posterior model probability structures using various choices of prior parameters. We generate survival times t_i, $i = 1, \ldots, n$, from a mixture of Weibull densities taking the form

$$f(t_i) = \psi f_1(t_i) + (1 - \psi) f_2(t_i) \; , \tag{37}$$

where

$$f_1(t_i) = 0.8\lambda_i(\lambda_i t_i)^{(0.8-1)} \; \exp(-(\lambda_i t_i)^{0.8}), f_2(t_i) = 2\lambda_i^2 t_i \; \exp(-(\lambda_i t_i)^2),$$

$0 \leq \psi \leq 1$, $n = 500$, $\lambda_i = \exp(2x_{i1} + 1.5x_{i2})$ and the $x_i = (x_{i1}, x_{i2})'$ are independent bivariate normal random vectors each with mean $\mu = (1, 1)'$ and covariance matrix

$\Sigma = \begin{pmatrix} .2 & .07 \\ .07 & .1 \end{pmatrix}$. The off-diagonal elements of Σ have been chosen so that the correlation between x_{i1} and x_{i2} is .5. In addition, 10% of the observations were randomly right censored. Two additional covariates (x_{i3}, x_{i4}) are independently randomly generated, each having a normal distribution with mean 1 and variance .25. Further, x_{i3} and x_{i4} are independent. Notice that $f_1(t_i)$ has a decreasing hazard function while $f_2(t_i)$ has an increasing hazard function. The full model ($k = 4$) thus contains four covariates $(x_1, \ldots, x_4)$, and the true model contains the covariates (x_1, x_2). To obtain the prior prediction y_0, we generate a new set of covariates that have the same distribution as the set corresponding to the t_i's. Then, using this new set of covariates, we generate another independent set of 500 survival times, denoted w_i, from model (37). For illustrative purposes, we consider the prior prediction $y_0 = (y_{01}, \ldots, y_{0n})'$ to be a perturbation of $w = (w_1, \ldots, w_n)'$, taking the form $y_{0i} = w_i + l_0\ \hat{\sigma}\ z_i$, $i = 1, \ldots, n$ where l_0 is a scalar multiple, $\hat{\sigma}$ is the standard deviation of the w_i's, and the z_i are *i.i.d.* truncated standard normal random variates. Once y_0 is specified, the formulas given in (12) and (13) are used to obtain the prior mean and precision matrix for $\pi(\beta^{(m)})$. For $\pi_0(\beta^{(m)})$, $W_{0,m}$ is taken as a diagonal matrix with the same diagonal elements as T_m in (13). Several choices of l_0, c_0, and (r_1, r_2), where $r_1 = ab$ and $r_2 = ab^{1/4}$, are considered to examine the behavior of the prior and posterior model probabilities. Since an intercept is not included in the model, we have a total of $\mathcal{K} = 15$ models in $\mathcal{M}$.

First, we fix $c_0 = .1$ and $(r_1, r_2) = (.1, .15)$. Table 17.1 gives the largest prior and posterior model probabilities based on different values of ψ and l_0. In Table 17.1, we use the following priors for $\pi(\Delta)$ and $\pi_0(\Delta)$. For $\pi(\Delta)$, we take $M = 60$, with each $\delta_i \sim G(s_i - s_{i-1}, .1)$, $i = 1, \ldots, M$. For $\pi_0(\Delta)$, we take $M_0 = 60$ with each $\delta_i \sim G(s_i - s_{i-1}, .001)$. We choose the subintervals $(s_{i-1}, s_i]$ to have equal numbers of failure or censored observations, i.e., s_i is chosen to be the $(i/M)th$ quantile of the t_i's. The prior model probabilities appear to be most sensitive to the quality of the prior prediction, that is, the choice of l_0 for each ψ.

Table 17.1. The Largest Prior and Posterior Model Probabilities for Various Values of ψ and l_0

ψ	l_0	model	$p(m)$	model	$p(m\|D)$
0.0	.01	(x_1, x_2)	.62	(x_1, x_2)	.74
	3.0	(x_1, x_2)	.31	(x_1, x_2)	.49
0.5	.01	(x_1, x_2)	.54	(x_1, x_2)	.67
	3.0	(x_1)	.27	(x_1, x_2)	.43
1.0	.01	(x_1, x_2)	.65	(x_1, x_2)	.71
	3.0	(x_1)	.18	(x_1, x_2)	.41

Table 17.1 shows that true model obtains the largest posterior probability regardless of l_0 or ψ. However, the posterior probabilities tend to decrease with poorer predictions (i.e., larger l_0). For $\psi = 1$ and $l_0 = 3$, the model with highest prior probability is the x_1 model, which has a posterior probability of nearly 0, while the true model, (x_1, x_2), has the second largest prior probability with a value of .14. Similar results are obtained for $\psi = 0.5$ and $l_0 = 3$. Thus it appears that the prior model probabilities are somewhat sensitive to choice of l_0, although one needs a fairly large l_0 to obtain a model other than the true model with the largest prior probability. We also note that the posterior model probabilities are not sensitive to the increasing hazard rate assumption implied by the EG process prior. When $\psi = .5$, we have a

non-monotonic hazard rate and the model with the largest posterior probability is (x_1, x_2) with value .67. The case $\psi = 1.0$ corresponds to a decreasing hazard rate, and still yields the (x_1, x_2) model with the largest posterior probability of .71.

We also calculated the simulation standard errors of the estimated posterior model probabilities. For all posterior model probabilities in Table 17.1, the simulation standard errors are between 0.01 and 0.04 with a simulation sample size of 5,000. A similar magnitude of standard errors was obtained for all of the remaining posterior model probabilities as well as for the other calculations below, and thus are not reported in the chapter.

Second, we fix $\psi = 0$ and vary l_0 and c_0. We use $l_0 = .01$ (i.e., a good prior prediction), and $c_0 = .1$. We also consider three different partitioning schemes for the time axis, that is, choices of the subintervals $(s_{i-1}, s_i]$. We choose the subintervals $(s_{i-1}, s_i]$ with (i) equal numbers of failures or censored observations; (ii) approximately equal lengths subject to the restriction that at least one failure or censored observation occurs in each interval; (iii) decreasing numbers of failures or censored observations. More specifically, in case (iii) we took s_i to be the $((1 - e^{(-j/M)})/(1 - e(-1)))th$ quantile of the t_i's. For all three partition schemes, the true model, (x_1, x_2) obtains the largest prior and posterior model probabilities. The largest prior probabilities are .62, .61, .61 and the largest posterior model probabilities are .74, .75, .72 for partition schemes (i), (ii), and (iii), respectively. Thus, the prior and posterior model probabilities do not appear to be too sensitive to the choices of the subintervals $(s_{i-1}, s_i]$. This is a comforting feature of our approach since it allows the investigator some flexibility in choosing the subintervals. For the remaining calculations, we will use the subintervals $(s_{i-1}, s_i]$ with equal numbers of failure or censored observations.

The third sensitivity analysis we conduct involves fixing $\psi = 0$, $l_0 = .01$, $c_0 = 1$, and varying (r_1, r_2). These results are summarized in Table 17.2.

Table 17.2. The Largest Prior and Posterior Model Probabilities for Various Choices of (r_1, r_2) with Fixed $\psi = 0$, $l_0 = 0.01$, and $c_0 = 1$

(r_1, r_2)	model	$p(m)$	model	$p(m\|D)$
(.1,.15)	(x_1, x_2)	.62	(x_1, x_2)	.70
(.1, .9)	(x_1, x_2)	.73	(x_1, x_2)	.78
(.5, .9)	(x_1, x_2)	.99	(x_1, x_2)	1.0

For each (r_1, r_2) given in Table 17.2, the model with the largest prior and posterior probability is the (x_1, x_2) model. We see that the prior model probability increases for the (x_1, x_2) model as r_1 and r_2 become larger in magnitude, as seen by $(r_1, r_2) = (0.5, 0.9)$. Using $c_0 = 0.1$ instead of $c_0 = 1$, we obtain results very similar to those in Table 17.2. The prior model probabilities are not very sensitive to changes in c_0 for this fixed value of $l_0 = .01$. Finally, when we use a poor prior prediction, the wrong model can obtain the largest prior and posterior probability. For example, when $\psi = 0$, $l_0 = 8$, $c_0 = 5$, and $(r_1, r_2) = (.10, .50)$, the model with the largest posterior probability is x_1, which has a prior probability of .13 and a posterior probability of .36. In this case, the model with the second largest posterior probability is the (x_1, x_3) model with prior probability of 0.07 and posterior probability of .24. The true model, (x_1, x_2), has the second largest prior probability of .13 and the fourth largest posterior probability of .10.

Next, we perform a sensitivity analysis for the prior parameters of $\pi(\Delta)$. We fix $\psi = 0$, $l_0 = 0.01$, $c_0 = .1$, $(r_1, r_2) = (.1, .15)$. The prior and posterior model probabilities were fairly robust when the shape parameter was kept fixed and the scale parameter was varied. Varying the scale parameter from .1 to .001 resulted in very little change in the posterior model probabilities. Changing the scale to $\lambda_i = .001$, the model with the largest posterior probability is (x_1, x_2) with prior probability of .62 and posterior probability of .73. Thus, as long as the shape parameter remains fixed, the results are not too sensitive to changes in the scale parameter. However, when we fix the scale parameter and vary the shape parameter, the results are sensitive. For example with $\psi = 0$, $l_0 = .01$, $c_0 = .1$, $(r_1, r_2) = (.1, .15)$, with a scale of $\lambda_i = .1$ and a shape parameter of $\alpha_i = 1$ for the ith interval, the (x_1, x_2) model obtains the largest prior probability (.62) but the full model (x_1, x_2, x_3, x_4) obtains the largest posterior probability (.99). However, if $\alpha_i = s_i - s_{i-1}$, the model with largest posterior probability is (x_1, x_2) with a value of .74. Further, when both the shape and the scale are changed such that their product remains constant, the results are less sensitive. For example, when $\psi = 0$, $l_0 = 1$, $(r_1, r_2) = (.1, .15)$, and $\alpha_i = \lambda_i = (.1(s_i - s_{i-1}))^{1/2}$, the model with the largest posterior probability is (x_1, x_2) with a value of .90. Similar phenomena were observed when doing a sensitivity analysis on $\pi_0(\Delta)$.

Finally, for $\psi = 0$, $l_0 = 0.01$ and $c_0 = .1$, we computed the posterior modes of the regression coefficients for the full model. These modes, along with the partial maximum likelihood estimates obtained by the PHREG procedure in SAS, are presented in Table 17.3. We mention that the PHREG procedure performs regression analysis of survival data based on Cox's partial likelihood. From Table 17.3, it can be seen that the Bayesian and non-Bayesian estimates are similar.

We have implemented the Gibbs sampler using the algorithm given in Section 3.2. The convergence of the Gibbs sampler was checked using several diagnostic procedures as recommended by Cowles and Carlin (1996). We ran 10 multiple chains with dispersed initial values and computed potential scale reductions (PSR's). PSR values close to 1 are indicative of convergence of the Markov chain to the target distribution (See Gelman and Rubin, 1992). We also computed the autocorrelation coefficients within chains to ascertain that it is "rapidly mixing" as discussed by Geyer (1992). For the full model, at 500 iterations, the PSR's (the 97.5 percentiles) are 1.03 (1.06), 1.06 (1.11), 1.01 (1.02), and 1.02 (1.03) for $\beta_1, \ldots, \beta_4$ respectively. We also found that the autocorrelations virtually disappear at lag 10 for all parameters. Therefore, the Gibbs sampler practically converges at 500 iterations. Since sampling from the posterior distribution is much cheaper than calculating posterior model probabilities, every $10th$ Gibbs iterate after convergence was used for estimation. 50,500 Gibbs iterates produced an estimation sample of size 5,000. Finally, we used the algorithm given in Section 3.1 to obtain the prior and posterior probabilities using samples from every $10th$ Gibbs iterate.

Table 17.3. Estimates of Regression Coefficients with Standard Errors in Parentheses

Parameter	Bayesian Estimate	SAS Output
β_1	1.926 (.140)	1.959 (.145)
β_2	1.550 (.182)	1.602 (.196)
β_3	0.021 (.086)	0.008 (.090)
β_4	0.046 (.088)	0.045 (.090)

5. Discussion

We have developed a semi-automatic method of doing Bayesian variable selection with informative priors for proportional hazards models. The computational algorithms discussed in Section 3 have proved to be reasonable to implement for small to moderate variable selection problems, say of the order 10-15 covariates. These methods have not been implemented for problems with more than 15 covariates. We note here, as in any variable subset selection technique, we view our procedure as a screening process for identifying a small set of reasonable models, which then can be investigated further. As is well known, there is rarely a single best model which describes a set of data. An important direction for future research is on elicitation of the prior parameters of the EG process prior. More sophisticated choices of $(\alpha(s), \lambda(s))$ which depend on y_0 are currently being investigated. The fruitful results obtained here have opened the door for further research of this issue.

Due to the inherent computational complexity of variable selection with the semiparametric proportional hazards model, a practical, tractable, and easy to interpret prior specification should be given in a closed form in which the user is able to easily specify a prior mean and covariance matrix. The priors given in (9) and (10) have been developed with this purpose. Having a prior that does not have a closed form, such as that given by (15), adds greatly to the computational complexity for this model and moves beyond the scope of the algorithms developed in Section 3. Thus, the analytical tractability of the prior is of practical and computational importance to us. Theoretical and computational properties of the prior in (15) is the subject of ongoing research.

References

Bondesson, L. (1982). On simulation from infinitely divisible distributions. *Advances in Applied Probability* **14**, 855-69.

Chen. M.-H. (1994). Importance-weighted marginal Bayesian posterior density estimation. *J. Am. Statist. Assoc.* **89**, 818-24.

Chen, M.-H. and Shao, Q.-M. (1997a). On Monte Carlo methods for estimating ratios of normalizing constants. *Ann. Statist.* **25**, 1563-1594.

Chen, M.-H. and Shao, Q.-M. (1997b). Estimating ratios of normalizing constants for densities with different dimensions. *Statistica Sinica* **7**, 607-630.

Chen, M.-H. and Shao, Q.-M. (1998). Monte Carlo methods on Bayesian analysis of constrained parameter problems. *Biometrika* **85**, to appear.

Chib, S. (1995). Marginal likelihood from the Gibbs output. *J. Am. Statist. Assoc.* **90**, 1313-21.

Clayton, D. G. (1991). A Monte Carlo method for Bayesian inference in frailty models. *Biometrics* **64**, 141-51.

Cowles, M. K. and Carlin, B. P. (1996). Markov chain Monte Carlo convergence diagnostics: a comparative review. *J. Am. Statist. Assoc.* **91**, 883-904.

Cox, D. R. (1972). Regression models and life tables, (with discussion). *J. R. Statist. Soc.*, **B 34**, 187-220.

Cox, D. R. (1975). Partial likelihood. *Biometrika* **62**, 269-76.

Damien, P., Laud, P. W. and Smith, A. F. M. (1995). Approximate random variate generation from infinitely divisible distributions. *J. R. Statist. Soc.* **B 57**, 547-564.

Dykstra, R. L. and Laud, P. W. (1981). A Bayesian nonparametric approach to reliability. *Ann. Statist.* **9**, 356-67.

Geisser, S. (1993). Predictive inference. New York: Chapman and Hall.

Gelfand, A. E. and Smith, A. F. M. (1990). Sampling-based approaches to calculating marginal densities. *J. Am. Statist. Assoc.* **85**, 398-409.

Gelman, A. and Meng, X.-L. (1994). Path sampling for computing normalizing constants: identities and theory. *Technical Report 377, Department of Statistics, The University of Chicago.*

Gelman, A. and Meng, X.-L. (1996). Simulating normalizing constants: from importance sampling to bridge sampling to path sampling. *Technical Report 440, Department of Statistics, The University of Chicago.*

Gelman, A. and Rubin, D. B. (1992). Inference from iterative simulation using multiple sequences. *Statistical Science* **7**, 457-511.

George, I. E., McCulloch, R. E. and Tsay, R. S. (1995). Two approaches to Bayesian model selections with applications. in *Bayesian Analysis in Econometrics and Statistics - Essays in Honor of Arnold Zellner*, eds. D. A. Berry, K. A. Chaloner and J. K. Geweke, 339-48.

Geyer, C. J. (1992). Practical Markov chain Monte Carlo (with discussion). *Statistical Science* **7**, 473-511.

Geyer, C. J. (1994). Estimating normalizing constants and reweighting mixtures in Markov chain Monte Carlo. *Revision of Technical Report No. 568, School of Statistics, University of Minnesota.*

Gilks, W. R. and Wild, P. (1992). Adaptive rejection sampling for Gibbs sampling. *Appl. Statist.* **41**, 337-48.

Gray, R. J. (1994). A Bayesian analysis of institutional effects in a multicenter cancer clinical trial. *Biometrics* **50**, 244-53.

Ibrahim, J. G. and Chen, M-H (1998). Prior distributions and Bayesian computation for proportional hazards models. *Sankhya*, **B 60**, 48-64.

Ibrahim, J. G. and Laud, P. W. (1994). A predictive approach to the analysis of designed experiments. *J. Am. Statist. Assoc.* **89**, 309-319.

Kalbfleisch, J. D. (1978). Non-parametric Bayesian analysis of survival time data. *J. R. Statist. Soc.* **B 40**, 214-221.

Kuo, L. and Smith, A. F. M. (1992). Bayesian computations in survival models via the Gibbs sampler. in *Survival Analysis: State of the Art*, eds. J. P. Klein and P. K. Goel, 11-24.

Laud, P. W. and Ibrahim, J. G. (1995). Predictive model selection. *J. R. Statist. Soc.* **B 57**, 247-62.

Laud, P. W. and Ibrahim, J. G. (1996). Predictive specification of prior model probabilities in variable selection. *Biometrika*, **83**, 267-274.

Laud, P. W., Smith, A. F. M. and Damien, P. (1996). Monte Carlo methods for approximating a posterior hazard rate process. *Statistics and Computing*, **6**, 77-83.

Meng, X. L. and Wong, W. H. (1996). Simulating ratios of normalizing constants via a simple identity: a theoretical exploration. *Statistica Sinica* **6**, 831-60.

Padgett, W. J. and Wei, L. J. (1980). Maximum likelihood estimation of a distribution function with increasing failure rate based on censored observations. *Biometrika* **67**, 470-74.

Raftery, A. E., Madigan, D. and Volinsky, C. T. (1995). Accounting for model uncertainty in survival analysis improves predictive performance. *Bayesian Statistics 5*, Eds. J. M Bernardo, J. O. Berger, A. P. Dawid and A. F. M. Smith, 323-50.

Skene, A. M. and Wakefield, J. C. (1990). Hierarchical models for multi-centre binary response studies. *Statistics in Medicine* **9**, 919-29.

Tsiatis, A. A. (1981). A large sample study of Cox's regression model. *Ann. Statist.* **9**, 93-108.

Zellner, A. (1986). On assessing prior distributions and Bayesian regression analysis with g-prior distributions. in *Studies in Bayesian Econometrics and Statistics*, eds. P. K. Goel and A. Zellner, New York: Elsevier, 233-43.

18

Bayesian Model Diagnostics for Correlated Binary Data

Dipak K. Dey
Ming-Hui Chen

ABSTRACT Bayesian methods are considered for the analysis of correlated binary data when each binary observation may have its own covariates. Several models, including different versions of logistic regression, multivariate probit and Student's t links are considered. Fully parametric classical approaches to these are intractable and thus Bayesian methods are pursued using sampling based approach through Markov chain Monte Carlo method. Several model diagnostics using simulation based approach are developed and implemented for model adequacy. the proposed methodology is implemented to study the voting behavior of residents of Troy, Michigan.

1. Introduction

In this chapter we consider an exact small sample Bayesian analysis of the models proposed by Prentice (1988) and also by Ochi and Prentice (1984). Our models can be classified into two groups. The first group of models generalize the binary logistic model to multivariate data by considering a particular parameterized representation for the correlations using the notion of random effects on the logistic regression structure or in terms of pairwise odds ratios. The other group of models are obtained by introducing a link function using an inverse cumulative distribution function (cdf). Multivariate probit (MVP) and multivariate t-link (MVT) models are obtained in this scenario. The MVP model was first introduced by Ashford and Sowden (1970) and studied further by Amemiya (1985). Recently, by introducing latent variables Chib and Greenberg (1998) analyzed the MVP model in a Bayesian framework.

The objective of this chapter is to explore different modeling strategies for the analysis of correlated binary data in a Bayesian perspective. As one entertains a collection of such models for a given data set one needs to address the problem of model adequacy. Specifically, we adopt the Markov chain Monte Carlo (MCMC) framework (e.g., Gelfand and Smith, 1990 and Tierney, 1994) to simulate the posterior distribution for proposed models. Instead of just taking formal Bayesian model adequacy criterion (as in Box, 1980), our approach is based on exploratory data analysis methods (e.g., Gelfand, Dey and Chang, 1992 and Dey, Gelfand, Swartz and Vlachos, 1995). We develop different diagnostic measures suitable for the binary data and apply them to model adequacy.

The remainder of the chapter is organized as follows. In Section 2, we consider different models and discuss how the likelihood is obtained for each model. Section 3 is devoted to the development of the prior distributions and the distribution theory involved for the posterior calculations. In Section 4, we discuss different methods involved in checking model adequacy. Section 5 is devoted to the application of our proposed methodologies to survey data on the voting behavior of residents of Troy, Michigan. Finally, Section 6 gives brief concluding remarks.

2. The Models

We first introduce some notations which will be used throughout the chapter. Suppose that we observe a binary (0-1) response Y_{ij} on the ith observations and jth variable and let $x_{ij} = (x_{ij1}, x_{ij2}, \ldots, x_{ijp_j})$ be the corresponding p_j-dimensional row regression vector for $i = 1, 2, \ldots, n$ and $j = 1, 2, \ldots, J$. (Note that x_{ij1} may be 1, which corresponds to an intercept.) Denote $Y_i = (Y_{i1}, Y_{i2}, \ldots, Y_{iJ})'$ and assume that Y_{i1}, Y_{i2}, ..., Y_{iJ} are dependent and Y_1, Y_2, ..., Y_n are independent. Let $y_i = (y_{i1}, y_{i2}, \ldots, y_{iJ})'$ and $y = (y_1, y_2, \ldots, y_n)$ be the observed data. Also corresponding to x_j, suppose $\beta_j = (\beta_{j1}, \beta_{j2}, \ldots, \beta_{jp_j})'$ is a p_j-dimensional column vector of regression coefficients for $j = 1, 2, \ldots, J$ and $\beta = (\beta_1', \beta_2', \ldots, \beta_J')'$.

Now we consider three different types of models for the above correlated binary data. The first two models, which are variants of random effects models, implicitly assume equicorrelation structure whereas the third model is based on latent variables where the correlation structure has more flexibility. There are other plausible models in the literature to incorporate correlation structure which are not pursued in this chapter.

2.1 Stratified and Mixture Models

In the spirit of the stratified and mixture models considered in Prentice (1988), we consider the following random effects binary logistic regression model for Y_{ij} given x_i as

$$P(Y_{ij} = y_{ij} | \alpha_i, \beta_j, x_{ij}) = \frac{\exp\{(\alpha_i + x_{ij}\beta_j) y_{ij}\}}{1 + \exp\{\alpha_i + x_{ij}\beta_j\}}, \tag{1}$$

where α_i denotes random effects on the ith observation, which captures the correlation among Y_{i1}, Y_{i2}, ..., Y_{iJ}. We take

$$\alpha_i \overset{i.i.d.}{\sim} N(0, \sigma_\alpha^2).$$

The above model is also known in the literature as a logit normal model.

2.2 Conditional Models

In certain studies there are natural ordering of the individuals within the block and the regression coefficients in the model may characterize dependencies of interest. Our second model is a symmetric conditional random effects binary logistic regression model (e.g., Prentice 1988), which is defined as

$$\begin{aligned} & P(Y_{i1} = y_{i1}, Y_{i2} = y_{i2}, \ldots, Y_{iJ} = y_{iJ} | \alpha_i, \beta, x_{i1}, \ldots, x_{iJ}) \\ = & \frac{\exp\{\sum_{j=0}^{\dot{y}_i - 1} \alpha_i(j) + \sum_{j=1}^{J} x_{ij}\beta_j y_{ij}\}}{\sum_{y_{i1}^*=0}^{1} \sum_{y_{i2}^*=0}^{1} \cdots \sum_{y_{iJ}^*=0}^{1} \exp\{\sum_{j=0}^{\dot{y}_i^* - 1} \alpha_i(j) + \sum_{j=1}^{J} x_{ij}\beta_j y_{ij}^*\}}, \end{aligned} \tag{2}$$

where $\dot{y}_i = \sum_{j=1}^{J} y_{ij}$ and $\dot{y}_i^* = \sum_{j=1}^{J} y_{ij}^*$. In (2), $\alpha_i(j)$ is chosen to be a polynomial in j. For example, the simplest choice of $\alpha_i(j) = \alpha_i$ gives

$$P(Y_{i1} = y_{i1}, Y_{i2} = y_{i2}, \ldots, Y_{iJ} = y_{iJ} | \alpha_i, \beta, x_{i1}, \ldots, x_{iJ})$$

$$= \frac{\exp\{\alpha_i \dot{y}_i + \sum_{j=1}^J x_{ij}\beta_j y_{ij}\}}{\sum_{y_{i1}^*=0}^1 \sum_{y_{i2}^*=0}^1 \cdots \sum_{y_{iJ}^*=0}^1 \exp\{\alpha_i \dot{y}_i^* + \sum_{j=1}^J x_{ij}\beta_j y_{ij}^*\}}, \tag{3}$$

where $\alpha_i \overset{i.i.d.}{\sim} N(0, \sigma_\alpha^2)$.

2.3 Multivariate Probit Models

Here, we consider multivariate probit (MVP) models; see also Chib and Greenberg (1998) for the Bayesian analysis of such models.

We introduce a J-dimensional latent variable $w_i = (w_{i1}, w_{i2}, \ldots, w_{iJ})'$ so that

$$Y_{ij} = \begin{cases} 1 & \text{if } w_{ij} > 0 \\ 0 & \text{if } w_{ij} \le 0 \end{cases} \tag{4}$$

and

$$w_i \sim N(x_i\beta, \Sigma), \tag{5}$$

where $x_i = diag(x_{i1}, x_{i2}, \ldots, x_{iJ})$. In (5), we take $\Sigma = (\rho_{jj^*})_{J\times J}$ to be a correlation matrix such that $\rho_{jj} = 1$ to ensure the identifiability of parameters. See Chib and Greenberg for the detailed discussions. The distribution of w_i determines the joint distribution of Y_i through (4) and the correlation matrix Σ captures the correlations among the Y_{ij}'s. More specifically, we have

$$\begin{aligned} &P(Y_{i1} = y_{i1}, Y_{i2} = y_{i2}, \ldots, Y_{iJ} = y_{iJ} | \beta, \Sigma, x_i) \\ &= \int_{A_{i1}} \int_{A_{i2}} \cdots \int_{A_{iJ}} \left(\frac{1}{\sqrt{2\pi}}\right)^J \frac{1}{|\Sigma|^{\frac{1}{2}}} \\ &\qquad \exp\left\{-\frac{1}{2}(w_i - x_i\beta)'\Sigma^{-1}(w_i - x_i\beta)\right\} dw_i, \end{aligned} \tag{6}$$

where

$$A_{ij} = \begin{cases} (-\infty, 0] & \text{if } y_{ij} = 0 \\ (0, \infty) & \text{if } y_{ij} = 1 \end{cases} \tag{7}$$

for $i = 1, 2, \ldots, n$.

2.4 Multivariate t-Link Models

A generalization of multivariate probit models is multivariate t-link models (MVT). Similar to the MVP models, we introduce a J-dimensional latent variable $w_i = (w_{i1}, w_{i2}, \ldots, w_{iJ})'$ so that

$$Y_{ij} = \begin{cases} 1 & \text{if } w_{ij} > 0 \\ 0 & \text{if } w_{ij} \le 0 \end{cases}$$

and

$$w_i \sim t_\nu(x_i\beta, \Sigma) \tag{8}$$

with probability density function

$$\begin{aligned} &\pi(w_i|\nu, x_i\beta, \Sigma) \\ &= \frac{\Gamma\left(\frac{1}{2}(\nu+J)\right)}{(\pi\nu)^{(1/2)J}\Gamma\left(\frac{1}{2}\nu\right)|\Sigma|^{1/2}} \left(1 + \nu^{-1}(w_i - x_i\beta)'\Sigma^{-1}(w_i - x_i\beta)\right)^{-\frac{1}{2}(\nu+J)}. \end{aligned} \tag{9}$$

(Note that one special case of (9) with $\nu = 1$ is termed as a multivariate Cauchy distribution and another special case of (9) with $\nu \to \infty$ is the multivariate normal distribution (5).) Thus, the joint distribution of Y_i is given by

$$\begin{aligned} & P(Y_{i1} = y_{i1}, Y_{i2} = y_{i2}, \ldots, Y_{iJ} = y_{iJ} | \beta, \Sigma, \nu, x_i) \\ = & \int_{A_{i1}} \int_{A_{i2}} \cdots \int_{A_{iJ}} \pi(w_i | \nu, x_i \beta, \Sigma) dw_i, \end{aligned} \tag{10}$$

where $\pi(w_i|\nu, x_i\beta, \Sigma)$ and A_{ij} are given in (9) and (7) respectively. For the ease of model complexity, we consider only fixed ν so that we can investigate different multivariate t-link models. A discrete prior on ν can be considered. However, this will bring an additional computational burden.

3. The Prior Distributions and Posterior Computations

In this section we present prior distributions for various models and develop algorithms to perform posterior computations for such models.

3.1 Prior Distributions

First, we choose the same prior distribution for the regression coefficient vector β for all four models presented in Section 2. That is,

$$\pi(\beta|\beta_0, B_0) \propto \exp\left\{-\frac{1}{2}(\beta - \beta_0)' B_0 (\beta - \beta_0)\right\}, \tag{11}$$

where B_0 is a precision matrix, β_0 is a location parameter vector, and both β_0 and B_0 are pre-specified. Typically, we choose $\beta_0 = 0$ and

$$B_0 = diag\left(B_{11}, B_{12}, \ldots, B_{1p_1}, B_{21}, B_{22}, \ldots, B_{2p_2}, \ldots, B_{J1}, B_{J2}, \ldots, B_{Jp_J}\right),$$

where B_{jl} is chosen to be small (e.g., $B_{jl} = 0.01$) so that a vague prior distribution for β is obtained, which ensures that the posterior is driven by the data.

Second, for the stratified and mixture models and the conditional models, we choose

$$\sigma_\alpha^2 \sim \mathcal{IG}(a, b), \tag{12}$$

where the density of the $\mathcal{IG}(a, b)$ is

$$\pi(\sigma_\alpha^2 | a, b) \propto (\sigma_\alpha^2)^{-(a+1)} e^{-b/\sigma_\alpha^2},$$

and a and b are chosen so that $E(\sigma_\alpha^2)$ is small (e.g., 0.001) and $Var(\sigma_\alpha^2)$ is large (e.g., 100) to ensure that the prior does not play a major role in the posterior.

Finally, for the multivariate probit and t-link models we denote

$$vec^*(\Sigma) = (\rho_{12}, \rho_{13}, \ldots, \rho_{J-1,J})'.$$

Then, analogous to Chib and Greenberg (1998), we choose

$$\begin{aligned} & \pi(vec^*(\Sigma)|\Sigma_0, G_0) \\ \propto & \exp\left\{-\frac{1}{2}(vec^*(\Sigma) - vec^*(\Sigma_0))' G_0 (vec^*(\Sigma) - vec^*(\Sigma_0))\right\} \end{aligned} \tag{13}$$

for $vec^*(\Sigma) \in A$ where Σ_0 is a $J \times J$ correlation matrix with all diagonal elements equal to one, G_0 is a $(J(J-1)/2) \times (J(J-1)/2)$ precision matrix, and the region A is a subset of the region $[-1,1]^{J(J-1)/2}$ that leads to a proper correlation matrix. As mentioned in Chib and Greenberg (1998) and also shown by Rousseeuw and Molenberghs (1994), the region A forms a convex solid body in the hypercube $[-1,1]^{J(J-1)/2}$. Note that both hyper parameters Σ_0 and G_0 are to be specified. The simplest choices of Σ_0 and G_0 are $\Sigma_0 = I_J$ and $G_0 = I_{J(J-1)/2}$, which are the J and $J(J-1)/2$ dimensional identity matrices.

3.2 Posterior Computations

We use Gibbs sampling (e.g, Geman and Geman, 1984 and Gelfand and Smith, 1990) to perform the posterior computation. We present necessary steps needed to perform the Gibbs sampling algorithms for all four models considered in Section 2 in turn.

Stratified and Mixture Models

To run the Gibbs sampler for the stratified and mixture models, we take all parameters from their respective conditionals as follows:

(a) For the ith random effect, let $\pi(\alpha_i|\beta, y_i)$ be the conditional density for α_i given the data and β. Then,

$$\pi(\alpha_i|\beta, y_i) \propto \prod_{j=1}^{J} \frac{\exp\{(\alpha_i + x_{ij}\beta_j)y_{ij}\}}{1+\exp\{\alpha_i + x_{ij}\beta_j\}}. \tag{14}$$

Since

$$\frac{d^2 \log \pi(\alpha_i|\beta, y_i)}{d\alpha_i^2} = -\sum_{j=1}^{J} \frac{\exp\{\alpha_i + x_{ij}\beta_j\}}{(1+\exp\{\alpha_i + x_{ij}\beta_j\})^2} < 0, \tag{15}$$

$\pi(\alpha_i|\beta, y_i)$ is log-concave. Therefore, we can use the adaptive rejection sampling algorithm of Gilks and Wild (1992) to generate α_i for $i = 1, 2, \ldots, n$.

(b) For the variance of the random effects α_i, we take

$$\sigma_\alpha^2 \mid \alpha \sim IG\left(\frac{n}{2} + a, b + \frac{1}{2}\sum_{i=1}^{n} \alpha_i^2\right), \tag{16}$$

where $\alpha = (\alpha_1, \ldots, \alpha_n)'$.

(c) For the regression coefficients β_{jl}'s, let $\pi(\beta|\alpha, y)$ be the conditional density for β given the data and random effects α. Assuming that we choose $\beta_0 = 0$ and B_0 is a diagonal precision matrix, then

$$\pi(\beta|\alpha, y) \propto \left[\prod_{i=1}^{n}\prod_{j=1}^{J} \frac{\exp\{(\alpha_i + x_{ij}\beta_j)y_{ij}\}}{1+\exp\{\alpha_i + x_{ij}\beta_j\}}\right] \cdot \left[\prod_{j=1}^{J}\prod_{l=1}^{p_j} \exp\{-\frac{1}{2}B_{jl}\beta_{jl}^2\}\right].$$

Similar to (14), it can be shown that $\pi(\beta|\alpha, y)$ is log-concave in each component of β and, moreover, it can be shown that $\pi(\beta|\alpha, y)$ is log-concave in the whole vector of β, i.e., the second derivative matrix $\left(\frac{\partial^2 \pi(\beta|\alpha,y)}{\partial\beta_{jl}\partial\beta_{j^*l^*}}\right)$ is non-positive definite and, therefore, we can again use the adaptive rejection sampling algorithm of Gilks and Wild to generate each component of β.

Conditional Models

For the sake of simplicity, we consider the conditional models given in (3). The conditional density for α_i is

$$\pi(\alpha_i|\beta, y_i) \propto \frac{\exp\{\alpha_i\dot{y}_i + \sum_{j=1}^J x_{ij}\beta_j y_{ij}\}}{\sum_{y_{i1}^*=0}^1 \sum_{y_{i2}^*=0}^1 \cdots \sum_{y_{iJ}^*=0}^1 \exp\{\alpha_i\dot{y}_i^* + \sum_{j=1}^J x_{ij}\beta_j y_{ij}^*\}}.$$

Similar to (15), it can be shown that $\pi(\alpha_i|\beta, y_i)$ is log-concave; therefore, we again use the algorithm of Gilks and Wild to generate α_i. The conditional density for σ_α^2 is the same as that given in (16); henceforth, it is trivial to sample σ_α^2 for its conditional distribution. Using the same prior for β as in the stratified and mixture models, the conditional density for β is

$$\begin{aligned} &\pi(\beta|\alpha, y) \\ &\propto \left[\prod_{i=1}^n \frac{\exp\{\alpha_i\dot{y}_i + \sum_{j=1}^J x_{ij}\beta_j y_{ij}\}}{\sum_{y_{i1}^*=0}^1 \sum_{y_{i2}^*=0}^1 \cdots \sum_{y_{iJ}^*=0}^1 \exp\{\alpha_i\dot{y}_i^* + \sum_{j=1}^J x_{ij}\beta_j y_{ij}^*\}}\right] \\ &\times \left[\prod_{j=1}^J \prod_{l=1}^{p_j} \exp\{-\frac{1}{2} B_{jl}\beta_{jl}^2\}\right]. \end{aligned}$$

Then,

$$\begin{aligned} \frac{\partial^2 \log \pi(\beta|\alpha, y)}{\partial \beta_{jl}^2} &= -\sum_{i=1}^n \left[\frac{\sum_{y_{i1}^*=0}^1 \sum_{y_{i2}^*=0}^1 \cdots \sum_{y_{iJ}^*=0}^1 x_{ijl}^2 {y^*}_{ij}^2 \exp\{\alpha_i\dot{y}_i^* + \sum_{j^*=1}^J x_{ij}\beta_{j^*} y_{ij^*}^*\}}{\sum_{y_{i1}^*=0}^1 \sum_{y_{i2}^*=0}^1 \cdots \sum_{y_{iJ}^*=0}^1 \exp\{\alpha_i\dot{y}_i^* + \sum_{j^*=1}^J x_{ij}\beta_{j^*} y_{ij^*}^*\}}\right. \\ &= \left. - \frac{\left(\sum_{y_{i1}^*=0}^1 \sum_{y_{i2}^*=0}^1 \cdots \sum_{y_{iJ}^*=0}^1 x_{ijl} {y^*}_{ij} \exp\{\alpha_i\dot{y}_i^* + \sum_{j^*=1}^J x_{ij}\beta_{j^*} y_{ij^*}^*\}\right)^2}{\left(\sum_{y_{i1}^*=0}^1 \sum_{y_{i2}^*=0}^1 \cdots \sum_{y_{iJ}^*=0}^1 \exp\{\alpha_i\dot{y}_i^* + \sum_{j^*=1}^J x_{ij}\beta_{j^*} y_{ij^*}^*\}\right)^2}\right] \\ &\quad - B_{jl}. \end{aligned}$$

Using the Cauchy-Schwarz inequality, we have

$$\begin{aligned} &\left(\sum_{y_{i1}^*=0}^1 \sum_{y_{i2}^*=0}^1 \cdots \sum_{y_{iJ}^*=0}^1 x_{ijl} {y^*}_{ij} \exp\{\alpha_i\dot{y}_i^* + \sum_{j^*=1}^J x_{ij}\beta_{j^*} y_{ij^*}^*\}\right)^2 \\ &\le \sum_{y_{i1}^*=0}^1 \sum_{y_{i2}^*=0}^1 \cdots \sum_{y_{iJ}^*=0}^1 x_{ijl}^2 {y^*}_{ij}^2 \exp\{\alpha_i\dot{y}_i^* + \sum_{j^*=1}^J x_{ij}\beta_{j^*} y_{ij^*}^*\} \\ &\quad \cdot \sum_{y_{i1}^*=0}^1 \sum_{y_{i2}^*=0}^1 \cdots \sum_{y_{iJ}^*=0}^1 \exp\{\alpha_i\dot{y}_i^* + \sum_{j^*=1}^J x_{ij}\beta_{j^*} y_{ij^*}^*\}. \end{aligned}$$

Thus, $\pi(\beta|\alpha, y)$ is log-concave in each component of β; therefore, again the algorithm of Gilks and Wild can be used to generate each component of β.

Multivariate Probit Models

To run the Gibbs sampler for the multivariate models, we need to sample β, w_i, and Σ from their respective conditional distributions. Let $\hat{\beta} = B^{-1}\left(B_0\beta_0 + \sum_{i=1}^n x_i' \Sigma^{-1} w_i\right)$ and $B = B_0 + \sum_{i=1}^n x_i'\Sigma^{-1}x_i$. Then, given the w_i and Σ, we have

$$\beta \mid w_1, w_2, \ldots, w_n, \Sigma, y \sim N\left(\hat{\beta}, B^{-1}\right).$$

From (6), it can be seen that the full conditional distribution of w_i is multivariate normal truncated to a region determined by y_i. More specifically,

$$\begin{aligned}&\pi(w_i \mid \beta, \Sigma, y_i)\\ &\propto \prod_{j=1}^{J}\left[1_{\{w_{ij}>0\}}1_{\{y_{ij}=1\}} + 1_{\{w_{ij}\le 0\}}1_{\{y_{ij}=0\}}\right]\\ &\quad \times \exp\left\{-\frac{1}{2}(w_i - x_i\beta)'\Sigma^{-1}(w_i - x_i\beta)\right\}.\end{aligned}$$

As suggested by Geweke (1991), we generate this truncated multivariate normal variate w_i using a cycle of J Gibbs steps through the components of w_i so that w_{ij} is sampled from the truncated normal $w_{ij} \mid z_{ij^*}, j^* \neq j, \beta, \Sigma, y_i$ respectively over an interval $(0, \infty)$ if $y_{ij} = 1$ or $(-\infty, 0]$ if $y_{ij} = 0$.

Finally, we consider sampling Σ from its conditional distribution. The conditional likelihood function $L(\Sigma|\beta, w, y)$, ignoring the normalizing constant, is

$$|\Sigma|^{-\frac{n}{2}} \exp\left\{-\frac{1}{2}\sum_{i=1}^{n}(w_i - x_i\beta)'\Sigma^{-1}(w_i - x_i\beta)\right\},$$

where $w = (w_1, w_2, \ldots, w_n)$ and $vec^*(\Sigma) \in A$. The full conditional density is proportional to $L(\Sigma|\beta, w, y)\pi(vec^*(\Sigma)|\Sigma_0, G_0)$. Because of the complexity of this conditional distribution, we use a Hastings algorithm (e.g., Metropolis *et al.* 1951, Hastings 1970, Tierney 1994) to generate Σ. Let Σ be the current value. Using an algorithm analogous to Chib and Greenberg (1998), we generate candidate values Σ^* by specifying a random walk chain

$$\Sigma^* = \Sigma + H,$$

where $H = (h_{ij})$ is an increment matrix with zeros on the diagonals and with means $E(h_{ij}) = 0$. Let λ be the least eigenvalue of Σ. Alternative to Chib and Greenberg's algorithm, we use a Metropolized hit-and-run algorithm (Chen and Schmeiser 1993) to simulate H. This algorithm operates as follows:

(i) generate an *i.i.d* $N(0,1)$ random variate sequence of $z_{12}, z_{13}, \ldots, z_{J-1,J}$;

(ii) generate d from $N(0, \sigma_d^2)$ truncated to $(-\frac{\lambda}{\sqrt{2}}, \frac{\lambda}{\sqrt{2}})$;

(iii) calculate

$$h_{ij} = \frac{dz_{ij}}{\left(\sum_{j=1}^{J-1}\sum_{l=j}^{J} z_{jl}^2\right)^{1/2}}$$

for $i < j$, $h_{ii} = 0$, and $h_{ij} = h_{ji}$ for $i > j$.

Note that in (ii), σ_d^2 is appropriately chosen so as to avoid excessive rejections in the Hastings algorithm. As discussed in Marsaglia and Olkin (1994), H generated in this manner guarantees that Σ^* is positive definite.

Let λ^* be the least eigenvalue of Σ^*. Following Chen and Schmeiser (1993), given the proposal value, a move to the point Σ^* is made with probability

$$\min\left\{\frac{L(\Sigma^*|\beta, w, y)\pi(vec^*(\Sigma^*)|\Sigma_0, G_0)\left(\Phi\left(\frac{\lambda^*}{\sqrt{2}\sigma_d}\right) - \Phi\left(-\frac{\lambda^*}{\sqrt{2}\sigma_d}\right)\right)}{L(\Sigma|\beta, w, y)\pi(vec^*(\Sigma)|\Sigma_0, G_0)\left(\Phi\left(\frac{\lambda}{\sqrt{2}\sigma_d}\right) - \Phi\left(-\frac{\lambda}{\sqrt{2}\sigma_d}\right)\right)}, 1\right\},$$

where $\Phi(\cdot)$ is the standard normal cumulative distribution function. Compared to the algorithm of Chib and Greenberg (1998), our Metropolized hit-and-run algorithm is more advantageous since the use of the hit-and-run algorithm ensures candidate correlation matrix Σ^* to be non-negative no matter which σ_d^2 is chosen.

Multivariate *t*-Link Models

It is well-known that a Multivariate t distribution is the Gamma mixture of normal distributions (e.g., see Johnson and Kotz 1976 or Bernardo and Smith 1994). From (9), we have

$$\begin{aligned}
&\pi(w_i|\nu, x_i\beta, \Sigma) \\
&= \int_0^\infty \left[\left(\frac{1}{\sqrt{2\pi}}\right)^J \frac{1}{|\Sigma/\xi_i|^{\frac{1}{2}}} \exp\left\{-\frac{1}{2}(w_i - x_i\beta)'\left(\frac{\Sigma}{\xi_i}\right)^{-1}(w_i - x_i\beta)\right\}\right] \\
&\quad \cdot \frac{1}{\Gamma\left(\frac{\nu}{2}\right)}\left(\frac{\nu}{2}\right)^{\nu/2} \xi_i^{\frac{\nu}{2}-1} e^{-\frac{\nu}{2}\xi_i} d\xi_i. \qquad (17)
\end{aligned}$$

To run the Gibbs sampler, we sample ξ_i, w_i, β, and Σ from their respective conditional distributions. Using (17), we independently generate

$$\xi_i \mid w_i, \beta, \Sigma \sim \mathcal{G}\left(\frac{\nu + J}{2}, \frac{1}{2}\left[\nu + (w_i - x_i\beta)'\Sigma^{-1}(w_i - x_i\beta)\right]\right),$$

where $\mathcal{G}(a, b)$ denotes a gamma distribution with density

$$g(\xi|a, b) \propto \xi^{a-1}e^{-b\xi}.$$

In a manner similar to the MVP models with obvious adjustments, we can generate w_i, β, and Σ given the ξ_i for the MVT models. For example, given ξ_i, w_i, and Σ, we have

$$\beta \mid \xi_i, w_i, i = 1, 2, \ldots, n, \Sigma, y \sim N\left(\hat{\beta}_t, B_t^{-1}\right),$$

where $\hat{\beta}_t = B_t^{-1}(B_0\beta_0 + \sum_{i=1}^n \xi_i x_i'\Sigma^{-1}w_i)$ and $B_t = B_0 + \sum_{i=1}^n \xi_i x_i'\Sigma^{-1}x_i$.

4. Model Adequacy for Correlated Binary Data

Once we have accomplished the first two steps of a Bayesian analysis, i.e., constructing probability models and computing the posterior distributions of all parameters of interest, using sampling based approach, it is natural to assess the fit of the models to the data and to own substantive knowledge. A good Bayesian analysis should include at least some checks of the adequacy of the fit of the model to the data. While model checking usually addresses the entire model specification (both likelihood and prior), model failures can happen in different phases, which include outliers, mean structure errors, dispersion misspecification, and inappropriate exchangeabilities in the hierarchical structure. In this section we consider different simulation based model checking approaches using certain discrepancy measure which is a function of data as well as parameters.

We first introduce the following notation which will be used in developing our model adequacy and selection procedures. Let θ be all parameters in the likelihood, e.g., $\theta = (\alpha_1, \ldots, \alpha_n, \beta_1, \ldots, \beta_J)$ for the stratified and mixture models, and δ be all the parameters in the prior (i.e., hyper parameters), e.g., $\delta = (\sigma_\alpha^2, a, b, \beta_0, B_0)$ for the stratified and mixture models. Also Let $f(y_i|\theta)$ be the joint distribution of $Y_i = (Y_{i1}, Y_{i2}, \ldots, Y_{iJ})'$ given θ and covariates $x_{i1}, \ldots, x_{iJ}$. Denote $p_{ij}(\theta) = E(Y_{ij}|\theta)$, which may be a function of all covariates as well, and $q_{ij}(\theta) = 1 - p_{ij}(\theta)$. Note that for the stratified and mixture models and the conditional models (3),

$$p_{ij}(\theta) = \frac{\exp\{\alpha_i + x_{ij}\beta_j\}}{1 + \exp\{\alpha_i + x_{ij}\beta_j\}},$$

for the MVP models $p_{ij}(\theta) = \Phi(x_{ij}\beta_j)$, and for the MVT models $p_{ij}(\theta) = t_\nu(x_{ij}\beta_j)$ where $t_\nu(\cdot)$ is the cdf of the t distribution with ν degrees of freedom which has the form

$$t_\nu(w) = \int_{-\infty}^{w} \frac{\Gamma\left(\frac{1}{2}(\nu+1)\right)}{(\pi\nu)^{(1/2)}\Gamma\left(\frac{1}{2}\nu\right)} \left(1 + \nu^{-1}u^2\right)^{-\frac{1}{2}(\nu+1)} du.$$

We denote $\pi(\theta|\delta)$ to be the prior distribution of θ given δ.

In order to introduce an appropriate discrepancy measure for the correlated binary data, we need to obtain the variance and covariance matrix of Y_i, which is denoted by $\Sigma_i^*(\theta) = (\sigma_{ijj^*}^*(\theta))_{J\times J}$. For the stratified and mixture models, $\sigma_{ijj^*}^*(\theta) = 0$ for $j \neq j^*$ and $\sigma_{ijj}^*(\theta) = p_{ij}(\theta)q_{ij}(\theta)$. For the conditional models (3),

$$\begin{aligned} \sigma_{ijj^*}^*(\theta) &= \frac{\exp\{2\alpha_i + x_{ij}\beta_j + x_{ij^*}\beta_j^*\}}{\sum_{y_{ij}^*=0}^{1}\sum_{y_{ij^*}^*=0}^{1} \exp\{\alpha_i(y_{ij}^* + y_{ij^*}^*) + x_{ij}\beta_j y_{ij}^* + x_{ij^*}\beta_j^* y_{ij^*}^*\}} \\ &\quad - p_{ij}(\theta)p_{ij^*}(\theta) \quad \text{for } j \neq j^* \end{aligned}$$

and $\sigma_{ijj}^*(\theta) = p_{ij}(\theta)q_{ij}(\theta)$. For the MVP models,

$$\begin{aligned} \sigma_{ijj^*}^*(\theta) &= \Phi(x_{ij}\beta_j) + \Phi(x_{ij^*}\beta_{j^*}) - 1 + \Phi\left(-x_{ij}\beta_j, -x_{ij^*}\beta_{j^*}; \rho_{jj^*}\right) \\ &\quad - \Phi(x_{ij}\beta_j)\Phi(x_{ij^*}\beta_{j^*}) \quad \text{for } j \neq j^*, \end{aligned}$$

where $\Phi(a^*, b^*; \rho)$ is the bivariate normal distribution function which is defined as

$$\Phi(a^*, b^*; \rho) = \frac{1}{2\pi\sqrt{1-\rho^2}} \int_{-\infty}^{a^*}\int_{-\infty}^{b^*} \exp\left\{-\frac{u^2 - 2\rho uv - v^2}{2(1-\rho^2)}\right\} dudv$$

and $\sigma_{ijj}^*(\theta) = \Phi(x_{ij}\beta_j)(1 - \Phi(x_{ij}\beta_j))$. For the MVT models,

$$\begin{aligned} \sigma_{ijj^*}^*(\theta) &= t_\nu(x_{ij}\beta_j) + t_\nu(x_{ij^*}\beta_{j^*}) - 1 + t_\nu\left(-x_{ij}\beta_j, -x_{ij^*}\beta_{j^*}; \rho_{jj^*}\right) \\ &\quad - t_\nu(x_{ij}\beta_j)t_\nu(x_{ij^*}\beta_{j^*}) \quad \text{for } j \neq j^*, \end{aligned}$$

where $t_\nu(a^*, b^*; \rho)$ is the bivariate t distribution function which is defined as

$$\begin{aligned} &t_\nu(a^*, b^*; \rho) \\ &= \frac{\Gamma\left(\frac{1}{2}(\nu+2)\right)}{(\pi\nu)\Gamma\left(\frac{1}{2}\nu\right)\sqrt{1-\rho^2}} \int_{-\infty}^{a^*}\int_{-\infty}^{b^*} \left(1 + \frac{u^2 - 2\rho uv - v^2}{\nu(1-\rho^2)}\right)^{-\frac{1}{2}(\nu+2)} dudv \end{aligned}$$

and $\sigma_{ijj}^*(\theta) = t_\nu(x_{ij}\beta_j)(1 - t_\nu(x_{ij}\beta_j))$. Note that $t_\nu(a^*, b^*; \rho)$ can be evaluated through the bivariate normal distribution function $\Phi(a^*, b^*; \rho)$, since we have the following relationship between $t_\nu(a^*, b^*; \rho)$ and $\Phi(a^*, b^*; \rho)$:

$$t_\nu(a^*, b^*; \rho) = \int_0^\infty \Phi(a^*\sqrt{\xi}, b^*\sqrt{\xi}; \rho) \frac{1}{\Gamma\left(\frac{\nu}{2}\right)} \left(\frac{\nu}{2}\right)^{\nu/2} \xi^{\frac{\nu}{2}-1} e^{-\frac{\nu}{2}\xi} d\xi. \tag{18}$$

In the earlier sections we present a collection of models and proceed with Bayesian inference. The issue at stake is which model is most appropriate for the given binary data as it is well-known that an improper model could lead to a misleading conclusion. In classical statistics, goodness of fit tests have been employed to check the plausibility of the model fit to the data. The classical goodness of fit test quantifies the extremeness of a particular discrepancy measure by calculating a tail area probability that the model under a specified null hypothesis is true. In the classical setup the test statistic and hence the tail area probability is a function of both the data as well as the unknown parameters which are specified only under the null hypothesis. Alternative to the classical goodness of fit test, we develop here the Bayesian model checking methods on the correlated binary data.

In order to incorporate the correlated nature of binary responses, we introduce the following observation-level Pearson residual discrepancy measure:

$$D_i(\theta) = (y_i - p_i(\theta))' \Sigma_i^{*-1} (y_i - p_i(\theta)), \tag{19}$$

where $p_i(\theta) = (p_{i1}(\theta), p_{i2}(\theta), \ldots, p_{iJ}(\theta))'$ for $i = 1, 2, \ldots, n$. Furthermore, if the overall performance for a model is of interest, we introduce the total Pearson residual discrepancy measure, which is

$$D(\theta) = \sum_{i=1}^{n} D_i(\theta). \tag{20}$$

Now, letting $d(data, \theta)$ be the generic notation for a discrepancy measure (e.g., an observation-level Pearson residual or the total Pearson residual), we propose the following two methods to perform the model adequacy study.

Method I: Posterior Predictive Comparison

This approach is motivated from Gelman, Meng and Stern (1996) where the technique for checking the fit of a model to data is to draw simulated values of a discrepancy measure from the posterior predictive distribution and compare these samples to the sample from the observed data.

Let y_{obs} be the observed data and y_{new} be the generated data. Let $f(\theta|y_{obs})$, $f(d(y_{obs}, \theta)|y_{obs})$ and $f(d(y_{new}, \theta)|y_{obs})$ be the posterior (predictive) distributions of θ, $d(y_{obs}, \theta)$, and $d(y_{new}, \theta)$ respectively. We generate $\theta^{(l)}$ from $f(\theta|y_{obs})$ (MCMC output) and $y^{(l)}$ from $f(y|\theta^{(l)})$ and calculate $d(y_{obs}, \theta^{(l)})$ and $d(y^{(l)}, \theta^{(l)})$ for $l = 1, 2, \ldots, B$. Then, $\{d(y_{obs}, \theta^{(l)}),\ 1 \le l \le B\}$ is a sample from $f(d(y_{obs}, \theta)|y_{obs})$ and $\{d(y^{(l)}, \theta^{(l)}),\ 1 \le l \le B\}$ is a sample from $f(d(y_{new}, \theta)|y_{obs})$.

We propose the following Bayesian exploratory data analysis to perform posterior predictive comparison.

(a) We overlay two box plots, i.e., box plot of $f(d(y_{obs}, \theta)|y_{obs})$ vs box plot of $f(d(y_{new}, \theta)|y_{obs})$. That is, we display these two box plots in a side-by-side fashion. If the model is adequate, the two box plots will be very much alike.

(b) For a given constant K, we calculate

$$P\left[\left|\frac{d(y_{obs}, \theta) - df}{\sqrt{2df}}\right| > K|y_{obs}\right] \text{ and } P\left[\left|\frac{d(y_{new}, \theta) - df}{\sqrt{2df}}\right| > K|y_{obs}\right], \tag{21}$$

where we take $df = nJ$ when $d(y_{new}, \theta) = D(\theta)$ and $df = J$ when $d(y_{new}, \theta) = D_i(\theta)$. Then we compare these two probabilities. If the model is appropriate, these two probabilities are comparable. Note that several different values of K, e.g., $K = 1, 2, 3$, will be tried.

(c) We calculate $P(d(y_{new}, \theta) \geq d(y_{obs}, \theta))$, which can be estimated by

$$\frac{1}{B} \sum_{l=1}^{B} 1_{\{d(y^{(l)}, \theta^{(l)}) \geq d(y_{obs}, \theta^{(l)})\}}$$

where $1_{\{d(y^{(l)}, \theta^{(l)}) \geq d(y_{obs}, \theta^{(l)})\}}$ is 1 if $d(y^{(l)}, \theta^{(l)}) \geq d(y_{obs}, \theta^{(l)})$ and 0 if otherwise. If the model is adequate and n is reasonably large, this probability will not be far away from one half. Whereas the model will be suspected when this probability is close to one or zero. For this case, $d(\cdot, \theta)$ may be chosen as an observation-level Pearson residual discrepancy or a total Pearson residual discrepancy.

Method II: Simulation Based Model Checking

This method is entirely simulation based and has been proposed in Dey *et al.* (1995). Like the posterior prior comparison method, this method only requires model specification and simulation draws from the posterior distribution under the model. This approach replicates a posterior of interest using data obtained under the model and enables us to observe the extent of variability in such a posterior. Then we compare this posterior obtained under the observed data with the different posterior replicates to determine whether the posterior obtained under the observed data is in agreement with the other posterior replicates. This enables us to surmise if the observed data came from the proposed model.

In short the approach indicated in Dey *et al.* (1995) compares posteriors generated from the observed data with the associated posteriors generated from the specific model. This method also uses a discrepancy measure as the model checking tool. The difference of this approach with the earlier method is that instead of using a single set of data under the model we replicate R (a large number) data sets. For each data set we obtain the posterior distribution of the discrepancy measure. In this way for a single discrepancy measure we have R posteriors obtained from the data generated under the model to which we can compare the posterior calculated from the observed data. The idea is that if the posterior obtained under the observed data fits among the other R posteriors then our model fits the data well for that particular discrepancy function. Due to the intensity of computation, we recommend a discrepancy measure $d(\cdot, \theta)$ to be chosen as the total Pearson residual discrepancy measure $D(\theta)$ in (20).

The advantage of using this method is that the variability among the posteriors allows us to judge better whether the posterior under the observed data fits the model.

The technical detail is given as follows. Similar to **Method I**, let $y^{(0)} = y_{obs}$ and $y^{(r)}$ be generated data at the rth replication for $r = 1, 2, \ldots, R$ (say, $R = 1,000$). More specifically, $y^{(r)}$ can be obtained by generating $\theta^{(r)}$ from $f(\theta|y_{obs})$ and hence generating $y^{(r)}$ from $f(y|\theta^{(r)})$ for $r = 1, 2, \ldots, R$. For each r, we generate $\theta^{(rj)}$ from the posterior $f(\theta|y^{(r)})$ and calculate $d(y^{(r)}, \theta^{(rj)})$ for $j = 1, 2, \ldots, B$ and $r = 0, 1, 2, \ldots R$. In this way, we obtain $R + 1$ samples $\{d(y^{(r)}, \theta^{(rj)}),\ j = 1, 2, \ldots, B\}$ from the posterior distribution $f(d(y^{(r)}, \theta)|y^{(r)})$ for $r = 0, 1, \ldots, R$. Using these $R + 1$ samples, we calculate 5–quantiles of each distribution, i.e., $q^{(r)} = \left(q_{0.05}^{(r)}, q_{0.25}^{(r)}, q_{0.5}^{(r)}, q_{0.75}^{(r)}, q_{0.95}^{(r)}\right)'$, $r = 0, 1, 2, \ldots, R$. Letting $\bar{q} = \frac{1}{R}\sum_{r=1}^{R} q^{(r)}$, we calculate $\xi_r = \left|\left|q^{(r)} - \bar{q}\right|\right|^2$, $r = 1, 2, \ldots, R$. Denote C to be, say, the 95th percentile of ξ_r's. Then we perform the following Monte Carlo test: if $\left|\left|q^{(0)} - \bar{q}\right|\right|^2 \leq C$, we conclude that the model is adequate. Note that choice of C is arbitrary, and is usually left to the user.

5. Voter Behavior Data example

To apply and illustrate our methodologies, we use the survey data on the voting behavior of 95 residents of Troy, Michigan given in Greene (1993) and further analyzed by Chib and Greenburg (1998). In this data set, the first decision (Y_{i1}) is whether to send at least one child to public school and the second (Y_{i2}) is whether to vote in favor of a school budget. As in Chib and Greenburg (1998), the covariates in x_{i1} are a constant, the natural logarithm of annual household income in dollars, and the natural logarithm of property taxes paid per year in dollars; and those in x_{i2} are a constant, annual household income in dollars, property taxes paid per year in dollars, and the number of years the resident has been living in Troy. Since each voter made two decisions, namely, Y_{i1} and Y_{i2}, then there is a natural correlation in the response.

We fit five models, i.e., a stratified and mixture model, a conditional model, a MVP model, two MVT models with $\nu = 1$ and $\nu = 8$, to this data set. For this data set, we have $n = 95$, $J = 2$, $\beta_1 = (\beta_{11}, \beta_{12}, \beta_{13})'$ and $\beta_2 = (\beta_{21}, \ldots, \beta_{24})'$. Since $J = 2$, we take $\sigma_\alpha^2 = 0$ for the stratified and mixture model and $\sigma_\alpha^2 = 1$ for the conditional model to ensure the identifiability of parameters. In our calculation, we used $\beta_0 = 0$, $B_0 = 0.01I_7$, $\Sigma_0 = I_2$, and $G_0 = 1$.

In our implementation of all five models, we first standardize covariates x_{i2}, x_{13}, x_{22}, x_{23}, and x_{24} to accelerate the convergence of Gibbs sampling. We observe that the standardization is very effective for this data set. We check the convergence of the Gibbs sampler using several diagnostic procedures as recommended by Cowles and Carlin (1996). We take 10 multiple chains with dispersed initial values and compute potential scale reductions (PSR's). (PSR values close to 1 are indicative of convergence of the Markov chain to the target distribution.) A detailed description of the PSR is given by Gelman and Rubin (1992). We compute the autocorrelation coefficients within chains to ascertain that it is "rapidly mixing" as discussed by Geyer (1992). Finally, we graph the sample paths from all chains to confirm at which iterate the convergence occurs. Using the above diagnostic methods, we find that for most cases the Gibbs sampler converges no later than 200 iterations. For instance, for the MVP model, for all parameters β_{ij} and ρ, the ranges of the PSRs are from 1.011 to 1.038 and the ranges for the 97.5 percentiles of the PSRs are from 1.017 to 1.075 at 200 iterations. Further we calculate the autocorrelations after the convergence using 5000 Gibbs iterates for each parameter and we find that the autocorrelations for all β_{ij}'s disappear at lag 5 and the autocorrelations for ρ are .787 at lag 1, .278 at lag 5, and .078 at lag 10. Note that the acceptance rate for the Metropolis step for generating ρ is about 0.62.

We use 10,000 "stationary" Gibbs iterates to perform all relative posterior computations. We apply the methodology proposed in Section 4 to check the model adequacy for all five models. We use the total Pearson residual discrepancy measure $D(\theta)$ in (20) as our model checking tool. For the voter behavior data example, the box plots of $f(d(y_{obs}, \theta)|y_{obs})$ and $f(d(y_{new}, \theta)|y_{obs})$ are displayed in Fig 18.1. The posterior predictive probabilities (21) with *df* = 190 for $K = 1, 2, 3$ are given in Table 18.1. The results from Figure 18.1 and Table 18.1 are consistent and it can be easily observed that the conditional model and the MVT model with $\nu = 1$ are less adequate than the other three. We also calculate the probabilities

$P(d(y_{new}, \theta) \geq d(y_{obs}, \theta))$ for all five models and the MVT model with $\nu = 1$ gives 0.088 while other four models give approximate 0.25. Again, the MVT model with $\nu = 1$ fails this type of model checking criterion. Finally we perform the simulation based model checking and all models pass the Monte Carlo test. For example, for the MVP and stratified and mixture models, the respective 95th percentile of ξ_r's are 5838.5 and 5776.0 and the respective observed distances $||q^{(0)} - \bar{q}||^2$ are 454.3 and 485.4. Note that the results for the stratified and mixed model and the MVT model with $\nu = 8$ are similar, which is consistent with the fact that t_8 is virtually the logistic distribution (e.g., Albert and Chib, 1993).

Table 18.1: Posterior Predictive Probabilities

	$P\left(\left\|\frac{D(\theta)-nJ}{\sqrt{2nJ}}\right\| > K \mid y_{obs}\right)$			$P\left(\left\|\frac{D(\theta)-nJ}{\sqrt{2nJ}}\right\| > K \mid y_{new}\right)$		
Model	$K = 1$	$K = 2$	$K = 3$	$K = 1$	$K = 2$	$K = 3$
Stratified and Mixture	0.607	0.363	0.204	0.424	0.130	0.036
Conditional	0.662	0.451	0.327	0.644	0.344	0.137
MVP	0.570	0.332	0.179	0.405	0.120	0.028
MVT ($\nu = 1$)	0.865	0.691	0.466	0.495	0.187	0.052
MVT ($\nu = 8$)	0.607	0.373	0.208	0.466	0.141	0.032

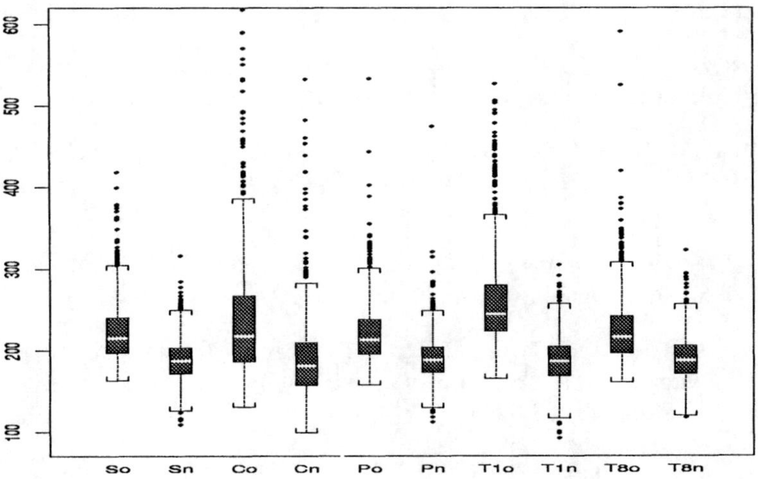

FIGURE 18.1. Distributions of Total Pearson Discrepancy Measures where S=Stratified and Mixture, C=Conditional, P=MVP, T1=MVT1 and T8=MVT8; o=observed and n=new.

6. Concluding Remarks

Correlated binary data often arise in experiments when two or more measurements are taken at one time for the same subjects or when repeated measurements are taken over time. If such correlation is ignored in the model, overstatement of the precision of parameter estimates results. We have considered several models in this paper to incorporate the correlation structure within the framework of multivariate generalized linear models.

In this chapter using Bayesian modeling technique, we have proposed two versions of logistic regression models with random effects component to incorporate dependencies and introduced latent variables to create multivariate probit and t-link models. Several model diagnostic criteria have been introduced for model adequacy, using discrepancy measures, and several graphical methods have been employed to get a visual representation of model diagnostics. Our calculations indicate that the conditional and multivariate Cauchy models are less adequate than the other three proposed models. Some possible extensions are to dynamic generalized linear models for correlated data, including time-varying coefficients, and polychotomous response model s which will be pursued in future work.

References

Albert, J.H. and Chib, S. (1993). Bayesian Analysis of Binary and Polychotomous Response Data. *Journal of the American Statistical Association, 88*, 669-679.

Amemiya, T. (1985). *Advanced Econometrics.* Boston: Harvard University Press.

Ashford, J.R. and Sowden, R.R. (1970). Multivariate Probit Analysis. *Biometrics, 26*, 535-546.

Bernardo, J.M. and Smith, A.F.M. (1994). *Bayesian Theory.* Wiley: New York.

Box, G.E.P. (1980). Sampling and Bayes's Inference in Scientific Modeling (with discussion). *Journal of the Royal Statistical Society, Series A, 143*, 383-430.

Chen, M.-H. and Schmeiser, B.W. (1993). Performance of the Gibbs, Hit-and-Run, and Metropolis Samplers. *The Journal of Computational and Graphical Statistics, 2*, 251-272.

Chib, S. and Greenberg, E. (1998). Bayesian Analysis of Multivariate Probit Models. *Biometrika, 85*, to appear.

Cowles, M.K. and Carlin, B.P. (1996). Markov Chain Monte Carlo Convergence Diagnostics: A Comparative Review. *Journal of the American Statistical Association, 91*, 883-904.

Dey, D.K., Gelfand, A.E., Swartz, T.B., and Vlachos, P.K. (1995). Simulation Based Model Checking for Hierarchical Models. *Technical Report #95-29, Department of Statistics, University of Connecticut.*

Gelfand, A.E., Dey, D.K. and Chang, H. (1992). Model Determination using Predictive Distributions with implementation via Sampling-Based Methods. *In: Bayesian Statistics, 4, (J. Bernardo, et al., eds.)*, Oxford University Press, Oxford, 147-158.

Gelfand, A. E. and Smith, A.F.M. (1990). Sampling Based Approaches to Calculating Marginal Densities. *Journal of the American Statistical Association, 85*, 398-409.

Gelman, A., Meng, X.L., and Stern, H.S. (1996). Posterior Predictive Assessment of Model Fitness via Realized Discrepancies (with discussion). *Statistica Sinica, 6*, 733-808.

Gelman, A. and Rubin, D.B. (1992). Inference from Iterative Simulation Using Multiple Sequences. *Statistical Science, 7*, 457-511.

Geman, S. and Geman, D. (1984). Stochastic Relaxation, Gibbs Distributions and the Bayesian Restoration of Images. *IEEE Transactions on Pattern Analysis and Machine Intelligence, 6*, 721-741.

Geweke, J. (1991). Efficient Simulation from the Multivariate Normal and Student-t Distributions Subject to Linear Constraints. *Computing Science and Statistics: Proceedings of the Twenty-Third Symposium on the Interface*, 571-578.

Geyer, C.J (1992). Practical Markov Chain Monte Carlo (with discussion). *Statistical Science, 7*, 473-511.

Gilks, W.R. and Wild, P. (1992). Adaptive Rejection Sampling for Gibbs Sampling. *Applied Statistics, 41*, 337-348.

Greene, W. (1983). *Econometric Analysis*, Second Edition. Macmillan: New York.

Hastings, W.K. (1970). Monte Carlo Sampling Methods Using Markov Chains and Their Applications. *Biometrika, 57*, 97-109.

Johnson, N.L. and Kotz, S. (1976). *Distributions in Statistics: Continuous Multivariate Distributions.* Wiley: New York.

Marsaglia, G. and Olkin, I. (1984). Generating Correlation Matrices. *SIAM Journal on Scientific and Statistical Computations, 5*, 470-475.

Metropolis, N., Rosenbluth, A.W., Rosenbluth, M.N., Teller, A.H. and Teller, E. (1953). Equations of state calculations by fast computing machines. *Journal of Chemical Physics, 21*, 1087-1092.

Ochi, Y. and Prentice, R.L. (1984). Likelihood Inference in a Correlated Probit Regression Model. *Biometrics, 71*, 531-543.

Prentice, R.L. (1988). Correlated Binary Regression with Covariate Specific to Each Binary Observation. *Biometrics, 44*, 1033-1048.

Rousseeuw, P. and Molenberghs, G. (1994). The Shape of Correlation Matrices. *American Statistician, 48*, 276-279.

Tierney, L. (1994). Markov Chains for Exploring Posterior Distributions (with discussions). *Annals of Statistics, 22*, 1701-1762.

Part VI

Challenging Approaches in GLMs

19

Bayesian Errors-in-Variables Modeling

Jon Wakefield
David Stephens

ABSTRACT Errors-in-variables models are relevant when explanatory variables are measured with error. These models fit very naturally into a Bayesian framework where the unobserved values of the explanatory variables are viewed as unknown parameters. In this paper we consider errors-in-variables modeling for generalized linear models. We present a general framework, review classical and Bayesian approaches and describe a Bayesian analysis of case-control data in which the association between respiratory disease and socio economic status is considered, the latter being inexactly measured.

1. Introduction

The conventional approach to regression modeling is to assume that explanatory variables are measured without error. The defining feature of an *errors-in-variables* problem is that rather than observing the explanatory variables without error, at least one is measured as an error-prone "surrogate". This situation is also sometimes referred to as a *measurement error* problem. Following Carroll, Ruppert and Stefanski (1995) we use X to denote the true value of the variable, W the surrogate and Z explanatory variables that are measured without error; Y will denote the response.

In this paper we consider generalized linear models in which:

$$\mathrm{E}[Y|X,Z] = \mu, \tag{1}$$

with monotonic link function

$$g(\mu) = \beta_0 + \beta_x X + \beta_z Z, \tag{2}$$

$\beta = (\beta_0, \beta_x, \beta_z)$ and Y from the exponential family. We let $h(\cdot)$ denote the inverse link, i.e. $\mu = h(\beta_0 + \beta_x X + \beta_z Z)$.

Now suppose that we observe W rather than X, then

$$\begin{aligned} \mathrm{E}[Y|W,Z] &= \mathrm{E}_{X|W,Z}\left[\mathrm{E}(Y|X,W,Z)\right] \\ &= \mathrm{E}_{X|W,Z}\left[\mathrm{E}(Y|X,Z)\right] \end{aligned} \tag{3}$$

if we assume that we have *non-differential* errors-in-variables, i.e.

$$p(Y|X,W,Z,\beta) = p(Y|X,Z,\beta).$$

In other words W adds nothing to the prediction of Y if X is known. Note that this is equivalent to

$$p(W|Y,X,Z,\beta) = p(W|X,Z,\beta). \tag{4}$$

One situation in which this will not be true is in epidemiological applications in which questionnaires are used to assess exposure and *recall bias* occurs (Clayton, 1992). In this case (4) is not true since individuals with the disease have a different measurement error model to individuals without the disease.

Now from (3),

$$\mathrm{E}[Y|W,Z] = \mathrm{E}_{X|W,Z}[\mu] = \mathrm{E}_{X|W,Z}\left[h(\beta_0 + \beta_x X + \beta_z Z)\right].$$

When we have an identity link function we obtain

$$\mathrm{E}[Y|W,Z] = \beta_0 + \beta_x \mathrm{E}[X|W,Z] + \beta_z Z, \tag{5}$$

i.e. a linear model in β. With a non-identity link function we have

$$\mathrm{E}[Y|W,Z] \neq h(\beta_0 + \beta_x \mathrm{E}[X|W,Z] + \beta_z Z).$$

Hence if the model is really given by (1) and (2) and we regress Y on (W, Z) we do not obtain a GLM with the same link and there is no obvious way of estimating β_x. This argument is relevant to the classical procedure of *regression calibration*, discussed in Section 3.2.

Clearly to analyze errors-in-variables problems we must consider the relationship between the true and surrogate responses given the known covariates, i.e. $p(W, X|Z)$. The *classical errors-in-variables* model considers

$$p(W,X|Z) = p(W|X,Z) \times p(X|Z),$$

whereas with the *Berkson errors-in-variables* model we have

$$p(W,X|Z) = p(X|W,Z) \times p(W|Z).$$

Since W is observed this approach only considers $p(X|W,Z)$.

In experimental studies the Berkson measurement error model will often be appropriate. For example when the level of a variable, W, is recorded from a machine setting; the true level X is then modeled as a function of W. Similarly in the protocol for collection of blood samples in a clinical trial a nominal time, W, will be specified and again the true value X will be modeled as a function of W. In observational studies the classical measurement error model is more typical though Berkson measurement errors do arise; Richardson and Deltour (1998) give as an example the situation in which ambient air pollution W is measured for a group of individuals, perhaps defined by a common area, and individual exposures X are required; a reasonable Berkson model may then be $\mathrm{E}[X|W] = W$.

A large amount of work on errors-in-variables has been motivated by epidemiological applications. In general, epidemiological studies consider observational data and exposure assessment is very difficult. In nutritional epidemiology for example a person's dietary intake is required and information may be obtained through non-perfect surrogates such as 24-hour recall, food-frequency questionnaires, or dietary records (e.g. Landin, Freedman and Carroll, 1995). Case-control studies have also been examined from an errors-in-variables perspective (e.g. Carroll, Gail and Lubin, 1993). In such applications it is conceptually convenient to consider the following separation for the classical errors-in-variables model (Clayton, 1992; Richardson and Gilks, 1993a,b):

$$\textbf{Disease Model}: \quad p(Y|X,Z,\beta). \tag{6}$$

$$\textbf{Measurement Model}: \quad p(W|X, Z, \theta). \tag{7}$$

$$\textbf{Exposure Model}: \quad p(X|Z, \phi). \tag{8}$$

Hence the measurement error model relating W to X and Z depends on parameters θ, and the exposure model for X given Z depends on parameters ϕ. The disease model may depend on additional nuisance parameters (such as variances) but for notational convenience we suppress dependence on these parameters. A Bayesian approach completes the above model by assigning prior distributions to the unknown parameters. Often these parameters will be assumed to be independent, i.e.

$$p(\beta, \theta, \phi) = p(\beta) \times p(\theta) \times p(\phi). \tag{9}$$

We note that in general the prior distribution for θ and ϕ in particular, may depend on Z, but for notational simplicity we suppress this possible dependence.

Richardson and Gilks (1993a,b) consider designs within this framework using graphical models. Unless there are previous data or other forms of information available to build the measurement error model, data on X and W *simultaneously* must be available for at least some experimental units (e.g. people); in this situation X is referred to as the *gold standard*. When Y is not available on the subset of individuals upon whom both W and X are measured, these data are referred to as *external*; otherwise they are *internal*. In the former situation the *transportability* (Carrol, Ruppert and Stefanski, 1995) of the model, i.e. whether the model from the external data is appropriate for the current study, is a crucial consideration. In other situations repeat measures on W may be available, or several measuring instruments may have been used. If X is never available (which is the case in nutritional epidemiology) then it is clear that the analysis is totally dependent on prior assumptions.

2. Illustrative Example: Case-control Study with Deprivation

The Small Area Variations in Air-quality and Health (SAVIAH) project (Briggs *et al.*, 1997) investigated the relationship between respiratory health and air pollution. In this paper we consider one of the study cities, the United Kingdom city of Huddersfield, in which a case-control study in children was carried out. The binary response was parentally-reported "wheezing or whistling in the chest in the past 12 months"; for brevity this will be referred to as "wheeze". In the original study there were 3326 children in total, of which 536 were cases. For illustration here we consider a random sample of 331 individuals (approximately 10% of the original data) including 53 cases.

A number of explanatory variables were available including a deprivation (socio-economic) score, the so-called Carstairs index (Carstairs and Morris, 1991), measured at the level of the decennial census-defined Enumeration District (ED). The Carstairs' index is the sum of four variables measured at the census at ED level, namely the percent of persons: with no car, in overcrowded housing, with household head in social class IV or V, and the percent of men unemployed. Each of these variables is then standardized across the whole country and the sum is taken as a measure of deprivation. High values of the score indicate deprived areas and low scores affluent areas. The ED's contain on average 400 individuals and within our study region there are a total of 427 containing between 0 and 35 study participants with a median of 7.

For each of the children in the study we have the location of the residence (in terms of Eastings and Northings), as well as the ED within which this location lies. Figure 19.1 shows the locations of the cases and the controls that we analyze.

The link between ill-health and deprivation is well-documented (e.g. Jolley, Jarman and Elliott, 1992). Clearly, in general, in studies of health it is better to possess individual-level data on variables such as diet, smoking status, etc. Only rarely are such data available, however. Often in small-area studies of health (e.g. Elliott and Wakefield, 1998) deprivation indices such as the Carstairs' index act as surrogates for unmeasured variables. The index may also be truly area-level (rather than individual-level) in the sense that it is measuring a characteristic of the area such as access to health services. We note that the relationship between health and deprivation is complex, and although the obvious causal link is for deprivation to cause ill-health, ill-health may cause deprivation (e.g. unemployment).

In the study we consider a number of explanatory variables are available. First let $p(X, Z) = \Pr(wheeze|X, Z)$. Here for illustration we assume that the logit of this probability depends on sex (girl/boy denoted Z_1), age (7/8–9 years, denoted Z_2), damp (without/with, denoted Z_3) and deprivation (denoted X) with interactions between deprivation and the main effects; Z_1, Z_2, Z_3 are assumed to be measured without error and X with error. A more complex analysis might also allow for the possibility that damp is measured with error also. The model is then given by:

$$\begin{aligned} \operatorname{logit} p(X, Z) &= \beta_0 + \beta_1 Z_1 + \beta_2 Z_2 + \beta_3 Z_3 + \beta_4 X \\ &+ \beta_5 X \times I(Z_1 = 1) + \beta_6 X \times I(Z_2 = 1) + \beta_7 X \times I(Z_3 = 1) \end{aligned}$$

where $I(\cdot)$ denotes the indicator function.

Here we may envisage two modeling strategies: we may view deprivation as an individual-level or an area-level variable. We describe both possibilities but then in Section 5 explicitly consider an individual-level analysis. Figure 19.2 shows, for the 108,423 ED's of England and Wales a histogram of Carstairs' scores. We note that a small proportion of ED's contain so few individuals that for confidentiality reasons they remain "unclassified" in terms of deprivation. These ED's have been removed in Figure 19.2.

When we treat deprivation as an area-level variable all individuals within a particular ED will be assigned the same score. For such an analysis we could assume that a classical errors-in-variables model were appropriate, i.e. if X_j denotes the deprivation score for ED j then we have

$$W_j = X_j + \epsilon_j^W,$$

where W_j is the observed surrogate and the ϵ_j are independent and identically distributed (i.i.d.) as $N(0, \sigma_W^2)$. Under this model we need to specify a prior distribution for each of the variables measured with error, in this case the deprivation score. We might consider constructing a prior distribution from the data of Figure 19.2 by using, say, a finite mixture-of-normals prior. Strictly we should use the "true" deprivation but this is an example in which no gold standard exists.

Recall from Section 1 that we need to specify a prior $p(X|Z)$ and we note that, for example, houses with damp (i.e. $Z_3 = 1$) will, in general, be in more deprived ED's. Consequently the use of the same prior for the effect of deprivation across all levels of the accurately-measured covariates is unrealistic. Further complications become evident, however, when one considers that within any one area we have a number of individuals, each with their own age, sex and damp variables. We have no information available to construct a more refined prior and, as noted above, the

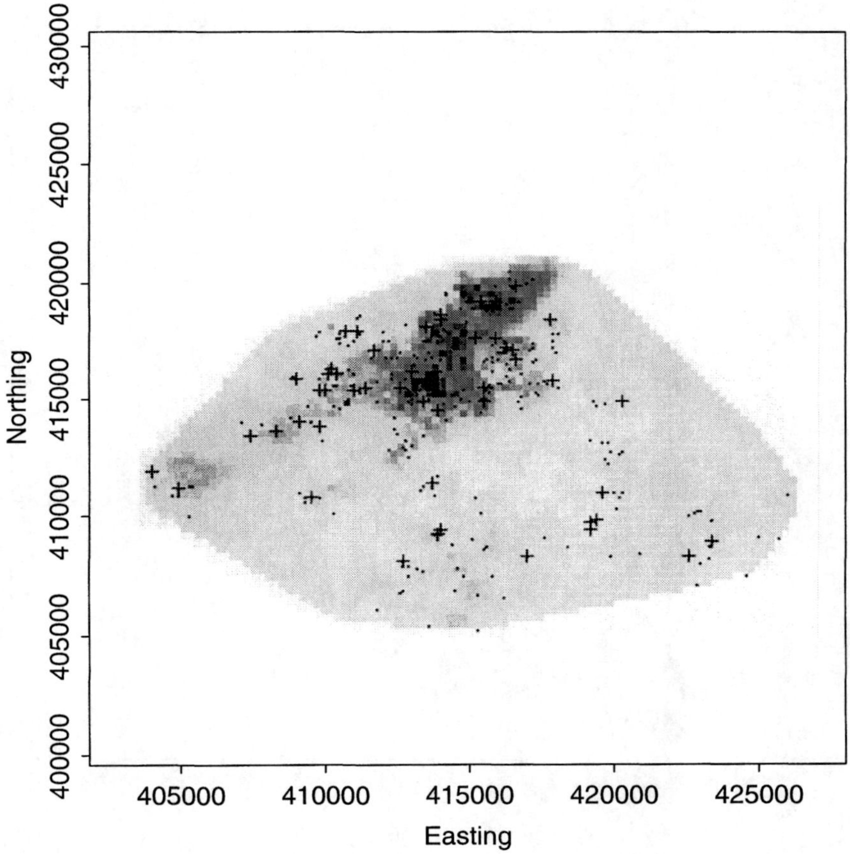

FIGURE 19.1. Spatial variation of deprivation score with a subset of cases (+) and controls (·); dark squares indicate deprived areas and light squares affluent areas.

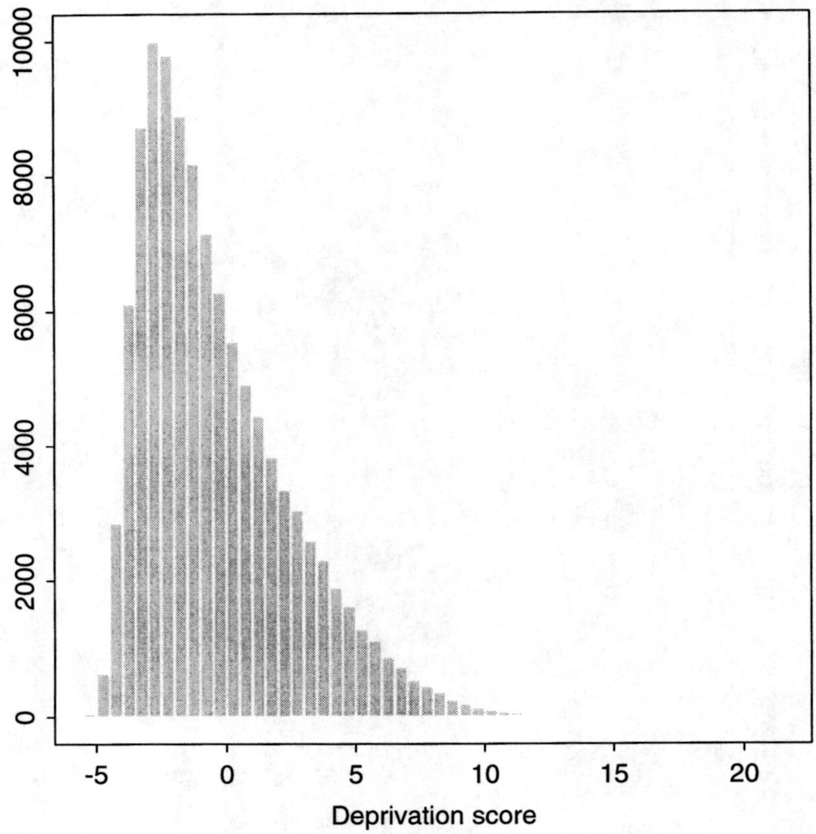

FIGURE 19.2. Histogram of enumeration district deprivation scores across England and Wales; high values indicate greatest deprivation.

assumption that the damp variable is measured without error is probably not realistic. An alternative therefore would be to specify a joint model for the distribution of damp and deprivation.

We also note that we would not expect the prior for the deprivation scores over all J areas to be of the form $\prod_{j=1}^{J} p(X_j|\phi)$ since there will be spatial dependence in the deprivation data; ED's that are geographically close will tend to have similar deprivation scores. This spatial dependence can clearly be seen in Figure 19.1. Bernardinelli *et al* (1997) carry out errors-in-variables modeling in a disease mapping context using a conditional autoregressive model for a spatially-varying explanatory variable that is assumed to be inexactly measured.

For the individual-level analysis that we present, we argue that a Berkson model is more appropriate since the observed surrogate is an *average* over the ED and so, denoting the deprivation of individual i by X_i, we have

$$X_i = W_i + \epsilon_i^X$$

where the ϵ_i^X are i.i.d. $N(0, \sigma_X^2)$. In this case, as commented in Section 1, we are not required to specify a prior distribution for X, but can examine the qualitative effect of varying the level of measurement error.

A major problem shared by this and many epidemiological examples is the lack of a gold standard. Hence the analyzes that we present in Section 5 can be viewed as a *sensitivity study*. We allow the measurement error standard deviations, σ_X, for the Berkson model to range from 1.0 to 3.0; the amount of variability depicted in Figure 19.2 indicates that this range of values is appropriate.

3. Classical approaches

3.1 Basic Formulation

Fuller (1987) and Carroll, Ruppert and Stefanski (1995) contain comprehensive accounts of the frequentist approach to errors-in-variables modeling (the latter also contains a chapter on Bayesian approaches).

Suppose first that we have normal linear model, i.e.

$$Y = \beta_0 + \beta_x X + \beta_z Z + \epsilon^Y,$$

with $\epsilon^Y \sim N(0, \sigma_Y^2)$. Then from (5) we see that regressing Y on (W, Z) can lead to bias in the estimation of both β_x and β_z. Specifically suppose that X and W are both scalar and are related via the classical errors-in-variables model:

$$W = \theta_0 + \theta_1 X + \epsilon^W,$$

with $\epsilon^W \sim N(0, \sigma_W^2)$. Here θ_0 and θ_1 represent *additive* and *multiplicative* bias terms, respectively. Now consider the exposure model

$$X = \mu^{X|Z} + \epsilon^{X|Z},$$

with $\epsilon^{X|Z} \sim N(0, \sigma_{X|Z}^2)$ and $(\epsilon^Y, \epsilon^W, \epsilon^{X|Z})$ all mutually independent. We further assume that (X, Z) are bivariate normal with mean vector and variance/covariance matrix:

$$\begin{bmatrix} \mu_X \\ \mu_Z \end{bmatrix}, \quad \begin{bmatrix} \sigma_X^2 & \sigma_{XZ} \\ \sigma_{XZ} & \sigma_Z^2 \end{bmatrix},$$

respectively. Then, using Bayes Theorem we obtain:

$$E[X|W,Z] = \lambda_0 + \lambda_1 W + \lambda_2 Z,$$

where

$$\begin{aligned}
\lambda_0 &= \frac{\mu_X \sigma_W^2 - \theta_0\theta_1\sigma_{X|Z}^2 + \mu_Z \sigma_{XZ}\sigma_W^2/\sigma_Z^2}{\theta_1^2\sigma_{X|Z}^2 + \sigma_W^2},\\
\lambda_1 &= \frac{\theta_1\sigma_{X|Z}^2}{\theta_1^2\sigma_{X|Z}^2 + \sigma_W^2},\\
\lambda_2 &= \frac{\sigma_{XZ}\sigma_W^2/\sigma_Z^2}{\theta_1^2\sigma_{X|Z}^2 + \sigma_W^2}.
\end{aligned}$$

Hence

$$E[Y|W,Z,\beta] = \beta_0^* + \beta_x^* W + \beta_z^* Z$$

where

$$\begin{aligned}
\beta_0^* &= \beta_0 + \beta_x\lambda_0\\
\beta_x^* &= \beta_x\lambda_1\\
&= \beta_x \times \frac{\theta_1\sigma_{X|Z}^2}{\theta_1^2\sigma_{X|Z}^2 + \sigma_W^2}\\
\beta_z^* &= \beta_z + \beta_x\lambda_2.
\end{aligned}$$

Using

$$\text{var}(Y|W,Z) = E_{X|W,Z}[\text{var}(Y|X,Z)] + \text{var}_{X|W,Z}(E[Y|X,Z]),$$

it is straightforward to derive

$$\text{var}(Y|W,Z) = \sigma_Y^2 + \beta_x^2 \frac{\sigma_W^2\sigma_{X|Z}^2}{\theta_1^2\sigma_{X|Z}^2 + \sigma_W^2}$$

showing that the error variance is inflated. We note the following:

- If we have $\theta_1 = 1$ and $\sigma_W^2 = 0$ but $\theta_0 \neq 0$, then we have no estimation bias in β_x and β_z due to θ_0 since we have simply relocated the X's and hence the slopes are unaffected.

- If $\theta_0 = 0$, $\theta_1 = 1$ (i.e. we have an unbiased surrogate) and there are no known covariates Z (and so $\sigma_{X|Z}^2 = \sigma_X^2$) then

  $$\beta_x^* = \beta_x\lambda_1$$

  where

  $$\lambda_1 = \frac{\sigma_X^2}{\sigma_X^2 + \sigma_W^2}$$

  and we obtain the often-quoted *attenuation* of the regression coefficient. The quantity λ_1 is sometimes referred to as the *reliability ratio*.

- If X and Z are correlated then there is bias in the estimation of β_z, as well as in β_x.

- If $\theta_1 \geq 1$ then we always have attenuation of β_x, for $\theta_1 < 1$ we may over estimate the size of the coefficient.

We end this section by noting that once we are outside the class of linear models very few results are available to suggest the effects of errors-in-variables. In particular the attenuation of estimated coefficients should not be assumed. For example, Brenner *et al* (1992) show that in ecological studies (in which group-level rather than individual-level data are analyzed), the effect of exposure misclassification is to *over-estimate* coefficients.

3.2 Modeling and Analysis: Classical Extensions and Procedures

We now briefly mention two of the more common extensions to the classical measurement error formulation described above.

Regression Calibration

In the regression calibration approach Y is regressed on $\mathrm{E}[X|W,Z]$, rather than X, thus producing a modified model for the observed data. A simple model for $\mathrm{E}[X|W,Z]$ (the so-called calibration model) is a linear regression. The regression calibration approach is straightforward to implement and has been applied to a range of models (including logistic regression for which it has been widely-used), see Carroll, Ruppert and Stefanski (1995), Chapter 3.

Simulation Extrapolation

Simulation extrapolation (SIMEX) is a procedure that can be used to assess the effect of measurement error on, for example, parameter estimates, by considering simulated replicate data sets derived from, the original data. The basic SIMEX principal is that by inflating the measurement error artificially (but in relation to the observed data) in a sequential fashion, and noting the varying effect on the estimated parameters, it is possible to extrapolate (backwards) to obtain estimates of parameters in the case of zero measurement error. Heuristic and more detailed mathematical justifications for the SIMEX approach are given in Carroll, Ruppert and Stefanski (1995), Chapter 4.

4. Bayesian Approaches

4.1 General Framework

As mentioned in the introduction it is very natural to view errors-in-variables models from a Bayesian perspective in which the unobserved exploratory variables X are treated as unknown parameters. Lindley and El-Sayyad (1968) were the first to consider a Bayesian approach and showed the difficulties that a likelihood approach encounters when the X values are viewed as parameters. This is to be expected since the usual regularity conditions are invalidated since the number of unknown parameters increases with the number of data points.

For ease of exposition we begin by considering the classical errors-in-variables model and the disease/measurement/exposure model described by Equations (6)–(8). The posterior distribution is then given by

$$p(X,\beta,\theta,\phi|Y,W,Z) = \frac{p(Y,W|X,\beta,\theta,\phi,Z) \times p(X,\beta,\theta,\phi|Z)}{p(Y,W|Z)}$$

$$\propto p(Y|X,Z,\beta) \times p(W|X,Z,\theta) \times p(X|Z,\phi) \times p(\beta,\theta,\phi), \tag{10}$$

using the conditional independencies implied by (6)-(8) (which include non-differential measurement error). As in (9) we will take independent priors, so that $p(\beta,\theta,\phi) = p(\beta) \times p(\theta) \times p(\phi)$. We re-iterate that, in general, the prior for θ and ϕ may depend on Z.

Similarly for the Berkson errors-in-variables model we have

$$p(X,\beta,\theta|Y,W,Z) \propto p(Y|X,Z,\beta) \times p(X|W,Z,\theta) \times p(\beta,\theta),$$

with $p(\beta,\theta) = p(\beta) \times p(\theta)$. There is no need to specify a prior for X when this model is used.

4.2 Implementation

As in many applications Markov chain Monte Carlo (MCMC) has greatly eased the analysis of errors-in-variables models. The full conditional distributions (Gilks, Richardson and Spiegelhalter, 1996) that are required for the classical errors-in-variables model are obtained from, (10):

$$\begin{aligned}
p(\beta|X,\theta,\phi,Y,W,Z) &= p(\beta|X,Y,Z) \\
&\propto p(Y|X,Z,\beta) \times p(\beta) \\
p(X|\beta,\theta,\phi,Y,W,Z) &= p(Y|X,Z,\beta) \times p(W|X,Z,\theta) \times p(X|Z,\phi) \\
p(\theta|\beta,X,\phi,Y,W,Z) &= p(\theta|X,W,Z) \\
&\propto p(W|X,Z,\theta) \times p(\theta) \\
p(\phi|X,\beta,\theta,Y,W,Z) &= p(\phi|X,Z) \\
&\propto p(X|Z,\phi) \times p(\phi).
\end{aligned}$$

This first conditional distribution corresponds to the posterior distribution that would have resulted if X had been observed. The conditional distributional for X involves all of the stages of the disease/measurement/exposure model, showing how information propagates between the different stages. The conditional distributions for the Berkson model follow similarly.

4.3 Previous Work

Richardson (1996) provides a review of Bayesian measurement error modeling. Racine-Poon, Weihs and Smith (1991) considered errors-in-variables for dilution errors in radioimmunoassay using a Berkson model and numerical integration for implementation. Schmid and Rosner (1993) used a Bayesian model to analyze epidemiological data in which alcohol consumption was modeled as a distribution that included a spike at zero to represent zero consumption.

The general parametric framework for epidemiological applications using MCMC for inference has been described by Richardson and Gilks (1993a,b). Fully parametric approaches using MCMC have been suggested for: generalized linear model (Stephens and Dellaportas, 1992); nonlinear models (Dellaportas and Stephens, 1995); ancillary risk factor models in epidemiology (Gilks and Richardson, 1992); disease mapping models, including a spatial clustering prior on the explanatory variables (Bernardinelli *et al*, 1997); ecological correlation studies with a Poisson regression model (Jordan *et al*, 1997); estimation of the dose-response relationship for atomic bomb survivors (Richardson and Deltour, 1998).

TABLE 19.1. Models for the wheeze data, logit $p(X_i, Z_i) = \phi_i$.

MODEL	Z_1	Z_2	Z_3	LINEAR PREDICTOR ϕ_i
1	0	0	0	$\beta_0 + \beta_4 X_i$
2	1	0	0	$\beta_0 + \beta_1 + (\beta_4 + \beta_5)X_i$
3	0	1	0	$\beta_0 + \beta_2 + (\beta_4 + \beta_6)X_i$
4	1	1	0	$\beta_0 + \beta_1 + \beta_2 + (\beta_4 + \beta_5 + \beta_6)X_i$
5	0	0	1	$\beta_0 + \beta_3 + (\beta_4 + \beta_7)X_i$
6	1	0	1	$\beta_0 + \beta_1 + \beta_3 + (\beta_4 + \beta_5 + \beta_7)X_i$
7	0	1	1	$\beta_0 + \beta_2 + \beta_3 + (\beta_4 + \beta_6 + \beta_7)X_i$
8	1	1	1	$\beta_0 + \beta_1 + \beta_2 + \beta_3 + (\beta_4 + \beta_5 + \beta_6 + \beta_7)X_i$

Dellaportas and Stephens (1995) and Walker and Wakefield (1998) consider errors-in-variables in second stage explanatory variables of a hierarchical model, the latter with a non-parametric distribution for the random effects in a nonlinear model. Another source of errors-in-variables in epidemiology is the consideration of inaccuracies in population *denominators* that are present due to, for example, underenumeration at the census and errors due to the need to construct counts for inter-censual years. Wakefield and Wallace (1999) consider an errors-in-variables approach to this problem. Bennett and Wakefield (1998) consider an errors-in-variables approach to population pharmacokinetic/pharmacodynamic modeling where the observed pharmacokinetic concentrations that act as explanatory variables for the pharmacodynamic response were modeled using a classical measurement error model and an informative prior derived from previous studies.

Muller and Roeder (1997) consider a non-parametric approach to errors-in-variables in case-control studies using Dirichlet processes. Mallick and Gelfand (1996) consider a semi-parametric approach for generalized linear models using incomplete beta functions.

5. Example revisited

We now present a number of analyzes that consider the effect of errors-in-variables in the deprivation score. In particular, we wish to study the qualitative effect of varying the amount of measurement error in the deprivation score. We therefore repeat the analysis several times with σ_X^2 in the Berkson model ranging from, 1.0 to 9.0, and inspect posterior summaries for the parameters of interest relative to the results obtained assuming zero measurement error.

The parameterization of the

SEX*DEP+AGE*DEP+DAMP*DEP

model used (in an obvious notation, see for example Wilkinson and Rogers, 1973), is as follows. In terms of parameters $\beta = (\beta_0, \beta_1, \ldots., \beta_8)$, we model the case/control random variable for individual i, Y_i, as having a Bernoulli distribution,

$$Y_i \sim Bernoulli\,\{p(X_i, Z_i)\}$$

where logit $p(X_i, Z_i) = \phi_i$, with ϕ_i and (Z_{1i}, Z_{2i}, Z_{3i}) is given in the table below. We have eight distinct models, each corresponding to a different combination of the explanatory variables sex (Z_1), age (Z_2) and damp (Z_3). Due to the interactions we have a different slope and intercept for each combination of the covariates.

The deprivation score for individual i is denoted X_i. The parameterization in the table is utilized to simplify likelihood computations. We make inference about the posterior via MCMC, all of the required conditional distributions follow well-known forms or are log-concave (for which the adaptive rejection algorithm of Gilks and Wild, 1992, can be used).

We report the results first in terms of posterior summaries of the slope and intercept pairs for each of the the eight models. The second summary we examine is the odds ratio for each of the main effects as a function of deprivation, in order to investigate the effect of the interactions. For age, sex and dampness, at deprivation level X, these odds ratios are given by $\exp(\beta_1 + \beta_5 X)$, $\exp(\beta_2 + \beta_6 X)$ and $\exp(\beta_3 + \beta_7 X)$, respectively. Posterior distributions for these functions are readily computed from the MCMC output.

Figure 19.3 depicts sample-based boxplot summaries (median, quartiles and 95% interval) for the slope and intercept parameters for models 1–8, with increasing measurement error left to right. The effect of varying measurement error is evident; as measurement error levels are increased, the posterior distributions exhibit increased variance, and location shift away from zero, i.e. attenuation is exhibited when measurement error is ignored. This qualitative behavior is in line with other comparable measurement error studies, but it is of interest to quantify the extent to which conclusions are altered as σ_X is varied over a plausible range.

Figure 19.4 depicts similar sample-based summaries for the main effects odds ratios, that is, odds ratios for boys versus girls, 8-9 year-olds versus 7 year olds, and for houses with damp versus houses without dampness, for deprivation scores ranging from -2.0 to 3.0. We see that boys are more likely to suffer from wheeze than girls, as are those in a damp house, and in both cases these relationships become more pronounced as deprivation increases. For age the relationship is less clear cut; the effect switches as a function of deprivation.

6. Conclusions and Discussion

Great care must be taken when errors-in-variables modeling is carried out; particularly in the case when no gold standard exists since the analysis is driven by assumptions that are uncheckable from the data being analyzed. If the errors-in-variables model is a bad approximation to the truth then spurious results will be obtained; if the posterior distributions of the true explanatory variables are far from the surrogates then this is a sign that a large amount of "feedback" has occurred suggesting that modeling assumptions are having a large effect. For this reason non- and semi-parametric methods are appealing in errors-in-variables modeling. A frequentist semi-parametric approach has recently been suggested by Spiegelman and Casella (1997). From a Bayesian perspective the modeling of the measurement error and exposure probability distributions can be achieved in a non-parametric fashion (using mixtures of Dirichlet processes or Polya tree formulations for example), or using the semi-parametric approach of Richardson and Green (1997), which is essentially a mixture of normals representation with variable number of components.

We finally note that although MCMC techniques allow the routine analysis of complex data structures they do not necessarily provide insight into the underlying structure of the problem since each application is a "one-off". For errors-in-variables specifically there is a need for more theoretical work in order to determine the effects of measurement error. For design in particular this would be highly desirable. Recently this problem has been examined by Devine and Smith (1998).

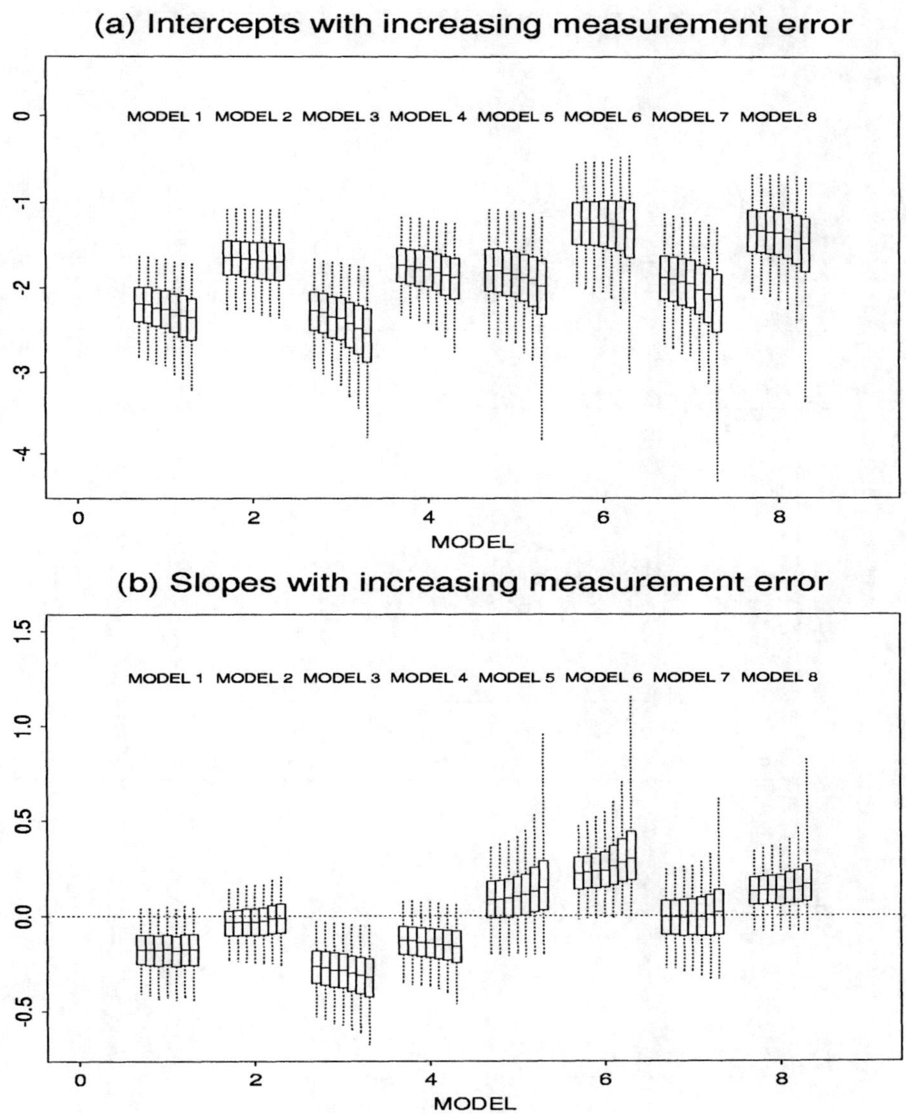

FIGURE 19.3. Boxplots of (a) intercept and (b) slope parameters for models 1–8 with increasing measurement error (left to right).

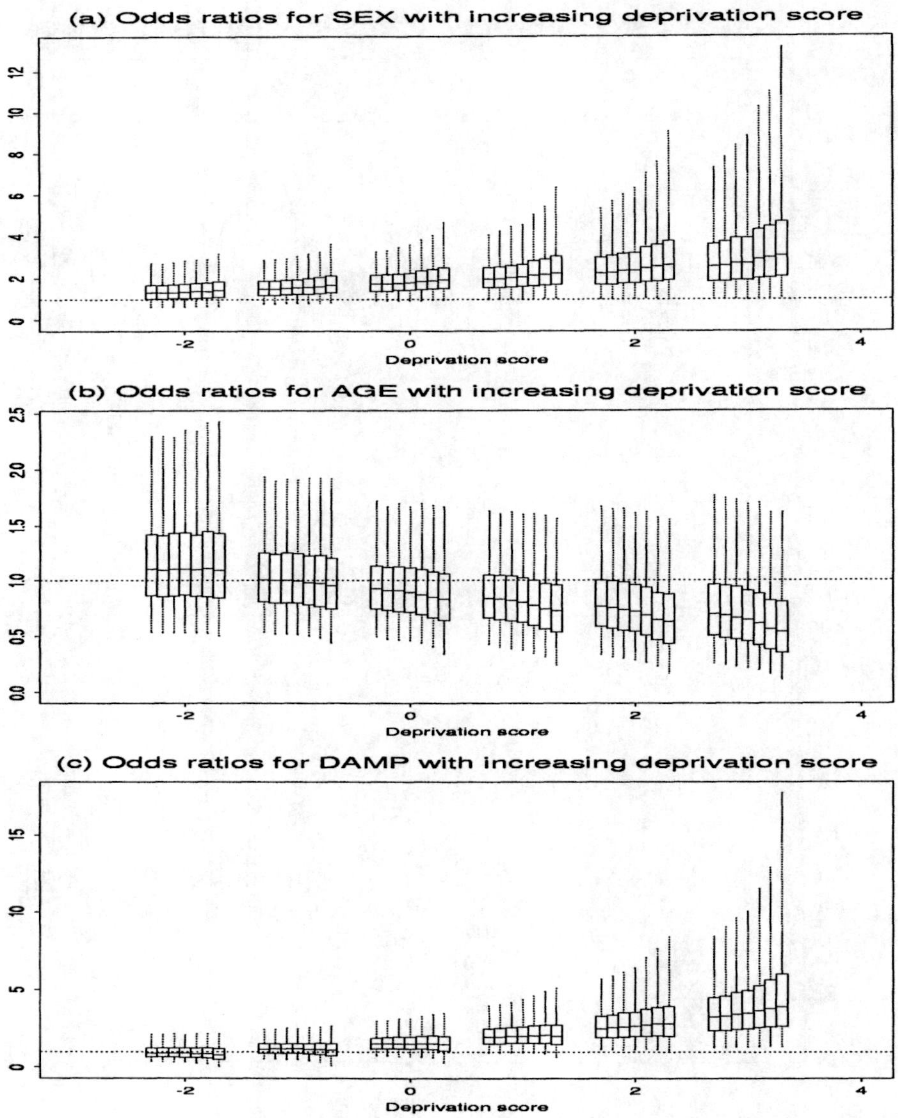

FIGURE 19.4. Posterior log-odds ratios for the SEX, AGE, and DAMP with increasing measurement error (left to right).

Acknowledgements

The authors would like to thank Dave Briggs of Nene College and Jeremy Bullard and Paul Elliott of Imperial College School of Medicine for supplying the data that were used in the example. The authors would also like to thank the Office for National Statistics who made the postcoded cancer data available for use. Population data came from the Estimating with Confidence project. This work is based on data provided with the support of the ESRC and JISC and uses census and boundary material which are copyright of the Crown, the Post Office and the ED-LINE Consortium.

References

Bennett, J.E. and Wakefield, J.C. (1998). The use of errors-in-variable modeling to link populations pharmacokinetic and pharmacodynamic data. Submitted to *Biometrics.*

Bernardinelli, L., Pascutto, C., Best, N.G., and Gilks, W.R. (1997). Disease mapping with errors in covariates. *Statistics in Medicine 16*, 741–752.

Brenner, H., Savitz, D.A., Jockel, K.-H. and Greenland, S. (1992). Effects of non-differential exposure misclassification in ecologic studies. *American Journal of Epidemiology 135*, 85–95.

Briggs, D.J., Collins, S., Elliott, P., Fischer, P., Kingham, S., Lebret, E., Pryl, K., Van Reeuwijk, H., Smallbone, K., and Van Der Veen, A. (1997). Mapping urban air pollution using GIS: a regression-based approach. *International Journal of Geographical Information Systems 11*, 699–718.

Carroll, R.J., Gail, M.H., and Lubin, J.H. (1993). Case-control studies with errors in covariates. *Journal of the American Statistical Society 88*, 185–99.

Carroll, R.J., Ruppert, D., and Stefanski, L.A. (1995). *Measurement Error in Nonlinear Models.* Chapman and Hall, London.

Carstairs, V. and Morris, R. (1991). *Deprivation and Health in Scotland.* Aberdeen University Press, Aberdeen.

Clayton, D.G. (1992). Models for the analysis of cohort and case-control studies with inaccurately measured exposures. In *Statistical Models for Longitudinal Studies of Health*, J.H. Dwyer, F. Manning, P. Lippert, and H. Hoffmeister (editors), pp. 301–331. Oxford University Press, Oxford.

Dellaportas, P. and Stephens, D.A. (1995). Bayesian analysis of errors-in-variables regression models. *Biometrics 51*, 1085–95.

Devine, O.J. and Smith, J.M. (1998). Estimating sample size for epidemiological studies: The impact of ignoring exposure measurement uncertainty. *Statistics in Medicine 17*, 1391–1402.

Elliott, P. and Wakefield, J.C. (1998). Small-area studies of environment and health. In *Statistics for the Environment 4: Health and the Environment*, V. Barnett, A. Stein and K.F. Turkman (editors), pp. 3–27, John Wiley, New York.

Fuller, W.A. (1987). *Measurement Error Models.* John Wiley and Sons, New York.

Gilks, W.R. and Richardson, S. (1992). Analysis of disease risks using ancillary risk factors, with application to job-exposure matrices. *Statistics in Medicine 11*, 1443–1463.

Gilks, W.R., Richardson, S., and Spigelhalter, D.J. (1996). *Markov Chain Monte Carlo in Practice.* Chapman and Hall, London.

Gilks, W.R. and Wild, P. (1992). Adaptive rejection sampling for Gibbs sampling. *Applied Statistics, 41*, 337–348.

Jolley, D., Jarman, B., and Elliott, P. (1992). Socio-economic confounding. In *Geographical and Environmental Epidemiology: Methods for Small-area Studies*, P. Elliott, J. Cuzick, D. English, and R. Stern (editors), pp. 115–24. Oxford University Press, Oxford.

Jordan, P., Brubacher, D., Tsugane, S. Tsubono, Y., Gey, K.F., and Moser, U. (1997). Modelling of mortality data from a multi-centre study in Japan by means of Poisson regression with errors in variables. *International Journal of Epidemiology 26*, 501–7.

Landin, R., Freedman, L.S. and Carroll, R.J. (1995). Adjusting for time trends when estimating the relationship between dietary intake obtained from, a food frequency questionnaire and true average intake. *Biometrics 51*, 169–181.

Lindley, D.V. and El-Sayyad, G.M. (1968). The Bayesian estimation of a linear functional relationship. *Journal of the Royal Statistical Society, Series B 30*, 190–202.

Mallick, B. and Gelfand, A.E. (1996). Semiparametric errors in variables models: a Bayesian approach. *Journal of Statistical Planning and Inference 52*, 307–321.

Muller, P. and Roeder, K. (1997). A Bayesian semiparametric model for case-control studies with errors in variables. *Biometrika 84*, 523–537.

Racine-Poon, A., Weihs, C., and Smith, A.F.M. (1991). Estimation of relative potency with sequential dilution errors in radioimmunoassay. *Biometrics 47*, 1235–1246.

Richardson, S. (1996). Measurement Error. In *Markov Chain Monte Carlo in Practice*, W.R. Gilks, S. Richardson, and D.J. Spiegelhalter (editors), pp. 401–417. Chapman and Hall, London.

Richardson, S. and Deltour, I. (1998). Bayesian modelling of measurement error problems with reference to the analysis of atomic-bomb survivor data. In *Statistics for the Environment 4: Health and the Environment*, V. Barnett, A. Stein and K.F. Turkman (editors), pp. 259–279, John Wiley, New York.

Richardson, S. and Gilks, W.R. (1993a). A Bayesian approach to measurement error problems in epidemiology using conditional independence models. *American Journal of Epidemiology 138*, 430–442.

Richardson, S. and Gilks, W.R. (1993b). Conditional independence models for epidemiological studies with covariate measurement error. *Statistics in Medicine 12*, 1703–1722.

Richardson, S. and Green, P.J. (1997). On Bayesian analysis of mixtures with an unknown number of components. *Journal of the Royal Statistical Society, Series B 59*, 731-792.

Schmid, C.H. and Rosner, B. (1993). A Bayesian approach to logistic regression models having measurement error following a mixture distribution. *Statistics in Medicine 12*, 1141–1153.

Spiegelman, D. and Casella, M. (1997). Fully parametric and semiparametric regression models for common events with covariate measurement error, in main study/validation study designs. *Biometrics 53*, 395–409.

Stephens, D.A. and Dellaportas, P. (1992). Bayesian analysis of generalized linear models with covariate measurement error. In *Bayesian Statistics 4*, J.M. Bernardo, J.O. Berger, A.P. Dawid, and A.F.M. Smith (editors), pp. 813–820. Oxford University Press, Oxford.

Wakefield, J.C. and Wallace, C. (1999). Implications of estimated counts for studies of environment and health in small-areas. To appear in *Office for National Statistics Occasional Series in Medical and Population Subjects.*

Walker, S.G. and Wakefield, J.C. (1998). Population models with a nonparametric random coefficient distribution. *Sankhya, Series B 60*, 196–212.

Wilkinson, G.N. and Rogers, C.E. (1973). Symbolic description of factorial models for analysis of variance. *Applied Statistics 22*, 392–9.

20

Bayesian Analysis of Compositional Data

Malini Iyengar
Dipak Dey

ABSTRACT Compositional data often result when raw data are normalized or when data is obtained as proportions of a certain heterogeneous quantity. These conditions are fairly common in geology, economics and biology. The result is therefore, a vector of such observations per specimen. The usual multivariate procedures are seldom adequate for the analysis of compositional data and there is a relative dearth of alternative techniques suitable for the same. In this chapter, parametric and semi-parametric methods to model such data is discussed and is illustrated on a real data set comprising sand, silt and clay compositions taken at various water depths in an Arctic lake. Simulation based approach is adopted to ascertain adequacy of the fit and several models are compared via measures based on posterior predictive distribution.

1. Introduction

Compositional data can be viewed as a non-negative vector with unit-sum constraint and therefore restricted to a simplex of appropriate dimension. A typical example of data of this nature would be compositional contents of alloys and ores. Analysis of such data also arises when dealing with synthesis of alloys or compounds i.e., in specifying a compositional range of contents to achieve a certain level of some desired physical property. In environmental studies, toxic contents in air reveal the extent of atmospheric pollution and analyzing these components helps in understanding the harmful effects of toxins on our health. In another case data on several life-forms in a particular region may be available as percentages and we may want to assess effect of time and other varying ecological conditions on their survival. Although not frequently identified as such, the scope for obtaining information compositionally is unlimited.

To begin with, Dirichlet distribution was used to explain compositional data. But, as Aitchison (1982) has commented, it turns out to be inadequate for the description of variability in this situation. One disadvantage with the Dirichlet environment is that the correlation structure of a Dirichlet composition is wholly negative thereby making it inappropriate to fit data with some positive correlations. Also, since every Dirichlet composition may be perceived as a composition formed from a basis of independent equi-scaled gamma distributed components, this class has a very strong intrinsic independence structure which makes it unsuitable for data with mild dependence. Now, occurrence of data on a simplex inhibits the usage of the existing plethora of multivariate analytical procedures. Aitchison and Shen (1985b) introduced logistic normal distribution and in the context of modeling compositional data, Aitchison first recognized the inherent difficulties and made a fundamental contribution in this area by primarily adopting logistic normal class to model such data. His modeling strategy was to transform the G component (composition) vector x to vector y in R^{G-1} through Additive Log Ratio (ALR) function (defined below).

Now, let S^{G-1} denote the $G-1$ dimensional simplex, so that $x \in S^{G-1}$. Aitchison's idea to model imprecision in observation x is as follows. If $\beta \in S^{G-1}$ represents the "true" unobserved vector of component proportions then, a perturbation operator "o" relates error term ζ in S^{G-1} with β as, $x = \beta o \zeta = (\frac{\beta_1\zeta_1}{\sum_1^G \beta_i\zeta_i} \cdots \frac{\beta_G\zeta_G}{\sum_1^G \beta_i\zeta_i})$. So that, $y = ALR(x)$ defined by $y = (\log(\frac{x_1}{x_G}), \ldots, \log(\frac{x_g}{x_G}))$, a (real) vector in R^{G-1} is then modeled in multivariate distributional framework. Aitchison (1986) assumed logistic normality of x or equivalently multivariate normality of y. We note that, under ALR transformation one automatically requires all component ratios to be positive, and especially so with the assumption of logistic normal density which is not defined on the boundary of the simplex. Further Aitchison and Shen (1985b) also demonstrated that logistic normal density is invariant to permutation of components and this in turn makes the model independent of order in which constituents are recorded. Aitchison also suggested using the Box-Cox transformations to extend the scope of analysis.

Rayens and Srinivasan (1991a, 1991b) adopted Aitchison's suggestion and incorporated Box-Cox transformations (see (2.1) for definition) as a generalization of ALR transform. But, they make strong structural assumptions on the covariance of the resulting transformed vector to ensure that classical inferential techniques are more tractable, as these procedures rely largely on the principle of maximizing likelihood. Their framework does not abide by the inherent constraints in such data and even under such assumptions introduction of covariates rapidly increases the level of complexity. Here we employ the Bayesian methods model such data using Box-Cox transformations. Let us begin by briefly examining this mode of analysis based on transformed quantities. Now, the LN class does not include Dirichlet distribution and it fails to model data with extreme independence properties. Also in transforming the composition vector to the real space, one has to sacrifice interpretability of parameter estimates in simplex. So, a subsequent question is whether we can model on S^g and if so are there other alternatives to the Dirichlet model? In other words, if we choose S^g as the surface to build the model then the idea is to investigate distributions on S^g that not only model complex dependence properties but can also entertain extreme independence. Aitchison (1985b) introduced the $A^g(\alpha, \beta)$ class (not discussed here) and this family includes both logistic normal and Dirichlet as a special case. However closed form expressions for this density exist only in these two special cases. Rayens and Srinivasan (1994) studied generalized Liouville distributions (GLD) on simplex in response to this but, in their treatment of the material it remains unclear as to how covariates can be included in the model. The GLD class is a definite improvement over Dirichlet class on simplex. It permits distributions that admit negative or mixed correlation. Further, (to quote Rayens and Srinivasan (1994)) the GLD family contains non-Dirichlet distributions with non-positive correlation and overcomes deficits in the $D^g(\alpha)$ class. Here we adopt a semiparametric Bayesian approach to model in the GLD framework.

We use a Bayesian paradigm to model such data in both parametric and semiparametric setup and illustrate with the aid of an example. Markov chain Monte Carlo (MCMC) methods (see Gelfand and Smith (1995) and Tierney (1994)) are employed to simulate from posterior distribution for proposed models and inferences are drawn from these simulated values. The material is organized as follows. In the next section we formulate a parametric model and in Section 3 we describe model determination. In section 4 we discuss a semiparametric model and the following section describes the estimation procedure for the two models. Section 6 provides a detailed analysis of results thus obtained. The last section is reserved for concluding remarks.

2. A Parametric Approach

The Box-Cox transformation was suggested by Aitchison as a means to develop more reliable models and this has been discussed by Rayens and Srinivasan (1991a, 1991b). We briefly describe the key ideas in their paper as follows. Let BC(ζ) denote the Box-Cox transformation (with parameter λ) of $\zeta \in S^g$ and as in Aitchson's model they let $x = \beta o \zeta$. Instead of modeling error term as logistic normal as was done by Aitchison (1982), Rayens and Srinivasan (1991b) assume that there exists $\lambda_i \in R$ so that,

$$\frac{(\zeta_i/\zeta_G)^{\lambda_i} - 1}{\lambda_i}$$

is $N(0, \sigma_i^2)$, $(1 \leq i \leq g)$. Note that they assume independence between the g components and estimations are performed maximizing likelihood. The independence assumption may not be appropriate owing to the nature of x i.e., x being subject to unit sum constraint is not composed of independent quantities. Further, the model is fit marginally due to this assumption.

We now adapt the above notation to subsequent treatment of the material. Consider a heterogeneous mixture with G components. Let $\beta_i = (\beta_{i1}, \beta_{i2}, \ldots, \beta_{iG})$ denote the parameter vector of constituent proportions and x_{ij} represent i^{th} observation for j^{th} constituent $(i = 1, \ldots, n;\ j = 1, \ldots, G)$ so that, $x_{ij} \geq 0$ and $\sum_{j=1}^{G} x_{ij} = 1$ $(\forall\ i = 1, \ldots, n)$. Here n denotes the number of composition vector samples. We note that β_i could denote the expected value of a composition vector under a Bayesian paradigm as opposed to a fixed "true" value in the frequentist case and may depend on covariate information (see subscript i). There is a natural unit sum constraint on the constituent vector $x_i = (x_{i1}, \ldots, x_{iG})$ so that, x_i has dimension $g = G - 1$ i.e., $x_i \in S^g$. Now, Aitchison's model depends on being able to fit compositional data under logistic normal class. Further, di! agnostic procedures proposed by Aitchison (1986) are dependent on the choice of divisor x_G even if the theory based on logistic normal distributions is not. In this regard, suppose we use another family of a general class of functions which also includes the ALR transformation we can increase the chances of being able to provide a better fit to such data. Now, consider a function $H(.)$ so that, $y_{ij} = H(\frac{x_{ij}}{x_{iG}})$, $j = 1, \ldots, g$; $i = 1, \ldots, n$ and let $y_i = (y_{i1}, \ldots, y_{ig})$, the $1 \times g$ vector of transformed composition. By using the ratio $\frac{x_{ij}}{x_{iG}}$ the dimensionality of simplex is effectively preserved and the function $H(.)$ is chosen so as to ensure that the resultant vector y_i has real components which can be modeled by a suitable likelihood. When we model $H(\frac{\beta_{ij}}{\beta_{iG}})$, the method acquires a GLIM like flavor with $H(.)$ as a link function. Next as a means to improve the scope of analytic tools! we adopt Aitchison's suggestion of studying the Box-Cox transformations as a natural extension of ALR.

Now, the Box-Cox transformation of x_i $(1 \leq i \leq n)$, is defined as follows,

$$y_{ij} = H(\frac{x_{ij}}{x_{iG}}) = \begin{cases} \frac{(x_{ij}/x_{iG})^{\lambda_j} - 1}{\lambda_j} & \text{if } \lambda_j \neq 0 \\ \log(\frac{x_{ij}}{x_{iG}}) & \text{otherwise} \end{cases} \tag{1}$$

where $\lambda_j \in R$ is an unknown parameter $(i = 1, \ldots, n$ and $j = 1, \ldots, g)$. This family includes ALR transformation as a special case (i.e., $\lambda_j = 0$ for all $j = 1, \ldots, g$). Here the positivity restriction on x_i reduces to having at least one component (x_{iG})

being uniformly positive. Now, we model y_i (whose elements are defined in (2.1)) as follows, for $1 \leq i \leq n$

$$Y_i = \alpha + z_i\theta + \epsilon_i \tag{2}$$

where z_i is a $1 \times p$ vector of covariates when i^{th} sample is observed, α is $1 \times g$ vector of intercept quantities and θ is $p \times g$ matrix of regression coefficients. Also $\epsilon_i = (\epsilon_{i1}, \ldots, \epsilon_{ig})$ denotes the model error vector. We assume that ϵ_i $(1 \leq i \leq n)$ are independent with the same distribution and model diagnostics are used to validate distributional assumptions of ϵ_i. We note that the dimension of this problem is at least $2g + g(g+1)/2$ and higher in presence of covariates. Now, estimates of β_i can be derived upon arriving at estimates of α and θ and this is also discussed here. As modeling assumptions are made more universal in nature it becomes practically infeasible to continue to depend on exact analytical solutions. It is in these circumstances that sampling based inferences can no longer be viewed as a surrogate method but rather a primary mode of extracting information for inference.

To illustrate the procedure we analyze data on sand, silt and clay compositions obtained at different water depths in an Arctic lake. The data originally appeared in Coakley and Rust (1968) and has been discussed in Aitchison (1986, page 359). Aitchison considers ALR transformation to fit the data and diagnostic measures examined by him do not ascertain goodness of fit.

3. Simulation Based Model Determination

The aspect of model determination comprises adequacy, selection and (possible) outlier detection. We rely largely on the predictive distribution to assess performance of the models as more often than not, models are developed for purposes of prediction and therefore it is only natural to study their predictive distributions. Further, well known procedures such as Bayes factor (when prior distribution is proper) may not be very informative especially when there are more than two models to compare. Also Bayes factors are optimal only under a 0-1 loss function i.e., when model choice is viewed as hypothesis testing. Moreover when comparing models, predictive distributions are comparable whereas posterior distributions are not so, it is more appropriate to examine predictive distributions under various models. Now by employing Bayesian inference we enjoy the availability of the predictive distribution and this in turn encourages a range of diagnostic studies. In our example we utilize the following simulation based approaches to evaluate the performance of models.

First, model adequacy is studied using the Gelman, Meng and Stern (1996) technique, where simulated values of a discrepancy measure $d(.,.)$ from predictive distribution are compared to the same from the data (i.e. y). Let γ denote all unknown parameters and y_{new}, the data generated from predictive distribution. Furthermore, let $f(\gamma \mid y)$, $f(d(y,\gamma) \mid y)$ and $f(d(y_{new},\gamma) \mid y)$ denote the posterior distributions of γ, $d(y,\gamma)$ and $d(y_{new},\gamma)$ respectively. Samples γ_l^*, are obtained from $f(\gamma \mid y)$ and y_l^* from $f(y \mid \gamma_l^*)$ for $l = 1, \ldots B$, where B represents number of iterates. From these samples $d(y,\gamma_l^*)$, $d(y_l^*,\gamma_l^*)$ are computed. Clearly $\{d(y,\gamma_l^*), l = 1,\ldots,B\}$ is a sample from $f(d(y,\gamma) \mid y)$ and $\{d(y_l^*,\gamma_l^*), l = 1,\ldots,B\}$ is a sample from $f(d(y_{new},\gamma) \mid y)$. Essentially, it is desired that the two discrepancy values, $d(y_{new},\gamma)$ and $d(y,\gamma)$ do not differ radically. One way to test this is would be by computing the probability $P(d(y_{new},\gamma) \geq d(y,\gamma))$, denoted henceforth by pr. This probability may not be analytically solvable and Monte Carlo methods can be used to estimate pr as follows,

$$\hat{pr} = \frac{1}{B}\sum_{l=1}^{B} I_{\{d(y_l^*,\gamma_l^*)\geq d(y,\gamma_l^*)\}} \tag{3}$$

The idea being that if the model is adequate, this probability would be about a half and the model would be judged fairly inadequate if this probability is either close to 1 or 0.

The second analysis for model criticism adapts the Gelfand, Dey and Chang (1992) results for a multivariate setup. They present four different techniques to assess performance of models, of which the discussion on Bayesian standardized residual is used. This would in particular address possible issue of outliers in the model. We describe it briefly as follows. Assume that new observations (i.e. y_{new}) are available for fixed values z of the covariate. From predictive distribution at these covariate values one can determine the mean m and covariance structure S. These values along with y_{new}, can be utilized to compute a standardized distance, r where $r^2 = (y_{new} - m)'S^{-1}(y_{new} - m)$. To be able to do this one however needs these new values which ideally are not available and to overcome this, one borrows from the concept of cross validation as explained under.

Let $y_{(i)}$ denote the data with information on ith sample deleted. Likewise let $z_{(i)}$ represent covariate information on all but the ith sample. The conditional predictive density of y_i given $y_{(i)}$, $z_{(i)}$ and z_i is known as Conditional Predictive Ordinate (CPO) (see Pettit and Young (1990)) and it captures support of ith observation for the model. Referring Gelfand, Dey and Chang (1992) it can be shown that

$$CPO_i = f(y_i \mid y_{(i)}, z_{(i)}, z_i) = \int f(y_i \mid \gamma, z_i) f(\gamma \mid y_{(i)}, z_{(i)}) d\gamma$$

where γ as before denotes all unknown parameters. Now, to estimate r^2 we need to compute mean and variance of this distribution. However, these closed form expressions are not always tractable and Monte Carlo estimates $\hat{CPO}_i$, $\hat{E}(Y_i \mid y_{(i)}, z_{(i)}, z_i)$, and $\hat{V}(Y_i \mid y_{(i)}, z_{(i)}, z_i)$ are used. See Gelfand, Dey and Chang (1992) for more details. Now based on the cross validation scheme, Bayesian standardized residual r_i $(i = 1, \ldots, n)$ can be derived from

$$r_i^2 = [y_i - E(Y_i \mid y_{(i)}, z_{(i)}, z_i)]'[V(Y_i \mid y_{(i)}, z_{(i)}, z_i)]^{-1}[y_i - E(Y_i \mid y_{(i)}, z_{(i)}, z_i)]$$

Here again B denotes number of iterates. Large values of r_i should increase concern about appropriateness of the model. Now, we obtain a Monte Carlo estimate of r_i^2 from

$$\hat{r}_i^2 = [y_i - \hat{E}(Y_i \mid y_{(i)}, z_{(i)}, z_i)]'\hat{V}(Y_i \mid y_{(i)}, z_{(i)}, z_i)[Y_i - \hat{E}(Y_i \mid y_{(i)}, z_{(i)}, z_i)] \tag{4}$$

These residuals are a visual means to compare models. These procedures may approve more than one model, in which case a summary statistic is sought to compare contending models. A certain measure of loss associated with competing models is elicited. Ideally, the measure is designed so as to quantify the loss attributed to the model's ability to predict. This is sorted by a predictive simulation based approach i.e., data from proposed models is simulated and they are compared with the aim of minimizing predictive loss. The method considered here involves simulation from predictive distribution. As before, let y_{new} denote data sampled from predictive distribution, $f(y_{new}, \gamma \mid y)$ and γ denote all unknown parameters in the model. In the present example we measure the discrepancy between simulated and observed data by the following loss function,

$$d = E(\| Y_{new} - y_{obs} \|^2 | \, y_{obs})$$

The natural choice is then to adopt the model which minimizes d. Now, if y_l^* denotes values simulated from predictive distribution, a simulation based approach on B iterations to estimate d is given by,

$$\hat{d} = \frac{1}{B} \sum_{l=1}^{B} \| y_l^* - y_l \|^2 \tag{5}$$

On arriving at a subset of desirable models, inference may be performed with information from simulations. To generate a joint confidence region for constituent proportions, convex hulls defined by Monte Carlo estimates of these proportions are peeled to reveal the inner most region of desired volume. Since components add to one, instead of considering all G constituents, information on any g are considered sufficient. The hulls so generated clearly do not depend on which constituent has been dropped because convex hulls created by g proportions is an intersection of convex hulls in G dimensions with their analog in g dimensions. This procedure may not generate an exact $100(1-f)\%$ $(0 < f < 1)$ convex credible region due to the fact that points are peeled on the basis of extremity of hulls they lie in and as a result of which deleted hulls may account for approximately $100f\%$ of points.

4. A Semiparametric Approach

We begin with an introduction to the GLD class and its properties. Let $u : R_+^g \to R_+$ be defined by $u(x_{-G}) = h\{(\frac{x_1}{q_1})^{b_1} + \cdots + (\frac{x_g}{q_g})^{b_g}\}$ where $h(\cdot)$ is continuous (almost everywhere) function from R_+ to R_+. Let $S = \{x_{-G} = (x_1, \ldots, x_g) : x_i \geq 0, \sum_{i=1}^{g} x_i \leq 1\}$ denote the simplex subset of S^g. The generalized Liouville family is defined as follows.

Definition 4.1: $x_{-G} \in S$ is said to have a GLD (with respect to Lebesgue measure) if the density of x_{-G} is of the form

$$f(x_{-G}) = A.u(x_{-G}) \prod_{i=1}^{g} x_i^{a_i - 1} \quad \text{for } x_{-G} \in S \text{ and } 0 \text{ otherwise}, \tag{6}$$

and is denoted by $L_g(h, \boldsymbol{a}, \boldsymbol{b}, \boldsymbol{q})$. Here $\boldsymbol{a} = (a_1, \ldots, a_g)$, $\boldsymbol{b} = (b_1, \ldots, b_g)$ and $\boldsymbol{q} = (q_1, \ldots, q_g)$ are vectors of positive elements. A is the normalizing factor and $u(x_{-G})$ is defined above. It is however sufficient for $(\frac{x_i}{q_i})^{b_i} \in R$, $\forall\, i = 1, \ldots, g$ and $\int_{S^g} h(x_{-G}) \prod_{i=1}^{g} x_i^{a_i - 1} dx_{-G} < \infty$. Clearly when $h(x_{-G}) = (1 - \sum_{i=1}^{g} x_i)^{a_G - 1}$, GLD corresponds to $D^g(\boldsymbol{a})$.

Now, Gupta and Richards (1987, 1992a, 1992b) have studied dependence properties of Liouville family in the case $\beta_i = q_i = 1$, $\forall\, i = 1, \ldots, g$. Rayens and Srinivasan (1991a) have demonstrated applicability of GLD class for compositional data analysis with some intuitive insight on the choice of $h(\cdot)$. However it remains unclear in their analysis, as to how one would incorporate covariates in their model. Typical choices discussed by Rayens and Srinivasan are $h(\tau) = constant$, $h(\tau) = \exp^{w(\tau)}$ with $w(\tau) = \eta \exp(-\tau)$ (with η an unknown parameter) etc. The final choice of $h(\cdot)$ is dictated by the desirable location of mode of the resulting density, L_g in the simplex, S. The density L_g can be inflated or depressed as a result of "pressure" applied by $h(\tau)$ on the curved surface $(\frac{x_1}{q_1})^{b_1} + \ldots + (\frac{x_g}{q_g})^{b_g} = \tau$ for each $0 < \tau < 1$. Now, the functional form of L_g (see (6)) is such that it leaves room for further creativity in choice of $h(\cdot)$ and it may be possible to include covariate effects in this part of L_g. Consider expressing $h(\tau) = g(h_0(\tau), \eta z)$ with z denoting covariate information (z could be a vector), η is an unknown parameter (vector) and $g(\cdot)$ represents the unknown function through which x is related to ηz. We note that this idea has a survival analysis flavor where the hazard function is modeled as above. Instead of intuitively assessing the nature of $h_0(\cdot)$, we use semiparametric Bayesian methods to estimate $h_0(\cdot)$ assuming that the form of $g(\cdot)$ is specified. First we partition the domain of $h_0(\cdot)$, the unit interval $[0, 1]$ into k arbitrary intervals by the partition $P = \{I_1, \ldots, I_k\}$. We also assume that $h_0(\cdot)$ has a piecewise constant form (see Breslow (1974)) i.e., $h_0(\tau) = t_i$ for $\tau \in I_i$. We use Dynamic methods for simultaneous estimation and smoothing of $h_0(\cdot)$. From a Bayesian perspective smoothing and estimation algorithms can be derived as posterior mode estimators. The estimators may be viewed as discrete spline smoothers. In the example, we consider $h(\cdot) = h_0(\cdot)^{\exp(\eta z)}$. The monotonic form $\exp(\eta z)$ is chosen for the sake of mathematical tractability and diagnostic procedures are used to detect any inadequacies in this choice. Gelfand and Mallick (1995) have dealt with estimation of $h_0(\cdot)$ in a different context with the functional form of $g(\cdot)$ unknown. However in this chapter, the accent is on demonstrating use of GLD class to fit compositional data with covariates and not so much the determination of $g(\cdot)$.

A first order regression model is used to determine $h_0(\tau)$ (see Fahrmeir (1994) for reference.) Now, let $h_0(\tau) = t_j$ for $\tau \in I_j$ and

$$\delta_j = \log t_j$$
$$\delta_j = \delta_{j-1} + e_j, \quad e_j \sim N(0, \sigma_e^2)$$

for $j = 1, \ldots, k$ with $t_0 = 1$. Further structure can be introduced in the evolution of δ_j, for example a negative slope corresponds to decaying effects or $\sigma_e^2 = 0$ would correspond to a constant slope. The above transition model allows a flexible structure for $h_0(\tau)$. The size of partition P is considered fixed here and the model is fit over several fixed sizes of P. Alternatively this quantity k may be taken random and Bayesian inference from posterior samples could be used as pointers for appropriate choices of k. The proposed model is fit with different sizes of the partition. We also fit the Dirichlet model for the sake of comparison and performance of these models is assessed via posterior predictive distributions. We use Conditional Predictive Ordinate (CPO) measures (see Pettit and Young (1990), Gelfand *et al* (1992)), along with Bayesian Information Criterion (BIC) (see Schwarz (1978)) to compare the models. Samples from posterior predictive distribution help compute conditional predictive ordinate values for the models. Now, the BIC criterion chooses the model that minimizes

$$-\log(L(x; \Omega)) + \frac{M}{2}\log(n)$$

where n is as defined earlier, Ω denotes all model parameters and M denotes the number of model parameters.

5. Posterior Distributions and Estimation

As an example we analyze sand, silt, clay compositions (see Coakley and Rust (1968) and Aitchison (1986 page 359.) Depth at which these sediment compositions are sampled, has also been recorded. Here $g = 2$, $n = 39$ and $p = 1$. On visually inspecting the data it becomes clear that sand content has a downward trend with an increase in depth whereas silt follows an upward trend with increasing depths of water. Clay quantities also consolidate steadily with depth. Five samples violate unit sum constraint by fractional amounts but, this irregularity does not pose a problem as simulated quantities obey built-in unit sum structure. The data set is examined under the above parametric and semiparametric framework.

Now, Equation (2.2) represents the relation between Y_i and unknown parameters. Flat tailed distributions that model outliers can be a good starting point and in this example an exponential power family is examined as a means to model data. This family has received considerable attention as it includes a wide spectrum of continuous and symmetric distributions and under certain conditions, members of this family can be expressed as scale mixtures of normals. It also includes the multivariate normal and multivariate double exponential class as a special case. Now, the pdf of (symmetric) MultiVariate Exponential Power family (MVEP) is given by,

$$EP(y_i \mid z_i\theta, \alpha, v) = c_g \, |\Sigma|^{-1/2} \exp -[c_0(y_i - \alpha - z_i\theta)'\Sigma^{-1}(y_i - \alpha - z_i\theta)]^v$$

with kurtosis parameter v such that, $1/2 \leq v \leq 1$, $c_0 = \frac{\Gamma(\frac{3}{2v})}{\Gamma(\frac{1}{2v})}$ and $c_g = \frac{v c_0^{g/2}\Gamma(\frac{g}{2})}{\Gamma(\frac{g}{2v})\pi^{g/2}}$. The kurtosis parameter v measures the degree of non-normality, when $v = 1/2$ multivariate double exponential family is realized and at $v = 1$, the familiar multivariate normal family is defined. When $1/2 \leq v \leq 1$, this family can be expressed as a scale mixture of multivariate normal density (see Choy (1995)). Under this distributional assumption on Y_i, we fit a hierarchical model. Here especially, as functional forms encountered do not appear to be analytically navigable we adopt a sampling routine based on the Metropolis algorithm (see Choy (1995)) to generate simulated values from posterior distribution and information thus gained enables inference on desired quantities. To simplify representation of the model, some more notation is given here. Let $\theta_{.j}$ denote the jth column of the $p \times g$ matrix θ and let $\lambda = (\lambda_1, \ldots, \lambda_g)$. Now, access to prior information is very desirable but may not always be possible and in this event non-informative priors or priors with suitable hyperparameters (so as to make the density sufficiently diffuse) may be used. The latter variety of priors are used in our analysis. For Σ (unknown and positive definite matrix) we use a Wishart prior $W_g(m, M)$ with $m \geq g$ for the density to be well defined. We note that, in our analysis no restriction is made on functional form for Σ. Now choosing m, the precision parameter small enough ensures that the Wishart prior has a large variance thereby making the density vague. Also this density for Σ serves as a conjugate prior facilitating the sampling procedure. We choose exponential family densities with large variance parameters as priors for θ, α and λ. Specifically we asssume that $\theta_{.j} \sim N(\mu 1, \sigma_\theta I_p)$ $(1 \leq j \leq g)$, $\alpha \sim N(\alpha_0 1, \sigma_\alpha I_g)$ and $\lambda \sim N(\lambda_0 1, \sigma_\lambda I_g)$ with hyperparameter $\mu \sim N(0, \tau)$. Here I_k denotes the $k \times k$ identity matrix. A one-one linear transformation of kurtosis parameter, v from $[1/2, 1]$ to $[0, 1]$ is made and $Beta(1/2, 1/2)$ density is used as a prior for this transformed parameter denoted by $h(v)$. Now the posterior distribution is proportional to the following complex product of likelihood and prior,

$$\begin{aligned} f(\theta, \alpha, \mu, \Sigma, \lambda, h(a) \mid y_i, z_i) \quad = \quad & EP(y_i \mid z_i\theta, \alpha, a) N_\theta(\mu 1, \sigma_\theta I_g) N_\alpha(\alpha_0 1, \sigma_\alpha I_g) N_\mu(\nu, \tau) \\ & \times N_\lambda(\lambda_0 1, \sigma_\lambda I_g) W_g(m, M) Beta(1/2, 1/2) \end{aligned}$$

The resulting posterior distribution for parameters are fairly complicated and we employ the Gibbs sampling routine (see Gelfand and Smith (1990)) and a Metropolis-

Hastings algorithm (see Metropolis *et al.* (1953)), a MCMC method to simulate from (posterior) distributions. We adopt the Odell and Feiveson (1966) procedure to sample from a Wishart density. The above steps are repeated to convergence and under mild regularity conditions, samples from complete conditionals approach samples from joint distribution for a sufficiently large number of iterations. Estimates are based on convergent simulated values from 50,000 iterations. To ascertain convergence, ten multiple chains with observations from every tenth iterate are taken around dispersed initial values and Potential Scale Reduction (PSR's) values (see Gelman and Rubin (1992)) are computed. When these values are close to one, convergence of the Markov chain to the target distribution is rendered imminent. All models considered shared this attribute with PSR values being nearly one in each case for all chains. We further subsampled the convergent chains and retained every fifth value to significantly reduce autocorrelation effects. Over the duration of analysis we tried several widely varying hyperparameters to ensure that inferences are data dependent. In results presented here the hyperparameters are $\sigma_\alpha = \sigma_\theta = \tau = 10$, $\sigma_\lambda = 1$, $\alpha_0 = \lambda_0 = \nu = 0$, $m = 5$ and $M = 10^7 I$. The MVEP model framework yields $\hat{v} = 0.95$, which is sufficiently close to multivariate normal case where $v = 1$. Subsequent models were fit under the simpler assumption that $Y_i \sim N(\alpha + z_i\theta, \Sigma)$. Now with kurtosis parameter v fixed at one, prior density for other model parameters are as in MVEP case. Here results are shown in the normal case only with the above hyperparameters for the sake of clarity.

Now, an estimate for $\beta_i = (\beta_{i1}, \ldots, \beta_{iG})$ at z_i when $\lambda = (\lambda_1, \ldots, \lambda_g) \neq 0$ is provided by

$$\hat{\beta_{ij}} = \frac{1}{B}\sum_{s=1}^{B} \frac{[\lambda_{js}^*(z_i\theta_{.js}^* + \alpha_{js}^*) + 1]^{1/\lambda_{js}^*}}{1 + [\lambda_{1s}^*(z_i\theta_{.1s}^* + \alpha_{1s}^*) + 1]^{1/\lambda_{1s}^*} + \cdots + [\lambda_{gs}^*(z_i\theta_{.gs}^* + \alpha_{gs}^*) + 1]^{1/\lambda_{gs}^*}}$$

for $1 \leq j \leq g$ and

$$\hat{\beta_{iG}} = \frac{1}{B}\sum_{s=1}^{B} \frac{1}{1 + [\lambda_{1s}^*(\theta_{.1s}^* z_j + \alpha_{1s}^*) + 1]^{1/\lambda_{1s}^*} + \cdots + [\lambda_{gs}^*(\theta_{.gs}^* z_j + \alpha_{gs}^*) + 1]^{1/\lambda_{gs}^*}}.$$

Here B represents number of iterations over which estimates are based and γ_s^* represents valuthe e of unknown γ at s^{th} simulation. Zero values in λ make the estimations simpler and can be deduced similarly.

We now examine estimation and other issues in the above semiparametric model. Let $p(\eta)$, $p(\boldsymbol{a})$, $p(\boldsymbol{b})$, $p(\boldsymbol{q})$ denote (independent) prior information for parameters. The product of likelihood and prior is of the form

$$L(x : \Omega) = L_g(h, \boldsymbol{a}, \boldsymbol{b}, \boldsymbol{q})p(\eta)p(\boldsymbol{a})p(\boldsymbol{b})p(\boldsymbol{q}) \tag{7}$$

Now, the posterior density of model parameters is proportional to the above product. We note that in this case A in (6) is unknown and dependent on the covariate z. We focus primarily on the likelihood when prior information may not be adequate. As mentioned in the earlier section we use diffuse priors in this event ensuring that inferences are data dependent. We employ Gaussian priors for a suitable transformation of the parameters with precision matrix close to 0 so that a sufficiently diffuse prior is achieved. Now, analytic resolution of posterior derived from (7) is infeasible. A simulation based approach using a Markov chain Monte Carlo algorithm to sample from the posterior distribution is implemented. In particular a Metropolis-Hastings (1970) version is used here with Gaussian proposal density for the parameters. As mentioned before since the parameters are assumed positive (see (6)) a transformation (logarithm) of parameters is sought to justify Gaussian prior assumptions for transformed parameters.

Now, moments of GLD class exist and can be determined if we compute integrals of the form,

$$\int_S h(\{(\frac{x_1}{q_1})^{b_1} + \cdots + (\frac{x_g}{q_g})^{b_g}\}) \prod_{i=1}^{g} x_i^{a_i - 1} dx_{-G}$$

In the particular case where $b_i = q_i = 1$, the above integral reduces to computing the one dimensional integral

$$\int_S h(x_1 + \cdots + x_g) \prod_{i=1}^{g} x_i^{a_i - 1} dx_{-G}$$

$$= \frac{\Gamma(a_1)\cdots\Gamma(a_g)}{\Gamma(a_1 + \cdots + a_g)} \int_0^1 h(\tau)\tau^{\sum_{i=1}^{g} a_i - 1} d\tau. \tag{8}$$

(see Edwards (1922), Whittaker and Watson (1952)). The integral on the right can be calculated by a number of known methods. Therefore in the case $b_i = q_i = 1$, covariance function of the density may be evaluated analytically depending on the nature of $h(\cdot)$. Now since $h(\cdot)$ is of an unknown piecewise constant form the integral in (8) reduces to a sum of integrals to be evaluated over the intervals in partition P i.e.,

$$\frac{\Gamma(a_1)\cdots\Gamma(a_g)}{\Gamma(a_1 + \cdots + a_g)} \sum_{i=1}^{k} \int_{I_i} \hat{t}_i^{\exp(\eta z)} \tau^{\sum_{i=1}^{g} a_i - 1} d\tau \tag{9}$$

Here $\hat{t}_i$ represent estimates for $h_0(\cdot)$ over the partition. Estimates of other parameters determined suitably can be used in (9) to compute the covariance. In the case some $b_i \neq 1$ or $q_i \neq 1$, Monte Carlo techniques are available to evaluate the covariance. This first entails being able to draw samples from GLD and in order to do so the following algorithm by Devroye (1986) can be used

i. Generate $(Y_1, \ldots, Y_{g-1}) \sim D^g(a_1/b_1, \ldots, a_g/b_g)$

ii. Generate $Y_g \sim L_1(h, (a_1/b_1, \ldots, a_g/b_g), 1, 1)$ independent of step (i.)

iii. Define $X_i = q_i[Y_i Y_g]^{1/b_i}$ for $1 \leq i \leq g - 1$

iv. Define $X_g = q_g[(1 - Y_1 - \cdots - Y_g)]^{1/b_g}$

v. $(X_1, \ldots, X_g)$ is $L_g(h, \boldsymbol{a}, \boldsymbol{b}, \boldsymbol{q})$ distributed

The GLD model is fit and results from the situation when $k = 4$, $k = 8$ with $P = \{[\frac{j-1}{k}, \frac{j}{k}); 1 \leq j \leq k\}$ and $\beta_i = q_i = 1$ for $1 \leq i \leq g$ are presented. Further, in the analysis we also assume that $h(\tau) = h_0(\tau)^{\exp(\theta z)}$. We also fit the $D^g(\alpha)$ class and note that this class has no covariate representation as in GLD case. About 100,000 iterations are run and one in every ten values is retained to minimize autocorrelation. The resulting chains are then tested for convergence via Raftery-Lewis (1992) diagnostic procedures. These diagnostics are satisfactory for all the chains, confirming convergence of samples to intended posterior distributions. The convergent chains are further sub-sampled with one in ten chosen to further mitigate autocorrelation effects, if any. Parameter estimates are based on the resulting samples.

6. Results

With regard to the example considered, some results are summarized in Tables 20.1 - 20.4. We present two cases in the multivariate normal setup viz., Aitchison's ALR transformation with $\lambda = 0$ (giving rise to logistic normal model) and the general Box-Cox representation (see (2.1)). Aitchison's ALR model in the absence of depth yields $\hat{\beta} = (17.79, 56.38, 25, 83)$ and in Bayesian paradigm this model results in $\hat{\beta} = (18.04, 55.96, 26.0)$ and $\hat{\alpha} = (-0.363, 0.783)$. In the absence of covariate, classical estimates for β is just the normalized version of geometric mean of the data in G categories (because of unit sum constraint on proportions). Next we consider models with covariate. Now, the Box-Cox transformed data are modeled in both classical and Bayesian setup and results are given in Table 20.1. Classical estimates presented in Table 20.1 are based on ideas to fit the model marginally (see Rayens and Srinivasan (1991b) as opposed to yielding (joint) g-dimensional multivariate normality (as in Bayesian inference). Consonance between the two methods of inference in this case is therefore not very satisfactory. Parameter estimates (not shown here) under the MVEP assumptions agree fairly well with Bayesian estimates of Table 20.1 for the normal model and this is anticipated since $\hat{v} = 0.95$ in MVEP case.

Now, consider diagnostic measures (see (3.1) and (3.3)) obtained for the five models in Table 20.2. On surveying $\hat{p}$ values, it is clear that models without covariate perform poorly in terms of explainability and these models can only be used as pointers. We also note that estimates of posterior predictive loss indicate a preference for the model with covariate and non-zero λ. The MVEP model yields $(\hat{p}, \hat{d}) = (0.5693, 3.79)$ and this model demonstrates an agreement with covariate assisted $\lambda \neq 0$ model on parameter estimates and performance diagnostics. However, complications involved in computations under MVEP framework do not offer any additional benefits when compared to its multivariate normal counterpart. Furthermore, estimates of v at 0.95 for the MVEP case goes one more step to instill faith in the subsequent multivariate normality assumption.

Convex credible regions (not shown here) for the proportions were generated at $z = 0$, lowest depth at 10.4 m, mean depth at 48.04m and finally at largest depth at 103.7m. Over the range of depth values considered it was evident that the model with non-zero λ gives rise to more compatible convex credible region i.e., for instance with increasing depth the decrease in sand content is readily noticeable. Also this model gives rise to convex credible regions of comparatively smaller volume. One more visual aid to judge adequacy is given by the Bayesian standardized residual plots (not shown here) for the four models seen in Table 20.2 under normality. Again models with covariate have smaller distances and among the three models with depth as an explanatory variable, the situation with non-zero λ is more attractive. Although $\hat{pr}$ values in models including covariate effect are comparable, observed versus predicted discrepancy estimates clearly indicate a preference for the case with non-zero λ. In both these cases we may want to perhaps include a non-linear function of covariate but, model adequacy considerations do not exhibit an oversight in this regard. Even if intercept term α at $(2.178, 0.618)$ dominates relation (2.2) over depth coefficient θ at $(-0.066, -0.005)$ there are distinct diagnostic advantages to models including covariate. At this stage there is sufficient evidence to conclude that the Box-Cox Bayesian model with covariate is better than the simpler (ALR) version with $\lambda = 0$. Now, credible regions are a visual aid to study behavior of data and unlike ternary diagrams used in the past by geologists, these convex credible regions may be created for any given value of z_i, regardless of dimension of z_i

TABLE 20.1. Non-zero λ with covariate

Parameter estimated	Classical inference	Bayesian method	95% Credible set
α_1	2.47	2.178	(1.433, 3.037)
α_2	0.77	0.618	(0.489, 0.766)
θ_1	-0.09	-0.066	(-0.083, -0.054)
θ_2	-0.009	-0.005	(-0.006, -0.003)
λ_1	-0.53	-0.339	(-0.502, -0.097)
λ_2	-0.009	-2.383	(-2.942, -1.527)

TABLE 20.2. Estimates of pr, d

Model	With covariate	Without covariate
$\lambda \neq 0$	0.4972, 3.09	0.4170, 9.10
$\lambda = 0$	0.5010, 4.90	0.7720, 10.79

in our model. However if g, the dimension of x_i is greater than three, it is not very easy to discriminate patterns in cloud of points in credible region plots. Now, Aitchison's (ALR) model without covariate does not yield satisfactory diagnostic measures i.e., his marginal tests do not support the claim to logistic normality. And further inclusion of a linear function of covariate also does not gain diagnostic support. Aitchison finally concludes that a logarithmic function of depth is possibly more suitable, and here he admits that even though logistic normality is not fully validated, it is more reasonable. Further, even this model produces unexplainable outliers (see Aitchison (1986) for details).

We now discuss some results for the semiparametric model. As mentioned before we fix $k = 4$ and $k = 8$ and compare these with the Dirichlet model. Estimates of $h_0(\cdot)$ for the two partitions are $\{1.07, 1.1, 1.13, 1.14\}$ and $\{1.0, 1.01, 1.02, 1.02, 1.03, 1.03, 1.03, 1.03\}$. These values indicate that $h_0(\cdot)$ is roughly constant on its domain. Table 20.3 contains estimates of $\boldsymbol{a}$, η for these two cases.

Table 20.3: Parameter Estimates for GLD case

Parameter	a_1	a_2	η
$k = 4$	0.575	1.952	-0.0003
$k = 8$	0.573	1.95	-0.00015

Since the two cases with $k = 4$ and $k = 8$ do not produce conflicting estimates, we retain the model with $k = 8$ for comparison and diagnostic procedures. In Table 20.4 estimates (expected value and credible set) for concentrations of sand and silt under GLD ($k = 8$) and $D^g(\alpha)$ models are given. We compute estimates for the GLD model when depth is 48.04 m (average value). However for the $D^g(\alpha)$ model these estimates do not include covariate values. Now, empirical estimates of covariance (sand, silt) turns out to be -0.02. For GLD case this is -0.02 (when depth is 48.04 m) but this quantity is -0.055 in the Dirichlet case. We calculate these values analytically from results in (9). The estimates from the GLD model are more compatible and this model also provides reasonable estimates of the covariance matrix. The results also indicate that the GLD family captures variability (in the data) better than Dirichlet class.

Table 20.4: GLD ($k = 8$, $z = 48.04$m) vs. $D^g(\boldsymbol{a})$

Model	E(Sand), 95% credible set	E(Silt), 95% credible set
GLD	0.162, (0.155, 0.169)	0.554, (0.540, 0.563)
$D^g(\boldsymbol{a})$	0.192, (0.181, 0.196)	0.561, (0.538, 0.562)

Next, we outline diagnostics for these models. Now, CPO values for the GLD model are much better than those for the $D^g(\boldsymbol{a})$ (not shown here). BIC values for GLD and $D^g(\alpha)$ models are 4.55 and 11.06 respectively. The diagnostic measures considered provide more support to GLD model than the Dirichlet model.

7. Conclusion

Compositional data can be modeled via the Box-Cox tranformation model and this approach has a GLIM like flavor. The GLD class can also be used to model compositional data on simplex and the density can be adapted to include the influence of independent variables on compositions. In the event of (at least two) missing observations in a composition in Bayesian framework, the unknown quantities may be treated as parameters with suitable priors and above computations can be adapted. It remains to be seen whether in general, inferences change when components are permuted under the Box-Cox transform. In this example however no major discrepancies in inferences (estimates of proportions, convex credible sets for β_i etc.) were detected.

References

Aitchison, J. (1982). The Statistical Analysis of Compositional Data.*Journal of Royal Statistical Society, B*, Vol 2, 139–177.

Aitchison, J. (1985a). A General class of Distributions on the Simplex. *Journal of Royal Statistical Society, B*, Vol. 47, 136–146.

Aitchison, J. (1986). The Statistical Analysis of Compositional Data. *Chapman and Hall.*

Aitchison, J. and Shen, S. M. (1985b). Logistic-Normal distributions: Some properties and uses. *Biometrika*, Vol. 47, 136–146.

Breslow, N. (1974). Covariate analysis of censored survival data. *Biometrics.* Vol. 30, 89–99.

Choy, S. T. B. (1995). Robust Bayesian Analysis Using Scale Mixture of Normals Distributions. *PhD Thesis, Department of Mathematics, Imperial College.*

Coakley, J. P. and Rust, B. R. (1968). Sedimentation in an Arctic lake. *Journal of Sedimentary Petrology*, Vol. 38, 1290–1300.

Devroye, L. (1986). Non-uniform random variate generation, *New York: Springer Verlag.*

Edwards, J. (1922). A treatise on the Integral Calculus, *New York: Macmillan*, Vol. II.

Fahrmeir, L. (1994). Dynamic modeling and penalized likelihood estimation for discrete time survival data", *Biometrika*, Vol. 81, 317–330.

Gelfand, A. E. and Dey, D. K. and Chang, H. (1992). Model determining using predictive distributions with implementation via sampling-based methods (with discussion), *Proceedings of the Fourth Valencia International Meeting on Bayesian Statistics, Oxford University Press*, eds. J. M. Bernardo and J.O. Berger and A.P. Dawid, 147–167.

Gelfand, A. E. and Mallick, B. K. (1995). Bayesian Analysis of Proportional Hazards Models Built from Monotone Functions, *Biometrics* Vol. 51, 843-853.

Gelfand, A. E. and Smith, A. F. M. (1990). Sampling based approaches to calculating marginal densities. *Journal of the American Statistical Association*, Vol. 85, 398–409.

Gelman, A., Meng, X. L. and Stern, H. S. (1996). Posterior Predictive Assessment of Model Fitness via Realized Discrepancies (with discussion). *Statistica Sinica* Vol. 6, 733-807.

Gelman, A. and Rubin, D. B. (1992). Inference from iterative simulation using multiple sequences. *Statistical Science*, Vol. 7, 457–511.

Gupta, R. D. and Richards, D. St. P. (1987). Multivariate Liouville Distributions, *Journal of Multivariate Analysis*, Vol. 43, 233–256.

Gupta, R. D. and Richards, D. St. P. (1992a). Multivariate Liouville Distributions, II.*Probability and Mathematical Statistics* Vol. 12, 291–309.

Gupta, R. D. and Richards, D. St. P. (1992b). Multivariate Liouville Distributions, III. *Journal of Multivariate Analysis* Vol. 43, 29–57.

Hastings, W. K. (1970). Monte Carlo sampling methods using Markov chains and their applications. *Biometrika*, Vol. 57, 97–109.

Metropolis, N., Rosenbluth, A. W., Rosenbluth, M. N. and Teller, A. H. (1953). Equations of state calculations by fast computing machines. *Journal of Chemical Physics*, Vol. 21, 1087–1091.

Odell, P. L. and Feiveson, A. H. (1966). A numerical procedure to generate a sample covariance matrix. *Journal of the American Statistical Association*, Vol. 61, 199–203.

Pettit, L.I. and Young, K. D. S. (1990). Measuring the effect of observation on Bayes factors. *Biometrika* Vol. 77, 455–466.

Raftery, A. E. and Lewis, S. M. (1992). How many iterations in the Gibbs sampler? *Proceedings of the Fourth Valencia International Meeting on Bayesian Statistics* Oxford University Press, ed. J.M. Bernardo and A.F.M. Smith and A.P. Dawid and J.O. Berger.

Rayens, W. S. and Srinivasan, C. (1991). Box-Cox Transformations in the Analysis of Compositional Data. *Journal of Chemometrics*, Vol. 5, 227–239.

Rayens, W. S. and Srinivasan, C. (1991). Estimation in Compositional Data Analysis. *Journal of Chemometrics* Vol. 5, 361–374.

Rayens, W. S. and Srinivasan, C. (1994). Dependence Properties of Generalized Liouville Distributions on the Simplex. *Journal of the American Statistical Association*, Vol. 89, 1465–1470.

Schwarz, G. (1978). Estimating the dimension of a model.*Annals of Statistics*, Vol. 6, 461–464.

Tierney, L. (1994). Markov Chains for Exploring Posterior Distributions (with discussion). *Annals of Statistics*, Vol. 22, 1701–1762.

Whittaker, E. T. and Watson, G. N. (1952). A course in modern analysis. *Cambridge: University Press.*

21

Classification Trees

D.G.T. Denison
B.K. Mallick

ABSTRACT The generalized linear model framework is often used in classification problems and the importance and effect of the predictor variables on the response is generally judged by examination of the relevant regression coefficients. This chapter describes classification trees which can also be used for performing classification but lead to more appealing 'rule-based' models which are simple to interpret. We utilise Markov chain Monte Carlo methods to formulate a Bayesian 'search' strategy to find good tree models which, in general, outperform the usual tree search methods.

1. Introduction

This chapter deals with the general classification problem. We wish to find a model, using a training set of data, which we can use to predict the (categorical) response of future observations given just their covariates. Using classical generalized linear model (GLM) theory for categorical data (Chapter 5, McCullagh and Nelder, 1989) is restrictive and the resulting model can be difficult to interpret. Due to these problems interest in performing classification using tree-based models has grown, especially by applied statisticians. The ease of interpretability and simple output of the model, especially its rule-based nature, is particularly appealing. This leads to definite classes being assigned to each datapoint rather than possible class probabilities. While this may not seem to incorporate the uncertainty in class assignments its definiteness is seen as important to decision makers (e.g. doctors, insurance salesmen) even though this may not appear completely reasonable to more theoretical statisticians.

Classification trees (Breiman *et al.*, 1984) aim to model the unknown true regression function g when the data come from the relationship

$$y_i = g(\mathbf{x}_i) \qquad i = 1, \ldots, n,$$

where the responses $y_i \in \{0, 1, \ldots, C\}$ refer to the category of the datapoint with predictors $\mathbf{x}_i$. Note that the most general classification problem assigns classes nominally so that the numbers given to the classes should not be thought of as *distances* between them. Also, classification trees can easily handle not just ordinal, but also categorical, predictor variables.

In the classification tree case we estimate g by the function $\widehat{g}$, which we can write as

$$\widehat{g}(X) = E(Y|X) = \sum_{i=1}^{k} \beta_i B_i(X), \tag{1}$$

where the B_i are the basis functions of the model, Y is the vector of responses and X the matrix of the predictor variables. Note that, unlike the standard GLM approach the responses are not transformed using a link function.

Nonparametric models use the data to find their form. In many cases these can be written as in equation (1.1), where the only differences between the models is in the form of the basis functions. In this chapter we will demonstrate the potential advantages that can be achieved using Bayesian methods to find these basis functions B_i. Classically this is done by performing a stepwise deterministic search of the model space induced by the basis functions (Breiman *et al.*, 1984) which we shall describe later. However, this only optimises the form of the basis function at each step of the algorithm in a 'greedy' manner and does not jointly 'train' them.

It has been shown (Denison, Mallick and Smith, 1998; Chipman, George and McCulloch, 1998) that better (in a predictive sense) classification trees can be found using a Bayesian search strategy. Unlike many of the chapters in this book where a full posterior distribution of the parameters of interest is recovered, often by simulation using Markov chain Monte Carlo algorithms, we use the same techniques just to find 'good' models knowing that generation from the full posterior is unlikely, except in the simplest of cases.

2. The Classification Tree Model

2.1 The Basis Functions

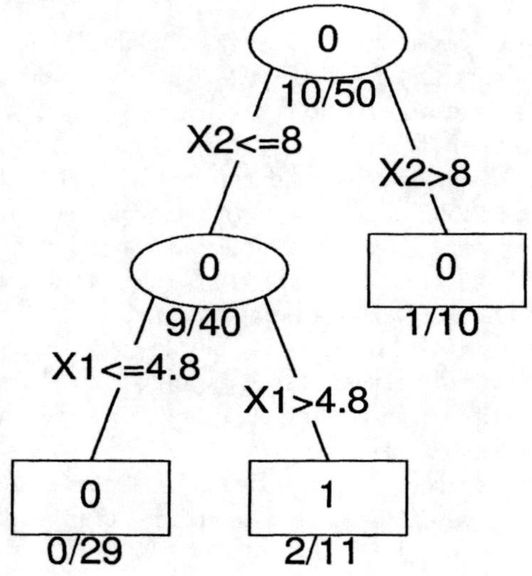

FIGURE 21.1. Example tree structure. The classification and the fraction of misclassified points is given inside and below each node, respectively.

In any nonparametric model the basis functions play an important role in determining both the predictive power and the interpretability of the model. We shall introduce the basis functions used in classification models via a binary regression example and the tree given in Fig. 1.

Suppose that there are 50 datapoints classified either as being in class 0 or 1. For illustration we chose 10 to be in class 1 with the other 40 in class 0. Associated with each datapoint are three predictor variables $X = (X_1, X_2, X_3)$. Suppose that X_1 and X_2 are ordinal variables taking continuous values on [0,10] and X_3 is nominal with 10 categories indexed $\{0, 1, \ldots, 9\}$. The aim is to find good basis functions, which is equivalent to finding good *splitting* questions (at the elliptical *splitting* nodes) involving X, in order to increases within-node homogeneity at the *terminal* nodes (represented by rectangular boxes). The class associated with a given terminal node is the one which has the most members in that node and if there are equal numbers between some classes then the assignment is done randomly between them.

In this example we can see that the tree has fitted well as the fraction of misclassified points, given below each node in Fig. 1, is small at the terminal nodes. It would be possible to carry on splitting the tree until there are no misclassified points but this would lead to the model overfitting so it is common to restrict the tree from allowing too few datapoints in any terminal node.

The basis functions which describe the tree structure in Fig. 1 are

$$\begin{aligned} B_1 &= H[X_2 \leq 8]H[X_1 \leq 4.8], \\ B_2 &= H[X_2 \leq 8]H[X_1 > 4.8], \\ B_3 &= H[X_2 > 8], \end{aligned}$$

where H is the Heaviside function which takes the value 1 when the statement is true and 0 otherwise. The corresponding bottom node assignments are $\boldsymbol{\beta} = (0, 1, 0)$. Note that X_3 was not used in the fit, this suggests that it is not relevant. However, if it was relevant because it is a categorical variable its basis functions would be of the form $H[X_3 \in \{0, 2, 5\}]H[\cdot]$.

As mentioned before, the problem of finding good classification trees is exactly the same as finding good questions to ask at each of the binary splitting nodes but we must also ensure that the tree does not overfit the data.

2.2 The Classical Approach

The usual approach to finding classification trees involves a deterministic search of the model space induced by the trees. In all but the simplest of cases this search cannot be undertaken exhaustively so restrictions in the manner it is undertaken are required. In common with other similar nonparametric methods (see Friedman (1991) for a review) an algorithm that incorporates stepwise addition followed by stepwise deletion is used. This approach is called *greedy* because it optimises the form of the tree at each step. Thus the basis functions are trained stepwise and not together which leads to many good models being missed in this type search. An outline of the algorithm is given below.

Classical Algorithm

1. *Start with a tree with no splitting nodes present.*
2. *Work out the lack-of-fit criterion when each possible split is made.*
3. *Execute the split which most decreases the lack-of-fit.*
4. *Repeat 2 and 3 until a large enough tree has been grown.*
5. *Find the lack-of-fit when each basis in the model is, in turn, deleted.*
6. *Delete the basis which most improves the fit.*
7. *Repeat 5 and 6 until the lack-of-fit criterion reaches a minimum.*

2.3 The Bayesian Approach

We propose a model which can be used to set up a probability distribution over the set of possible trees.

Any binary tree-based model can be uniquely defined by the positions of the splitting nodes present, together with the variables and points where these variables are split. We define these variables respectively as s_i^{pos}, s_i^{var} and s_i^{rule}. The only other unknown parameter is the number of splitting nodes in the model. When there are k terminal nodes in the tree we can show that there are always $k-1$ split nodes so we also use k to define the tree structures.

The tree in Fig. 1 is useful for helping to understand the role of the variables $\theta^{(k)} = (s_1^{pos}, s_1^{var}, s_1^{rule}, \ldots, s_{k-1}^{rule})$. In this case we know $k = 3$ and we write $\theta = (1, 2, 8, 2, 1, 4.8)$ as we define $s^{pos} = 1$ for the splitting position at the root node and $s^{pos} = 2^\ell$ at the left-most possible split position at the ℓth level below the root node. The other possible split positions are labelled consecutively, left to right, along each level of the tree.

To specify a Bayesian model we must assign priors to the unknown variables. We put a Poisson prior (restricted to be greater than 0) with parameter λ over k with uniform priors over the other unknowns to set up a prior probability distribution for the model. These priors are chosen reflecting the knowledge that we know little of the structure *a priori* except that we want to penalise large trees more than smaller ones, hence the Poisson prior over the number of terminal nodes.

Assuming that each datapoint comes from a multinomial distribution then we can write down the likelihood of the data

$$p(Y|X, k, \theta^{(k)}) = \prod_{i=1}^{k} \left\{ H(n_i \geq T_{min}) \prod_{j=0}^{C} (p_{ij})^{m_{ij}} \right\},$$

where n_i is the number of datapoints in the ith terminal node with m_{ij} the number of datapoints in class j and T_{min} is the minimum number of datapoints allowed in a terminal node. Henceforth we shall take $T_{min} = 5$. By standard analytical simplification when using the reference Dirichlet prior for the p_{ij} we find that we can integrate our the p_{ij} from the unknown parameters (Chipman *et al.*, 1998) in the model giving the marginal likelihood (assuming all the nodes have more than T_{min} datapoints) as

$$p(Y|X, k, \theta^{(k)}) = \prod_{i=1}^{k} \left\{ \frac{\prod_j m_{ij}!}{(n_i + C)!} \right\}. \tag{2}$$

Now we can construct a sampler over the varying dimensional parameter space using the reversible jump MCMC algorithm proposed by Green (1995). We must combine move types, some within a dimension and some between dimensions, to form a hybrid sampler (Tierney, 1994). Each move type is randomly chosen according to a defined proposal structure which depends on the number of terminal nodes currently in the model. The move types we choose are: splitting a terminal node (a birth step); recombining adjacent terminal nodes (a death step); changing a splitting variable at a node and changing just the rule of a split at a node. The last two moves types do not change the dimension of the model so we can use a simple Metropolis-Hastings step (Metropolis *et al.*, 1953; Hastings, 1970) to work out the acceptance probability of the proposed model in these cases.

Bayesian Algorithm

1. *Start with a tree with no splitting nodes present.*
2. *Set k equal to the number of terminal nodes in the present tree.*
3. *Attempt to perform either a birth, death, change of split variable or change of split rule move type (the probability that each move type is attempted is usually a function of k).*
4. *Repeat 2 until a sufficient sized sample of trees has been collected, ignoring the initial burn-in period when the log marginal likelihood of the trees varies wildly.*

Unfortunately the hierarchical nature of the classification tree basis functions (proposed new basis functions depend on the bases already in the model) leads to problems in simulating from the posterior distribution of $(k, \theta^{(k)})$. When the tree in the chain has grown to a reasonable size proposed changes in splitting questions near the top of the tree are unlikely to be accepted. This is because changing a question here alters many basis functions which affect lower branches in the structure. As all these bases have been trained together changing many of them significantly in a random way gives poor acceptance rates. This problem is exacerbated because, after large well-fitting trees have been found, the chain spends little time at structures with very few nodes and these are the only times when changes in the top nodes are likely. Thus sampling from the full posterior distribution of tree structures currently seems infeasible so we must content ourselves with performing only a stochastic search for good tree structures using our algorithm.

One way to view the generated sample of trees is that it comes from an approximate posterior distribution of tree structures conditional on the top few splitting nodes. So, due to the importance of the top few splits, to produce a sample of 'good' structures we restrict the tree from growing more than some small number of terminal nodes (e.g. 6) at the beginning of the burn-in period. Thus reasonable initial splits are found and the tree does not go quickly down poor 'blind alleys'. For a fuller exposition of the methodology refer to Denison *et al.* (1998).

3. Real data example

We illustrate the Bayesian classification tree methodology on the breast cancer data studied in Breiman (1996) and Chipman *et al.* (1998). The aim is to predict whether a patient has a benign (B) or malignant (M) tumour based on nine predictor variables which are all numeric and take values on $\{1, \ldots, 10\}$. Of the 699 observations, 16 had missing values and these were removed before the analysis was undertaken. Note that the data can be obtained from the Irvine repository database at the University of California (`ftp://ftp.ics.uci.edu/pub/machine-learning-databases`) and it was originally studied in Wolberg and Mangasarian (1990).

We took the independent sample of trees to be every 5th model visited from the last 300,000 iterations of the chain after an initial 10,000 burn-in iterations. We took the Poisson mean for the number of terminal nodes, the only user-set parameter in our model, to be 2.7 to penalise complex trees (for explanation see Denison, Smith and Mallick, 1998).

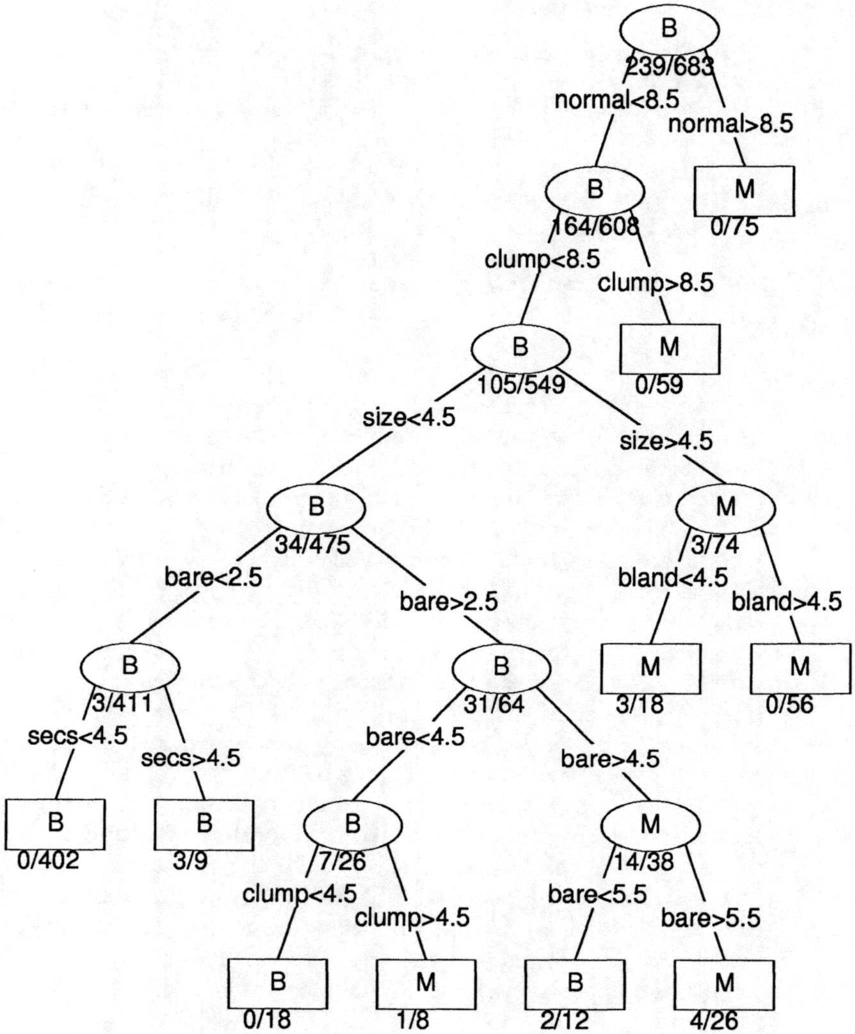

FIGURE 21.2. The tree structure with the largest LML using the breast cancer dataset and the Bayesian search for classification trees. It has 10 terminal nodes and a misclassification rate of 13/683.

Comparing the trees in the generated sample is not a trivial problem and many different approaches could be taken (see Chipman *et al.* (1998) and related discussion). We do not discuss this point further in this paper and just use the log marginal likelihood (LML), which can be easily found from (1.2), for comparison.

In Fig. 2 we display the tree structure from the generated sample which has the highest LML. It has a very low misclassification rate (13 points out of 683) and only 6 of the 9 predictors are used for the classification.

Table 1 gives more details of our analysis of the dataset. It is clear that although we do not sample from the true posterior distribution many different trees with varying numbers of nodes are visited. Also the misclassification rates of our trees compare very well with those found using the 'greedy' method. Using the Splus

Table 21.1. Analysis of the Breast Cancer dataset. Posterior weights along with the largest LMLs of the trees with each number of terminal nodes. The number of misclassified points corresponds to the trees with the largest LML.

Number of Bottom nodes	Posterior Weight	LML	Number misclass.
< 8	0.002	-70.1	17
8	0.041	-65.7	15
9	0.147	-63.4	13
10	0.244	-61.7	13
11	0.248	-61.8	13
12	0.180	-61.6	12
13	0.097	-62.0	10
14	0.031	-63.4	12
> 14	0.010	-65.3	10

function `tree` (Clark and Pregibon, 1992) we fitted greedy trees to the same data and found that all the trees with 11 or fewer terminal nodes misclassified at least 22 points and for tree with between 12-14 terminal nodes 20 points were still misclassified. This demonstrates that the Bayesian method tends to also can find more parsimonious representations of the data than the greedy approach.

4. Discussion

Although classification trees have obvious appeal due to their simple representation through tree diagrams, they do not truly reflect our underlying beliefs about the data. They do not produce continuous models and splits are made perpendicular to the coordinate axes in the predictor space. Some current research involves generalising these tree models, using a Bayesian framework, to overcome some of these difficulties. In particular, the Bayesian partition model (Holmes, Denison and Mallick, 1998) allows the predictor space to be split up with more flexible region boundaries and can be thought of as having many splitting questions at the root node off which many terminal nodes are linked. This makes the exploration of the full posterior distribution of the model structure possible, in contrast to the Bayesian classification tree, because of the lack of hierarchy in the model.

References

Breiman, L., Friedman, J.H., Olshen, R. and Stone, C.J. (1984). *Classification and Regression Trees.* Belmont, CA: Wadsworth.

Breiman, L. (1996). Bagging predictors. *Machine Learning*, **24**, 123-140.

Chipman, H., George, E.I. and McCulloch, R.E. (1998). Bayesian CART model search (with discussion). *J. Am. Statist. Assoc.*, **93**, 937-960.

Clark, L.A. and Pregibon, D. (1992). Tree based models. In *Statistical Models in S*, (Ed. J.M. Chambers and T.J. Hastie), pp. 377-420. Pacific Grove, CA: Wadsworth.

Denison, D.G.T., Mallick, B.K. and Smith, A.F.M. (1998). A Bayesian CART algorithm. *Biometrika*, **85**, 363-377.

Denison, D.G.T., Smith, A.F.M. and Mallick, B.K. (1998). Comment on Bayesian CART model search by H. Chipman, E.I. George and R.E. McCulloch. *J. Am. Statist. Assoc.*, **93**, 954-957.

Friedman, J.H. (1991). Multivariate adaptive regression splines. *The Annals of Statistics*, **19**, 1-141.

Green, P.J. (1995). Reversible jump Markov chain Monte Carlo computation and Bayesian model determination. *Biometrika*, **82**, 711-732.

Hastings, W.K. (1970). Monte Carlo sampling methods using Markov chains and their applications. *Biometrika*, **57**, 97-109.

Holmes, C.C., Denison, D.G.T. and Mallick, B.K. (1998). Bayesian partitioning for classification and regression. *Technical report.* Imperial College, London.

McCullagh, P. and Nelder, J.A. (1989). *Generalized Linear Models.* London: Chapman and Hall.

Metropolis, N., Rosenbluth, A.W., Rosenbluth, M.N., Teller, A.H. and Teller, E. (1953). Equations of state calculations by fast computing machines. *J. Chem. Phys.*, **21**, 1087-1091.

Tierney, L. (1994). Markov chains for exploring posterior distributions (with discussion). *Ann. Statist.*, **22**, 1701-1762.

Wolberg, W.H. and Mangasarian, O.L. (1990). Multisurface method of pattern separation for medical diagnosis applied to breast cytology. *Proceedings of the National Academy of Sciences*, **87**, 9193-9196.

22

Modeling and Inference for Point-Referenced Binary Spatial Data

Alan E. Gelfand
Nalini Ravishanker
Mark D. Ecker

ABSTRACT The problem of modeling and analyzing point referenced binary spatial data is addressed. We formulate a hierarchical model introducing spatial effects at the second stage. Rather than capturing the second stage spatial association using a Markov random field specification, we employ a second order stationary Gaussian spatial process. In fact, we introduce a convenient latent Gaussian spatial process making our observed data a realization of an indicator process on this latent process. Convenient analytic results ensue. Working with a Gaussian process specification introduces the need for matrix inversion to implement likelihood or Bayesian inference. When the number of sites is large, high dimensional inversion is required which is slow (or possibly infeasible) and subject to inaccuracy and instability. We propose an alternative approach replacing inversion with simulation by introducing a suitable importance sampling density. We illustrate with the analysis of a set of repeat sales of residential properties in Baton Rouge, Louisiana.

1. Introduction

We consider the problem of modeling and analyzing binary response data collected at sampling locations in a fixed region D. A critical objective of the analysis is to assess spatial association between the observations. Let $\{\boldsymbol{s}_i : \boldsymbol{s}_i \in D \subseteq R^d\}$ for $i = 1, \ldots, n$ denote a set of spatial locations, let $Y(\boldsymbol{s}_i)$ denote the binary response at site $\boldsymbol{s}_i$ and let $\boldsymbol{Y} = (Y(\boldsymbol{s}_1), Y(\boldsymbol{s}_2), \ldots, Y(\boldsymbol{s}_n))'$. Associated with each site $\boldsymbol{s}_i$ will be a vector $\boldsymbol{X}(\boldsymbol{s}_i)$ which can reflect both spatial and nonspatial information associated with the site.

Our approach is to define a binary stochastic process, $Y(\boldsymbol{s})$, over D. This requires the specification of the joint distribution of $Y(\boldsymbol{s}_1), Y(\boldsymbol{s}_2), \ldots, Y(\boldsymbol{s}_n)$ for any n and for any set of locations $\boldsymbol{s}_1, \boldsymbol{s}_2, \ldots, \boldsymbol{s}_n$ given n.

Formally, our modeling approach is included within that described in Diggle, Tawn and Moyeed (1998). That is, we conceptualize a latent stationary Gaussian spatial process $W(\boldsymbol{s})$ over D such that $\boldsymbol{W} = (W(\boldsymbol{s}_1), W(\boldsymbol{s}_2), \ldots, W(\boldsymbol{s}_n))' \sim N(\boldsymbol{0}, \sigma^2 H(\delta))$ where $(H(\delta))_{ij} = \rho(\boldsymbol{s}_i - \boldsymbol{s}_j; \delta)$, ρ being a valid correlation function in R^2. In addition, we introduce a vector of regression coefficients, $\boldsymbol{\beta}$, associated with $\boldsymbol{X}(\boldsymbol{s}_i)$. Then, we model the $Y(\boldsymbol{s}_i)$ as conditionally independent given $\boldsymbol{\beta}$ and $\boldsymbol{W}$. In fact, we assume

$$\Pr(Y(\boldsymbol{s}_i) = 1 \mid \boldsymbol{X}(\boldsymbol{s}_i), \boldsymbol{\beta}, \boldsymbol{W}) = h(\boldsymbol{X}(\boldsymbol{s}_i)'\boldsymbol{\beta} + W(\boldsymbol{s}_i)) \tag{1}$$

for a specified link function h.

To obtain several useful analytic results and to facilitate computations, we envision an underlying $Z(\boldsymbol{s}_i) = \boldsymbol{X}(\boldsymbol{s}_i)'\boldsymbol{\beta} + W(\boldsymbol{s}_i) + \epsilon(\boldsymbol{s}_i)$ where the $\epsilon(\boldsymbol{s}_i)$ are *iid* $N(0, \tau^2)$. That is, $\epsilon(\boldsymbol{s})$ is a white noise or measurement error process introduced apart from the spatial process $W(\mathbf{s})$ (see, e.g., Cressie, 1993 or Diggle, Liang and Zeger, 1994). Then, we set $Y(\mathbf{s}_i) = 1$ or 0 according to whether $Z(\mathbf{s}_i) > 0$ or < 0, respectively. That is, $Y(\mathbf{s})$ is an indicator process associated with $Z(\mathbf{s})$. As with the $Y(\mathbf{s}_i)$, the $Z(\mathbf{s}_i)$ are conditionally independent given $\boldsymbol{\beta}$ and $\boldsymbol{W}$.

Considering the $Y(\mathbf{s}_i)$'s or the $Z(\mathbf{s}_i)$'s, we have formulated a hierarchical model with a conditionally independent first stage and a model for the $\boldsymbol{\beta}$ and for the spatial effects provided at the second stage. The parameters of the $W(\mathbf{s})$ process, σ^2 and δ, become hyperparameters requiring a third modeling stage. The advantages to recognizing this hierarchical structure when $\boldsymbol{W}$ is modeled through a Gaussian process appear to have only been recently noticed in the literature, see, e.g., Diggle, Tawn and Moyeed (1998) or Ecker and Gelfand (1997). This contrasts with the use of Markov random field specifications for the $W(\mathbf{s}_i)$ which have a longer history and consequently, a richer literature. See, for example, Besag, York and Mollié (1991) or Clayton and Bernardinelli (1992).

This is not surprising. The latter approach models the conditional distribution $W(\mathbf{s}_i)|W(\mathbf{s}_j), j \neq i$ or equivalently, the inverse of the covariance matrix (Besag, 1974; Bernardinelli and Montomoli, 1992) and is well suited for likelihood or Bayesian inference and in particular for simulation based model fitting using Gibbs sampling (Gelfand and Smith, 1990). Possible disadvantages are the fact that the covariance matrix is typically not of full rank; the density for $\boldsymbol{W}$ is not proper much less Gaussian. Even if the inverse covariance matrix is proper, off diagonal entries measure conditional associations between pairs $W(\mathbf{s}_i)$ and $W(\mathbf{s}_j)$ given $W(\mathbf{s}_k), k \neq i, j$ usually through an arbitrarily chosen neighborhood structure. Unconditional association between $W(\mathbf{s}_i)$ and $W(\mathbf{s}_j)$ comes from the resulting covariance matrix and need not be monotonically decreasing with distance (Besag and Kooperberg, 1995). Furthermore, there is no associated spatial process $W(\mathbf{s})$ on D.

Modeling spatial dependence continuously (perhaps monotonically) by analogy with classical geostatistics may be more appropriate in many applications. We may be interested in the variogram, the sill, the nugget, the range, possible anisotropy, kriging, etc. Unfortunately, modeling the covariance matrix requires matrix inversion when evaluating the density for $\boldsymbol{W}$ which is needed for likelihood or Bayesian inference. Such inversion can be slow, perhaps infeasible, as n increases. More importantly, rounding error resulting from the numerous arithmetic operations becomes an issue; we can not be confident in the accuracy or stability of the resulting inverse. This problem arises in the more general setting of Diggle, Tawn and Moyeed (1998), who propose an arbitrary generalized linear model at the first stage for the $Y(\mathbf{s}_i)$. They do not seem to acknowledge it, perhaps because they use at most moderate sample sizes. In fact, they employ a Gibbs sampler to fit their models, which would appear to exacerbate the inversion problem

Recognition of this difficulty accounts for the preponderance of non-model based work in the case of observed $Z(\mathbf{s}_i)$'s. Variogram modeling becomes a curve fitting exercise. Kriging, including universal kriging, becomes a constrained optimization problem. Unfortunately, in the absence of a stochastic specification for the Z's, inference is limited to point estimation. In the case of binary response data, such *nonparametric* approaches are modified to accommodate indicator tranformations of the data. Indicator variograms (Cressie, 1993) and indicator kriging (Journel, 1983) have been proposed though theoretical justification is weak (Cressie, 1993; Papritz and Moyeed, 1997).

De Oliveira (1997) also models $Y(\mathbf{s}_i) = 1$ or 0 according to whether $Z(\mathbf{s}_i) > 0$ or < 0, calling $Y(\mathbf{s})$ a clipped Gaussian field. He does not, however, introduce the white noise process $\epsilon(\mathbf{s})$. As a result, no conditional independence for the $Y(\mathbf{s})$ is obtained; the joint distribution of $\boldsymbol{Y}$ arises as an n dimensional integration of the density of $\boldsymbol{W}$ over a subset of R^n. De Oliveira (1997) introduces latent $Z(\boldsymbol{s}_i)$'s in order to run a Gibbs sampler to avoid such integration. This requires sampling the full conditional distributions for the $Z(\boldsymbol{s}_i)$'s which again introduces repeated matrix inversion.

Hence, one additional contribution of this paper is to suggest that matrix inversion can be replaced with suitable simulation using importance sampling. Matrix inversion is an order n^3 operation. In its most polished form, our approach reduces computation to order n, making it feasible to provide approximate Bayesian inference for larger spatial datasets than previously considered in the literature. Indeed, because our approach runs much faster than performing inversions, it may be preferable even for moderate n.

Returning to the binary spatial process example, questions of interest involve the strength of spatial association between sites. This can be investigated in several ways. Focusing on a pair of sites $\boldsymbol{s}_i$ and $\boldsymbol{s}_j$ results in a 2×2 table for the possible outcomes of the pair $(Y(\boldsymbol{s}_i), Y(\boldsymbol{s}_j))$. The four cell probabilities can be used to compute covariances (or correlations) as well as odds ratios (or log odds ratios). As functions of the model parameters, these quantities are random variables having posterior distributions. These posteriors can be used to infer (e.g., using the median together with lower and upper quantiles) about strength of association. Also, there is a latent variogram, in fact as we shall see, a renormalized variogram associated with $W(\boldsymbol{s}) + \epsilon(\boldsymbol{s})$, which again has a posterior distribution that can be obtained to learn about the association in the latent Gaussian process. Another question of interest involves $\Pr(Y(\boldsymbol{s}_0) = 1)$ where $\boldsymbol{s}_0$ may be an observed site or a new site of interest. How does this probability change as levels of $\boldsymbol{X}(\boldsymbol{s}_0)$ are changed? For a given $\boldsymbol{X}(\boldsymbol{s}_0)$, the posterior distribution of this probability can be obtained so that the effect of change in $\boldsymbol{X}(\boldsymbol{s}_0)$ can be assessed.

We illustrate our approach with an application to a set of 857 residential properties in Baton Rouge, Louisiana, where the $\boldsymbol{s}_i$ are geographical coordinates for the ith house. All properties in the dataset are "repeat sales", that is, they sold once and then were resold during a ten year period of observation. $Y(\boldsymbol{s}_i) = 1$ if the property resold within four years, $Y(\boldsymbol{s}_i) = 0$ if not. We are looking to identify whether there is spatial association in the pattern of quicker resales (≤ 4 years) compared with the slower resales (> 4 years). The $\boldsymbol{X}(\boldsymbol{s}_i)$ record house characteristics anticipated to be important in property sales and resales: age of the house at the time of sale, living area, other area (total square footage less living area) and the number of bathrooms.

The format of the paper is the following. In Section 2, we provide modeling details including several useful analytic calculations. We ultimately arrive at the full hierarchical model employed in fitting the foregoing data. In Section 3, we summarize our simulation-based approach to avoid matrix inversion, extending Gelfand and Ravishanker (1998). Section 4 provides the analysis of the repeat sales real estate data. Finally, in Section 5, we offer a few related remarks.

2. Modeling Details

To formalize modeling details, assume that $\boldsymbol{Z} = (Z(\boldsymbol{s}_1), Z(\boldsymbol{s}_2), \ldots, Z(\boldsymbol{s}_n))'$ is a realization from a spatial Gaussian process of the form

$$\boldsymbol{Z} = \boldsymbol{X}\boldsymbol{\beta} + \boldsymbol{W} + \boldsymbol{\epsilon} \tag{2}$$

where $\boldsymbol{X}$ is $n \times p$ with ith row $\boldsymbol{X}(\boldsymbol{s}_i)$; $\boldsymbol{\beta}$ is $p \times 1$; $\boldsymbol{W}$ is $n \times n$ from a second order stationary Gaussian process, i.e., $\boldsymbol{W} \sim N(\boldsymbol{0}, \sigma^2 H(\delta))$ where $(H(\delta))_{ij} = \text{Corr}(W(\boldsymbol{s}_i), W(\boldsymbol{s}_j)) = \rho(\boldsymbol{s}_i - \boldsymbol{s}_j; \delta)$, ρ a valid 2-dimensional correlation function parameterized by δ; and $\boldsymbol{\epsilon}$ is $n \times 1$ distributed $N(\boldsymbol{0}, \tau^2 I)$. Then,

$$\boldsymbol{Z}|\boldsymbol{\beta}, \boldsymbol{W} \sim N(\boldsymbol{X}\boldsymbol{\beta} + \boldsymbol{W}, \tau^2 I). \tag{3}$$

Let $2\gamma(\boldsymbol{s}_i - \boldsymbol{s}_j)$ be the detrended variogram (see Ecker and Gelfand, 1997) associated with $\boldsymbol{Z}$, i.e.,

$$\begin{aligned} 2\gamma(\boldsymbol{s}_i - \boldsymbol{s}_j) &= \text{Var}((Z(\boldsymbol{s}_i) - \boldsymbol{X}(\boldsymbol{s}_i)'\boldsymbol{\beta}) - (Z(\boldsymbol{s}_j) - \boldsymbol{X}(\boldsymbol{s}_j)'\boldsymbol{\beta})) \\ &= 2\{\tau^2 + \sigma^2(1 - \rho(\boldsymbol{s}_i - \boldsymbol{s}_j; \delta))\}. \end{aligned} \tag{4}$$

$\boldsymbol{Z}$ is not observed, but rather $\boldsymbol{Y} = (Y(\boldsymbol{s}_1), Y(\boldsymbol{s}_2), \ldots, Y(\boldsymbol{s}_n))'$ is, where $Y(\boldsymbol{s}_i) = 1$ if $Z(\boldsymbol{s}_i) > 0$ and $Y(\boldsymbol{s}_i) = 0$ if $Z(\boldsymbol{s}_i) \leq 0$. Given $\boldsymbol{Y}$, we seek to infer about the process parameters $\boldsymbol{\beta}, \delta, \sigma^2, \tau^2$ as well as the probability that $\boldsymbol{Y}(\boldsymbol{s}_0) = 1$ where $\boldsymbol{s}_0$ may be an observed or a future site and also the association between $Y(\boldsymbol{s}_i)$ and $Y(\boldsymbol{s}_j)$.

Note that the $Y(\boldsymbol{s}_i)$ are conditionally independent Bernoulli random variables given $\boldsymbol{\beta}$ and $\boldsymbol{W}$. In fact,

$$\begin{aligned} \Pr(Y(\boldsymbol{s}_i) = 1 \mid \boldsymbol{\beta}, \boldsymbol{W}) &= \Pr(Z(\boldsymbol{s}_i) > 0 \mid \boldsymbol{\beta}, \boldsymbol{W}) \\ &= 1 - \Phi(-\frac{1}{\tau}(\boldsymbol{X}(\boldsymbol{s}_i)'\boldsymbol{\beta} + W(\boldsymbol{s}_i))) \\ &= \Phi(\frac{1}{\tau}(\boldsymbol{X}(\boldsymbol{s}_i)'\boldsymbol{\beta} + \sigma V(\boldsymbol{s}_i))) \end{aligned} \tag{5}$$

where $\boldsymbol{W} = \sigma \boldsymbol{V}$; i.e., $\boldsymbol{V} \sim N(\boldsymbol{0}, H(\delta))$. Thus, we see that, e.g., τ is not identifiable; if we multiply $\boldsymbol{\beta}$, σ and τ all by a constant c, (5) is unchanged. Without loss of generality, we can set $\tau^2 = 1$ or equivalently we can work with the renormalized variogram by dividing (4) by τ^2.

Marginalizing (5) over $V_i = V(\boldsymbol{s}_i) \sim N(0, 1)$, we obtain

$$\begin{aligned} \Pr(Y(\boldsymbol{s}_i) = 1 \mid \boldsymbol{\beta}, \sigma^2) &= \int_{-\infty}^{\infty} \phi(V_i) \int_{-\infty}^{\boldsymbol{X}(\boldsymbol{s}_i)'\boldsymbol{\beta} + \sigma V_i} \phi(U_i)\, dU_i\, dV_i \\ &= \Pr(U_i < \boldsymbol{X}(\boldsymbol{s}_i)'\boldsymbol{\beta} + \sigma V_i \mid \boldsymbol{\beta}, \sigma^2) \\ &= \Pr(U_i - \sigma V_i < \boldsymbol{X}(\boldsymbol{s}_i)'\boldsymbol{\beta} \mid \boldsymbol{\beta}, \sigma^2) \\ &= \Phi\left(\frac{\boldsymbol{X}(\boldsymbol{s}_i)'\boldsymbol{\beta}}{\sqrt{1 + \sigma^2}}\right). \end{aligned} \tag{6}$$

We seek to study the behavior of (6) as levels of $\boldsymbol{X}(\boldsymbol{s}_i)$ are varied. From the Bayesian perspective, with a prior specification for $\boldsymbol{\beta}$ and σ^2, (6) has a posterior distribution given $\boldsymbol{Y}$ for each $\boldsymbol{X}(\boldsymbol{s}_i)$. Hence, we can compare the posteriors as we change $\boldsymbol{X}(\boldsymbol{s}_i)$.

Just as (6) provides $E(Y(\boldsymbol{s}_i) \mid \boldsymbol{\beta}, \sigma^2)$, we obtain

$$\begin{aligned} \text{Var}(Y(\boldsymbol{s}_i) \mid \boldsymbol{\beta}, \sigma^2) &= E(Y(\boldsymbol{s}_i) \mid \boldsymbol{\beta}, \sigma^2) - (E(Y(\boldsymbol{s}_i) \mid \boldsymbol{\beta}, \sigma^2))^2 \\ &= \Phi\left(\frac{\boldsymbol{X}(\boldsymbol{s}_i)'\boldsymbol{\beta}}{\sqrt{1+\sigma^2}}\right) - \Phi^2\left(\frac{\boldsymbol{X}(\boldsymbol{s}_i)'\boldsymbol{\beta}}{\sqrt{1+\sigma^2}}\right) \\ &= \Phi\left(\frac{\boldsymbol{X}(\boldsymbol{s}_i)'\boldsymbol{\beta}}{\sqrt{1+\sigma^2}}\right) \cdot \Phi\left(-\frac{\boldsymbol{X}(\boldsymbol{s}_i)'\boldsymbol{\beta}}{\sqrt{1+\sigma^2}}\right). \end{aligned} \tag{7}$$

Turning to the association structure

$$\text{Cov}(Y(\boldsymbol{s}_i), Y(\boldsymbol{s}_j) \mid \boldsymbol{\beta}, \sigma^2, \delta) = \Pr(Y(\boldsymbol{s}_i) = 1, Y(\boldsymbol{s}_j) = 1 \mid \boldsymbol{\beta}, \sigma^2, \delta)$$

$$-\Pr(Y(\boldsymbol{s}_i) = 1 \mid \boldsymbol{\beta}, \sigma^2, \delta) \cdot \Pr(Y(\boldsymbol{s}_j) = 1 \mid \boldsymbol{\beta}, \sigma^2, \delta). \tag{8}$$

Employing the same reasoning that led to (6), we obtain

$$\Pr(Y(\boldsymbol{s}_i) = 1, Y(\boldsymbol{s}_j) = 1 \mid \boldsymbol{\beta}, \sigma^2, \delta) \; =$$

$$\Pr(U_i - \sigma V_i < \boldsymbol{X}(\boldsymbol{s}_i)'\boldsymbol{\beta}, U_j - \sigma V_j < \boldsymbol{X}(\boldsymbol{s}_j)'\boldsymbol{\beta}). \tag{9}$$

Since the joint distribution of $T_i = U_i - \sigma V_i$ and $T_j = U_j - \sigma V_j$ is bivariate normal with mean 0 and covariance matrix

$$\begin{bmatrix} 1+\sigma^2 & \sigma^2\rho(\boldsymbol{s}_i - \boldsymbol{s}_j;\delta) \\ \sigma^2\rho(\boldsymbol{s}_i - \boldsymbol{s}_j;\delta) & 1+\sigma^2 \end{bmatrix},$$

given $\boldsymbol{\beta}$, σ^2 and δ, (9) can be readily calculated. Though an analytic form for (8) is messy, a Monte Carlo integration by drawing pairs (T_i, T_j) is routine. Using (9) and (6) provides (8). Expressions analogous to (9) can be obtained for $\Pr(Y(\boldsymbol{s}_i) = 1, Y(\boldsymbol{s}_j) = 0 \mid \boldsymbol{\beta}, \sigma^2, \delta)$, $\Pr(Y(\boldsymbol{s}_i) = 0, Y(\boldsymbol{s}_j) = 1 \mid \boldsymbol{\beta}, \sigma^2, \delta)$ and $\Pr(Y(\boldsymbol{s}_i) = 0, Y(\boldsymbol{s}_j) = 0 \mid \boldsymbol{\beta}, \sigma^2, \delta)$ by again using Monte Carlo integration. In fact, since $\Pr(Y(\boldsymbol{s}_i) = 1, Y(\boldsymbol{s}_j) = 0 \mid \boldsymbol{\beta}, \sigma^2, \delta) = \Pr(Y(\boldsymbol{s}_i) = 1 \mid \boldsymbol{\beta}, \sigma^2) - \Pr(Y(\boldsymbol{s}_i) = 1, Y(\boldsymbol{s}_j) = 1 \mid \boldsymbol{\beta}, \sigma^2, \delta)$, the left side probability can be calculated using (9) and (6). Hence, once (6) has been obtained for each i, the four joint probabilities for $(Y(\boldsymbol{s}_i), Y(\boldsymbol{s}_j))$ can be computed using a single Monte Carlo integration. As a result, odds ratios and log odds ratios associated with the joint distribution of $(Y(\boldsymbol{s}_i), Y(\boldsymbol{s}_j))$ can be calculated.

In the case where no covariates are included, i.e., $\boldsymbol{X}(\boldsymbol{s}_j)'\boldsymbol{\beta} \equiv \mu$, for all i, (6) becomes $\Phi\left(\frac{\mu}{\sqrt{1+\sigma^2}}\right)$ and (7) becomes $\Phi\left(\frac{\mu}{\sqrt{1+\sigma^2}}\right) - \Phi^2\left(\frac{\mu}{\sqrt{1+\sigma^2}}\right)$. Also, suppose we calculate the variogram associated with the $Y(\boldsymbol{s})$ process

$$\mathrm{Var}(Y(\boldsymbol{s}_i) - Y(\boldsymbol{s}_j) \mid \mu, \sigma^2) =$$

$$2\{(\mathrm{Var}\, Y(\boldsymbol{s}_i) \mid \mu, \sigma^2) \; - \; \mathrm{Cov}((Y(\boldsymbol{s}_i), Y(\boldsymbol{s}_j) \mid \mu, \sigma^2, \delta)\}$$

$$= 2\left\{\Phi\left(\frac{\mu}{\sqrt{1+\sigma^2}}\right) \; - \; E(\Phi(\mu + \sigma V_i)\; \Phi(\mu + \sigma V_j))\right\}. \tag{10}$$

From the argument below (9), the expectation in (10) depends upon $\boldsymbol{s}_i$ and $\boldsymbol{s}_j$ only through $\boldsymbol{s}_i - \boldsymbol{s}_j$ so that the variogram is stationary. It is an indicator variogram in the sense of Cressie (1993, p. 282).

When σ is small, to a first order approximation, $\Phi(\mu + \sigma V) \approx \Phi(\mu) + \phi(\mu)\sigma V$. Inserting this approximation into the calculation of $\mathrm{Cov}(Y(\boldsymbol{s}_i), Y(\boldsymbol{s}_j) \mid \mu, \sigma^2, \delta)$, we obtain $\phi^2(\mu)\sigma^2\rho(\boldsymbol{s}_i - \boldsymbol{s}_j;\delta)$. That is, for a given μ, the spatial association in the $Y(\boldsymbol{s})$ process is roughly $\phi^2(\mu)$ times the spatial association in the $W(\boldsymbol{s})$ process.

To complete the hierarchical Bayesian model specification requires a prior distribution for $\boldsymbol{\beta}$, σ^2 and δ. Taking these to be flat, proper Inverse Gamma and proper Gamma, respectively provides the full distributional specification

$$f(\boldsymbol{Y} \mid \boldsymbol{\beta}, \boldsymbol{W}) \;\cdot\; f(\boldsymbol{W} \mid \sigma^2, \delta) \;\cdot\; f(\boldsymbol{\beta}) \cdot f(\sigma^2) \cdot f(\delta) \tag{11}$$

where

$$f(\boldsymbol{Y} \mid \boldsymbol{\beta}, \boldsymbol{W}) = \prod_{i=1}^{n} (\Phi(\boldsymbol{X}(\boldsymbol{s}_i)'\boldsymbol{\beta} + W_i))^{Y(\boldsymbol{s}_i)} \;\cdot\; (\Phi(-(\boldsymbol{X}(\boldsymbol{s}_i)'\boldsymbol{\beta} + W_i)))^{1-Y(\boldsymbol{s}_i)}$$

with $W_i = W(\boldsymbol{s}_i)$. Lastly, the posterior for $\boldsymbol{\beta}$, σ^2, δ and $\boldsymbol{W}$ given $\boldsymbol{Y}$ is proportional to (11).

Models of the form (2), hence the associated model for $\boldsymbol{Y}$, are such that the data can not fully separate spatial variability and pure heterogeneity (white noise). Prior specifications strongly influence the inference. This is well known in the case where the distribution for $\boldsymbol{W}$ is given through a Markov random field (Clayton and Bernardinelli, 1992, Bernardinelli and Montomoli, 1992) but seems less appreciated for the case where $\boldsymbol{W}$ is a stationary Gaussian process. As a result, according to the priors on σ^2 and τ^2, we can tell a primarily spatial or primarily heterogeneity story, respectively. The above papers offer empirical guidelines with regard to providing priors which balance these two sources of variation. In our case with $\tau^2 = 1$, it is much simpler to give a *neutral* specification. We take σ^2 to follow an inverse Gamma distribution with mean 1 and infinite variance.

3. Computational Issues

We propose the use of simulation to fit the model in (11). That is, we seek to draw samples from the posterior $f(\boldsymbol{\beta}, \sigma^2, \delta, \boldsymbol{W} \mid \boldsymbol{Y})$. We find the widely used Markov Chain Monte Carlo approach (see, e.g., Gilks, Richardson and Spiegelhalter, 1995) to be unattractive here, particularly for moderate to large n. This is due to the fact that, in evaluating (11), we must calculate $f(\boldsymbol{W} \mid \sigma^2, \delta)$ which requires $H^{-1}(\delta)$ (and $|H(\delta)|$). Hence, for each δ, a high dimensional matrix inversion is required which, due to the enormous number of arithmetic operations, may become very slow and, with cumulative rounding error, inaccurate. Moreover, to implement thousands of iterations may become infeasible within realistic time constraints.

Following Gelfand and Ravishanker (1998), we propose to use a noniterative Monte Carlo approach with a suitably selected importance sampling density (ISD). This ISD enables both Monte Carlo integration for expectations (as in Geweke, 1989) and Monte Carlo sampling for other distributional features (as in Smith and Gelfand, 1992). The ISD is of very high dimension ($n + p + 2$, if δ is one-dimensional) and thus would require an astronomical amount of sampling to learn about $f(\boldsymbol{\beta}, \sigma^2, \delta, \boldsymbol{W} \mid \boldsymbol{Y})$. However, concern lies only in the $p+2$ dimensional posterior $f(\boldsymbol{\beta}, \sigma^2, \delta \mid \boldsymbol{Y})$. In fact, as a result of marginalizing over $\boldsymbol{W}$, all of the quantities of interest in the previous section are functions solely of $\boldsymbol{\beta}$, σ^2 and δ. Samples from $f(\boldsymbol{\beta}, \sigma^2, \delta \mid \boldsymbol{Y})$ provide samples from the posteriors for any of these quantities, enabling inference regarding these quantities. For the posterior $f(\boldsymbol{\beta}, \sigma^2, \delta \mid \boldsymbol{Y})$, a much smaller number of draws may be sufficient. Of course, this number will grow with n, increasing run times, but there is no feasible alternative.

We propose an ISD of the form

$$g_s(\boldsymbol{\beta}, \sigma^2, \delta, \boldsymbol{W}; \boldsymbol{Y}) = g_s(\boldsymbol{\beta} \mid \boldsymbol{W}; \boldsymbol{Y}) \cdot g_s(\boldsymbol{W} \mid \sigma^2, \delta; \boldsymbol{Y}) \cdot g_s(\sigma^2, \delta; \boldsymbol{Y}). \qquad (12)$$

We now describe each of the components on the right hand side of (12). With a flat prior on $\boldsymbol{\beta}$, we take $g_s(\boldsymbol{\beta} \mid \boldsymbol{W}; \boldsymbol{Y})$ to be $N(\boldsymbol{\beta} \mid \hat{\boldsymbol{\beta}}_{\boldsymbol{W}}, \hat{\Sigma}_{\hat{\boldsymbol{\beta}}_{\boldsymbol{W}}})$ where $\hat{\boldsymbol{\beta}}_{\boldsymbol{W}}$ and $\hat{\Sigma}_{\hat{\boldsymbol{\beta}}_{\boldsymbol{W}}}$ is the maximum likelihood estimate and associated asymptotic covariance arising, at a given $\boldsymbol{W}$, from $f(\boldsymbol{Y} \mid \boldsymbol{\beta}, \boldsymbol{W})$ whose form appears below (11). Since $-\hat{\Sigma}^{-1}_{\hat{\boldsymbol{\beta}}_{\boldsymbol{W}}}$ is the $p \times p$ Hessian for $\boldsymbol{\beta}$ at a given $\boldsymbol{W}$, it can be computed explicitly and hence $g_s(\boldsymbol{\beta} \mid \boldsymbol{W}; \boldsymbol{Y})$ is easily evaluated. As a p-variate normal distribution, it is routinely sampled.

An alternative would be to sample $f(\boldsymbol{\beta} \mid \boldsymbol{Y}, \boldsymbol{W})$, the full conditional distribution for $\boldsymbol{\beta}$. Since $f(\boldsymbol{Y} \mid \boldsymbol{\beta}, \boldsymbol{W})$ is log concave in $\boldsymbol{\beta}$ (Wedderburn, 1976), a rejection algorithm can be designed using bounding hyperplanes.

Next let $g_s(\boldsymbol{W} \mid \sigma^2, \delta; \boldsymbol{Y})$ be $f(\boldsymbol{W} \mid \sigma^2, \delta)$ and $g_s(\sigma^2, \delta; \boldsymbol{Y}) = f(\sigma^2)f(\delta)$. In certain cases, we may be able to refine these choices, as discussed in Gelfand and Ravishanker (1998). Then the sampling weights, dividing (11) by (12), become

$$\begin{aligned} w(\boldsymbol{\beta}, \sigma^2, \delta, \boldsymbol{W}; \boldsymbol{Y}) &= \frac{f(\boldsymbol{Y} \mid \boldsymbol{\beta}, \boldsymbol{W}) \cdot f(\boldsymbol{W} \mid \sigma^2, \delta) \cdot f(\sigma^2) \cdot f(\delta)}{N(\boldsymbol{\beta} \mid \hat{\boldsymbol{\beta}}_{\boldsymbol{W}}, \hat{\Sigma}_{\hat{\boldsymbol{\beta}}_{\boldsymbol{W}}}) \cdot f(\boldsymbol{W} \mid \sigma^2, \delta) \cdot f(\sigma^2) \cdot f(\delta)} \\ &= \frac{f(\boldsymbol{Y} \mid \boldsymbol{\beta}, \boldsymbol{W})}{N(\boldsymbol{\beta} \mid \hat{\boldsymbol{\beta}}_{\boldsymbol{W}}, \hat{\Sigma}_{\hat{\boldsymbol{\beta}}_{\boldsymbol{W}}})}, \end{aligned} \tag{13}$$

free of σ^2 and δ. Note that calculation of (13) requires no $n \times n$ matrix inversion. Our choice for $g_s(\boldsymbol{W} \mid \sigma^2, \delta; \boldsymbol{Y})$ has annihilated $f(\boldsymbol{W} \mid \sigma^2, \delta)$; we replace matrix inversion with sampling from an n dimensional normal. If $f(\boldsymbol{\beta} \mid \boldsymbol{W}, \boldsymbol{Y})$ replaces the normal density for $\boldsymbol{\beta}$ in the denominator of (13), $w(\boldsymbol{\beta}, \sigma^2, \delta, \boldsymbol{W}; \boldsymbol{Y})$ becomes $\int f(\boldsymbol{Y} \mid \boldsymbol{\beta}, \boldsymbol{W})\, d\boldsymbol{\beta}$, free of $\boldsymbol{\beta}$ as well as σ^2 and δ.

Draws $(\sigma_\ell^{2*}, \delta_\ell^*, \boldsymbol{W}_\ell^*, \boldsymbol{\beta}_\ell^*)$, $\ell = 1, \ldots, L$, from $g_s(\boldsymbol{\beta}, \sigma^2, \delta, \boldsymbol{W}; \boldsymbol{Y})$ yield weights w_ℓ^* using (13) which may be normalized to probabilities $q_\ell = w_\ell^* / \sum w_\ell^*$. How do we use these random draws and weights to infer about $f(\boldsymbol{\beta}, \sigma^2, \delta \mid \boldsymbol{Y})$? For $f(\boldsymbol{\beta} \mid \boldsymbol{Y})$, we note that

$$\frac{f(\boldsymbol{\beta} \mid \boldsymbol{Y})}{1} = \frac{\int f(\boldsymbol{\beta}' \mid \boldsymbol{W}, \boldsymbol{Y}) \cdot f(\boldsymbol{W} \mid \boldsymbol{Y})\, d\boldsymbol{W}}{\int f(\boldsymbol{\beta}' \mid \boldsymbol{W}, \boldsymbol{Y}) \cdot f(\boldsymbol{W} \mid \boldsymbol{Y})\, d\boldsymbol{\beta}' d\boldsymbol{W}}$$

$$= \frac{\int f(\boldsymbol{\beta} \mid \boldsymbol{W}, \boldsymbol{Y}) \cdot \frac{f(\boldsymbol{\beta}', \sigma^2, \delta, \boldsymbol{W} \mid \boldsymbol{Y})}{g_s(\boldsymbol{\beta}', \sigma^2, \delta, \boldsymbol{W}; \boldsymbol{Y})} \cdot g_s(\boldsymbol{\beta}', \sigma^2, \delta, \boldsymbol{W}; \boldsymbol{Y})\, d\boldsymbol{\beta}' d\sigma^2 d\delta d\boldsymbol{W}}{\int \frac{f(\boldsymbol{\beta}', \sigma^2, \delta, \boldsymbol{W} \mid \boldsymbol{Y})}{g_s(\boldsymbol{\beta}', \sigma^2, \delta, \boldsymbol{W}; \boldsymbol{Y})} \cdot g_s(\boldsymbol{\beta}', \sigma^2, \delta, \boldsymbol{W}; \boldsymbol{Y})\, d\boldsymbol{\beta}' d\sigma^2 d\delta d\boldsymbol{W}}$$

$$= \frac{\int f(\boldsymbol{\beta} \mid \boldsymbol{W}, \boldsymbol{Y}) \cdot w(\boldsymbol{\beta}', \sigma^2, \delta, \boldsymbol{W}; \boldsymbol{Y}) \cdot g_s(\boldsymbol{\beta}', \sigma^2, \delta, \boldsymbol{W}; \boldsymbol{Y})\, d\boldsymbol{\beta}' d\sigma^2 d\delta d\boldsymbol{W}}{\int w(\boldsymbol{\beta}', \sigma^2, \delta, \boldsymbol{W}; \boldsymbol{Y}) \cdot g_s(\boldsymbol{\beta}', \sigma^2, \delta, \boldsymbol{W}; \boldsymbol{Y})\, d\boldsymbol{\beta}' d\sigma^2 d\delta d\boldsymbol{W}}.$$

Hence a Monte Carlo integration for $f(\boldsymbol{\beta} \mid \boldsymbol{Y})$ is the mixture distribution

$$\hat{f}(\boldsymbol{\beta} \mid \boldsymbol{Y}) = \sum f(\boldsymbol{\beta} \mid \boldsymbol{W}_\ell^*, \boldsymbol{Y}) \cdot q_\ell \tag{14}$$

Expression (14) is awkward to work with since, for each $\boldsymbol{W}_\ell^*$, (11) only provides $f(\boldsymbol{Y} \mid \boldsymbol{\beta}, \boldsymbol{W}_\ell^*)$ explicitly. The normalizing constant for $f(\boldsymbol{\beta} \mid \boldsymbol{W}_\ell^*, \boldsymbol{Y})$ requires a p-dimensional integration. In the case where no covariates are included, $(p = 1)$, this is no problem. However, in general it is easier to run a Gibbs sampler at each $\boldsymbol{W}_\ell^*$, using adaptive rejection sampling (Gilks and Wild, 1992) since $f(\boldsymbol{\beta} \mid \boldsymbol{W}_\ell^*, \boldsymbol{Y})$, equivalently $f(\boldsymbol{Y} \mid \boldsymbol{\beta}, \boldsymbol{W}_\ell^*)$, is log concave in each component of $\boldsymbol{\beta}$.

If a few weights dominate with regard to the q_ℓ, this will be preferable to the discrete distribution placing mass q_ℓ at $\boldsymbol{\beta}_\ell^*$. Taking the approximation one step further, we might replace $f(\boldsymbol{\beta} \mid \boldsymbol{W}_\ell^*, \boldsymbol{Y})$ in (14) with $N(\boldsymbol{\beta} \mid \boldsymbol{\beta}_{\boldsymbol{W}_\ell^*}, \Sigma_{\hat{\boldsymbol{\beta}}_{\boldsymbol{W}_\ell^*}})$.

Similarly, for $f(\sigma^2, \delta \mid \boldsymbol{Y})$, if a few weights dominate, the discrete distribution in $(\sigma_\ell^{2*}, \delta_\ell^*)$ will be unsatisfactory. The mass may be smoothed for each ℓ using a bivariate t-distribution with low degrees of freedom for $\epsilon_1 = \log \sigma^2, \epsilon_2 = \log \delta$.

To determine the t-distribution requires a location vector and a dispersion matrix. The location vector will be $\epsilon_{1\ell} = \log \sigma_\ell^{2*}, \epsilon_{2\ell} = \log \delta_\ell^*$. We can obtain a dispersion matrix using a parametric bootstrap. Generate a sample of $\boldsymbol{W}$'s from $N(\mathbf{0}, \sigma_\ell^{2*} H(\delta_\ell^*))$. For each $\boldsymbol{W}$, estimate σ^2 and δ, hence ϵ_1 and ϵ_2. Use the sample of (ϵ_1, ϵ_2)'s to create a sample covariance matrix.

The ISD approach still requires drawing $\boldsymbol{W}$ from $N(\mathbf{0}, \sigma^2 H(\delta))$ which in turn requires triangulation (Cholesky decomposition) of $H(\delta)$, an order n^2 operation. Though cheaper than inversion (an order n^3 operation) and more stable, repeated sampling of $\boldsymbol{W}$ as δ changes requires repeated triangulation of $H(\delta)$, which may still result in very long run times if n is large.

We propose a first-order dependent approximation to sampling from $N(\mathbf{0}, \sigma^2 H(\delta))$. Suppose $\boldsymbol{W}$ is blocked as $(\boldsymbol{W}_1, \boldsymbol{W}_2, \ldots, \boldsymbol{W}_K)$ with $\boldsymbol{W}_i$, $n_i \times 1$ and $\sum n_i = n$. Partition $H(\delta)$ accordingly, i.e., suppressing δ, the $n_i \times n_j$ matrix $H_{ij} = \text{Cov}(\boldsymbol{W}_i, \boldsymbol{W}_j)$. Consider the joint distribution for $\boldsymbol{W}$ created as $\boldsymbol{W}_1 \sim N(\mathbf{0}, \sigma^2 H_{11})$ and $\boldsymbol{W}_i \mid \boldsymbol{W}_{i-1} \sim N(H_{i,i-1} H_{i-1,i-1}^{-1} \boldsymbol{W}_{i-1}, \sigma^2 (H_{ii} - H_{i,i-1} H_{i-1,i-1}^{-1} H_{i-1,i}))$, $i = 1, \ldots, k$. It is straightforward to check that the resulting distribution for $\boldsymbol{W}$ is again an n dimensional multivariate normal, that the mean for $\boldsymbol{W}$ is $\mathbf{0}$ and that the covariance matrix is $\sigma^2 \tilde{H}(\delta)$ where $\tilde{H}_{ii} = H_{ii}$, $\tilde{H}_{i,i+1} = H_{i,i+1}$, and, for $i < j$, $\tilde{H}_{ij} = H_{i,i+1}\ H_{i+1,i+1}^{-1}\ H_{i+1,i+2} \ldots, H_{j-1,j-1}^{-1}\ H_{j-1,j}$. In other words, $\tilde{H}$ agrees with H on the main and first off block-diagonals.

But then, a natural suggestion emerges to strengthen the quality of the approximation. By permuting the components of $\boldsymbol{W}$, with corresponding permutation of the rows and columns of H, we can consider arbitrary blocking for $\boldsymbol{W}$. Suppose, for example, that we partition the region D into vertical strips defined by creating intervals on the x axis. Then each site will fall into one and only one strip and roughly, pairs of sites from adjacent strips, will, *a priori*, be more strongly associated than pairs from non-adjacent strips. In fact, association should tend to decline as strips become farther apart. For the corresponding H, entries in H_{ij} will tend to grow smoother as $|i - j|$ grows larger. In fact, this will also be true for $\tilde{H}_{ij}$ enhancing the quality of the approximation. In a sense, this approach offers a solution, in the simulation context, to the well-known problem of how to define and piece together subregions to infer about the spatial association in the overall region.

The first order approximation is obviously not unique. In particular, we could employ horizontal strips rather than vertical ones. Possibly the strips could be constructed by the orientation of D. Also, the number of and width of individual strips is flexible. For us, the objective would be to achieve roughly equal n_i across the strips. Then, the advantage to the approximation is apparent. In sampling $\boldsymbol{W}$, we never need work with a matrix larger than max $n_i \times$ max n_i. Choosing k large enough, we can ensure that max $n_i \times$ max n_i is small enough so that the required inversion and triangulation to do the sampling runs very quickly. In the example of section 4, we have $n = 857$ with $k = 10$ and max $n_i = 92$.

4. An Illustration

We examine a dataset consisting of 857 residential properties in Baton Rouge, Louisiana. The locational coordinates are plotted in Figure 22.1. The shape of the region encourages partitioning using horizontal strips. Again, the $\boldsymbol{s}_i$ are the geographical coordinates of the ith house. All properties in the dataset are repeat sales. That is, during a ten year period of observation, they sold once and then were subsequently resold. We let $Y(\boldsymbol{s}_i) = 1$ if the property resold within four years and $Y(\boldsymbol{s}_i) = 0$ if not. The proportion of $Y(\boldsymbol{s}_i) = 1$ in the sample is 45%. We seek to identify whether there is spatial association in the pattern of quicker (≤ 4 years)

FIGURE 22.1. Home Locations in Baton Rouge, Louisiana.

TABLE 22.1. Posterior Mean, Standard Deviation, 2.5% ,50% and 97.5% quantiles of Model Parameters

	Post. Mean	Post. StDev.	2.5% quantile	Post. Median	97.5% quantile
β_0	-0.115	0.053	-0.225	-0.115	-0.015
β_1	0.224	0.081	0.062	0.221	0.381
β_2	0.554	0.293	-0.031	0.564	1.146
β_3	0.093	0.154	-0.201	0.094	0.394
β_4	-0.254	0.304	-0.853	-0.259	0.374
σ^2	0.212	0.034	0.152	0.208	0.284
δ	14.18	2.12	10.45	14.02	18.83
r	0.216	0.033	0.159	0.214	0.285

resales relative to the slower (> 4 years) resales. Four covariates, reflecting house characteristics, are employed at each site. $X_1(\boldsymbol{s}_i)$ is the age of the house at the time of first sale, $X_2(\boldsymbol{s}_i)$ reflects the living area of the house, $X_3(\boldsymbol{s}_i)$ represents other area (apart from living area), while $X_4(\boldsymbol{s}_i)$ refers to the number of bathrooms per house. All variables were standardized. Hence, $\boldsymbol{\beta}$ is 5×1, including an intercept term. For illustration, we adopt $\rho(\boldsymbol{s}_i - \boldsymbol{s}_j; \delta) = \exp(-\delta d_{ij})$ where d_{ij} is the Euclidean distance between sites $\boldsymbol{s}_i$ and $\boldsymbol{s}_j$. The remainder of the model specification takes the form (11) with $f(\sigma^2) = IG(2, 1)$ and $f(\delta) = Gamma(2, 0.04)$. The former has mean 1 with infinite variance following the discussion at the end of seStion 2. The later has mean 50 (with very large variance) which corresponds to a prior guess of 0.06 degrees for the range (see next paragraph), roughly 4 to 5 miles.

The model is fitted using the approach of Section 3.

Table 22.1 summarizes the posteriors for $\boldsymbol{\beta}$, σ^2, and δ using the mean, standard deviation, the median and the lower and upper 0.025 quantiles. The latent detrended variogram (4) with $\tau^2 = 1$ is a function of σ^2, δ, and the Euclidean distance (d) between sites. We plot the posterior mean, $2E(1 + \sigma^2(1 - \exp(-\delta d)) \mid \boldsymbol{Y})$ versus d in Figure 22.2. The range is defined to be the distance r such that $0.05 = \exp(-\delta r)$, i.e., $r = -\log(0.05)/\delta$. Its posterior is also summarized in Table 22.1. Finally, a plot of the observed binary response at each spatial location is not very insightful regarding spatial patterns. Instead, using (6), we can obtain the posterior mean of $\Pr(Y(\boldsymbol{s}_i) = 1 \mid \boldsymbol{\beta}, \sigma^2)$ for each site.

A greyscale plot of these expected probabilities, using S-Plus, is given in Figure 22.3, smoothing the spatial pattern in the raw data.

We attempt some intrepretation of these figures and the table. First, it is the case that the southern part of the city has evolved as the more desirable area to live in. There is more real estate activity in this area and it is expected that the bulk of the resales would be in this area, as Figure 22.1 reveals. Turning to the coefficients, a standard logistic regression of Y on X_1, X_2, X_3 and X_4 yields significant positive age effect ($P = 0.003$) and a nearly significant positive living area effect ($P = 0.091$). These results concur with Table 22.1. (Magnitudes need not align due to the implicit scaling in setting $\tau^2 = 1$).

The implication is that the probability of quicker resale increases in age and living area. To clarify, the southwestern part of the city is near Louisiana State University and is dominated, residentially, by professionals. Hence, it would be expected to

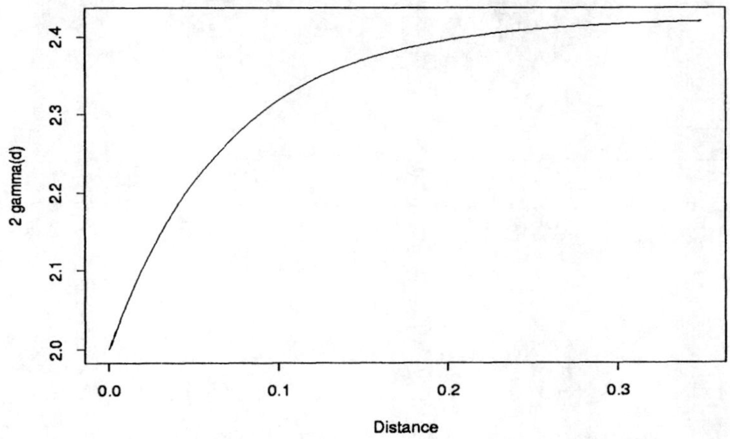

FIGURE 22.2. Plot of Estimated Latent Detrended Variogram.

exhibit more rapid turnover than other areas.

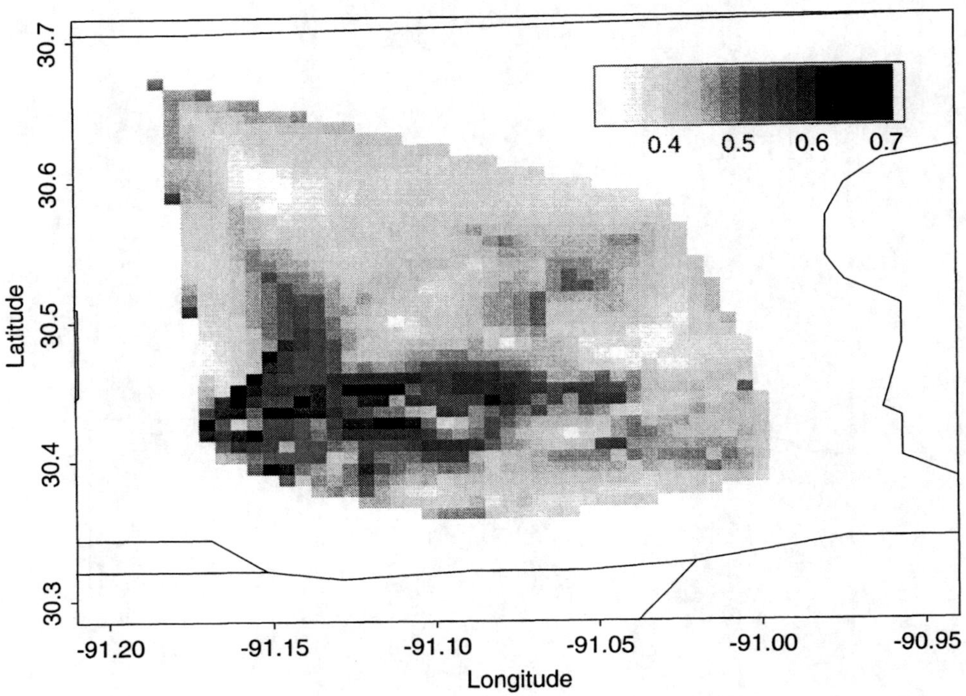

FIGURE 22.3. Greyscale plot of $E\{Pr(Y(s_i) = 1|\beta, \sigma^2)\}$ versus s_i.

Figure 22.3 supports this, showing highest expected probabilities in the southwest. But also, homes in this area tend to be older and larger helping to explain the positive β_1 and β_2.

The magnitude of σ^2 (relative to τ^2) shows that starting from a neutral specification with regard to source of variation, spatial variability emerges as roughly one-fifth of the white noise variability. The choice of covariates may have explained much of the spatial hetergeneity. Still, the spatial component is consequential, as the detrended variogram reveals. Moreover, within this component, association declines slowly with distance; the estimated range of roughly 0.2 is approximately 14 miles.

5. Related Remarks

We have considered the problem of modeling and analyzing point referenced binary spatial data. We have formulated a hierarchical model introducing spatial effects at the second stage in the spirit of familiar Markov random field approaches.

However, we capture the second stage spatial association using an, arguably, more direct and natural second order stationary Gaussian process. In the fitting of such models, we have indicated how the problem of repeated high dimensional matrix inversion can be replaced by simulation. Such simulation enables full inference regarding all parameters and parametric functions of interest.

Link functions h in (1) besides the probit can be adopted. However, the attractive connection with the latent $Z(\boldsymbol{s})$ process is lost and thus so are the convenient calculations in (6)-(9). A related question concerns whether $W(\boldsymbol{s})$ can be modeled with an alternative to the Gaussian process. For instance, we might propose a heavier tailed *t-process* such that $\boldsymbol{W} \sim t_r(\mathbf{0}, \sigma^2 H(\delta))$, a multivariate t distribution with r degrees of freedom, location vector $\mathbf{0}$ and dispersion matrix $\sigma^2 H(\delta)$. Unfortunately, with just a single data vector $\boldsymbol{Y}$, i.e., no replication, it will not be possible to distinguish a t-process from a Gaussian one.

Acknowledgements

The work of the first author was supported in part by NSF grant DMS 96-25383. The authors thank C.F. Sirmans for valuable discussions.

References

Bernardinelli, L. and Montomoli C. (1992). Empirical Bayes versus Fully Bayesian Analysis of Geographical Variation in Disease Risk. *Statistics in Medicine.* **11**, pp 983-1007.

Besag, J.E. (1974). Spatial Interaction and the Statistical Analysis of Lattice Systems (with discussion). *Journal of the Royal Statistical Society, Series B.* **36**, pp 192-236.

Besag, J.E. and Kooperberg, K.L. (1995). On Conditional and Intrinsic Autoregressions. *Biometrika.* **82**, pp 733-746.

Besag, J.E., York, J.C. and Mollié, A. (1991). Bayesian Image Restoration, with two Applications in Spatial Statistics (with discussion). *Annals of the Institute of Statistical Mathematics.* **43**, pp 1-59.

Clayton, D.G. and Bernardinelli, L. (1992). Bayesian Methods for Mapping Disease Risk. In *Small Area Studies in Geographical and Environmental Epidemiology*, J. Cuzick and P. Elliott (eds). Oxford: Oxford University Press.

Cressie, N. (1993). *Statistics for Spatial Data.* New York: John Wiley and Sons.

De Oliveira, V. (1997). Bayesian Prediction of Clipped Gaussian Random Fields. Technical Report, Department of Mathematics, University of Maryland.

Diggle, P.J., Liang, K-L. and Zeger, S.L. (1994). *Analysis of Longitudinal Data.* New York: Oxford Science Publications.

Diggle, P.J., Tawn, J.A. and Moyeed, R.A. (1998). Model-based Geostatistics (with discussion). *Applied Statistics.* (**47**(3), pp 299-350.

Ecker, M.D. and Gelfand, A.E. (1997). Spatial Modeling and Prediction under Range Anisotropy. Technical Report 97-26, Department of Statistics, University of Connecticut.

Gelfand, A.E. and Ravishanker, N. (1998). Inference for Bayesian Variogram Models with Large Data Sets. Technical Report 98-18, Department of Statistics, University of Connecticut.

Gelfand, A.E. and Smith, A.F.M. (1990). Sampling Based Approaches to Calculating Marginal Densities. *Journal of American Statistical Association* **85**, pp 398-409.

Geweke, J. (1989). Bayesian Inference in Econometric Models using Monte Carlo Integration. *Econometrica.* **57**, pp 1317-1339.

Gilks, W.R., Richardson, S. and Spiegelhalter, D.L. (1996). *Markov Chain Monte Carlo in Practice.* New York: Chapman and Hall.

Gilks, W.R. and Wild P. (1992). Adaptive rejection sampling for Gibbs sampling. *Applied Statistics.* **41**, pp 337-348.

Journel, A.G. (1983). Nonparametric Estimation of Spatial Distributions. *Mathematical Geology.* **15**, pp 445-468.

Papritz, A. and Moyeed, R.A. (1997). Empirical Validation of Linear and Nonlinear Methods for Spatial Point Prediction. Technical Report, Department of Statistics, Lancaster University.

Smith, A.F.M. and Gelfand, A.E. (1992). Bayesian statistics without tears: A sampling-resampling perspective. *The American Statistician.* **46**(2), pp 84-88.

Wedderburn, R.W.M. (1976). On the Existence and Uniqueness of the Maximum Likelihood Estimates for Certain Generalized Linear Models. *Biometrika.* **63**, pp 27-32.

Whittaker, J. (1990). *Graphical Models in Applied Multivariate Statistics.* Chichester: John Wiley and Sons.

23

Bayesian Graphical Models and Software for GLMs

Nicky Best
Andrew Thomas

ABSTRACT In this chapter, we describe two useful tools for the Bayesian analysis of generalized linear models (GLM's) and their extensions. The first is a graphical modelling approach for representing the conditional independence assumptions and qualitative structure underlying a statistical model. The second is the WinBUGS statistical software, which implements a Markov chain Monte Carlo approach to Bayesian inference. Graphical models form the central construct of both the statistical model and the software by providing a direct link between the model description and the computational solutions to the associated inference problem. We describe how these parallels are exploited by the WinBUGS software and show how this leads to a readily extensible programming environment for complex Bayesian modeling. The remainder of the chapter offers practical guidelines for implementing Bayesian GLM's using the WinBUGS software. Particular emphasis is placed on the flexibility of both the Bayesian graphical modeling approach and the WinBUGS program to extend the standard GLM framework by accommodating additional sources of complexity such as correlated data structures, overdispersion, measurement error, missing data, outlying observations and so on.

1. Bayesian Graphical Models and Conditional Independence Structures

Graphical representations of dependencies between variables in a statistical model have been used to express the underlying problem structure in a wide variety of applications. Spiegelhalter (1998) cites examples including path analysis diagrams, structural equation models, Bayesian networks, classical conditional independence graphs, neural networks and Bayesian graphical models. Interest in the latter has grown rapidly in recent years due to the recognition that the formal properties of such graphs provide a direct link to the Markov Chain Monte Carlo (MCMC) computational algorithms used in complex Bayesian inference.

A Bayesian graphical model is simply a pictorial representation of the conditional independence assumptions underlying a statistical model. In general, these may be expressed in the form of a *directed acyclic graph* (DAG) in which all edges are directed and no cycles are permitted. Figure 23.1 shows a simple example of a DAG for the generalized linear model (GLM) described in Section 2. Each quantity in the model is represented by a node in the graph: by convention, ellipses represent random variables (both observed data and unknown parameters, missing data, latent variables and so on) and rectangles (if any) represent fixed constants. Repetitive structures (for example, measurements of the random variable `Y[i]` on different units `i=1`, ..., `I`) are known as "plates" and are represented by a large thick-edged rectangle enclosing the repeated nodes. Directed edges (arrows) between nodes indicate dependencies: solid edges represent probabilistic relationships and hollow edges represent deterministic relationships. For example, in Figure 23.1, `Y[i]` is a stochastic node which depends probabilistically on `mu[i]` and `phi`, whilst

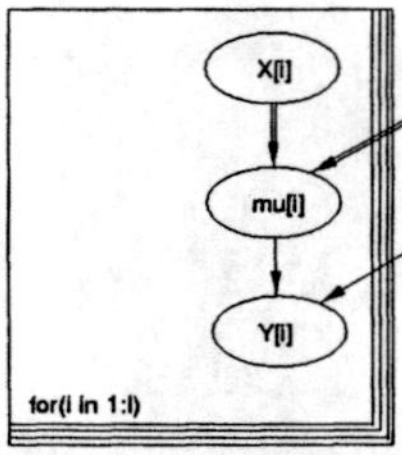

FIGURE 23.1. DAG for a simple GLM

`mu[i]` is a logical node which may be expressed as a deterministic function of `X[i]` and `beta`. Assuming for now that the quantities in the graph represent the genes of individuals in a family tree, and ignoring deterministic edges, we may regard `mu[i]` and `phi` as the genes of a pair of "parents" with a 'child' whose genes are represented by `Y[i]`. The conditional independence structure implied by a DAG follows naturally from the Mendelian inheritance laws associated with this genetic analogy. The genetic properties of a generic individual v depend only on the genes of its parent(s). Conditional on knowing the parent genes, no other individual can provide information about v's genes unless that individual is a "descendant" of v. This property may be expressed formally as follows

$$v \perp\!\!\!\perp \text{non-descendants}[v]|\text{parents}[v] \qquad (1)$$

where $\perp\!\!\!\perp$ denotes "is conditionally independent of." It can also be shown (Lauritzen *et al.* , 1990) that (1) is equivalent to assuming that the joint distribution of all quantities $V = \{v\}$ in the graph has a simple factorization

$$p(V) = \prod_{v \in V} p(v|\text{parents}[v]) \qquad (2)$$

1.1 Computation on Bayesian Graphical Models

Bayesian inference involves specification of a full probability model to describe the joint distribution of the observed data D (e.g. `Y[i]` and `X[i]` in Figure 23.1) and unobserved quantities Θ (e.g. `beta` and `phi` in Figure 23.1). Note that logical nodes such as `mu[i]` in Figure 23.1 are included to simplify the graph, but are collapsed over when identifying probabilistic relationships. According to the factorization theorem (??) for DAGs, we need only specify the parent-child conditional distributions $p(v|\text{parents}[v])$ for each node $v \in V = \{D, \Theta\}$ in order to fully define the model. Conditioning on the known value of D by applying Bayes theorem will then yield a posterior density $p(\Theta|D)$. For most real-life applications, Θ will be high dimensional and scientific interest will focus on the *marginal* posterior distribution of individual components $\theta_k \in \Theta$. This involves integrating the joint posterior density $p(\Theta|D)$ over the remaining unknowns $\theta_{j \neq k} \in \Theta$. Such integrations are usually complex and best carried out using sampling-based methods such as Markov Chain Monte Carlo (MCMC) algorithms. MCMC methods involve simulation of values from a Markov Chain whose stationary distribution is the required posterior density. Inference is based on data-analytic summaries of the sampled values, such as means and quantiles, using the principle of Monte Carlo integration.

Detailed accounts of MCMC methods are provided elsewhere (Smith and Roberts, 1993; Brooks, 1998). The key point here is that MCMC algorithms such as the Gibbs sampler involve simulation from the full conditional distribution of each unknown quantity v given all the other terms in the model $V\backslash v$. It is easily shown that the factorization theorem (??) for DAGs leads to the following form for the full conditional distribution

$$\begin{aligned} p(v|V\backslash v) &\propto \text{terms in factorization of } p(V) \text{ containing } v \\ &= p(v|\text{parents}[v]) \prod_{v \in \text{parents}[w]} p(w|\text{parents}[w]) \end{aligned}$$

By exploiting conditional independence structures, the graph thus provides a direct link between the statistical model and the computational algorithms required for inference. This leads to the crucial ideas behind the BUGS software (described below), namely specification of full probability models in terms of parent-child relationships and *local* computation on full conditional distributions.

1.2 Constructing Software from Graphical Models

The conceptual design of the BUGS software is based on constructing an internal representation of the graph describing the joint probability model. The current version of BUGS (WinBUGS) has been developed for Microsoft Windows within the Black Box environment using the computer language Component Pascal. A component-orientated philosophy has been adopted. This is a new software engineering approach which aims to create fully extensible modular systems. The software consists of a number of components which implement the nodes in a DAG. Each component has a well-defined interface describing the implemented entities which can be used in other components. These are linked together at load time or run time to create a series of connected objects (conceptually similar to the parent-child links in a DAG) required to execute the model. An example of a WinBUGS component is given in the Appendix.

2. Implementing GLM's Using WinBUGS

In this section we illustrate the principles of model specification and implementation for a simple Bayesian GLM using the WinBUGS software. More complex GLMs are considered in Sections 3-6. The reader is also referred to the Classic BUGS manual (Spiegelhalter *et al.*, 1996) and WinBUGS online documentation for additional information.

A standard GLM consists of 3 components:

1. A random component specifying the probability distribution of the response variable Y_i for units $i = 1, \ldots, I$. In general, this will depend on a *mean* parameter μ_i and on a global *dispersion* parameter ϕ.

2. A deterministic component specifying a linear function of covariates X_i upon which the response Y_i is assumed to depend. This is termed the *linear predictor*, denoted $\eta_i = \beta' X_i$.

3. A *link function* $g(.)$ that relates the linear predictor to the mean μ_i of the response variable: $\mu_i = g^{-1}(\eta_i) = g^{-1}(\beta' X_i)$.

The WinBUGS software provides a graphical interface called DoodleBUGS which enables direct specification of the model as a DAG. Figure 23.1 shows how the structure of a standard GLM may be expressed using DoodleBUGS (see Section 1.) for a description of the graph notation and interpretation). Model specification is then completed by

1. Specifying the form of the probability distribution (likelihood) for the response variable `Y[i]`. DoodleBUGS provides a menu of distributions from which the density for a stochastic node may be selected. Common choices include *Bernoulli* or *binomial* (binary responses), *Poisson* (count data), *normal* or *Student t* (continuous responses), *Weibull* or *gamma* (survival times).

2. Specifying the deterministic function of covariates and regression coefficients which equate to the mean `mu[i]`. In DoodleBUGS, logical nodes have a menu of link functions associated with them, plus a text-entry value field where the mathematical expression for the linear predictor should be typed. Common choices are are the *logit* link $g(\mu_i) = \log \frac{\mu_i}{1-\mu_i}$ (Bernoulli or binomial GLMs), the *log* link $g(\mu_i) = \log \mu_i$ (Poisson or Weibull GLMs) and the *identity* link (normal or t GLMs). Other options include the *probit* and *complementary log-log* links.

3. Assigning prior distributions for all "founder" nodes (i.e. stochastic nodes with no arrows pointing to them), namely the regression coefficient `beta` and the dispersion parameter `phi`. Again, these are selected from the menu of probability distributions in DoodleBUGS, and suitable choices are discussed in Section 7.2. Note that for Bernoulli, binomial and Poisson GLMs, the dispersion parameter `phi` is not required and so may be deleted from the graph.

The DoodleBUGS graph now describes a full probability model for a standard GLM. This may be passed to the WinBUGS program where the structures underlying the graphical description of the model are parsed into a tree form. Observed nodes in the graph are then identified by loading the *data file*: these are the nodes which will be conditioned on when carrying out Bayesian posterior inference (see Section 1.1). WinBUGS *compiles* the model by traversing the tree to create a series of connected objects (see Appendix) which form a low-level executable version of the Bayesian graphical model. Each object has a set of attributes containing relevant *local* information such as its parents and children in the graph and a suite of algorithms to implement the probability distribution or link function implied by the associated edges in the graph. This information is used to construct full conditional distributions needed for the MCMC simulation using the factorization theorem for DAGs. Once compilation is complete, the user must assign *initial values* for each stochastic node, either by loading a separate file of values and/or by requesting that WinBUGS automatically generate values by forward sampling from the prior distribution.

The MCMC sampler may now be invoked using the *Update* menu in WinBUGS to initiate the required number of iterations. Posterior inference is achieved by using the *Statistics* menu to request summary statistics for the samples generated. A variety of graphical outputs are also available, including trace plots, autocorrelation plots and density plots.

For more complex models, particularly those with large numbers of parameters and/or multivariate nodes, formal specification via the DoodleBUGS graphical interface can become tricky and most users prefer to specify the model directly using the text-based BUGS language. This has a syntax similar to the Splus language and follows reasonably intuitively from the equation-based representation of a statistical model. Nonetheless, the graphical representation of a full probability model can provide a valuable aid to model elaboration as illustrated in Sections 3.-6., and remains the key to its underlying implementation in WinBUGS.

3. GLMs with Non-canonical Links

For scientific reasons, or to explore sensitivity to different assumptions, non-canonical link functions are often assumed for GLMs. Specification of such models in WinBUGS is achieved by simply selecting the desired function from the available links. However, care must be taken when the chosen link allows for values of μ_i outside the valid range. For example, consider the analysis of data from the Solomon-Wynne experiment. The experiment involves applying an electric shock to a dog 10 seconds after a light stimulus. The dog can avoid the shock by jumping out of the cage after the light, in which case a success is recorded; if the dog fails to jump and receives the shock a failure is recorded. A plausible model is to suppose that a dog learns from previous experiments, with the probability of success depending on the number of previous successes x_{ij} and failures $j - x_{ij}$, i.e. $\mu_{ij} = \theta_1^{x_{ij}} \theta_2^{j-x_{ij}}$. This is equivalent to the following log-linear model

$$\log \mu_{ij} = \beta_1 x_{ij} + \beta_2 (j - x_{ij})$$

giving a Bernoulli GLM with a non-canocnical log link. This is trivial to implement in WinBUGS. However, prior distributions for the regression coefficients should reflect the fact that μ_{ij} is a probability i.e. $0 < e^{\beta_1 x_{ij} + \beta_2 (j - x_{ij})} < 1$. Hence constraints should be imposed to ensure β_1 and β_2 are negative. This is easily achieved in WinBUGS using the following notation to denote right-truncated normal prior distributions with upper bound of, say, -0.000001 for β_1 and β_2

```
beta[1] ~ dnorm(0, 0.001) I(,-0.000001)
beta[2] ~ dnorm(0, 0.001) I(,-0.000001)
```

Implementation of user-defined links in WinBUGS is difficult and involves writing a new software component to execute the specific function (see Section 8.) and the Appendix for further details). Alternatively, non-standard links may be implemented by transforming the linear predictor using the inverse link function, although strictly speaking, the model is no longer a GLM. The latter approach is used for the multinomial-logistic model described in Section 5.

4. Generalized Linear Mixed Models (GLMMs)

In many applications the variance of the response Y_i exceeds the nominal variance for the assumed probability distribution. This phenomenon is termed overdispersion. Overdispersion can arise in a number of ways, but is most commonly assumed to result from clustering in the population. That is, the response sample is actually drawn from a population consisting of many subgroups. For example, repeat observations made on a sample of individuals may be clustered within subjects. Households or neighborhoods in a geographical study of disease rates represent another potential source of clustering. Overdispersion may also be induced if the response depends on an unobserved covariate, with responses sharing the same unknown covariate value forming the clusters.

There are various methods for dealing with overdispersion in GLMs (see the chapter by Dey and Ravishanker (this volume) for a discussion). From a Bayesian perspective, one of the most natural approaches is to introduce *random effects* to model heterogeneity in the responses. Such models are termed generalized linear mixed models (GLMMs) or hierarchical GLMs.

4.1 *Exchangeable Random Effects*

The simplest form of random effects model introduces an additional parameter λ_i into the linear predictor for each response unit i. The $\lambda_i, i = 1, ..., I$ are assumed independent and exchangeable and are modeled with a mean zero normal population distribution with unknown variance σ^2. The latter is often termed a hyperparameter, for which a hyperprior distribution must be specified. For computational convenience, WinBUGS parameterizes the normal distribution in terms of the mean and *precision* $\tau = \sigma^{-2}$. The conjugate prior for a normal precision parameter is the gamma distribution, and a gamma(0.001, 0.001) is often chosen as a proper but diffuse hyperprior for τ (see Section 7.2).

Figure 23.2 *(a)* shows how a simple extension of the DoodleBUGS graph for a standard GLM allows for the inclusion of exchangeable random effects. The population distribution for `lambda[i]` and the hyperprior for `tau` are selected from the menu of probability distributions as before. Note that the user is not restricted to the assumption of normality for the random effects. A particular heavy-tailed population distributions, such as a Student t with small degrees of freedom or a double exponential (Laplace) distribution, may be appropriate if outlying observations are suspected.

The strong conditional independence structure exhibited by GLMMs with exchangeable random effects makes them remarkably easy to estimate via MCMC methods such as those implemented in WinBUGS. However, readers should be aware of the potential for nonconvergence and poor mixing of the sampler in certain situations, particularly when data are sparse (for example, binomial data with small denominator) and vague hyperprior distributions are used. Specifying an informative prior for the random effects precision (see Section 7.2) can often remedy the situation.

4.2 *Correlated Random Effects*

There are many situations where independent exchangeable random effects do not adequately capture the pattern of variation in the data. This is particularly so for temporal or spatial data where proximity in time or space may lead to correlated responses. Correlation amongst the random effects in a GLMM may be modeled in a variety of ways. These include direct specification of a multivariate normal prior with suitably parameterized covariance matrix for the joint distribution of $\lambda = \{\lambda_1, \ldots, \lambda_I\}$, or specification of an autoregressive (AR) or conditional autoregressive (CAR) prior for the conditional distribution of each random effect given the remaining effects and an overall precision parameters τ, i.e. $p(\lambda_i|\lambda_{j\neq i}, \tau)$. The reader is referred to the chapter by Sun, Speckman and Tsutakawa (this volume) for further details.

The graphical model underlying a GLMM with conditionally specified correlated random effects is more complex than the DAGs discussed so far. Links between the individual random effects are *undirected* due to the correlation structure. This leads to a *chain graph* and the rules which WinBUGS uses to construct full conditional distributions from nodes in a DAG no longer apply (Spiegelhalter, Thomas and Best, 1996). (Note that it is possible to "trick" WinBUGS into implementing

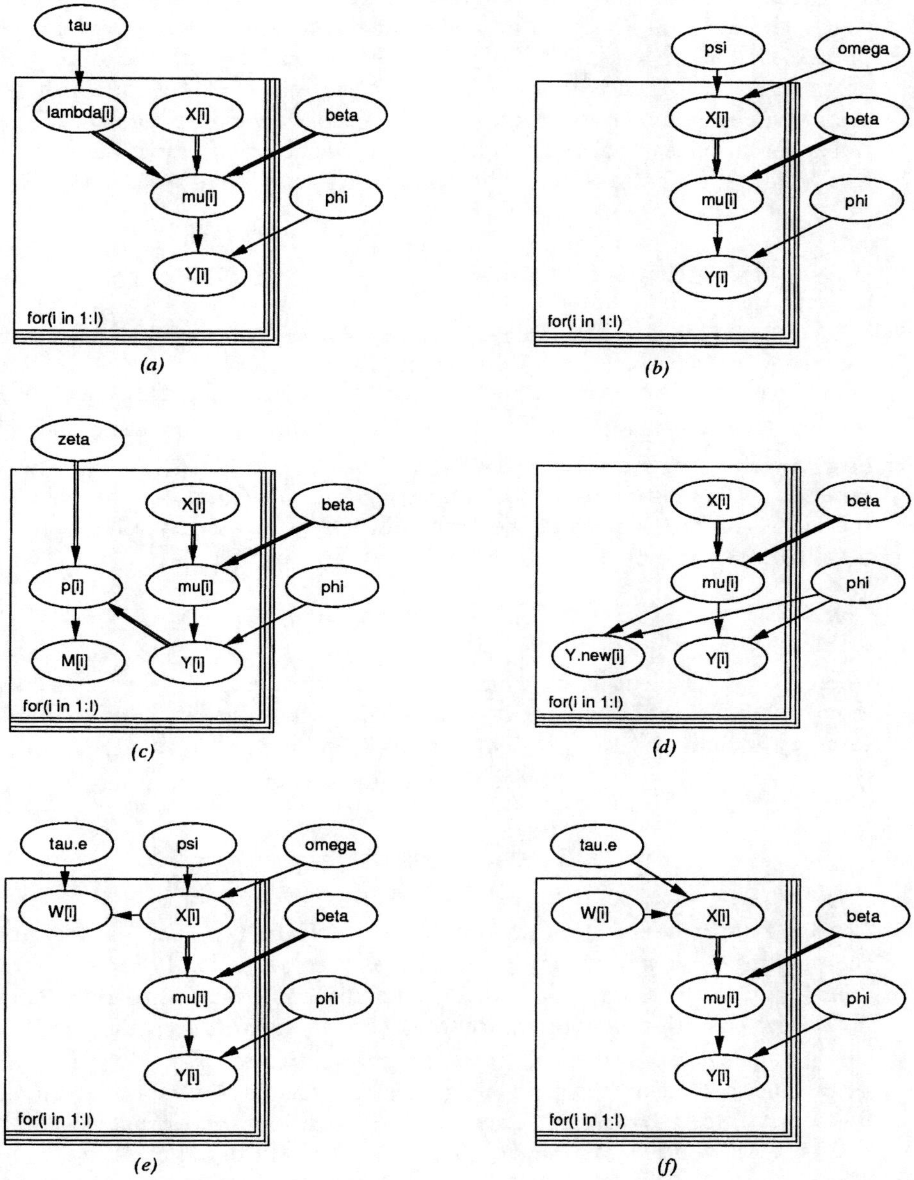

FIGURE 23.2. Extensions to the DAG for a simple GLM: *(a)* GLMM with exchangeable random effects; *(b)* missing covariate data; *(c)* non-ignorable missing response data; *(d)* prediction; *(e)* classical covariate measurement error; *(f)* Berkson covariate measurement error.

the correct model for a chain graph by explicitly specifying the required full conditional distributions in the model description — see Spiegelhalter, Thomas and Best (1996) for details). If the correlated random effects are modeled *jointly* using a multivariate prior however, it is possible to treat the vector $\lambda = \{\lambda_1, \ldots, \lambda_I\}$ as a single multivariate node in the graph. The graph thus reverts to a simple DAG with acyclic directed links between all nodes. At the time of writing, it is not possible to implement such models in WinBUGS since version 1.1.1 only handles multivariate normal nodes with unstructured (inverse) covariance matrices for which a Wishart prior must be assumed. However, future versions are intended to allow fully parameterized covariance matrices and will include special density functions corresponding to the joint distributions underlying certain AR and CAR processes.

5. Polytomous Responses

Polytomous data comprise responses which are restricted to one of a fixed set of possible values or categories. The likelihood is usually taken to be $Y_i \sim$ Multinomial $(N_i, \mu_{i1}, \mu_{i2}, \ldots, \mu_{iK})$ where Y_i is a vector of counts $(Y_{i1}, Y_{i2}, \ldots, Y_{iK})$ for each of the K possible outcome categories, with $N_i = \sum_{k=1}^{K} y_{ik}$. The parameter μ_{ik} represents the probability of observing the kth category for unit i, with $\sum_{k=1}^{K} \mu_{ik} = 1$. A standard way to parameterize the multinomial GLM is by extending the logistic link function to obtain the logarithm of the ratio of the probability of each category relative to that of a baseline category, say $k = 1$. That is

$$\log \frac{\mu_{ik}}{\mu_{i1}} = \eta_{ik} = \beta_k' X_i \tag{3}$$

where $\beta_1 = 0$ for identifiability and β_k $(k > 1)$ represents the effect of a unit change in covariate X_i on the probability of observing a response in category k relative to category 1. Since $\sum_{k=1}^{K} \mu_{ik} = 1$ and from (3) $\mu_{ik} = \mu_{i1} e^{\eta_{ik}}$, then $\sum_{k=1}^{K} \mu_{i1} e^{\eta_{ik}} = 1$ giving $\mu_{i1} = 1/\sum_{k=1}^{K} e^{\eta_{ik}}$. Hence (3) is equivalent to

$$\mu_{ik} = g^{-1}(\eta_{ik}) = \frac{e^{\beta_k' X_i}}{\sum_{j=1}^{K} e^{\beta_j' X_i}} \tag{4}$$

The latter parameterization must be used in WinBUGS since the transformation $g(\mu_{ik}) = \log \frac{\mu_{ik}}{\mu_{i1}}$ is not explicitly available as a link function in the software.

An alternative parameterization for the multinomial GLM is to use a Poisson likelihood and condition on appropriate totals (Baker, 1995). This will generally be more efficient since it avoids the time-consuming evaluation of $\sum_{j=1}^{K} e^{\beta_j' X_i}$ in (4). Suppose we assume that the data were in fact generated by as Poisson random variables $Y_{ik} \sim$ Poisson(θ_{ik}) for $k = 1, \ldots, K$. Then *conditional* on the total count $N_i = \sum_{k=1}^{K} Y_{ik}$, the joint distribution of $Y_i = (Y_{i1}, Y_{i2}, \ldots, Y_{iK})$ is Multinomial$(N_i, \mu_{i1}, \mu_{i2}, \ldots, \mu_{iK})$ with $\mu_{ik} = \theta_{ik}/\sum_{j=1}^{K} \theta_{ij}$. We incorporate the constraint on the multinomial total into the Poisson model using the following 'trick'. Let

$$\theta_{ik} = \theta_{i1} e^{\beta_k' X_i} \tag{5}$$

Then the Poisson likelihood is for the joint distribution of Y_i is

$$\prod_{k=1}^{K} \theta_{ik}^{Y_{ik}} e^{-\theta_{ik}} = \theta_{i1}^{N_i} e^{\sum_{k=1}^{K} Y_{ik} \beta_k' X_i} e^{-\theta_{i1} \sum_{k=1}^{K} e^{\beta_k' X_i}}$$

Assuming a Gamma(a,b) prior for θ_{i1} and integrating gives a marginal likelihood for Y_i of

$$e^{\sum_{k=1}^{K} Y_{ik}\beta_k' X_i} \int \theta_{i1}^{N_i} e^{-\theta_{i1}\sum_{k=1}^{K} e^{\beta_k' X_i}} \theta_{i1}^{a-1} e^{-b\theta_{i1}} d\theta_{i1} \quad \propto \quad \frac{e^{\sum_{k=1}^{K} Y_{ik}\beta_k' X_i}}{\left[\sum_{k=1}^{K} e^{\beta_k' X_i} + b\right]^{N_i+a}}$$

As $a, b \to 0$ the correct multinomial likelihood is obtained. A Gamma(0,0) prior on θ_{i1} is improper and hence is not allowed in WinBUGS (see Section 7.2). However this is formally equivalent to a uniform prior on $\log \theta_{i1}$, so that we may approximate the above model by replacing (5) with

$$\log \theta_{ik} \quad = \quad \lambda_i + \beta_k' X_i$$

and give λ_i a proper but locally uniform prior such as a normal with very large variance.

5.1 Ordered Categories

Ordinal data are a special case of polytomous data for which the categories have a natural ordering, such as "low", "medium" and "high". We wish to incorporate such structure into the model, but in this context it does not make sense to talk of "distance" or "spacing" between pairs of response categories. One solution is to assume a multinomial GLM but to define the linear predictor as a function of the *cumulative* probabilities $\pi_{ik} = \sum_{j \le k} \mu_{ij}$ instead of the individual category probabilities μ_{ij} (Agresti, 1989). For example

$$Y_i \quad \sim \quad \text{Multinomial}(N_i, \mu_{i1}, \mu_{i2}, ..., \mu_{iK}) \qquad (6)$$
$$g(\pi_{ik}) \quad = \quad \delta_k - \beta\prime X_i \qquad (7)$$

The unknown parameters δ_k may be interpreted as cutpoints on the scale of an underlying continuous latent variable Z_i. If the true but unobserved value of Z_i lies in the interval $[\delta_{k-1}, \delta_k)$, where $\delta_0 = -\infty$, $\delta_K = \infty$ and $\delta_1 < \ldots < \delta_{K-1}$, then category k will be recorded. Various distributions may be assumed for Z_i, the most popular of which is the logistic distribution with mean $\beta' X_i$ and scale $= 1$. This corresponds to assuming a logistic link function $g(.)$ in (7). Other common choices for $g(.)$ include the complementary log-log link (corresponding to the assumption of an extreme value distribution for the latent variable Z_i) or the probit link (corresponding to the assumption of a Normal distribution for Z_i).

Implementation of cumulative logit models is straightforward using WinBUGS (see Best *et al.* (1996) for an example). In addition to specifying equations (6) and (7), the user must include deterministic expressions for the individual category probabilities $\mu_{ik} = \pi_{ik} - \pi_{i,k-1}$ (where $\mu_1 = \pi_1$ and $\mu_K = 1 - \pi_{k-1}$) and prior distributions for β and each unknown cutpoint δ_k. Order constraints must be imposed on the latter using the `I(,)` notation to denote a bounded distribution in WinBUGS

```
delta[1] ~ dnorm(0, 1.0E-6) I(         , delta[2])
delta[2] ~ dnorm(0, 1.0E-6) I(delta[1], delta[3]) .....
```

Note that the model presented by Best *et al.* (1996) uses a categorical distribution for the response likelihood. This is formally equivalent to a multinomial with $N_i = 1$, and is appropriate, for example, when repeated ordered categorical responses are available for each subject, and an individual level analysis with time-dependent covariates is required.

6. Adding Complexity in GLMs/GLMMs

6.1 Missing Data

Missing data are handled very naturally within the Bayesian framework, being treated in the same way as all other unknown quantities in the model. Posterior inference is based on the joint distribution of parameters β and missing data Y_{miss} given the modeling assumptions and the observed data Y_{obs}. In practice, this involves simulating values from the conditional distribution of Y_{miss} given Y_{obs} and the current values of β, followed by simulation from the conditional distribution of β given the complete data $Y = \{Y_{obs}, Y_{miss}\}$, where Y_{miss} take on their currently estimated values. The marginal posterior for β involves integrating the joint posterior $p(\beta, Y_{miss}|Y_{obs})$ over the distribution of missing values. The posterior estimates of β are thus fully adjusted for uncertainty due to the missing data.

The missing value code in WinBUGS is **NA**: whenever the software encounters this code in a data file, it will automatically treat that node as an unknown quantity and generate values from the appropriate full conditional during MCMC simulation. If the missing values occur in the response variable, the structure of the graph remains unchanged from the complete data case and WinBUGS simply generates samples of Y_{miss} from $p(Y|\beta)$, the likelihood distribution specified for the response variable. If missing *covariates* X_{miss} are involved, the graph must be extended to include an exposure distribution for X, since all unknown stochastic quantities must be assigned a probability distribution within a fully Bayesian model. Suitable exposure distributions for continuous covariates include a normal distribution with mean ψ and precision ω (see Figure 23.2 *(b)*). The hyperparameters ψ and ω may be fixed *a priori* or assigned (possibly) vague hyperpriors (the observed covariate values X_{obs} will contain information by which to identify ψ and ω). Suitable exposure distributions for binary or polytomous covariates include Bernoulli or categorical distributions respectively.

6.2 Informative Missing Data

Rubin (1976) discusses a number of mechanisms by which missing data may arise in practice. Let M_i be a binary indicator taking value 0 if $Y_i \in Y_{obs}$ and 1 if $Y_i \in Y_{miss}$ and consider the following models for the missing data mechanism

$$p(M|Y,\xi) = p(M|\xi) \tag{8}$$

$$p(M|Y,\xi) = p(M|Y_{obs},\xi) \tag{9}$$

$$p(M|Y,\xi) = p(M|Y_{obs},Y_{miss},\xi) \tag{10}$$

where ξ are nuisance parameters controlling the probability of non-repsonse. Model (8) represents data *missing completely at random* (MCAR), in which the the probability of non-response is constant (ξ). Model (9) represents data *missing at random* (MAR), in which the probability of non-response may depend on *observed* data. For example, subjects in a longitudinal study who record an extreme response on one occasion may be more likely to dropout at the next occasion than subjects with more moderate responses. The probability of non-response may also depend on observed covariate values under the MAR mechanism. Model (10) represents an *informative missing data* mechanism. Here, the probability of non-response depends directly on the value of the response which would have been recorded had it been measured. For example, in a study of depression in the elderly, death may represent an informative dropout mechanism. Depression scores cannot be measured for those individuals who have died but are likely to be higher than in the surviving subjects since severe depression is a risk factor for death.

Under MCAR or MAR missing data mechanisms, it can be shown that the marginal posterior distribution of the parameters of interest does not depend on the pattern of missing data i.e. $p(\beta|Y_{obs}, M) = p(\beta|Y_{obs})$. The missing data mechanism is said to be ignorable and there is no need to include an explicit model for M in the analysis. The automatic handling of missing response data in WinBUGS assumes such an ignorable mechanism.

If an informative mechanism is suspected, the user must augment the data set with the missing data indicator M and explicitly define $p(M|Y_{miss}, \xi)$ in the model specification. Two frequently used models for an informative missing data process are the pattern-mixture model (Little, 1993) and the selection model (Diggle and Kenward, 1994) . Best *et al.* (1996) present an example of how to implement the latter using the BUGS software; the general formulation is as follows

$$\begin{aligned} M_i &\sim \text{Bernoulli}(p_i) \\ \text{logit}(p_i) &= \zeta_0 + \zeta_1 Y_i \end{aligned} \tag{11}$$

where Y_i will be missing whenever $M_i = 1$. The corresponding DoodleBUGS graph is shown in Figure 23.2 *(c)*. Note that the linear predictor (11) may also depend on observed responses and covariates. An informative prior is generally required for ζ_1, and the user is advised to investigate sensitivity of the resulting inference to different assumptions for this prior and to the assumed model for missing data mechanism.

6.3 Prediction

The posterior predictive distribution for an observable Y is given by

$$p(Y_{i,new}|Y) = \int p(Y_{i,new}, \beta|Y)d\beta = \int p(Y_{i,new}|\beta)p(\beta|Y)d\beta$$

where $p(Y_{i,new}|\beta)$ is the likelihood for a future response $Y_{i,new}$ and $p(\beta|Y)$ is the posterior distribution of β given the observed data Y. Samples from the posterior predictive distribution are easily obtained by generating draws from the posterior distribution of β and then simulating new values $Y_{i,new}$ by forward sampling from the likelihood conditional on the current value of β. These samples may be obtained in WinBUGS by simply adding a new node `y.new[i]` to the graph with the same parents as the observed response data (see Figure 23.2 *(d)*).

6.4 Covariate Measurement Error

Measurement error on covariates is widespread in practice, and may lead to serious bias in the estimated regression coefficients if ignored. Readers are referred to the chapter by Wakefield and Stephens (this volume) for a detailed discussion of this problem.

Classical measurement error

Let X_i denote the true covariate of interest, but suppose we only observe an imprecise measurement W_i. The classical measurement error model for a continuous covariate assumes that the observed value depends on the true value according to some probability model with error variance σ_ϵ^2. For example, $W_i \sim \text{Normal}(X_i, \tau_\epsilon)$, where $\tau_\epsilon = \sigma_\epsilon^{-2}$. Both X_i and τ_ϵ are unknown quantities which must be assigned prior distributions in a full probability model. The exposure distribution for the true covariate X_i is typically assumed to be normal with unknown mean ψ and precision ω which are assigned vague hyperpriors. The prior for τ_ϵ must be informative about the expected precision of W_i as a proxy for X_i, since this parameter is not identifiable from the observed data. However, if a relevant calibration sample (containing measurements of *both* W_i and X_i for some units i) or repeated proxy measurements W_{ij} $(j = 1, 2, \ldots)$ for each X_i are included in the model, then these extra data do contain information about the measurement precision and so a more diffuse prior may be assumed for τ_ϵ. The DoodleBUGS graph for a GLM with classical covariate measurement error is shown in Figure 23.2 *(e)*. Note that the structure of this model is similar to that for the model with *missing* covariate data (Figure 23.2 *(b)*) except for the addition of "children" for X_i in the form of the imprecisely observed covariate values W_i.

Classical measurement error on a binary covariate may be modelled as

$$\begin{aligned} W_i &\sim \text{Bernoulli}(p_i) \\ \text{logit}(p_i) &= \alpha_0 + \alpha_1 X_i \end{aligned}$$

where $e^{\alpha_0} = \Pr(W_i = 1|X_i = 0)$ and $e^{\alpha_0+\alpha_1} = \Pr(W_i = 1|X_i = 1)$. Again, informative priors must be specified for α_0 and α_1 unless a relevant calibration sample or repeated measurements W_{ij} are available.

Berkson measurement error

Berkson measurement error models assume that the true covariate depends on the observed value. Such errors typically occur in experimental studies, where W_i are the nominal values of pre-set design points such as time of measurement or machine setting, and X_i are the actual measurement times or values output by the machine. The DoodleBUGS graph for Berkson measurement error in a GLM is shown in Figure 23.2 *(f)*. Note that the direction of the arrow between X_i and W_i is the reverse of the classical measurement error model (Figure 23.2 *(e)*). Hence there is no need to specify a separate exposure distribution for the unknown true covariate X_i since under Berkson error this depends directly on the observed value W_i. However, an informative prior or calibration sample are still required for the precision parameter τ_ϵ.

7. General Advice on Modeling Using WinBUGS

In this section we consider some general issues concerning the specification and implementation of Bayesian models using WinBUGS. Further discussion of these and related topics can be found in a recent paper on MCMC methods and their application by Brooks (1998).

7.1 Parameterization

Samples generated using MCMC algorithms can sometimes exhibit poor mixing. That is, the sampler does not move rapidly throughout the support of the target distribution. Poor mixing may be identified by examining plots of the sample trace and autocorrelation functions in WinBUGS for evidence of high within-chain correlations. This may lead to slow convergence and reduce the efficiency of the Markov chain for posterior estimation (see Section 7.3 for further discussion of these issues).

Poor mixing often occurs when variables in the model are nearly collinear. Reparameterization may reduce these correlations, and a number of simple techniques are discussed by Gilks and Roberts (1996). In particular, we recommend that covariates appearing in the linear predictor of a GLM or GLMM should always be standardized about their sample mean

$$g(\mu_i) = \beta_0 + \beta_1(X_{i1} - \overline{X}_1) + \beta_2(X_{i2} - \overline{X}_2)$$

Scaling covariates by the sample standard deviation may also be appropriate. This leads to approximate posterior orthogonality between the regression coefficients $\{\beta_k\}$.

Gelfand, Sahu and Carlin (1995) recommend a strategy known as *hierarchical centring* to reduce posterior correlations amongst random effects in a GLMM. This involves introducing extra layers into the hierarchical model to provide a "better behaved posterior surface". For example, consider a GLMM with linear predictor $g(\mu_i) = \beta_0 + \beta_1 X_i + \lambda_i$ where $\lambda_i \sim \text{Normal}(0, \tau)$. The full hierarchically centred parameterization is

$$g(\mu_i) = \lambda_i; \qquad \lambda_i \sim \text{Normal}(\theta_i, \tau); \qquad \theta_i = \beta_0 + \beta_1' X_i$$

A partial hierarchically centred parameterisation of the same model is

$$g(\mu_i) = \lambda_i + \beta_1' X_i; \qquad \lambda_i \sim \text{Normal}(\beta_0, \tau)$$

In general, hierarchical centring will work best for random effects which have a large posterior variance. However, the theory underlying selection of the most efficient parameterization is difficult and problem-specific, and users are recommended to experiment with different forms of full-, partial- and non-hierarchically centred random effects to find the best parameterization for a given application.

Poor mixing may also arise as a direct consequence of the random walk behaviour exhibited by the Gibbs sampling algorithm. Neal (1998) proposes an "overrelaxed" variant of the Gibbs sampler which generates new values that are negatively correlated with the current values. This tends to cause sampled values to alternate from one side of the conditional mean to the other at consecutive updates. The net result is sample trajectories which move in a consistent direction (subject to some random deviations and reflection from the tails of the distribution), thus covering the sample space more efficiently. WinBUGS provides an option for the user to request overrelaxed sampling where possible: the sample time per iteration will increase but within-chain correlations will generally be lower. Hence fewer iterations may be necessary to achieve convergence and the desired efficiency of the chain for posterior inference.

Over-parameterized models

An over-parameteriszd model results in some parameters being nonidentifiable in the likelihood and in the prior. However, it is still feasible (but potentially dangerous) to run an MCMC sampler for such a model provided that posterior inference is only based on identifiable functions of the parameters. Since there is no notion of convergence for nonidentifiable terms in the model, examination of trace plots and other diagnostics is only meaningful for the identifiable functions. For example, consider the following parameterizations for the linear predictor of a GLM with covariate X_i representing a 3-level factor:

(i) Explicit aliasing

$$\begin{aligned} g(\mu_i) &= \beta_0 + \beta_{X_i} \quad X_i = 1,2,3 \\ \beta_1 &= 0;\ \beta_0, \beta_2, \beta_3 \sim \text{Independent vague priors} \end{aligned}$$

(ii) Identifiable contrasts

$$\begin{aligned} g(\mu_i) &= \beta_0^* + \beta_{X_i}^* \quad X_i = 1,2,3 \\ \beta_0^*, \beta_1^*, \beta_2^*, \beta_3^* &\sim \text{Independent vague priors} \\ \beta_0 &= \beta_0^* + \frac{1}{3}\sum_{k=1}^{3} \beta_k^*;\ \beta_2 = \beta_2^* - \beta_1^*;\ \beta_3 = \beta_3^* - \beta_1^* \end{aligned}$$

Parameterization *(i)* employs the usual corner-point constraint to alias one of the factor effects (β_1) to zero. The remaining factor coefficients are identifiable directly and measure the effect of level $k > 1$ of factor X relative to level $k = 1$. Parameterization *(ii)* imposes no constraints on the factor coefficients. These are no longer identified since adding a constant to each β_k^* and subtracting the same constant from β_0^* does not affect the probability. However, constrasts such as $\beta_2^* - \beta_1^*$ are identified since the arbitrary constant cancels.

Both models may be implemented in WinBUGS: it is the user's responsibility to check for over-parameterization and to ensure that posterior inference is based only on identifiable quantities.

7.2 Prior Specification

WinBUGS requires that a full probability model is defined and so does not allow improper prior distributions. This precludes the use of many standard "reference" or "non-informative" prior distributions. Here we offer some brief advice on how to choose proper priors for the unknown parameters of a GLM or GLMM in WinBUGS. Further details can be found in the Classic BUGS manual (Spiegelhalter *et al.*, 1996).

Lack of prior knowledge for a location parameter such as a regression coefficient may be expressed by assuming a locally uniform prior. The most common choice is a mean-zero normal distribution with large variance, s^2. A rule-of-thumb is to set s^2 at least one to two orders of magnitude greater than the expected value of the regression coefficient. (Recall that WinBUGS parameterizes the normal distribution in terms of the mean and *precision*, so that the actual prior specified in the code will have very small precision $= s^{-2}$).

Specification of a proper prior to express lack of knowledge for the precision of the random effects in a GLMM requires care. A widely used option is to a adopt a gamma(ϵ, ϵ) prior where ϵ is a very small number such as 0.001. This represents a "just proper" form of the improper gamma(0,0) prior which is formally equivalent to assuming a uniform distribution on the log of the scale parameter $\sigma = \tau^{-\frac{1}{2}}$.

It is often better to specify a moderately informative proper prior for the random effects precision parameter. GLMMs typically lead to inference concerning relative risk or odds ratios in which interpretation of the variance of a random effect is independent of context. For example, a normal random effect with standard deviation $\sigma = 0.59$ implies that 95% of subjects with identical covariates will have log odds (e.g. binomial logistic model) or log relative risk (e.g. Poisson log link model with offset) within a range of width $2 \times 1.96 \times 0.59 = 2.3 = \log 10$; that is, the 97.5% quantile of the distribution of odds ratios or relative risks for identical subjects will be one order of magnitude greater than the 2.5% quantile. This interpretation provides a mechanism by which to choose a suitable informative prior for the random effects precision parameter τ in many GLMM applications. For example, suppose we thought it plausible that there was roughly one order of magnitude difference between the odds or relative risk of the response of interest for subjects with identical observed covariates. This implies that $(2 \times 1.96/\log 10)^2 = 2.90$ is a reasonable guess for τ. Suppose also that we would be surprised to find more than 2 orders of magnitude difference between all but the most extreme 5% of such subjects. This gives an "low" value for τ (i.e. an upper value for the variance) of $(2 \times 1.96/\log 100)^2 = 0.73$. A gamma(3,1) distribution has a mean of 3 and 96% probability of exceeding 0.73 and hence might be an appropriate prior for τ in this context.

7.3 Convergence and Posterior Sample Size

Two of the most important implementation problems in any practical application of MCMC methods are (i) determining the number of initial iterations to discard as "burn-in" and (ii) deciding how long the simulations should be run following convergence.

To avoid potential bias due to the influence of arbitrary starting values, an initial portion of the simulation should be discarded. This is known as the "burn-in" and represents samples generated during an initial transient phase before the simulation has converged (at least approximately) to the posterior target distribution. Formal methods for assessing convergence (and hence the length of the required burn-in) have been the topic of a large and somewhat controversial literature. However, no method has yet provided a global and foolproof diagnostic (see Cowles and Carlin (1996) and Robert (1996) for reviews). We recommend a strategy of examining several different diagnostics before deciding on an appropriate "burn-in" period for a given model. WinBUGS provides a number of informal graphical diagnostics such as plots of the sample trace, autocorrelation function and running mean and 95% quantiles. At the time of writing, no formal diagnostics are implemented in WinBUGS version 1.1.1. However future versions of the software are intended to include a facility for running multiple simulations using different starting values in order to calculate the diagnostic proposed by Gelman and Rubin (1992) based on classical analysis of variance of the between and within chain variation.

Determining when to stop the simulation after convergence has been achieved depends on the level of precision required for posterior inference. The asymptotic variance of the Monte Carlo estimator (i.e. sample average) of a posterior expectation $E_\pi(f(\theta))$ is given by σ^2/N where N is the sample size and σ^2 is a positive constant depending on the posterior variance of $f(\theta)$ and the sample autocorrelation. Various methods exist for estimating σ^2; WinBUGS uses the batch means method outlined by Roberts (1996) , and reports the quantity $\sigma/\sqrt{N}$ in the column headed *Monte Carlo error* of the summary statistics table for each monitored node in the model. The simulation should be run until this error term is suitably small (say, 5% or 1% of the estimated posterior mean). However, care must be taken to ensure that N is large enough for the batch means approximation of σ^2 to be valid: users are thus recommended to run a minimum of 1000 iterations following the "burn-in" period in order to obtain a reliable estimate of the Monte Carlo error in WinBUGS.

7.4 Model Checking

Methods for Bayesian model criticism have been widely debated since the introduction of MCMC methods into mainstream statistics opened the way for practical application of Bayesian inference to real-life problems. Many purist still argue that the Bayes factor, based on the ratio of the marginal distributions of the observed data under two competing models, is the only criterion necessary for model comparison. However, there are serious problems with the interpretation and computation of the Bayes factor (see Gelfand (1996) for a brief discussion) which have led many authors to suggest alternative approaches. These include Bayesian model averaging (Raftery, Madigan and Hoeting, 1997) , cross-validation procedures (Gelfand, Dey and Chang, 1992) and criteria based on the log-likelihood distribution (Gelfand and Ghosh, 1998; Spiegelhalter, Best and Carlin, 1998).

WinBUGS provides no formal tools for model checking or comparison. However, the user is at liberty to implement many of the methods proposed in the literature. Section 9.3 of the Classic BUGS manual (Spiegelhalter *et al.*, 1996) outlines how some of these may be achieved, and the chapter by Dellaportas, Forster and Ntzoufras (this volume) provides further illustrations of Bayesian variable selection methods which may be implemented in WinBUGS. The posterior predictive model checks discussed in the chapters by Dey and Ravishanker (this volume) and Albert and Ghosh (this volume) may be implemented by generating samples Y_{new} from the posterior predictive distribution as described in Section 6.3. A deterministic node set equal to the log likelihood (calculated explicitly by the user) may also be included in the WinBUGS model specification. Posterior samples of this node may then be used to compute global model comparison criteria such as the *expected predictive deviance* (Gelfand and Ghosh, 1998) or the *deviance information criterion* (Spiegelhalter, Best and Carlin, 1998).

8. Extending the WinBUGS Software

The aim of previous sections has been to provide a flavor of some of the models and problems that can be handled using the graphical modeling approach in WinBUGS, and to illustrate the flexibility of the software to extend standard GLMs and accommodate almost arbitrary complexity. With some thought, most of the models discussed in other chapters of this book could be implemented using the existing WinBUGS program. However, many of these involve sophisticated, state-of-the art model design and analytic techniques, and their implementation in WinBUGS may prove difficult and inefficient compared with using the authors' problem-specific computer code. By its nature, routine statistical software such as WinBUGS must lag behind cutting-edge statistical methodology: it is neither feasible nor desirable to implement every new algorithm and model that becomes available. Nonetheless, the ability of researchers to utilize standard features of the WinBUGS package, such as the model specification language, input/output facilities and error checking, during their own methodological development work would be highly advantageous.

As already discussed, WinBUGS is constructed using a graphical model framework. This facilitates software extensibility in much the same way as the graph underlying a statistical model may be extended. That is, by exploiting "conditional independencies" between the components which form the building blocks of the software graph. Inclusion of new methods and applications is achieved by writing extra components which simply either "plug in" to relevant slots in existing modules or make use of existing modules without requiring any part of the software to be recompiled. This feature represents a key strength of the WinBUGS design, and both facilitates the ongoing development of the core program and provides the structure which will enable users to write customised "add-on" routines to link into and exploit existing facilities within the software. We anticipate that such "add-ons" will cover three main requirements: (i) new logical functions; (ii) new probability distributions; and (iii) new sampling algorithms. An example outlining the steps involved in writing a component to implement a new logical function is given in the Appendix. Interested readers should contact the authors for further information.

The WinBUGS software, documentation and worked examples are freely available over the World Wide Web from `http://www.mrc-bsu.cam.ac.uk`.

section*Appendix: Software components

WinBUGS builds an internal representation of the graphical model out of software components (objects). The nodes of the graph are represented by objects of type "Node" and the edges of the graph by pointers to node objects. The nodes of the graph can be classified into two families: type "Stochastic" and type "Logical". Stochastic nodes are quite complex and will not be discussed further. Here we consider the simpler logical node.

The component interface for a logical node describes the list of implemented entities (methods) that can be used in other components. These fall into two groups: methods used to build the graphical model (*Set*, *Check*, *CanEvaluate*, *Map*, *Optimize*, *IsConstant*, *Parents* and *Path*) and methods used for the MCMC inference (*Value*, *MonitoredValue*, *LogValue*, *LogitLogQ*). The *Set* method is used to set up any hidden attributes of the node such as its parents; the *Check* method can implement various consistency checks on the graphical model; the *CanEvaluate* method is a predicate which decides if the parents of a logical node have been initialized; the *Map* method allows the software to classify the functional form of the logical node and hence optimize the sampling methods used; the *Optimize* method allows a logical node to replace itself by a more efficient version; the *IsConstant* method is a predicate that tests whether the logical node is a function of only fixed value nodes; the *Parents* method finds all parents of the logical node that are of type "Stochastic" and are not data; the *Path* method is a predicate that tests if there is a path between two nodes; the *Value* method calculates the value of all the parents of a logical node and then calculates the logical expression associated with the node using these values; the *MonitoredValue* method is an optimization of the *Value* method used when monitoring nodes during the simulation; the *LogValue* and *LogitLogQ* methods calculate functions of the value of a logical node.

To create a new type of logical node, an extension of the type "Logical" is made, in which each of the above methods must be implemented. For example, suppose we wish to create a component to implement the solution of Keplers equation $x = l + e\sin(x)$, which would be used to define a logical node in the BUGS language viz. `x <- kepler(l, e)`. A new object must be made, called type "Kepler" (say), which is a pointer to a node of type "Logical". Each method required for a node of type "Logical" must then be defined for the new type "Kepler" together with any additional procedures required to implement the node. Methods are defined by writing `PROCEDURE` statements. For example, the *Value* method for the Kepler object is

```
PROCEDURE (keppler:Kepler) Value(): REAL;
VAR
  I, e: REAL;
BEGIN
  I := kepler.I.Value(); e := kepler.e.Value();
  RETURN SolveKepler(I, e)
END Value
```

where `SolveKepler` is a new procedure which solves the Keplers equations. Having defined all the required methods and new procedures, the component interface is then written. This is quite simple and describes which of the entites implemented in the Kepler component can be used by other components

```
DEFINITION GraphKepler;
  IMPORT GraphNodes;
  VAR
    dir-: GraphNodes.Directory;
  PROCEDURE Install;
  PROCEDURE SetDir (d: GraphNodes.Directory);
END GraphKepler.
```

Here, the entity `dir` can be used to create a "Kepler node". The component interface is encoded in a machine readable format called a symbol file. This file allows the consistency of the component interfaces to be checked both at compile time and link time, thus improving the reliability of the software.

References

Agresti, A. (1989). A survey of models for repeated ordered categorical response data. *Statistics in Medicine*, **8**, 1209–24.

Albert, J. and Ghosh, M. (1999). Item response modeling. In *Generalized linear models: a Bayesian perspective*, (ed. D. Dey, S. Ghosh, and B. Mallick). Marcel Dekker, New York.

Baker, S. (1995). The multinomial-Poisson transformation. *The Statistician*, **43**, 495–504.

Best, N. G., Spiegelhalter, D. J., Thomas, A., and Brayne, C. E. G. (1996). Bayesian analysis of realistically complex models. *Journal of the Royal Statistical Society, Series A*, **159**, 323–42.

Brooks, S. (1998). Markov chain Monte Carlo method and its application. *The Statistician*, **47**, 69–100.

Cowles, M. and Carlin, B. (1996). Markov chain Monte Carlo convergence diagnostics: a comparative review. *J Am Statist Assoc*, **91**, 883–904.

Dellaportas, P., Forster, J., and Ntzoufras, I. (1998). Bayesian variable selection using the Gibbs sampler. In *Generalized linear models: a Bayesian perspective*, (ed. D. Dey, S. Ghosh, and B. Mallick). Marcel Dekker, New York.

Dey, D. and Ravishanker, N. (1998). Bayesian approaches for overdispersion in generalized linear models. In *Generalized linear models: a Bayesian perspective*, (ed. D. Dey, S. Ghosh, and B. Mallick). Marcel Dekker, New York.

Diggle, P. J. and Kenward, M. G. (1994). Informative dropout in longitudinal data analysis. *Applied Statistics*, **43**, 49–94.

Gelfand, A. (1996). Model determination using sampling-based methods. In *Markov chain Monte Carlo in practice*, (ed. W. R. Gilks, S. Richardson, and D. J. Spiegelhalter), pp. 145–61. Chapman and Hall.

Gelfand, A., Dey, D., and Chang, H. (1992). Model determination using predictive distributions with implementation via sampling-based methods. In *Bayesian statistics 4*, (ed. J. M. Bernardo, J. O. Berger, A. P. Dawid, and A. F. M. Smith), pp. 147–68. Oxford University Press.

Gelfand, A. E. and Ghosh, S. (1998). Model choice: a minimum posterior predictive loss approach. *Biometrika.* **85**, pp.1-11.

Gelfand, A. E., Sahu, S. K., and Carlin, B. P. (1995). Efficient parameterisations for generalized linear models. In *Bayesian Statistics 5.* Clarendon Press, Oxford, UK.

Gelman, A. and Rubin, D. B. (1992). Inference from iterative simulation using multiple sequences. *Statistical Science*, **7**, 457–72.

Gilks, W. R. and Roberts, G. O. (1996). Strategies for improving MCMC. In *Markov chain Monte Carlo in practice*, (ed. W. R. Gilks, S. Richardson, and D. J. Spiegelhalter), pp. 89–114. Chapman and Hall.

Lauritzen, S. L., Dawid, A., Larsen, B., and Leimer, H. (1990). Independence properties of directed Markov fields. *Networks*, **20**, 49–505.

Little, R. (1993). Pattern-mixture models for multivariate incomplete data. *J. Am. Statist. Assoc.*, **88**, 125–34.

Neal, R. M. (1998). Suppressing random walks in Markov chain Monte Carlo using ordered overrelaxation. In *Learning in graphical models*, (ed. M. Jordan), p. (to appear). Kluwer Academic Press.

Raftery, A. L., Madigan, D., and Hoeting, J. (1997). Bayesian model averaging for linear regression models. *J Amer Statist Assoc*, **92**, 179–91.

Robert, C. (1996). Convergence assessments for Markov chain Monte Carlo methods. *Statistical Science*, **10**, 231–53.

Roberts, G. (1996). Markov chain concepts related to sampling algorithms. In *Markov chain Monte Carlo in practice*, (ed. W. R. Gilks, S. Richardson, and D. J. Spiegelhalter), pp. 45–57. Chapman and Hall.

Rubin, D. B. (1976). Inference and missing data. *Biometrika*, **63**, 581–92.

Smith, A. F. M. and Roberts, G. O. (1993). Bayesian computation via the Gibbs sampler and related Markov chain Monte Carlo methods (with discussion). *J Roy Statist Soc B*, **55**, 3–24.

Spiegelhalter, D., Best, N., and Carlin, B. (1998). Bayesian deviance, the effective number of parameters, and the comparison of arbitrarily complex models. Technical report, MRC Biostatistics Unit, Cambridge.

Spiegelhalter, D. J. (1998). Bayesian graphical modeling: a case-study in monitoring health outcomes. *Applied Statistics*, **47**, 115–33.

Spiegelhalter, D. J., Thomas, A., and Best, N. G. (1996*a*). Computation on Bayesian graphical models. In *Bayesian Statistics 5*, (ed. J. M. Bernardo, J. O. Berger, A. P. Dawid, and A. F. M. Smith), pp. 407–25. Clarendon Press, Oxford.

Spiegelhalter, D. J., Thomas, A., Best, N. G., and Gilks, W. R. (1996*b*). *BUGS 0.5 Bayesian inference using Gibbs sampling manual (version ii).* Medical Research Council Biostatistics Unit, Cambridge.

Sun, D., Speckman, P., and Tsutakawa, R. (1999). Random effects in generalized linear mixed models (GLMMs). In *Generalized linear models: a Bayesian perspective*, (ed. D. Dey, S. Ghosh, and B. Mallick). Marcel Dekker, New York.

Wakefield, J. and Stephens, D. (199). Bayesian errors-in-variables modeling. In *Generalized linear models: a Bayesian perspective*, (ed. D. Dey, S. Ghosh, and B. Mallick). Marcel Dekker, New York.

Index